Mathematics
for Electronics
Second Edition

Mathematics
for Electronics
Second Edition

Forrest L. Barker
Los Angeles City College

Gershon J. Wheeler
Ohlone College and
West Valley College

The Benjamin/Cummings Publishing Company, Inc.
Menlo Park, California ● Reading, Massachusetts
London ● Amsterdam ● Don Mills, Ontario ● Sydney

The cover shows the foil pattern for an Optical Synthesizer, printed by permission of *Radio Electronics*. © Gernsback Publications Inc. 1977.

Sponsoring editor: Adrian Perenon
Production editor: Ruth Cottrell
Book designer: Donna Davis
Cover designer: Michael Rogondino
Artist: M. Patricia Rogondino

ISBN 0-8053-0340-5
  BCDEFGHIJ-AL-798

The Benjamin/Cummings Publishing Company, Inc.
2727 Sand Hill Road
Menlo Park, California 94025

PREFACE

Many significant changes have taken place in the world since the first edition of *Mathematics for Electronics* was published, and some of them have had a dramatic impact upon the teaching of electronics. This second edition has been designed to respond to those changes in a way that will help to assure effective learning for the electronics student of the late 1970s.

A SPECIALIZED TEXT

The value of a specialized textbook on mathematics for electronics is well established. Experience has amply demonstrated that students of electronics are best served by an appropriate selection of mathematical material, which not only *includes* certain well-chosen topics but also deliberately *leaves out* others. Furthermore, the immediate application of mathematical tools to electronics problems not only increases motivation to learn the mathematics, but also aids comprehension of both the mathematics and electronics theory.

LEVEL OF THE TEXT

This book has been designed for a two- or three-term sequence for students of electronics technology and allied curricula at the community college and technical institute level. It is also suitable for use in high school electronics programs. Finally, it incorporates sufficient rigor to serve those technology students who desire to transfer to four-year level engineering technology programs.

SCOPE OF THE TEXT

The overall objective of the text is to enable electronics students to develop skill with the mathematical tools they will need. They will combine learning electronics theory with practice, and they will perform tasks required by industry. The approach utilized to achieve this objective relies heavily on worked-out examples; a generous quantity are provided.

Examples are clearly identified and set apart. They are carefully designed and presented in a style that demonstrates how to perform the mathematical operation being developed, and/or how to apply it to electronics.

More than 3000 graded exercises, carefully designed to further the understanding of processes and sharpen performance skills, provide the practice that is essential for the development of confidence in performing a new skill. Answers to all odd-numbered exercises are provided at the back of the book and the answers to even-numbered exercises are included in a *Solutions Manual* available from the publisher.

The mathematical topics presented here are a limited selection from algebra and trigonometry. The order of presentation of the topics from algebra follows, generally, that of typical introductory algebra texts, although some differences in order were deemed worthwhile. For example, a relatively complete development of exponents, including powers-of-ten notation, appears earlier than usual because of the extensive, early use of powers-of-ten in electronics calculations and unit conversions.

One of the more momentous of recent developments affecting electronics instruction is itself a product of the electronics industry—the hand-held electronic calculator. Comparative experience in electronics instruction has demonstrated that dramatic gains in learning occur when students use an electronic calculator instead of the slide rule, gains that go far beyond the obvious one of time saved in performing individual calculations. A student's early success in evaluating the formulas of electronics, with the aid of a calculator, appears to generate confidence in his or her mathematical ability. This, in turn, leads to an increased interest in mathematics and the motivation to learn more about it. Furthermore, the typical student is sufficiently motivated, by the novelty of the calculator, no doubt, to learn to use it with very little help from the instructor. Hence, nearly all of the time (and textbook space) previously required for slide rule instruction can now be devoted to other topics.

As this is being written (1977), completely adequate calculators are available for about the same cost as a moderately priced pair of shoes, and prices are still dropping. These inexpensive calculators include such special features as scientific notation, direct and inverse trigonometric functions, reciprocal function, change sign, etc. Considering the gains from using an electronic calculator, it is a good buy and within the reach of virtually every electronics student. Indeed, the student who, in the past, would have experienced great difficulty in learning to use a slide rule, cannot afford now to try to learn electronics without a calculator!

NEW IN THE SECOND EDITION

In this edition, then, all references to the slide rule have been omitted. The text incorporates presentations on the calculator and its use. These are included primarily to encourage students to use the calculator, to learn to use it effectively and efficiently, and to take advantage of some of the special features found on the typical slide-rule calculator.

Other changes incorporated in this second edition include the following features:

1. Special student aids, such as memory joggers, step-by-step procedural outlines, and key formulas and mathematical relationships are highlighted with boxes for easy identification and review.

2. Most of the presentations on algebra have been completely redesigned to improve teachability and match more closely the needs of electronics students with regard to concurrent theory course.

3. Derivations have been drastically shortened, simplified, or eliminated. Emphasis on the how-to aspects of electronics mathematics has been increased and may additional worked-out examples are included. These changes are in response to the dominant learning habits of today's students.

4. Those presentations that have not been rewritten completely have been slimmed and trimmed to improve their efficiency.

5. A few sections have been eliminated completely, but only those which a national survey by the publisher revealed were seldom used.

6. The number of exercises has been increased significantly. As in the first edition, exercises progress from the simple to the more difficult. Thus instructors are able, by the choice of assigned exercises, to determine the depth to which each topic will be covered in their classrooms.

ACKNOWLEDGMENTS

The comments and suggestions of many persons, both students and professors, have been considered in the preparation of this second edition. For their specific suggestions and many helpful critical comments we would like to acknowledge with sincere gratitude the contributions of the following professors: Thomas C. Power, Santa Rosa Junior College, Santa Rosa, California; Floyd A. Lambert, Institute of Electronic Science, Texas A & M University, College Station, Texas; J. L. Williams, Washtenaw Community College, Ann Arbor, Michigan; Eddy E. Pollock, W6LC, Cabrillo College, Aptos, California; L. W. Churchman, Allan Hancock College, Santa Maria, California; Melvin H. Smith, El Paso Community College, Colorado Springs, Colorado; and Alexander W. Avtgis, Wentworth Institute and College of Technology, Boston, Massachusetts.

Finally, and most importantly, we wish to thank Mrs. Carolyn Berbower Barker for her typing of the complete manuscript and the many helpful suggestions concerning its preparation.

F. Barker
Burbank, CA.
G. Wheeler
Mountain View, CA
December 1977

Mathematics
for Electronics
Second Edition

TABLE OF CONTENTS

TABLE OF CONTENTS

TABLE OF CONTENTS

TABLE OF CONTENTS

INTRODUCTION

TO THE STUDENT

The fact that you are reading this probably reflects your interest in learning about electronics, not mathematics. Electronics is a most fascinating subject and one which grows more exciting each day, primarily because of its apparent capability for producing an endless stream of miracles: radio, television, computers, communications satellites, close-up color pictures of the surface of Mars, and so on. However, as a student of electronics it is important for you to be aware that these truly marvelous inventions are possible only because of the science of mathematics, which has been developed over the past several thousand years. Without our ability to generalize the relationships between the various elements of an electric circuit with mathematical symbols, we would not be able to design new circuits and predict their behavior—the necessary steps of electronic invention.

Yes, mathematics is an essential element to the study and practice of electronics at any level. But this need not be a handicap to you, even if you have been turned off by mathematics earlier in your educational career and have experienced it as a difficult subject. Many students, motivated to study the required mathematics because of their interest in electronics, have discovered that mathematics can be fascinating as well as of great practical value. In many cases, discovering that mathematics is "not so bad" has led students to continue their education to levels far beyond those anticipated when they began their study of electronics.

One of the more recent miracles of the world of electronics is the miniature electronic calculator—a development that has virtually revolutionized the study of electronics, as well as of many other subjects. These calculators not only eliminate much of the tedium of the numerous calculations required in electronics but, of greater importance, also eliminate many difficult learning steps previously required for developing a skill in using the slide rule. Your first step, then, in preparing to master a course in which you will be using this text, should be to select and purchase a calculator with features appropriate to the work in electronics.

© 1976 by NEA, Inc.

" 'Gentlemen,' I said, 'We're going to stick with our line of slide rules. These electronic pocket calculators are just a fad . . .' "

Reprinted by permission of NEA.

BUYING AN ELECTRONIC CALCULATOR

Calculators are available with many different sets of features and levels of quality, with price tags to match. For best results, you should only consider the purchase of calculators that have the following functions or capabilities:

1. The four basic functions, $+$, $-$, $\times$, $\div$.

2. Powers-of-ten notation, usually indicated by the presence of a key labeled   EE .

3. Squaring function, x^2 .

4. Square root function, $\sqrt{x}$.

5. Direct trigonometric functions, SIN , COS , and TAN .

6. Inverse trigonometric functions, SIN⁻ , COS⁻ , and TAN⁻ , or sometimes simply, ARC .

7. Reciprocal function, $1/x$.

8. Change sign function +/- .

9. Common logarithm functions, $\text{LOG } X$ and 10^x .

10. Natural logarithm functions, $\ln X$ and e^x .

11. An ac converter/charger.

Other very strongly recommended but not absolutely essential features include the following:

12. Decimal degree-to-radian angle conversion, $d\text{-}r$.

13. $\rightarrow P$, $\rightarrow R$, facility for making conversions between rectangular and polar coordinates as used in the study of alternating current.

14. y^x , $\sqrt[x]{y}$, facility for raising any number to any power, or finding any root of any number.

15. M , $M+$, $M-$, memory capabilities, for temporarily storing numbers, or accumulating sums or differences.

16. (,) , or [(,)] , facility for performing parenthetical (out of the normal sequence) operations.

When you go to purchase a calculator, it is generally safe to let your pocketbook be your guide. For example, calculators with at least the first set of features in the preceding list are available, as this is being written (1977), for about the same cost as a pair of shoes, and prices are still dropping. A bargain calculator, if it has the desired features, is usually a genuine bargain. On the other hand, a higher-quality, higher-priced calculator can be expected to experience fewer breakdowns, last longer, and receive better service from the manufacturer, as well as have more functions. However, a calculator with too many functions, for example, a programmable calculator, would probably be a hindrance to you when you are beginning your study of electronics.

HOW TO STUDY *MATHEMATICS FOR ELECTRONICS*

The mathematics that you will be learning in this text is mostly of a how-to-do-it nature, that is, you will be learning the skills of evaluating equations and formulas, performing operations with algebraic expressions, writing and solving equations, and so on. Learning mathematical skill is very similar to learning any other skill—bicycling, swimming, driving a car, operating a computer—and consists of three basic steps:

1. Observe (study) a demonstration of the way a particular action is performed.

2. Carefully perform the action yourself, usually slowly. For best results, do it under the supervision of someone who can correct any errors in your performance.

3. Practice your performance correctly until you develop confidence. Speed follows.

You will be performing Step 1 when you use this text to study and follow, very carefully, each step of the worked-out examples that are specially designed and presented in a style to demonstrate specific skills. You will be performing Step 2, learning a skill, if you cover the solution of an example and perform the steps of the solution yourself, checking your solution later. Finally, you will be performing Step 3, which leads to confidence and speed, if you work a generous number of the exercises provided in the text. You can find the answers to all odd-numbered exercises at the back of the book.

1-2-3 Study Guide A 1-2-3 Guide to the things you should do, then, to assure yourself of success in learning the mathematics you will need in electronics, may be summarized as follows:

1. **Choose a quiet place and a regular time to study daily.**

2. Have an electronic calculator with appropriate functions (see the preceding list), and, of course, paper and pencil (or pen).

3. Read over a complete section, quickly, to get an overall view of the operations that you are to learn.

4. Return to the beginning of the section and study, very carefully, the worked-out examples, paying close attention to (a) the overall logic, (b) each step, and (c) the format used.

5. Cover the solutions and work the examples yourself on a separate sheet of paper. Organize your written work using a neat, orderly format (see the format of the example in the text). Check your solutions against those in the text.

6. Next work several of the exercises found at the end of the section. Check your answers against those at the back of the book. If you do not get a correct answer, check your method and try to determine the reason for your incorrect answer. Try working the exercise again.

7. Continue to practice your new skill, working exercises and checking your answers, until you feel confident of your ability to work problems of that particular type.

8. Go on to the next section and repeat Steps 2–7.

An example of a calculator which is appropriate for the calculations of electronics. (Courtesy of Hewlett Packard Company)

Chapter 1
Introduction to Algebra: Signed Numbers

Signed numbers are useful in many areas of our everyday lives, even in the sport of golf where a signed number may be used to indicate the number of strokes above or below par. (Andy Mercado/Jeroboam, Inc.)

1-1 THE ALGEBRA-ELECTRONICS CONNECTION

When we read the first few paragraphs of a typical basic electronics theory textbook, we find something like the following:

All matter contains two basic particles of electricity—the *electron*, the unit of *negative* electric charge, and the *proton*, the unit of *positive* electric charge.

The symbol for electric charge is Q or q, and the practical unit of charge is the *coulomb* (abbreviated C) named after Charles A. Coulomb (1736–1806), a French physicist who was a pioneer in investigating the effects of electric charge.

If we collect 6.25×10^{18} electrons together in one place we have 1 coulomb of charge. Symbolically, $Q = 1$ C.

We have immediately begun to use algebra in electronics! Let us examine some of the connections between algebra and electronics in this preliminary statement from atomic theory.

Using letters (for example, Q or q) to represent numbers is a fundamental characteristic of algebra. We call letters used in this way *literal numbers*. (Numerals such as 1, 15, and 5.68 are *explicit numbers*.) The expression $Q = 1$ C is an *equation*. Equations are very important in electronics and learning how to work with them is a major part of the study of algebra.

The number 6.25×10^{18} is an example of *scientific notation*. Note that this notation includes the multiplication sign $\times$; it is therefore a *product*. It has two parts called *factors*: the first part (6.25) is called the *mantissa* of the notation and the second part (10^{18}) is called the *characteristic*.

Scientific notation is sometimes written $6.25(10^{18})$. In algebra, writing two quantities without a space between them implies multiplication; that is, $4Q$ means $4 \times Q$, ab means $a \times b$, $6.25(10^{18})$ means 6.25×10^{18}, and so forth. Algebra also uses a dot $\cdot$ to indicate multiplication: for example, $ab = a \cdot b = a \times b$; $10 \cdot 10 \cdot 10 = 10 \times 10 \times 10$.

The notation 10^{18} from our example, read "ten to the eighteenth power," means "take 10 as a factor 18 times"; that is, $10^{18} = 10 \cdot 10 \cdot 10 \cdot 10 \cdot 10 \cdot 10 \cdot 10 \cdot 10 \cdot 10 \cdot 10 \cdot 10 \cdot 10 \cdot 10 \cdot 10 \cdot 10 \cdot 10 \cdot 10 \cdot 10$, or $10^{18} = 1,000,000,000,000,000,000$ and $6.25 \times$

$10^{18} = 6,250,000,000,000,000,000$. In this notation, or in any like it, the number 10 is called the *base* and 18 is called the *exponent*. Any number (except zero) can be used as a base and any number (including zero) can be used as an exponent.

Example 1   Read the following expressions and state their meaning.

(a) 10^2

(a) 10^2 is read "ten to the second power" or "ten squared" and means "take 10 as a factor two times" or $10^2 = 10 \cdot 10 = 100$.

(b) 2^5

(b) 2^5 is read "two to the fifth power" and means "take two as a factor five times" or $2^5 = 2 \cdot 2 \cdot 2 \cdot 2 \cdot 2 = 32$.

(c) x^3

(c) x^3 is read "x to the third power" or "x cubed" and means "take x as a factor three times" or $x^3 = x \cdot x \cdot x$.

(d) y^a

(d) y^a is read "y to the a power" and means "take y as a factor a times."

(e) 3.14×10^5

(e) 3.14×10^5 is read "three point one four times ten to the fifth power" and means "$3.14 \times 100,000 = 314,000$." (The exponent tells us the number of places we should move the decimal point to the right.)

(f) 6.51×10^{-6}

(f) 6.51×10^{-6} is read "six point five one times ten to the minus six" and means "$6.51 \times 10^{-6} = 0.00000651$." (The negative sign in front of the exponent means that we must move the decimal point to the left to find the equivalent decimal number.)

You will be required to use scientific and powers-of-ten notation beginning with the first chapter of your electronics theory textbook. Operations with numbers in this form require that you perform operations with exponents, the subject of Chapter 2 in this book. Exponents, however, may be positive or negative numbers. *Signed numbers* and their operations are the subject of this first chapter.

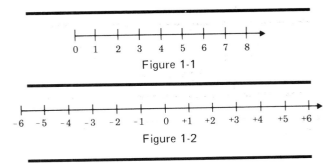

Figure 1-1

Figure 1-2

Exercises 1-1 In Exercises 1–16 state the numbers as they would be read and give their meaning (see Example 1).

1. 2.75×10^3
2. 3.81×10^{15}
3. 4.96×10^{18}
4. 8.19×10^{28}
5. 3.16×10^{-3}
6. 4.76×10^{-8}
7. 5.97×10^{-16}
8. 1.75×10^{-24}
9. 10^3
10. 10^7
11. 2^3
12. 2^5
13. x^2
14. y^5
15. x^a
16. y^6

1-2 SIGNED NUMBERS

We read in an electronics theory textbook that the mass of an electron is 9.108×10^{-28} grams (g). In the notation 10^{-28}, read "ten to the negative twenty-eight," we are confronted with a signed (negative) number. What is the meaning of signed numbers and how do we work with them? A graphic device called a *number line* will help us answer this question.

Number Line In arithmetic we worked with the set of numbers,

$$0, 1, 2, 3, 4, \ldots$$

known as *whole numbers*. These are called *integers* in algebra. (The three dots in the listing indicate that the set is not completely represented; it continues indefinitely.) We represent these numbers by equally spaced marks on a straight line (Figure 1-1), called a *number line*. The line starts at the mark labeled 0, called the *origin*. The arrowhead indicates the direction in which the numbers are getting larger.

The numbers used in algebra include another complete set called *negative numbers*, and the whole numbers of arithmetic become the *positive integers*. Together these are known as *signed numbers*. Signed numbers permit us to express a quality other than magnitude, such as direction as in these examples: $-50°C$ and $+27°C$, -300 ft elevation (300 ft below sea level) and $+5000$ ft (5000 ft above sea level), and so on. Negative numbers also provide a way of indicating the result of subtracting a larger number from a smaller number; for example, $6 - 8 = -2$. Signed numbers also include fractions. A portion of the set of signed numbers is depicted on the number line in Figure 1-2.

Reading Signed Numbers

-5 is read "minus five" or "negative five."

$+15$ is read "positive fifteen," "plus fifteen," or "fifteen."

36 is read "thirty-six," or if we want to emphasize that it is positive, "plus thirty-six."

When a number is positive, we usually omit saying plus or positive. Often we do not write the $+$ sign.

Example 2 Express the following ideas using signed numbers.

(a) 21° above zero, 10° below zero (a) $+21°, -10°$

(b) an electron charge of 2 C ($-$ particles), a proton charge of 3 C ($+$ particles) (b) -2 C, $+3$ C

(c) receive $50, pay $10 (c) $+\$50, -\$10$

Symbols of Relations Let a and b represent any two signed numbers or points on the number line. If a and b are at the same point, the relation is expressed as $a = b$, which is read "a is equal to b." If a and b are not at the same point, then $a \neq b$, which is read "a is not equal to b." Other relations include the following:

$a < b$ is read "a is less than b" and means that a is to the left of b on the number line; for example, $-5 < 2$.

$a > b$ is read "a is greater than b" and means that a is to the right of b on the number line; for example, $0 > -7$.

$a \leq b$ is read "a is less than or equal to b" and means that a is less than b or in some instances may be equal to b.

$a \geq b$ is read "a is equal to or greater than b."

$a \leq x \leq b$ is read "a is less than or equal to x is less than or equal to b" and is used to prescribe a range of values for x.

The use of the slash line, $\neq$, $\not>$, and so forth, puts a "not" in the meaning of the symbol.

MEMORY JOGGER

small end points to smaller value

larger end for larger value

Example 3   Using appropriate symbols, state the relations of the following.

(a) -15 and 3 (a) $-15 < 3$
(b) -5 and -15 (b) $-5 > -15$
(c) -550 C and 25 C   (c) -550 C < 25 C (Remember: C is the abbreviation for coulomb, the unit of electric charge.)
(d) $\$2$ and $50\cent$ (d) $\$2 > 50\cent$

Example 4   Read (express in words) the following.

(a) $5 > -3$ (a) five is greater than minus three
(b) $x \leq y$ (b) x is less than or equal to y

(c) $Q_1 > Q_2$ (c) Q one is greater than Q two (or Q sub one is greater than Q sub two)

(In the notation Q_1 and Q_2 the 1 and 2 are called *subscripts*. Subscripts permit us to use the same literal number to represent two or more different quantities.)

Example 5   Given: x and y are integers, $-5 \leq x < 3$, and $y > 3$.

(a) Show range of x and y on the number line.

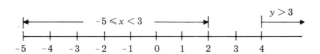

(b) List all possible values of x.   (b) x may be $-5, -4, -3, -2, -1, 0, 1,$ and 2
(c) Compare possible values of x and y.   (c) $y > x$

Magnitude and Absolute Value   A signed number consists of two parts:

$$\text{sign} \underset{\nearrow}{} -5 \underset{\nwarrow}{} \text{number part}$$

the *sign* and the *number part*. The number part of the signed number indicates the *magnitude* of the number; it is also known as its *absolute value*. The symbol $|\ \ |$ is used to indicate the absolute value; for example, $|-5| = 5$, $|+5| = 5$.

Example 6   Express the following symbolically.

(a) absolute value of -10 (a) $|-10|$
(b) magnitude of -50 C   (b) $|-50$ C$|$
(c) magnitude of $-x$ (c) $|-x|$

Example 7   Read (express in words) the following.

(a) $|-355|$ (a) absolute value of minus three hundred fifty-five

(b) $|12|$ (b) absolute value of twelve

(c) $|-Q|$ (c) magnitude of $-Q$

Example 8 State the equivalent of the following.

(a) $|-23|$ (a) $|-23| = 23$

(b) $-|+15|$ (b) $-|+15| = -15$

(c) $|-45\ C|$ (c) $|-45\ C| = 45\ C$

(d) $|-x|$ if $x = 2$ (d) $|-x| = |-2| = 2$

(e) $-|-y|$ if $y = 3$ (e) $-|-y| = -|-3| = -3$

Example 9 Using appropriate symbols, $<$, $>$, $=$, and so forth, write comparisons of the following sets of quantities.

(a) $-3, |-12|$ (a) $-3 < |-12|$

(b) $|-4|, 0$ (b) $|-4| > 0$

(c) $|-8\ C|, 3\ C$ (c) $|-8\ C| > 3\ C$

(d) $-|-x|, 4$ if $x = 8$ (d) $-|-x| = -|-8| = -8$ therefore $-|-x| < 4$

Example 10 Express the following ranges with symbols.

(a) x is greater than or equal to -4 but less than 5   (a) $-4 \leqslant x < 5$

(b) the charge Q is between -45 C and 15 C   (b) $-45\ C < Q < 15\ C$

(c) the charge q is between -15 C and 15 C, inclusive (may be equal to either extreme value)   (c) $-15\ C \leqslant q \leqslant 15\ C$

Exercises 1-2 In Exercises 1–6 express the stated ideas using signed numbers (see Example 2).

1. $59°$ below zero; $75°$ above zero

2. 250 ft below sea level; 4900 ft above sea level

3. receive $150; pay $75

4. electric charges of 5 C of electrons ($-$); 4 C of protons ($+$)

5. a gain of 15 yd; a loss of 7 yd

6. a space capsule increases speed by 4500 meters per second (m/s); reduces speed by 3600 m/s

In Exercises 7–24 use the symbols $<$ or $>$ to compare the given values (see Example 9).

7. $-3, 2$ 8. $-20, 3$ 9. $-2, -15$

10. $-3, -25$ 11. $0, -15$ 12. $0, 11$

13. $-100, -40$ 14. $-20, -60$

15. $|-5|, 3$ 16. $12, |-20|$

17. $|-18|, |-30|$ 18. $25, |+30|$

19. $|x|, 4$ (if $x = -5$)

20. $|-y|, 15$ (if $y = 22$)   21. $-5\ C, 3\ C$

22. $|-8\ C|, 3\ C$

23. Q_1, Q_2 (if $Q_1 = -4\ C, Q_2 = -9\ C$)

24. $|Q_1|, Q_2$ (if $Q_1 = -50\ C, Q_2 = 40\ C$)

In Exercises 25–30 x is an integer. (a) Show range of x on a number line. (b) List the possible values of x permitted by the stated ranges (see Example 5).

25. $-2 \leqslant x \leqslant 2$ 26. $-4 < x \leqslant 3$

27. $-2 < x < 0$ 28. $0 < x < 5$

29. $-6 \leqslant x \leqslant 2$ 30. $-15 < x < -10$

1-3 ADDITION OF SIGNED NUMBERS

Operations with signed numbers are performed according to definite rules which we state without proof. Memorize these rules so that you can recall them quickly and easily. Study the worked-out examples and observe carefully how the rules are applied.

RULE FOR ADDITION OF SIGNED NUMBERS

1. For numbers with like signs (all + or all $-$):
   a. Add the number parts.
   b. Use the common sign with the sum.
2. For numbers with unlike signs:
   a. Subtract the absolute values (number parts), smaller from larger.
   b. Use the sign of number with larger absolute value with the result.

Example 11

(a) Add: 4 and 5
(a) Both signs are positive; we add and show the sum as positive. $4 + 5 = 9$.

(b) Add: -7 and -2
(b) Both signs are negative; we add the number parts, $7 + 2 = 9$, and use the common sign $-$; therefore $-7 + (-2) = -9$. (We use parentheses to keep the number sign $-$ separated from the sign of operation $+$.)

(c) Add: -15 and 8
(c) Signs are unlike; we subtract absolute values, $15 - 8 = 7$, and use the sign of the larger with the result, $-15 + 8 = -7$.

(d) Add: 14 and -5
(d) Signs are unlike; we subtract absolute values, $14 - 5 = 9$, and use the sign of the larger with the result, $14 + (-5) = 9$.

Example 12 Add the signed numbers.

(a) $5 + 12$ (a) $5 + 12 = 17$

(b) $-8 + (-6)$ (b) $-8 + (-6) = -14$

(c) $-25 + 12$ (c) $-25 + 12 = -13$
                        $(25 - 12 = 13$ and sign of larger is $-$.)

(d) $-20 + 35$ (d) $-20 + 35 = 15$
                        $(35 - 20 = 15$ and sign of larger is $+$.)

(e) $\begin{array}{r} 12 \\ 3 \\ \hline \end{array}$ (e) $\begin{array}{r} 12 \\ 3 \\ \hline 15 \end{array}$

(f) $\begin{array}{r} -\ 3 \\ -15 \\ \hline \end{array}$ (f) $\begin{array}{r} -\ 3 \\ -15 \\ \hline -18 \end{array}$

(g) $\begin{array}{r} 13 \\ -28 \\ \hline \end{array}$ (g) $\begin{array}{r} 13 \\ -28 \\ \hline -15 \end{array}$

(h) $\begin{array}{r} -7 \\ 9 \\ \hline \end{array}$ (h) $\begin{array}{r} -7 \\ 9 \\ \hline 2 \end{array}$

Exercises 1-3 In Exercises 1–30 add the signed numbers (see Examples 11 and 12).

1. 4 and 7   2. 16 and 7
3. 3.7 and 5.6   4. 13.6 and 12.9
5. -6 and -5 6. -15 and -2
7. -4 and -23 8. -2.5 and -6.7
9. -9 and 3   10. -3.8 and 7.3
11. -5 and 7   12. -8 and 4

13. -12 and 3   14. -6.5 and 3.7
15. -8 and 7   16. $\begin{array}{r} -13.7 \\ 2.5 \\ \hline \end{array}$

17. $\begin{array}{r} 3 \\ -7 \\ \hline \end{array}$ 18. $\begin{array}{r} 6 \\ -2 \\ \hline \end{array}$

19. $\begin{array}{r} 12 \\ -5 \\ \hline \end{array}$ 20. $\begin{array}{r} 6.5 \\ -3.2 \\ \hline \end{array}$

21. $\begin{array}{r} 15.7 \\ 6.5 \\ \hline \end{array}$ 22. $\begin{array}{r} -4.85 \\ -2.53 \\ \hline \end{array}$

23. $\begin{array}{r} 18.98 \\ -6.54 \\ \hline \end{array}$ 24. $3.75 + (-4.69)$

25. $-5.63 + (-3.58)$ 26. $-6.73 + 2.98$
27. $-15.17 + 29.59$ 28. $33.65 + (-48.69)$
29. $4.63 + (-74.96)$ 30. $48.94 + (-6.32)$

In Exercises 31–34 apply the rules for addition of signed numbers to obtain solutions.

31. A charge of 7 C is to be added to a charge of -3 C. What is total charge in coulombs?

32. The charge Q_1 is to be added to Q_2. If $Q_1 = 15$ C and $Q_2 = -27$ C, what is the sum of the charges?

33. If $Q_T = Q_1 + Q_2$, $Q_1 = -15$ C, and $Q_2 = -3$ C, what is Q_T?

34. Refer to Exercise 33. Determine Q_T when $Q_1 = -30$ C and $Q_2 = 17$ C.

1-4 SUBTRACTION OF SIGNED NUMBERS

RULE FOR SUBTRACTION OF SIGNED NUMBERS

To subtract one signed number from a second signed number
1. Change the sign of the number being subtracted. (This number is called the subtrahend.)
2. Add the signed numbers obtained using the rules for addition of signed numbers (see Section 1-3).

Study the following examples, noting carefully how the rules are applied.

Example 13

(a) Subtract: 4 from 7

(a) The 4 is being subtracted, therefore we change its sign and then add as in Section 1-3.

$$7 - (+4) = 7 + (-4) = 3$$

(b) Subtract: 8 from 3

(b) Subtracting 8, we change its sign and add.

$$3 - (+8) = 3 + (-8) = -5$$

(c) Subtract: 4 from -15

(c) We change the sign of 4 and add.

$$-15 - (+4) = -15 + (-4)$$
$$= -19$$

(d) Subtract: -5 from 7

(d) We change the sign of -5 to + and add.

$$7 - (-5) = 7 + 5 = 12$$

(e) Subtract: -8 from -3

(e) We change the sign of -8 to + and add.

$$-3 - (-8) = -3 + (+8)$$
$$= 5$$

Example 14 Perform the indicated subtractions.

(a) $-8 - (+5)$

(a) $-8 - (+5) = -8 + (-5)$
    $= -13$

(b) $12 - 25$

(b) $12 - 25 = 12 + (-25)$
    $= -13$

(c) $\begin{array}{r} -10 \\ -7 \\ \hline \end{array}$

(c) $\begin{array}{r} -10 \\ -7 \\ \hline \end{array} = \begin{array}{r} +-10 \\ +7 \\ \hline -3 \end{array}$

(d) $\begin{array}{r} -8 \\ +3 \\ \hline \end{array}$

(d) $\begin{array}{r} -8 \\ +3 \\ \hline \end{array} = \begin{array}{r} + -8 \\ -3 \\ \hline -11 \end{array}$

Example 15

(a) A body has a charge $Q_T = -45$ C. A charge of electrons $Q_e = -15$ C is removed from the body. What is Q_T after Q_e is removed?

$$-45 - (-15) = -45 + 15 = -30 \text{ C}$$

(b) A body is charged so that $Q_T = -57$ C. A charge of protons $Q_p = 24$ C is removed. Find Q_T.

$$-57 - (24) = -57 + (-24) = -81 \text{ C}$$

Exercises 1-4 In Exercises 1–14 perform the indicated subtractions (see Examples 13 and 14).

1. $5 - (+2)$ 2. $9 - 3$ 3. $3 - 8$
4. $2 - 9$ 5. $-9 - (+3)$ 6. $-3 - (+8)$
7. $-5 - (+3)$ 8. $-6 - (+2)$ 9. $-2 - 6$
10. $6 - (-3)$ 11. $9 - (-11)$
12. $3 - (-5)$ 13. $-12 - (-5)$
14. $-15 - (-18)$

In Exercises 15–30 subtract the lower number from the upper number (see Example 14).

15. $\begin{array}{r} 12 \\ -15 \\ \hline \end{array}$ 16. $\begin{array}{r} -8 \\ -19 \\ \hline \end{array}$ 17. $\begin{array}{r} -12 \\ -3 \\ \hline \end{array}$ 18. $\begin{array}{r} 3 \\ 12 \\ \hline \end{array}$

19. $\begin{array}{r} 15 \\ -8 \\ \hline \end{array}$ 20. $\begin{array}{r} -6 \\ -3 \\ \hline \end{array}$ 21. $\begin{array}{r} 0 \\ 5 \\ \hline \end{array}$ 22. $\begin{array}{r} 0 \\ -7 \\ \hline \end{array}$

23. $\begin{array}{r} -1 \\ -8 \\ \hline \end{array}$ 24. $\begin{array}{r} -2.9 \\ 3.6 \\ \hline \end{array}$ 25. $\begin{array}{r} -5.73 \\ 2.36 \\ \hline \end{array}$

26. $\begin{array}{r} 15.92 \\ -3.68 \\ \hline \end{array}$ 27. $\begin{array}{r} -2.93 \\ 15.63 \\ \hline \end{array}$ 28. $\begin{array}{r} -4.61 \\ -2.15 \\ \hline \end{array}$

29. $\begin{array}{r} 27.35 \\ -42.69 \\ \hline \end{array}$ 30. $\begin{array}{r} -16.37 \\ 2.59 \\ \hline \end{array}$

In Exercises 31–35 use the rules for subtraction of signed numbers and the relation $Q_T = Q_1 - Q_2$ to find Q_T (see Example 15).

31. $Q_1 = 12$ C, $Q_2 = -.5$ C

32. $Q_1 = 8$ C, $Q_2 = 23$ C

33. $Q_1 = -15$ C, $Q_2 = 35$ C

34. $Q_1 = -50$ C, $Q_2 = -20$ C

35. $Q_1 = -48$ C, $Q_2 = -64$ C

1-5 ALGEBRAIC SUMS

In the previous sections we have presented operations that were explicitly the addition or subtraction of signed numbers. In many of our problems, however, we will deal with expressions such as

$$3 - 7 - 5 = ?$$

and so on. Are we to subtract +7 and +5 or are we to add -7 and -5? The answer is that either approach gives the same correct result. Let us consider the expression as an addition.

$$3 + (-7) + (-5) = -4 + (-5) = -9$$

(We recall from arithmetic that terms may be added in any order, for example, $2 + 3 = 5$,

3 + 2 = 5—the *commutative law*. Also, for addition, terms may be grouped in any order, 2 + 3 + 4 = 5 + 4 = 9, 2 + 7 = 9—the *associative law*.)

If we consider the expression to mean 3 – (+7) – (+5), we evaluate it as

$$3 - (+7) = 3 + (-7) = -4$$

and

$$-4 - (+5) = -4 + (-5) = -9$$

Although either approach gives the correct answer, the most common practice is to consider all such expressions as indicating *algebraic addition*. That is, *all signs are considered to be signs of quality and all operations are to be addition*. According to this practice our example is interpreted as

$$3 - 7 - 5 = 3 + (-7) + (-5) = -9$$

This interpretation also applies equally well to expressions with only two terms.

$$3 - 5 = 3 + (-5) = -2$$

$$-4 - 2 = -4 + (-2) = -6$$

$$-5 + 3 = -5 + (+3) = -2$$

Using this interpretation, we call a result an *algebraic sum* and the process *algebraic summation*. We state the rule for finding the algebraic sum of three or more terms.

RULE FOR FINDING THE ALGEBRAIC SUM OF THREE OR MORE TERMS

1. **Find the separate totals of positive terms and negative terms.**
2. **Combine the two totals using rules for addition of two signed numbers.**

Example 16   Evaluate the given expressions.

(a) 3 + 6 – 2   (a) 3 + 6 – 2 = 9 – 2 = 7
                      (9 = 3 + 6)

(b) –3 + 6 + 2   (b) –3 + 6 + 2 = –3 + 8 = 5
                      (8 = 6 + 2)

(c) 3 – 6 – 2   (c) 3 – 6 – 2 – 4 =
    – 4   3 – 12 = –9
                      Note: –12 = (–6) +
                      (–2) + (–4)

(d) –3 – 6 + 2   (d) We find the total of the
    + 4 – 5   negatives,

$$(-3) + (-6) + (-5)$$
$$= -14$$

and of the positives, 2 + 4 = 6. Then – 14 + 6 = – 8.

(e) 5 – 2 + 7   (e) Follow the same pro-
    – 3 + 1   cedure as in (d).
    – 5 + 2

$$(-2) + (-3) + (-5) = -10$$
$$5 + 7 + 1 + 2 = 15$$
$$-10 + 15 = 5$$

Example 17   The total charge Q_T of a body is the algebraic sum of individual charges Q_1, Q_2, Q_3, and so on:

$$Q_T = Q_1 + Q_2 + Q_3 + \cdots$$

Determine Q_T when $Q_1 = -15$ C, $Q_2 = 45$ C, $Q_3 = -29$ C, $Q_4 = -36$ C, and $Q_5 = 69$ C.

$$Q_T = -15 + 45 - 29 - 36 + 69 = 34 \text{ C}$$

("net" charge is positive).

Algebraic summation can be performed conveniently on slide-rule calculators. On the calculator we consider all signs to be signs of operation since the calculator enters all numbers as positive. Consult the instruction manual for your calculator for explicit instructions for combining signed numbers.

Exercises 1-5   In Exercises 1–10 evaluate the given expressions (see Example 16). Perform the operations, first mentally, or with pencil and paper. Then check your answers by performing operations on your calculator.

1. 3 – 7 – 4   2. –2 + 8 + 5
3. –3 – 8 – 4   4. –5 + 2 + 6
5. 2 + 8 – 7 – 4   6. –3 + 7 – 8 – 5
7. 1 + 9 + 2 – 5   8. –1 – 2 – 3 + 5
9. –2 – 8 + 5 – 4 + 9
10. 1 + 12 – 5 + 4 – 6

In Exercises 11–20 total the columns.

11.	12.	13.	14.
2	–7	9	–2
–7	–8	–8	–5
5	+2	4	–3
		–7	–6

15. -9   16.   1   17.   1.5   18. -8.3
      8   -15   -2.3   -7.5
      2   -12   -6.3   5.6
     -7   3   -1.9   -3.8
     -6   - 2   3.8   -5.9

19.   1.85   20. -0.06
     -2.79   -0.78
      0.68   0.69
     -2.75   0.57
     -3.69   0.65

In Exercises 21-25 find Q_T using the relation $Q_T = Q_1 + Q_2 + Q_3 + \cdots$. The values given are for Q_1, Q_2, Q_3, and so forth (see Example 17).

21. -12 C, 15 C, -30 C
22. 45 C, -36 C, -70 C, 65 C
23. 14 C, -50 C, -70 C, 37 C, -15 C, 75 C
24. 3.5 C, -6.8 C, 8.7 C, -15.6 C, -36.9 C
25. -0.69 C, 0.75 C, 0.09 C, -1.95 C, -0.37 C

1-6 MULTIPLICATION AND DIVISION OF SIGNED NUMBERS

RULE FOR MULTIPLYING ONE SIGNED NUMBER BY ANOTHER

1. Multiply the absolute values (number parts).
2. The product is positive if both numbers have the same sign (+ or −). Use a negative sign with the product if numbers have opposite signs.

MEMORY JOGGER

$+ \cdot + = +$
$- \cdot - = +$
$+ \cdot - = -$
$- \cdot + = -$

Learn this rule so that you can recall it easily. Study the examples demonstrating the rule. We introduce here another algebraic practice; multiplication is indicated simply by writing parentheses side by side: $(-3)(-7)$ means -3 times -7.

Example 18 Find the indicated products.

(a) $(+3)(+5)$

(a) $(+3)(+5) =$
$+(3 \cdot 5) = 15$

(b) $(-3)(-7)$

(b) $(-3)(-7) =$
$+(3 \cdot 7) = 21$
The product is positive because the numbers have the same sign.

(c) $(-4)(+5)$

(c) $(-4)(+5) =$
$-(4 \cdot 5) = -20$
The product is negative because the numbers have opposite signs.

(d) The charge Q_1 is $6Q_2$ ($Q_1 = 6Q_2$). Find Q_1 if $Q_2 = -7$ C.

(d) $Q_1 = 6Q_2 = 6(-7) =$
$-(6 \cdot 7) = -42$ C

RULE FOR DIVIDING ONE SIGNED NUMBER BY ANOTHER

1. Divide the absolute values.
2. The quotient is positive if both numbers have the same sign (+ or −). Use a negative sign with the quotient if the numbers have opposite signs.

MEMORY JOGGER

$\dfrac{+}{+} = +, \quad \dfrac{-}{-} = +, \quad \dfrac{+}{-} = -, \quad \dfrac{-}{+} = -$

We demonstrate the application of the rule. Find the solutions to the sample exercises yourself, first mentally and then with your calculator. Remember, to enter a negative number you must use the "change sign" key, (+/−). In algebra division is indicated in several ways: $a \div b$, $\dfrac{a}{b}$, or a/b.

Example 19 Perform the indicated division operations.

(a) $+6 \div (+3)$

(a) $\dfrac{+6}{+3} = 2$
Numbers have the same sign.

(b) $\dfrac{-15}{-3}$

(b) $\dfrac{-15}{-3} = 5$
Numbers have like signs.

(c) $21/-7$

(c) $21/-7 = -3$
Numbers have opposite signs.

(d) A charge Q_T is to be divided into 6 equal charges, Q.

$Q = \dfrac{Q_T}{6}$

Find Q if $Q_T = -48$ C

(d) $Q = \dfrac{Q_T}{6} = \dfrac{-48}{6}$
$= -8$ C

More Than Two Factors Sometimes we encounter multiplication or division operations in which there are more than two factors involved. We state the rule that applies.

RULE FOR MULTIPLICATION OF SIGNED NUMBERS

Find the product of absolute values of factors. Use a negative sign with the product if number of negative factors is odd, that is, 1, 3, 5, 7, ...

RULE FOR DIVISION OF SIGNED NUMBERS

Find the quotient using absolute values of factors. Use a negative sign with the quotient if number of negative factors, counting both numerator and denominator, is odd.

Study the following examples carefully. Please note that in division problems we can cancel common factors in the numerator and the denominator to simplify the work. A calculator is a great time saver with this type of problem.

Example 20 Find the indicated products.

(a) $(-3)(+5)(-2)$

(a) $(-3)(+5)(-2) =$
$+(3 \cdot 5 \cdot 2) = 30$

(b) $(-1)(-4)(+2)(-3)$

(b) $(-1)(-4)(+2)(-3) =$
$-(1 \cdot 4 \cdot 2 \cdot 3) = -24$

(c) $(-3)(-2)(-4)(-1)$

(c) $(-3)(-2)(-4)(-1) =$
$+(3 \cdot 2 \cdot 4 \cdot 1) = 24$

(d) $(+3)(+2)(+4)(+1)$

(d) $(+3)(+2)(+4)(+1) =$
$+(3 \cdot 2 \cdot 4 \cdot 1) = 24$

Example 21 Perform the indicated operations.

(a) $\dfrac{(-6)(-5)}{3}$

(a) $\dfrac{(-6)(-5)}{3} =$
$+\dfrac{\overset{2}{\cancel{6}} \cdot 5}{\cancel{3}} = \dfrac{10}{1} = 10$

(b) $\dfrac{(24)(-9)}{(2)(-4)(-6)}$

(b) $\dfrac{(24)(-9)}{(2)(-4)(-6)} =$
$-\dfrac{\overset{1}{\cancel{24}} \cdot 9}{2 \cdot \cancel{4} \cdot \cancel{6}} = -\dfrac{9}{2}$
$= -4.5$

(c) $\dfrac{(-15)(-7)(+6)}{(5)(-36)}$

(c) $\dfrac{(-15)(-7)(+6)}{(5)(-36)} =$
$-\dfrac{\cancel{15} \cdot 7 \cdot \cancel{6}}{\cancel{5} \cdot \cancel{36}} =$
$-\dfrac{7}{2} = -3.5$

Exercises 1-6 In Exercises 1–50 perform the indicated operations (see Examples 18, 19, 20, and 21).

1. $(+3)(+4)$ 2. $(+6)(+2)$ 3. $(+4)(-3)$
4. $(+2)(-8)$ 5. $(-5)(+2)$ 6. $(-6)(+3)$
7. $(-7)(-3)$ 8. $(-6)(-4)$ 9. $+15 \div (+3)$
10. $+12 \div (+4)$ 11. $9/-3$ 12. $8/-4$
13. $10/-2$ 14. $18/-6$ 15. $\dfrac{-20}{5}$
16. $\dfrac{-28}{4}$ 17. $\dfrac{-24}{8}$ 18. $\dfrac{-28}{7}$ 19. $\dfrac{-4}{-2}$
20. $\dfrac{-6}{-3}$ 21. $\dfrac{-8}{-2}$ 22. $\dfrac{-15}{-5}$ 23. $\dfrac{5}{-1}$
24. $\dfrac{-7}{-1}$ 25. $\dfrac{0}{-5}$ 26. $\dfrac{0}{12}$ 27. $(2)(3)(4)$
28. $(+3)(+2)(+8)$ 29. $(-2)(+3)(+6)$
30. $(+5)(-3)(+4)$ 31. $(-3)(+4)(-2)$

32. $(+1)(-6)(-3)$ 33. $(-2)(-10)(-3)$

34. $(-1)(-12)(-2)$ 35. $(-2)(1)(3)(6)$

36. $(1)(-2)(3)(-4)$ 37. $(-10)(1)(0)(-6)$

38. $(-1)(0)(-3)(-6)$

39. $(-1)(-2)(3)(4)(-10)$

40. $(-4)(-3)(-2)(-1)(-10)$ 41. $\dfrac{2 \cdot 6}{-3}$

42. $\dfrac{5 \cdot 4}{2}$ 43. $\dfrac{-40}{5 \cdot 2}$ 44. $\dfrac{-30}{(-6)(2)}$

45. $\dfrac{(-3)(15)(-6)}{(4)(-12)}$ 46. $\dfrac{(-4)(-24)(-6)}{(12)(-8)}$

47. $\dfrac{(-48)(24)}{(-6)(-4)(18)}$ 48. $\dfrac{(-6)(64)}{(18)(16)(-2)}$

49. $\dfrac{(-2)(-3)(-49)(4)}{(-7)(14)(6)}$

50. $\dfrac{(60)(-15)(22)(14)}{(7)(12)(25)(4)}$

In Exercises 51–54 find Q_T using the relation $Q_T = nQ$, where n is the number of equal charges combined to make Q_T.

51. $Q = -5$ C, $n = 6$ 52. $Q = 15$ C, $n = 4$

53. $Q = -15.7$ C, $n = 6$ 54. $Q = -0.74$ C, $n = 12$

In Exercises 55–58 find Q using the relation $Q = Q_T/n$.

55. $Q_T = 75$ C, $n = 5$ 56. $Q_T = -36$ C, $n = 4$

57. $Q_T = -57$ C, $n = 12$ 58. $Q_T = -0.67$ C, $n = 14$

Chapter 2

Exponents and Powers of Ten: Concepts, Operations, and Applications

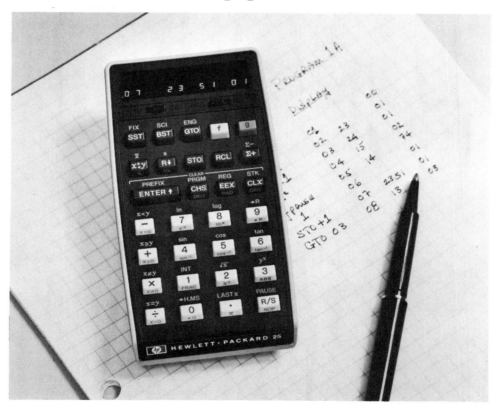

Exponents and their applications are very important in the calculations of technical disciplines. Note that the illustrated calculator has keys labeled $[y^x]$, $[x^2]$, $[\sqrt{x}]$, $[e^x]$, and $[10^x]$. The keys labeled [FIX], [SCI], and [ENG] provide for entry and readout of data in fixed decimal notation, scientific notation, or engineering notation $(10^{-3}, 10^{-6}, 10^3, 10^6,$ and so forth), respectively. (Courtesy of Hewlett Packard Company)

In Section 1-1 we stated that an exponent is a number which indicates the number of times another number, called the base, is taken as a factor. For example,

$$\underbrace{2 \cdot 2 \cdot 2 \cdot 2 \cdot 2}_{\substack{2 \text{ is a factor} \\ 5 \text{ times}}} = 2^5 \qquad \underbrace{x \cdot x \cdot x}_{\substack{x \text{ is a factor} \\ 3 \text{ times}}} = x^3$$

$$\underbrace{i \cdot i \cdot i \cdot i}_{\substack{i \text{ is a factor} \\ 4 \text{ times}}} = i^4$$

An application of exponents which will be of direct and immediate benefit to you in your study of electronics is that of *powers of ten*, the subject of the next section.

2-1 POWERS OF TEN

A brief review of our decimal number system will aid you in understanding powers of ten.

Digits The decimal number system works because of a few basic ideas. The first of these ideas is the use of *ten characters* or *symbols* to represent numbers. These ten characters, the familiar symbols, 0, 1, 2, 3, 4, 5, 6, 7, 8, 9, are called *digits*. Each, taken by itself, represents a specific quantity. The system is called a *decimal system* (deci = 10) or a *base-10 system* because it is designed around ten digits.

Place Value In order to count and represent very large or very small quantities with only ten characters the system must include some plan which enables a digit to represent more than one value. That plan provides for what is called *place value*: the location of a particular digit in a multidigit number determines the total value which that digit represents. For example, 5 simply means five; but 50 means five times ten or five 10's, 500 means five times one hundred or five 100's, and so on.

Each place in a multidigit number, then, has a value or name, its *place value* or *place name*. Because our system is based on ten, the place values are related to each other by powers of ten. Here are examples of powers of ten and their relation to the place values.

Power of ten		Decimal equivalent		Name
10^{-6}	=	0.000001	=	one millionth
10^{-5}	=	0.00001	=	one hundred-thousandth
10^{-4}	=	0.0001	=	one ten-thousandth
10^{-3}	=	0.001	=	one thousandth
10^{-2}	=	0.01	=	one hundredth
10^{-1}	=	0.1	=	one tenth
10^{0}	=	1	=	one unit
10^{1}	=	10	=	ten
10^{2}	=	100	=	one hundred
10^{3}	=	1000	=	one thousand
10^{4}	=	10,000	=	ten thousand
10^{5}	=	100,000	=	one hundred-thousand
10^{6}	=	1,000,000	=	one million

etc.

(Powers of ten are read: 10^{-6}, "ten to the minus six," 10^{2}, "ten to the two" or "ten squared," 10^{5}, "ten to the fifth," etc.)

A table, showing how powers of ten and place names relate to the way we write numbers, is also useful. See Figure 2-1.

10^6	10^5	10^4	10^3	10^2	10^1	10^0	.	10^{-1}	10^{-2}	10^{-3}	10^{-4}	10^{-5}	10^{-6}
millions	hundred-thousands	ten-thousands	thousands	hundreds	tens	units	decimal point	tenths	hundredths	thousandths	ten-thousandths	hundred-thousandths	millionths

Figure 2-1

Let us consider further the relation between place value and powers of ten in the way we write numbers. Let us take a portion of the table of Figure 2-1, for example, and place the digit 6 in the *thousands* position, that is, 6___. To show, without question, the location intended for the 6 we write zeros in the positions between it and the decimal point; thus 6000. We know from arithmetic that we read this "six thousand." We now want to realize that it also means 6×1000 or 6×10^3. This, in turn, points the way to representing any number with powers of ten. If 6000 can be represented as 6×10^3, then 5360 can be represented as 5.36×10^3, and so on.

Scientific Notation The number 5360 is said to be in *common notation (CN)*. When a number is written in the form 5.36×10^3, it is said to be in *scientific notation (SN)*. This notation consists of *two parts* which are *factors*. The first factor contains the significant digits of the original number (that is, 5, 3, and 6 in our example); it is called the *mantissa* (Mn). The second part is an appropriate power of ten (*PT*) factor; it is called the *characteristic:*

$$5360 \;=\; \underbrace{5.36}_{\text{mantissa}} \times \underbrace{10^3}_{\text{characteristic}}$$

$$\underbrace{\text{number in } CN}_{} \qquad \underbrace{\text{equivalent in } SN}_{}$$

Here is a 1-2-3 procedure for converting from common notation *CN* to scientific notation *SN*.

To convert a number given in common notation CN to its equivalent in scientific notation SN:

1. Rewrite the number placing the decimal point (DP) immediately to the right of the first nonzero digit. Write the 10 factor.

    $$12345.67 \longrightarrow 1.234567 \times 10$$

2. Count the number of places the decimal point was moved. Write this integer as the number part of the exponent of the ten factor.

    $$12345.67 \longrightarrow 1.234567 \times 10^4$$
    $$\underbrace{4\;3\;2\;1}$$

    DP moved 4 places, "4" becomes number part of exponent

3. Note the direction that the decimal point was moved and affix the sign of the exponent. If DP was moved:
    (a) LEFT, exponent is POSITIVE and no sign need be written (+ may be written).
    (b) RIGHT, write – in front of exponent.

    $$36750.7 = 3.67507 \times 10^4 \quad (\text{or } 10^{+4})$$
    $$\underbrace{4\;3\;2\;1}$$
    $\longleftarrow$ left, hence exponent is 4 (or +4)
    $$0.00367507 = 3.67507 \times 10^{-3}$$
    $$\underbrace{1\;2\;3}$$
    $\longrightarrow$ right, –3 is exponent

Example 1 Convert the following numbers in *CN* to their equivalents in *SN*.

(a) 3560

(a) $3\underbrace{560}_{3\,2\,1} = 3.56 \times 10^3$

(b) 56

(b) $5\underbrace{6}_{1} = 5.6 \times 10$

(when exponent is 1 it is not written)

(c) 754,000

(c) $7\underbrace{54,000}_{5\,4\,3\,2\,1} = 7.54 \times 10^5$

(d) 0.000972

(d) $0.000\underbrace{972}_{1\,2\,3\,4} = 9.72 \times 10^{-4}$

(e) 3.8

(e) $3.8 = 3.8$ (Since $1 \leqslant 3.8 < 10$, this number is already in *SN*. Sometimes we write it 3.8×10^0.)

MEMORY JOGGER

Common Notation $\longrightarrow$ Scientific Notation
If DP moved left, exponent is + ($\overset{+}{\longleftarrow}$.).
If DP moved right, exponent is – (. $\overset{-}{\longrightarrow}$).

Powers-of-Ten Notation Scientific notation is a very precisely defined format: the decimal point *must always* be written immediately to the right of the first nonzero digit. Or, $1 \leqslant \text{Mn} < 10$, "the mantissa is always a number between 1 and 10." Scientific notation is used for the final form of the solution of a problem or to give data in textbooks. In solving problems in electronics, however, it may be convenient to use a form of power-of-ten notation in which the mantissa is not between one and ten. For purposes of the problem it may be more convenient to work with a form like one of the following:

56×10^4 or 0.78×10^{-6} or 375×10^3, etc.

These are simply called numbers in *powers-of-ten notation (PTN)*. Converting to *PTN* is very similar to converting to *SN*.

To convert a number given in CN to its equivalent in PTN:
1. Rewrite the number and include the factor 10. Place the DP in the desired location.
2. Count the number of places the DP was moved. Write this integer as the number part of the exponent of the ten factor.
3. Note the direction the DP was moved and affix appropriate sign to exponent:
   (a) DP moved left, exponent is + ($\xleftarrow{+}$.).
   (b) DP moved right, exponent is – (. $\xrightarrow{-}$).

To convert a number in PTN (or SN) to its equivalent in CN:
1. Rewrite the Mn leaving out the decimal point but placing a caret (‸) at the previous location of the DP (or simply make a mental note of the location of the DP),

$$123.456 \times 10^2 \longrightarrow 123\underset{\wedge}{456}$$

2. Prepare to count off decimal places to determine new location of DP:
   If exponent is POSITIVE, prepare to count RIGHT.
   If exponent is NEGATIVE, prepare to count LEFT.
3. Count places, from original location of DP, in direction determined in Step 2, until count equals number part of exponent. Write DP at that position.

$$123.456 \times 10^2 = 12345.6$$
$$\underset{\underset{\xrightarrow{}}{1\,2}}{}$$

$$123.456 \times 10^{-5} = 0.00123456$$
$$\underset{\underset{\xleftarrow{}}{5\,4\,3\,2\,1}}{}$$

Example 2 Convert the following to *PTN*, with decimal point located as specified.

(a) 169800 with mantissa between 10 and 100

(a) $169800 =$
$\underset{4\,3\,2\,1}{}$
$\xleftarrow{}$
16.98×10^4

(b) 0.0000375 with DP to the left of first nonzero digit

(b) $0.0000375 =$
$\underset{1\,2\,3\,4}{}$
$\xrightarrow{}$
0.375×10^{-4}

(c) 910,000 with *DP* at point marked with caret (^)

(c) $910,000 =$
91×10^4

(d) 1,500,000 with *DP* at point marked with caret

(d) $1,500,000 =$
150×10^4

(e) 0.00000417 with $100 < \text{Mn} < 1000$

(e) $0.00000417 =$
417×10^{-8}

(f) 5 with $1000 < \text{Mn} < 10,000$

(f) $5.000 =$
5000×10^{-3}

Example 3 Convert the following to *CN*.

(a) 6.28×10^5
(a) $6.28 \times 10^5 = 628,000$
$\quad {}_5\llcorner\!\!\xrightarrow{}$

(b) 3.651×10^{-3}
(b) $3.651 \times 10^{-3} = 0.003651$
$\quad \xleftarrow{}\!\lrcorner_3$

(c) 485×10^3
(c) $485 \times 10^3 = 485,000$
$\quad {}_3\llcorner\!\!\xrightarrow{}$

(d) 0.00692×10^2
(d) $0.00692 \times 10^2 = 0.692$
$\quad {}_2\llcorner\!\!\xrightarrow{}$

(e) 375×10^{-4}
(e) $375 \times 10^{-4} = 0.0375$
$\quad \xleftarrow{}\!\lrcorner_4$

(f) 0.0593×10^{-2}
(f) $0.0593 \times 10^{-2} = 0.000593$
$\quad \xleftarrow{}\!\lrcorner_2$

Converting from Scientific Notation or Powers-of-Ten Notation to Common Notation Converting a number expressed with powers of ten to common decimal form requires that we reverse the steps outlined above.

MEMORY JOGGER

PTN (or SN) ⟶ CN

1. Sign of exponent + (or no sign), move DP RIGHT ⟶.
2. Sign of exponent –, move DP LEFT ⟵.

Exercise 2-1A In Exercises 1–6 convert the given numbers to numbers expressed in SN.

1. (a) 45
   (b) 450
   (c) 4500
   (d) 4.5
   (e) 450,000
   (f) 4,500,000

2. (a) 10.5
   (b) 105
   (c) 1.05
   (d) 1050
   (e) 10,500
   (f) 10,500,000

3. (a) 0.077
   (b) 0.77
   (c) 0.0077
   (d) 0.00077
   (e) 7.7
   (f) 0.0000077

4. (a) 0.00053
   (b) 0.0000053
   (c) 0.53
   (d) 5.3
   (e) 0.000000053
   (f) 0.00000000000053

5. (a) 8.13
   (b) 0.00813
   (c) 8,130,000
   (d) 813,000,000
   (e) 0.0000000813
   (f) 81.3

6. (a) 0.0000376
   (b) 3760
   (c) 37.6
   (d) 3.76
   (e) 376,000,000
   (f) 376,000,000,000

In Exercises 7–9 convert the given numbers to their equivalent in *PTN* with the decimal point of the mantissa located at position marked with a caret (ˏ) (see Example 2).

7. (a) 36ˏ84   (b) 368ˏ400   (c) ˏ3684
   (d) 0.036ˏ84   (e) 0.000ˏ3684
   (f) 0.03684ˏ

8. (a) ˏ37.0   (b) 37.00ˏ   (c) ˏ0037.0
   (d) 0.00ˏ37   e) 0.000ˏ37
   (f) 0.03700ˏ

9. (a) 0.000ˏ901   (b) 0.09010ˏ
   (c) 0.90ˏ1   (d) 90ˏ10   (e) 901ˏ000
   (f) ˏ90100

10. Convert the given numbers to their equivalent in *PTN* with mantissas between 10 and 100 (10 < Mn < 100).
    (a) 2750   (b) 36,800   (c) 0.00573
    (d) 0.000619   (e) 0.0253   (f) 230

11. Convert the given numbers to their equivalent in *PTN* with 100 < Mn < 1000.
    (a) 31,200   (b) 0.0463   (c) 5720
    (d) 0.000695   (e) 0.5970
    (f) 6,800,000

12. Convert the given numbers to their equivalent in *PTN* with 0.1 < Mn < 1.0.
    (a) 3790   (b) 0.00972   (c) 0.00563
    (d) 365,000   (e) 1,500,000   (f) 2.357

In Exercises 13–24 convert the given numbers to *CN*.

13. (a) 3.56×10^3
    (b) 3.56×10^5
    (c) 3.56×10^{-2}
    (d) 3.56×10^{-6}
    (e) 3.56×10^{-9}
    (f) 3.56×10^{12}

14. (a) 1.09×10^{-1}
    (b) 1.09×10
    (c) 1.09×10^0
    (d) 1.09×10^7
    (e) 1.09×10^{-4}
    (f) 1.09×10^{-7}

15. (a) 9.12×10^{-6}
    (b) 9.12×10^6
    (c) 9.12×10^{-5}
    (d) 9.12×10^{15}
    (e) 9.12×10^2
    (f) 9.12×10^{-3}

16. (a) 4.35×10^0
    (b) 4.35×10
    (c) 4.35×10^{-1}
    (d) 4.35×10^{-7}
    (e) 4.35×10^{-13}
    (f) 4.35×10^3

17. (a) 8.88×10^{-5}
    (b) 8.88×10^5
    (c) 8.88×10^4
    (d) 8.88×10^7
    (e) 8.88×10^{-4}
    (f) 8.88×10^{-7}

18. (a) 2.13×10^{-1}
    (b) 2.13×10^{-9}
    (c) 2.13×10^{-2}
    (d) 2.13×10^{11}
    (e) 2.13×10^4
    (f) 2.13×10^2

19. (a) 628×10^2 (b) 628×10^{-4}
    (c) 6280×10^2 (d) 62.8×10^{-3}
    (e) 0.628×10^{-2} (f) 0.00628×10^4

20. (a) 3005×10^{-2} (b) 0.3005×10^4
    (c) 0.003005×10^{-1}
    (d) 0.0003005×10^5
    (e) $30,050 \times 10^{-3}$
    (f) $300,500 \times 10^{-2}$

21. (a) 17.8×10 (b) 17.8×10^{-1}
    (c) 1780×10^2 (d) 0.0178×10^{-1}
    (e) 0.00178×10^{-3}
    (f) 0.000178×10^6

22. (a) 36.8×10 (b) 4750×10^{-3}
    (c) $378{,}000 \times 10^2$
    (d) 0.00573×10^2 (e) 0.0617×10^3
    (f) 0.000309×10^2
23. (a) 0.00603×10^{-2} (b) 75×10^5
    (c) 67.3×10^4 (d) $20{,}500 \times 10$
    (e) 3160×10^{-5} (f) 0.761×10^{-2}
24. (a) 376×10^4 (b) $43{,}700 \times 10^{-3}$
    (c) 12.5×10^5 (d) 0.527×10^4
    (e) 0.657×10^{-2} (f) 23.7×10^3

Specified Characteristic It is frequently necessary to convert a number to its equivalent in *PTN* with a specified exponent in the characteristic (ten factor). For example, when working with multiple and submultiple units (Section 2-2) we must frequently convert numbers to their equivalent in *PTN* but with characteristics in which the exponents are multiples of 3, or -3. The number we start with may be in *CN* or *PTN*. Here is a 1-2-3 procedure for converting from *CN* to *PTN* with specified characteristic.

Example 4 Convert the following numbers to their equivalents in *PTN* with characteristics as specified.

(a) $560{,}000$ (to Ch $= 10^3$)
(a) $560{,}000 = 560.000 \times 10^3$
               $\longleftarrow 3$

(b) 0.00075 (to Ch $= 10^{-6}$)
(b) $0.00075 = 000750. \times 10^{-6} = 750 \times 10^{-6}$
            $6\ \longrightarrow$

(c) 4700 (to Ch $= 10^{-3}$)
(c) $4700 = 4700000. \times 10^{-3}$
            $3\ \longrightarrow$

(d) 0.0056 (to Ch $= 10^3$)
(d) $0.0056 = 0.0000056 \times 10^3$
            $\longleftarrow\ 3$

When it is necessary to convert a number in *PTN* or *SN* to the equivalent in *PTN* with a different, but specified, characteristic, great care must be exercised to avoid making an error. The following 1-2-3 procedure outlines the steps necessary.

To convert a number in CN to its equivalent in PTN with specified characteristic:
1. Rewrite the number temporarily omitting the decimal point but mentally noting its location in the given number. Write the desired characteristic.
2. Prepare to count off decimal places to determine new location of DP:

   If exponent of desired characteristic is POSITIVE, prepare to count LEFT.
   If exponent of desired characteristic is NEGATIVE, prepare to count RIGHT.

3. Count places beginning at previous position of DP, in direction determined in Step 2, until count equals number part of exponent. Write DP at that position.

To convert a number in PTN/SN to its equivalent in PTN with a specified characteristic:
1. Rewrite the mantissa and the new, specified characteristic. Temporarily omit the DP from the mantissa.
2. Prepare to count off decimal places to determine new location of DP:

   If new characteristic is LARGER than old, prepare to move DP to LEFT.
   If new characteristic is SMALLER than old, prepare to move DP to RIGHT. (Remember: $10^{-6} < 10^{-3}$, $10^3 < 10^6$, etc.)

3. Find the difference in absolute values of new and old exponents. Count off decimal places, starting from previous location of DP, in direction determined in Step 2, until count equals difference (in exponents).

Example 5 Convert the following numbers to their equivalents in *PTN* with the specified characteristics.

(a) 1.5×10^6 (to Ch = 10^3)

(a) $1.5 \times 10^6 =$
1500×10^3
3⌐→

(b) 72×10^3 (to Ch = 10^6)

(b) $72 \times 10^3 =$
0.072×10^6
←⌐3

(c) 0.0025×10^{-3} (to Ch = 10^{-6})

(c) $0.0025 \times 10^{-3} =$
0002.5×10^{-6}
3⌐→
($10^{-6} < 10^{-3}$, hence move *DP* RIGHT)

(d) 755×10^{-6} (to Ch = 10^{-3})

(d) $755 \times 10^{-6} =$
0.755×10^{-3}
←⌐3
($10^{-3} > 10^{-6}$, move *DP* LEFT)

MEMORY JOGGER
CONVERTING TO SPECIFIED CHARACTERISTIC

If new characteristic is smaller than old, new mantissa must be LARGER than old. If new characteristic is LARGER than old, new mantissa must be smaller than old.

Exercises 2-1B In Exercises 25–28 convert the numbers given in *CN* to their equivalents in *PTN* with characteristics as specified in the parentheses to the right of the numbers (see Example 4).

25. (a) 4570 (to Ch = 10^3)
 (b) 45,700 (to Ch = 10^6)
 (c) 457,000 (to Ch = 10^3)
 (d) 0.0457 (to Ch = 10^{-3})
 (e) 0.00457 (to Ch = 10^{-6})
 (f) 0.000000457 (to Ch = 10^{-9})
26. (a) 560,000 (to Ch = 10^6)
 (b) 560,000 (to Ch = 10^3)
 (c) 560,000,000 (to Ch = 10^6)
 (d) 0.00056 (to Ch = 10^{-3})
 (e) 0.00056 (to Ch = 10^{-6})
 (f) 0.00056 (to Ch = 10^{-9})

27. (a) 810 (to Ch = 10^{-3})
 (b) 81000 (to Ch = 10^3)
 (c) 8100 (to Ch = 10^6)
 (d) 0.810 (to Ch = 10^{-3})
 (e) 0.0081 (to Ch = 10^{-6})
 (f) 0.000081 (to Ch = 10^{-9})
28. (a) 0.00048 (to Ch = 10^{-6})
 (b) 0.480 (to Ch = 10^{-6})
 (c) 4.8 (to Ch = 10^3)
 (d) 48 (to Ch = 10^3)
 (e) 4800 (to Ch = 10^6)
 (f) 480,000,000 (to Ch = 10^9)

In Exercises 29–36 convert the numbers given in *PTN* to their equivalents in *PTN* with new characteristics as specified in the parentheses to the right of the numbers (see Example 5).

29. (a) 673×10^3 (to Ch = 10^6)
 (b) 67.3×10^4 (to Ch = 10^6)
 (c) 67.3×10^4 (to Ch = 10^3)
 (d) 0.673×10^{-3} (to Ch = 10^{-6})
 (e) 0.0673×10^{-6} (to Ch = 10^{-9})
 (f) 673×10^{-6} (to Ch = 10^{-3})
30. (a) 0.455×10^6 (to Ch = 10^3)
 (b) 455×10^6 (to Ch = 10^9)
 (c) 455×10^{-9} (to Ch = 10^{-6})
 (d) $455,000 \times 10^{-9}$ (to Ch = 10^{-3})
 (e) 0.000455×10^{-3} (to Ch = 10^{-9})
 (f) 0.000455×10^{-6} (to Ch = 10^{-12})
31. (a) 703×10 (to Ch = 10^3)
 (b) 703×10^2 (to Ch = 10^6)
 (c) 7.03×10^{-2} (to Ch = 10^{-3})
 (d) 70.3×10^{-4} (to Ch = 10^{-6})
 (e) 0.703×10^{-7} (to Ch = 10^{-9})
 (f) 7.03×10^{-11} (to Ch = 10^{-9})
32. (a) 747×10^{-1} (to Ch = 10^3)
 (b) 747×10^{-2} (to Ch = 10^{-4})
 (c) 747×10^2 (to Ch = 10^3)
 (d) 0.0747×10^5 (to Ch = 10^3)
 (e) 0.00747×10^9 (to Ch = 10^6)
 (f) 74.7×10^{-3} (to Ch = 10^{-6})
33. (a) 755×10^{-5} (to Ch = 10^{-3})
 (b) 0.00755 (to Ch = 10^{-6})
 (c) 755,000 (to Ch = 10^6)
 (d) 0.755×10^6 (to Ch = 10^3)

34. (a) 10.7×10^6 (to Ch = 10^3)
    (b) $1,570,000,000$ (to Ch = 10^{12})
    (c) 0.000367 (to Ch = 10^{-9})
    (d) 759×10^{-5} (to Ch = 10^{-6})
35. (a) 5400 (to Ch = 10^5)
    (b) 74.04×10^4 (to Ch = 10^6)
    (c) 63.05×10^{-5} (to Ch = 10^{-6})
    (d) 0.0000472 (to Ch = 10^{-3})
36. (a) 0.0319×10^{-4} (to Ch = 10^{-9})
    (b) 0.00913×10^{-7} (to Ch = 10^{-12})
    (c) 40500×10^{-13} (to Ch = 10^{-6})
    (d) 0.00517×10^{13} (to Ch = 10^6)

2-2 UNITS AND UNIT CONVERSIONS

Most, if not all, the quantities we use to describe the properties and responses of electric/electronic circuits, as well as a host of other phenomena in our world, are measurable. If we can measure something, we agree upon a *standard amount*—a *unit*—of that something, and then assign a *name* to that unit. A set of such amounts and names is right now (1978) in the process of being adopted worldwide. This updated and standardized set of units was agreed on by an international conference in 1960 and is called Systéme International d'Unités (International System of Units), or officially SI, for short. Listed here (Table 2-1) are the most frequently used electrical units and their abbreviations. Also listed are the letter symbols most frequently used to represent the associated variables.

Many units have proven to be of inconvenient size for electronics. For such units the numbers that arise out of actual circuits are typically either very large or very small. For example, resistance values in electronics are commonly greater than 1000 Ω, current values are of the order of 0.001 A, and frequently less. Very large and very small numbers make calculations cumbersome and communications about them awkward. To ease this burden we use *multiple* and *submultiple* units.

A multiple or submultiple unit is made up of the basic unit name and a prefix, for example,

milliampere (mA),
Prefix Unit   Abbreviation
        name   of submultiple unit

kilohm (kΩ)
Prefix Unit   Abbreviation
        name   of multiple unit

Each prefix, which may be used with any unit, corresponds to a specific power-of-ten

Table 2-1

Quantity	Variable symbol	Electrical unit name	Standard abbreviation of unit
potential difference	V, v	volt	V
current	I, i	ampere	A
energy	W	joule	J
power	P	watt	W
resistance	R, r	ohm	Ω
conductance	$G, g\ (1/R)$	siemens (formerly mho)	S (formerly ℧)
capacitance	C	farad	F
inductance	L	henry	H
reactance	X	ohm	Ω
impedance	Z	ohm	Ω
frequency	$f\ (1/T)$	hertz (cycles per second)	Hz
time	t	seconds	s
period	$T\ (1/f)$	seconds	s
wavelength	λ	meters	m

Table 2-2

Prefix	Standard abbreviation (or symbol)	Equivalent power-of-ten factor	Example of typical application
exa	E	10^{18}	EHz (10^{18} hertz)
peta	P	10^{15}	PHz (10^{15} hertz)
tera	*T*	10^{12}	*Thz (10^{12} hertz)*
giga	*G*	10^{9}	*GHz (10^{9} hertz)*
mega	*M*	10^{6}	*MΩ (10^{6} ohm)*
kilo	*k*	10^{3}	*kΩ (10^{3} ohm)*
—	—	$10^{0} = 1$	*basic unit*
milli	*m*	10^{-3}	*mA (10^{-3} ampere)*
micro	*μ**	10^{-6}	*μS (10^{-6} siemens)*
nano	*n*	10^{-9}	*ns (10^{-9} seconds)*
pico	*p*	10^{-12}	*pF (10^{-12} farads)*
femto	f	10^{-15}	fs (10^{-15} seconds)
atto	a	10^{-18}	as (10^{-18} seconds)

*Greek letter mu.

factor. The prefix and its equivalent factor can quite literally be exchanged, one for the other, in the writing of the unit. For example, the equivalent factor of "milli" is 10^{-3}; therefore

$$25 \text{ mA} = 25 \times 10^{-3} \text{ A}$$

or

$$450 \times 10^{-3} \text{ A} = 450 \text{ mA}$$

Table 2-2 lists the common prefixes, their equivalent power-of-ten factors, and their standard abbreviations with typical application. The prefixes most frequently used are printed in italics. We cannot stress too strongly the value of memorizing the prefixes and their power-of-ten equivalents in this table.

MEMORY JOGGERS

1. All the prefixes listed in Table 2-2 (there are other prefixes which are used in fields other than electronics) have equivalent power-of-ten factors whose exponents are multiples of three (in converting between units we will move the decimal point multiples-of-three places). (At least one calculator manufacturer refers to numbers expressed with such powers of ten—10^{-6}, 10^{-3}, 10^{3}, and so forth—as *engineering notation* because these factors are so common in engineering calculations.)

2. Abbreviations for all multiple units, except *kilo*, are capital letters; abbreviations for all submultiple units are lower-case letters. (Multiple unit = *large* unit = *large letter*)

Converting from Basic Units to Prefixed Units To be successful in learning and practicing electronics we must be able to convert, quickly and accurately, between the various forms of units.

To convert from a basic unit to a multiple or submultiple unit (a "prefixed" unit):

1. Determine the power-of-ten factor which corresponds to the prefix of the desired unit (see Table 2-2).
2. Convert the given basic-unit value to an expression containing the power-of-ten factor of the desired prefixed unit (see Section 2-1).
3. Replace power-of-ten multiplier with the prefix of the desired unit.

Example 6 Convert the given basic-unit values to their equivalent in the unit specified in parentheses.

(a) 0.025 A (mA)
(a) 0.025 A = 25×10^{-3} A = 25 mA

(b) 27000 Ω (kΩ)
(b) 27000 Ω = 27×10^{3} Ω = 27 kΩ

(c) 6,800,000 Ω (MΩ)
(c) 6,800,000 Ω = 6.8 × 10⁶ Ω = 6.8 MΩ

(d) 2.2 × 10⁻¹²F (pF)
(d) 2.2 × 10⁻¹² F = 2.2 pF

(e) 36 × 10⁴ Ω (kΩ)
(e) 36 × 10⁴ Ω = 360 × 10³ Ω = 360 kΩ

(f) 57 × 10⁻⁴ A (mA)
(f) 57 × 10⁻⁴ A = 5.7 × 10⁻³ A = 5.7 mA

(g) 57 × 10⁻⁴ A (μA)
(g) 57 × 10⁻⁴ A = 5700 × 10⁻⁶ A = 5700 μA

> To change from one prefixed unit to another prefixed unit:
> 1. Replace the first prefix with its equivalent power-of-ten multiplier (Table 2-2).
> 2. Convert the number to an expression containing a power-of-ten multiplier equal to that of the desired prefix (see Section 2-1 and Table 2-2).
> 3. Replace power-of-ten multiplier with the desired prefix.

The procedure for converting from a prefixed unit to a basic unit is very direct.

> To convert from a prefixed unit to a basic unit:
> 1. Replace the prefix with its power-of-ten equivalent (see Table 2-2).
> 2. Convert the number obtained to SN or CN, as desired.

The steps required for this type of conversion are demonstrated in Example 8. Study each example carefully until you are sure you understand the procedural steps and can follow their implementation.

Example 7 Convert the following quantities to basic units.

(a) 35 mA   (a) 35 mA = 35 × 10⁻³ A = 3.5 × 10⁻² A or 0.035 A

(b) 47 kΩ   (b) 47 kΩ = 47 × 10³ Ω = 4.7 × 10⁴ Ω or 47,000 Ω

(c) 0.33 MΩ   (c) 0.33 MΩ = 0.33 × 10⁶ Ω = 3.3 × 10⁵ Ω or 330,000 Ω

(d) 7.56 GHz   (d) 7.56 GHz = 7.56 × 10⁹ Hz

(e) 27 μV   (e) 27 μV = 27 × 10⁻⁶ V = 2.7 × 10⁻⁵ V

Example 8 Convert the following to the unit specified in parentheses.

(a) 0.015 mA (μA)
(a) 0.015 mA = 0.015 × 10⁻³ A = 15 × 10⁻⁶ A = 15 μA

(b) 4400 μs (ms)
(b) 4400 μs = 4400 × 10⁻⁶ s = 4.4 × 10⁻³ s = 4.4 ms

(c) 150 pF (μF)
(c) 150 pF = 150 × 10⁻¹² F = 0.00015 × 10⁻⁶ F = 0.00015 μF

(d) 0.68 MΩ (kΩ)
(d) 0.68 MΩ = 0.68 × 10⁶ Ω = 680 × 10³ Ω = 680 kΩ

(e) 1540 kHz (MHz)
(e) 1540 kHz = 1540 × 10³ Hz = 1.54 × 10⁶ Hz = 1.54 MHz

The third conversion procedure, which we find we need to perform frequently in electronics, is that of changing from one prefixed unit to another prefixed unit.

Exercises 2-2 In Exercises 1 and 2 convert the given basic-unit values to corresponding "milli" units.

1. (a) 0.065 A   (b) 0.256 s
   (c) 0.375 H   (d) 0.037 S
   (e) 1.568 A   (f) 0.000739 s
   (g) 1.037 V   (h) 0.0049 V
   (i) 0.00000425 V
   (j) 32.5 A

2. (a) 0.019 A   (b) 0.483 s
   (c) 0.419 H   (d) 0.565 V
   (e) 0.000043 V   (f) 0.077 s

(g) 6.397 A   (h) 0.059 H

(i) 0.0000129 S

(j) 21.65 V

In Exercises 3 and 4 convert the given basic-unit values to "kilo" units.

3. (a) 5600 Ω (b) 750 V
   (c) 12,000 Hz   (d) 680 Ω
   (e) 330,000 V   (f) 1,010,000 Hz
   (g) 1600 m   (h) 72 Ω
   (i) 2200 Ω (j) 980 Hz

4. (a) 1500 Hz   (b) 33,000 Ω
   (c) 2400 V   (d) 28,000 m
   (e) 640,000 Hz   (f) 480 Ω
   (g) 47,000 V   (h) 1800 Ω
   (i) 455,000 Hz   (j) 91 Ω

In Exercises 5 and 6 convert the given basic-unit values to "Mega" units.

5. (a) 455,000 Hz   (b) 2,700,000 Ω
   (c) 770,000 V   (d) 680,000 Ω
   (e) 10,700,000 Hz
   (f) 68,000 Ω (g) 48,000,000 Hz
   (h) 150,000 Ω (i) 800,000,000 Hz
   (j) 410,000 Ω

6. (a) 56,000 Hz   (b) 620,000 Ω
   (c) 7,500,000 Ω (d) 570,000 Hz
   (e) 88,000,000 Hz
   (f) 5,100,000 Ω (g) 3,579,545 Hz
   (h) 920,000 Hz   (i) 130,000 Ω
   (j) 685,793 Ω

In Exercises 7 and 8 convert the given basic-unit values to "micro" units.

7. (a) 0.0000056 F
   (b) 0.000458 S
   (c) 0.00000057 s
   (d) 0.002755 H
   (e) 0.000015 F
   (f) 0.000985 H
   (g) 0.00000000015 F
   (h) 0.0000877 S
   (i) 0.000078 s
   (j) 0.03278 s

8. (a) 0.0000237 H
   (b) 0.000000045 F
   (c) 0.000565 s

(d) 0.00637 S

(e) 0.000000075 F

(f) 0.000084 H

(g) 0.0000968 s

(h) 0.495 H

(i) 0.0369 s

(j) 0.00000246 F

In Exercises 9 and 10 convert the given basic-unit values to convenient prefixed-unit values of your choosing.

9. (a) 27,000,000,000 Hz
   (b) 0.00000000015 F
   (c) 0.00000000000000045 m
   (d) 500,000,000,000,000,000 Hz
   (e) 0.0000000025 W
   (f) 15,000,000,000,000 Hz
   (g) 8.0×10^{17} Hz
   (h) 5.5×10^{-11} F
   (i) 1.8×10^{-19} m
   (j) 7.0×10^{13} Hz

10. (a) 0.0000000000002 F
    (b) 3,000,000,000,000,000 Hz
    (c) 50,000,000,000,000 Hz
    (d) 0.0000000000000004 m
    (e) 0.0000000000006 s
    (f) 7.0×10^{17} Hz
    (g) 6.5×10^{-13} W
    (h) 8.0×10^{19} Hz
    (i) 3.5×10^{14} Hz
    (j) 5.5×10^{-14} m

In Exercises 11–14 convert the given prefixed-unit values to basic-unit values. Express numbers in either *CN* or *SN*.

11. (a) 15 mA   (b) 370 kΩ (c) 53 μs
    (d) 71 MΩ (e) 960 ns
    (f) 840 GHz   (g) 675 μH
    (h) 0.48 MΩ (i) 0.029 kΩ
    (j) 0.16 fW

12. (a) 455 kHz   (b) 65 mV
    (c) 84 MHz   (d) 91 ns   (e) 75 pF
    (f) 50 GHz   (g) 35 fW   (h) 19 mA
    (i) 0.68 kΩ (j) 8.2 μW

13. (a) 10.7 MHz   (b) 0.015 μF
    (c) 0.041 MΩ (d) 720 kΩ

(e) 3.7 mS   (f) 3.6 ns   (g) 5 fW

(h) 8.2 μs   (i) 0.64 MHz

(j) 1.5 EHz

14. (a) 4.5 THz   (b) 27 pF   (c) 3.3 kΩ

(d) 56 kΩ (e) 680 kΩ (f) 250 μH

(g) 2.7 ns   (h) 4.3 aW   (i) 5.5 mA

(j) 7.5 μA

In Exercises 15–20 convert the given unit values to new units, as specified in the parentheses.

15. (a) 1500 μA (mA)

(b) 455 kHz (MHz)

(c) 0.015 μF (pF)

(d) 3.6 ns (μs)

(e) 0.48 MΩ (kΩ)

(f) 50 GHz (MHz)

(g) 675 μH (mH)

(h) 370 kΩ (MΩ)

(i) 0.16 fW (aW)

(j) 53 μS (mS)

16. (a) 91 ns (ps)

(b) 19 mA (A)

(c) 840 GHz (THz)

(d) 8.2 μW (mW)

(e) 0.041 MΩ (kΩ)

(f) 560 kΩ (MΩ)

(g) 450 mV (V)

(h) 685 mA (A)

17. (a) 4.5 THz (GHz)

(b) 27 pF (μF)

(c) 3.3 kΩ (Ω)

(d) 680 kΩ (MΩ)

(e) 56 kΩ (Ω)

(f) 250 μH (mH)

(g) 2.7 ns (ps)

(h) 2.7 ns (μs)

(i) 4.3 aW (fW)

(j) 5.5 mA (μA)

18. (a) 375 μA (mA)

(b) 4500 mA (A)

(c) 2 pF (μF)

(d) 0.455 MHz (kHz)

(e) 10.7 MHz (Hz)

(f) 250 mH (H)

(g) 0.375 mA (μA)

(h) 1.6 MΩ (kΩ)

(i) 3.9 kΩ (MΩ)

(j) 37 μW (mW)

19. (a) 200 mA (A)

(b) 200 μV (mV)

(c) 480 kΩ (MΩ)

(d) 1200 pF (μF)

(e) 12,000 ns (μs)

(f) 25,000 MHz (GHz)

(g) 0.0045 ma (μA)

(h) 0.0065 MΩ (Ω)

(i) 3.7×10^3 kΩ (MΩ)

(j) 1.5×10^{-3} mA (μA)

20. (a) 0.0035 mA (μA)

(b) 0.0048 MΩ (kΩ)

(c) 5600 kΩ (MΩ)

(d) 2500 μA (mA)

(e) 1500 mW (W)

(f) 1800 ns (μs)

(g) 950 kW (MW)

(h) 1200 Ω (MΩ)

(i) 2.5×10^4 kΩ (MΩ)

(j) 3.6×10^5 MHz (THz)

2-3 RULES FOR EXPONENTS: MULTIPLICATION AND DIVISION

Multiplication with Exponents We state the rule for multiplication of numbers with the same (common) base.

> **To find the product of two or more powers of a common base, add the exponents.**
>
> $$a^u \cdot a^v = a^{u+v} \qquad a \neq 0$$

Example 9   Simplify the given expressions using the rule for products of powers of a common base.

(a) $10^3 \cdot 10^2$

(a) $10^3 \cdot 10^2 = 10^{3+2} = 10^5$
Proof: $10^3 \cdot 10^2 =$
$\underbrace{(10 \cdot 10 \cdot 10)} \cdot \underbrace{(10 \cdot 10)}$
10 is a factor 5 times
$= 10^5$

(b) $10^6 \cdot 10^{-2}$

(b) $10^6 \cdot 10^{-2} = 10^{6+(-2)} =$
10^4 (Review rule for addition of signed numbers in Section 1-3.)

(c) $2^4 \times 2^5$
$\times 2^2$

(c) $2^4 \times 2^5 \times 2^2 = 2^{11}$
Proof: $2^4 \times 2^5 \times 2^2 =$
$\underbrace{2 \cdot 2 \cdot 2 \cdot 2}_{2^4} \times$
$\underbrace{2 \cdot 2 \cdot 2 \cdot 2 \cdot 2}_{2^5} \times$
$\underbrace{2 \cdot 2}_{2^2} = 2^{11}$

(d) $a^3 \cdot a^4 \cdot a^{-2}$

(d) $a^3 \cdot a^4 \cdot a^{-2} =$
$a^{3+4+(-2)} = a^5$

(e) $b^7 \cdot b^{-3}$

(e) $b^7 \cdot b^{-3} = b^{7+(-3)} = b^4$

(f) $x^p \cdot x^q$

(f) $x^p \cdot x^q = x^{p+q}$

(g) $2 \times 10^3 \times$
3×10^5

In (g) we must use two basic properties of multiplication which apply to algebra as well as to arithmetic:

1. *Multiplication can be performed in any order—the Commutative Law.*

   $3 \times 2 = 6, \quad 2 \times 3 = 6$

2. *Factors may be grouped (associated) in any fashion—the Associative Law.*

   $2 \times 3 \times 4 = 6 \times 4$ or 2×12 or 8×3

   In each case the result is 24.

(g) $2 \times 10^3 \times 3 \times 10^5 =$
$2 \cdot 3 \cdot 10^3 \cdot 10^5 =$
$6 \times 10^{3+5} = 6 \times 10^8$

(h) 3.68×10^{-2}
$\times 2.13 \times$
10^5

(h) $3.68 \times 10^{-2} \times 2.13 \times 10^5 = 3.68 \cdot 2.13 \cdot 10^{-2} \cdot 10^5 = 7.8384 \times 10^{-2+5} = 7.8384 \times 10^3 = 7838.4$

Division with Exponents

To find the quotient of two powers of a common base, subtract the exponent of the power in the denominator (the divisor) from the exponent of the power in the numerator (the dividend). (See Section 1-4 for rules for subtraction of signed numbers.)

$$\frac{a^u}{a^v} = a^{u-v} \qquad \frac{10^x}{10^y} = 10^{x-y}$$

Example 10   Simplify the given expressions by applying the rule for the quotient of two powers of the same base.

(a) $\dfrac{10^5}{10^2}$

(a) $\dfrac{10^5}{10^2} = \dfrac{10 \cdot 10 \cdot 10 \cdot 10 \cdot 10}{10 \cdot 10} = 10^3$
$(= 10^{5-2})$

(b) $\dfrac{10^5}{10^{-3}}$

(b) $\dfrac{10^5}{10^{-3}} = 10^{5-(-3)} = 10^{5+3} = 10^8$

(c) $\dfrac{10^{-4}}{10^3}$

(c) $\dfrac{10^{-4}}{10^3} = 10^{-4-3} = 10^{-7}$

(d) $\dfrac{10^{-5}}{10^{-4}}$

(d) $\dfrac{10^{-5}}{10^{-4}} = 10^{-5-(-4)} = 10^{-5+4} = 10^{-1}$

(e) $\dfrac{x^5}{x^3}$

(e) $\dfrac{x^5}{x^3} = x^{5-3} = x^2$

(f) $\dfrac{a^5 b^3}{a^4 b^{-1}}$

(f) $\dfrac{a^5 b^3}{a^4 b^{-1}} = a^{5-4} b^{3-(-1)} = ab^4$

(g) $\dfrac{8 \times 10^5}{2 \times 10^2}$

(g) $\dfrac{8 \times 10^5}{2 \times 10^2} = \dfrac{8}{2} \times 10^{5-2} = 4 \times 10^3$

(h) $\dfrac{18 \times 10^{-3}}{3 \times 10^4}$

(h) $\dfrac{18 \times 10^{-3}}{3 \times 10^4} = \dfrac{18}{3} \times 10^{-3-4} = 6 \times 10^{-7}$

Meaning of Zero as an Exponent When a division operation involves the division of the power of a base by itself, the result is 1 because any number divided by itself is 1; $N/N = 1$. Let us examine the implication. Consider the example

$$\frac{10^4}{10^4} = 10^{4-4} = 10^0 = 1$$

From this we generalize:

Any number (except zero) to the zero power is 1.

$$N^0 = 1$$

Example 11 Simplify the following.

(a) 10^0 (a) $10^0 = 1$

(b) x^0 (b) $x^0 = 1$

(c) $(4{,}569 \times 693{,}752)^0$ (c) $(4{,}569 \times 693{,}752)^0 = 1$

The Negative Exponent Let us consider this example, $\dfrac{10^3}{10^7}$:

$$\frac{10^3}{10^7} = \frac{\cancel{10} \cdot \cancel{10} \cdot \cancel{10}}{\cancel{10} \cdot \cancel{10} \cdot \cancel{10} \cdot 10 \cdot 10 \cdot 10 \cdot 10} = \frac{1}{10^4}$$

But

$$\frac{10^3}{10^7} = 10^{3-7} = 10^{-4}$$

That is,

$$10^{-4} = \frac{1}{10^4}$$

Or, in general

$$X^u = \frac{1}{X^{-u}}, \qquad X^{-u} = \frac{1}{X^u}$$

Any base (except zero) raised to any power (except zero) is equal to one divided by that base, to the negative of the original power.

$$A^u = \frac{1}{A^{-u}} \qquad A^{-u} = \frac{1}{A^u}$$

This property of exponents has a very important practical application in division operations involving powers of ten.

Example 12 Find the value of the indicated operations.

(a) $\dfrac{28 \times 10^{-7}}{4 \times 10^3}$

(a) This can be evaluated as in Example 10:

$$\frac{28 \times 10^{-7}}{4 \times 10^3} = 7 \times 10^{-7-(+3)} = 7 \times 10^{-10}$$

Some users prefer to move the power(s) of ten from the denominator to the numerator, using the property of the negative exponent, as follows:

$$\frac{28 \times 10^{-7}}{4 \times 10^3} = \frac{28 \times 10^{-7} \times 10^{-3}}{4}$$

$$= 7 \times 10^{-7+(-3)} = 7 \times 10^{-10}$$

(b) $\dfrac{24 \times 10^3 \times 12 \times 10^{-5}}{15 \times 10^{-2} \times 6 \times 10^{-4}}$

(b) $\dfrac{24 \times 10^3 \times 12 \times 10^{-5}}{15 \times 10^{-2} \times 6 \times 10^{-4}}$

$$= \frac{24 \times 12 \times 10^3 \times 10^{-5} \times 10^2 \times 10^4}{15 \times 6}$$

$$= 3.2 \times 10^{(3-5+2+4)} = 3.2 \times 10^4$$

Exercises 2-3 In Exercises 1–34 simplify the given expressions by using the rules for exponents as stated in this section. Express results in exponential notation (with exponents). (Study Examples 9, 10, and 11 carefully.)

1. $10^5 \cdot 10^2$ 2. $10^9 \cdot 10^5$

3. $10^{-6} \cdot 10^3$ 4. $10^5 \cdot 10^{-9}$

5. $10^{-3} \cdot 10^{-7}$ 6. $10^{-5} \cdot 10^{-1}$

7. $2^3 \cdot 2^2$ 8. $2^{-6} \cdot 2^4$ 9. $x^3 \cdot x^2$

10. $x^5 \cdot x^{-7}$ 11. $10^4 \cdot 10^{-3} \cdot 10^2$

12. $10^{-5} \cdot 10^7 \cdot 10^{-4} \cdot 10^3$ 13. $10^x \cdot 10^y$

14. $X^a \cdot X^b$ 15. $10^5/10^2$ 16. $10^4/10^7$

17. $10^{-5}/10^2$ 18. $10^{-9}/10^{-15}$

19. $\dfrac{10^4 \cdot 10^{-6}}{10^{-3}}$ 20. $\dfrac{10^{-5} \cdot 10^3 \cdot 10^{-4}}{10^{-7} \cdot 10^6}$

21. $\dfrac{a^6}{a^4}$ 22. $\dfrac{a^3}{a^7}$ 23. $\dfrac{a^{-2}}{a^5}$ 24. $\dfrac{b^{-9}}{b^2}$

25. $\dfrac{I^2 R}{IR}$ 26. $\dfrac{V^2 G^{-1}}{VG^{-1}}$ 27. $\dfrac{10^6}{10^6}$

28. 10^0 29. $\dfrac{X^{-3}}{X^{-3}}$

30. $(35{,}000 \times 2{,}568{,}397)^0$

31. $\dfrac{1}{10^5} = 10^x$, $x = ?$

32. $10^{-3} = \dfrac{1}{10^x}$, $x = ?$

33. $10^6 = \dfrac{1}{10^x}$, $x = ?$

34. $10^{-5} = \dfrac{1}{10^x}$, $x = ?$

In Exercises 35–45 perform the indicated operations mentally. Express results in both *SN* and *CN*.

35. $2 \times 10^2 \times 4 \times 10^3$

36. $1.5 \times 10^{-5} \times 2 \times 10^6$

37. $2 \times 10^2 \times 2 \times 10^6 \times 3 \times 10^{-4}$

38. $3 \times 10^{-6} \times 4 \times 10^{-2} \times 2 \times 10^5$

39. $\dfrac{12 \times 10^6}{2 \times 10^3}$ 40. $\dfrac{56 \times 10^{-4}}{8 \times 10^4}$

41. $\dfrac{24 \times 10^{-2}}{8 \times 10^{-6}}$ 42. $\dfrac{32 \times 10^2}{16 \times 10^{-6}}$

43. $\dfrac{2 \times 10^4 \times 6 \times 10^{-3}}{4 \times 10^{-5}}$

44. $\dfrac{12 \times 10^{10}}{2 \times 10^4 \times 3 \times 10^5}$

45. $\dfrac{3 \times 10^{-5} \times 16 \times 10^{12}}{4 \times 10^3 \times 6 \times 10^{-6}}$

In Exercises 46–52 use your calculator as an aid in finding the result of the indicated operations. Study the instruction manual carefully to learn how to enter exponents for powers of ten. Express your results in both *SN* and *CN*.

46. $1.29 \times 10^3 \times 4.86 \times 10^5$

47. $3.86 \times 10^5 \times 2.98 \times 10^{-7}$

48. $8.97 \times 10^{-12} \times 1.05 \times 10^3 \times 6.89 \times 10^4$

49. $\dfrac{2.69 \times 10^8}{8.97 \times 10^2}$ 50. $\dfrac{5.95 \times 10^{-6}}{3.27 \times 10^4}$

51. $\dfrac{3.09 \times 10}{4.17 \times 10^{-3}}$ 52. $\dfrac{0.516 \times 10^{-15}}{0.0713 \times 10^{-12}}$

2-4 ADDITION AND SUBTRACTION INVOLVING POWERS OF TEN

When we want to add 5 hundred dollars ($500) and 6 thousand dollars ($6000), we do not add 5 and 6 and get 11 dollars, or 11 hundred dollars, or even 11 thousand dollars; rather, we add as follows:

```
$  500        $  5 hundred
   6000   or    60 hundred
 $6500        $65 hundred
```

Similarly, if we need to add 5×10^2 and 6×10^3 we must not add 5, which represents a quantity of 10^2s and 6, which represents 10^3s. Addition is meaningful only when we add like quantities. In order to add 5×10^2 and 6×10^3 correctly, we must either change 5×10^2 to 0.5×10^3, or 6×10^3 to 60×10^2, or both to expressions with the same power-of-ten multiplier, or to common notation:

(a) 0.5×10^3 (b) 5×10^2 (c) 500
    6.0×10^3 60×10^2 6000
    6.5×10^3 65×10^2 6500

The same reasoning applies to subtraction of expressions in *PTN*. We summarize these observations in the following rule.

To add (or subtract) numbers in PTN, either:
1. Convert numbers to expressions containing the same power-of-ten multipliers.
2. Then add (or subtract) the numeral factors to obtain the numeral factor of result.
3. Include common power-of-ten multiplier in result. Or:
1. Convert numbers to expressions in common decimal notation.
2. Add (or subtract) according to rules for decimal numbers.

Example 13 Perform the indicated operations.

(a) Add 3.5×10^3 and 0.56×10^6

(a) $\begin{array}{r} 3.5 \times 10^3 \\ 560.0 \times 10^3 \\ \hline 563.5 \times 10^3 \end{array}$

(b) Add 4.56×10^{-2} and -53.7×10^{-3}

(b) $\begin{array}{r} 45.6 \times 10^{-3} \\ -53.7 \times 10^{-3} \\ \hline -8.1 \times 10^{-3} \end{array}$

(c) Subtract -570×10^{-7} from 89.3×10^{-6}

(c) $\begin{array}{r} 89.3 \times 10^{-6} \\ 57.0 \times 10^{-6} \\ \hline 146.3 \times 10^{-6} \end{array}$

(d) Add $56 \text{ k}\Omega$ and $0.68 \text{ M}\Omega$

(d) $\begin{array}{r} 56 \text{ k}\Omega \\ 680 \text{ k}\Omega \\ \hline 736 \text{ k}\Omega \end{array}$

We note that quantities with units must be in identical units.

(e) Add $475 \text{ }\mu\text{A}$ and 1.853 mA

(e) $\begin{array}{r} 1.853 \text{ mA} \\ .475 \text{ mA} \\ \hline 2.328 \text{ mA} \end{array}$

Exercises 2-4 In Exercises 1–4 find the sums of the given quantities.

1. (a) 3.5×10^2, 4.7×10^3
   (b) 6.8×10^{-4}, -3.9×10^{-3}
   (c) 67×10^2, 4.3×10^4, 580×10
   (d) 3.8×10^{-2}, -93×10^{-3}, 795×10^{-4}

2. (a) 7.8×10^6, -83×10^5
   (b) 42×10^{-9}, 0.16×10^{-7}
   (c) 485, -0.197×10^5, -2.13×10^3
   (d) 0.436×10^{-5}, 25.6×10^{-7}, -678×10^{-9}

3. (a) $61 \text{ k}\Omega$, $0.048 \text{ M}\Omega$
   (b) $475 \text{ }\mu\text{A}$, 1.57 mA
   (c) $3300 \text{ }\Omega$, $0.47 \text{ M}\Omega$, $560 \text{ k}\Omega$, $3.9 \times 10^4 \text{ }\Omega$
   (d) -0.019 V, 2.7 mV, $-3.6 \times 10^3 \text{ }\mu\text{V}$, $-9.6 \times 10^{-3} \text{ V}$

4. (a) 453 mS, $15.9 \times 10^{-3} \text{ S}$
   (b) $27{,}000 \text{ pF}$, $0.01 \text{ }\mu\text{F}$
   (c) $27 \text{ k}\Omega$, $0.091 \text{ M}\Omega$, $3.9 \times 10^5 \text{ }\Omega$, $72{,}000 \text{ }\Omega$
   (d) $-3500 \text{ }\mu\text{A}$, 27.5 mA, $-4.15 \times 10^{-3} \text{ A}$, $7.58 \times 10^6 \text{ nA}$

In Exercises 5–8 find the indicated differences.

5. (a) $48 \times 10^3 - 3.3 \times 10^4$
   (b) $450 \times 10^{-6} - 2.50 \times 10^{-3}$
   (c) $-2.9 \times 10^6 - (-870 \times 10^4)$
   (d) $-6900 \times 10^{-9} - 2.1 \times 10^{-6}$

6. (a) $59 \times 10^2 - 1600$
   (b) $0.00042 - 0.68 \times 10^{-4}$
   (c) $-179 \times 10^4 - 48.3 \times 10^6$
   (d) $0.483 - 2.75 \times 10^{-2}$

7. (a) $33 \text{ k}\Omega - 27{,}000 \text{ }\Omega$
   (b) $5.5 \text{ mA} - 1500 \text{ }\mu\text{A}$
   (c) $4500 \text{ }\mu\text{V} - 2.7 \text{ mV}$
   (d) $1500 \text{ mS} - 0.75 \text{ S}$

8. (a) $15 \text{ }\mu\text{s} - 4300 \text{ ns}$
   (b) $1.6 \text{ M}\Omega - 680{,}000 \text{ }\Omega$
   (c) $33 \text{ kV} - 9500 \text{ V}$
   (d) $450{,}000 \text{ pF} - 0.1 \text{ }\mu\text{F}$

2-5 APPLICATION: OHM'S LAW

Evaluation A common task in algebra (and electronics) is *to evaluate* an equation or formula. To evaluate literally means "to find the value of," but practically it means simply that we substitute known explicit numbers for literal numbers in equations or formulas and then perform the indicated operations of multiplication, division, or whatever.

The most frequently performed evaluations in practical electronics are probably those involving the three forms of Ohm's law and the power formula:

$$R = \frac{V}{I} \qquad I = \frac{V}{R} \qquad V = IR \qquad P = VI$$

A knowledge of multiple and submultiple units and powers-of-ten notation and a skill in performing operations with this notation are indispensable tools for such evaluations. The following suggestions for organizing your evaluations, if followed carefully, will help to ensure that you have a maximum of correct evaluations and a minimum of errors.

1. Follow a consistent format for evaluations; for example:
   (a) Write the correct formula.
   (b) Rewrite the formula with known values substituted *in basic units* (convert prefixes to equivalent powers of ten) for the literal numbers of the formula.
2. Perform the indicated operations (with a calculator, if more convenient). Repeat all calculator evaluations until the same answer is obtained at least twice.
3. Write down the answer as a basic unit, using powers-of-ten notation, if appropriate.
4. Finally, rewrite the answer using a convenient prefixed unit. Identify the answer by underlining or enclosing in a box ☐ .

The following worked-out examples demonstrate the use of this recommended procedure. Study them carefully.

Example 14   Find the desired electrical quantities by evaluating the appropriate formulas. Express results in convenient prefixed units.

(a) $V = 12$ V, $I = 2.5$ mA. Find R.

$$R = \frac{V}{I} \qquad R = \frac{12}{2.5 \times 10^{-3}} = 4.8 \times 10^3 \ \Omega$$

$$\boxed{R = 4.8 \text{ k}\Omega}$$

(b) $V = 27$ mV, $R = 15$ kΩ. Find I.

$$I = \frac{V}{R} \qquad I = \frac{27 \times 10^{-3}}{15 \times 10^3} = 1.8 \times 10^{-6} \ \text{A}$$

$$\boxed{I = 1.8 \ \mu\text{A}}$$

(c) $I = 45 \ \mu$A, $R = 72$ kΩ. Find V.

$$V = IR \qquad V = 45 \times 10^{-6} \times 72 \times 10^3$$

$$= 3240 \times 10^{-3} \ \text{V} \qquad \boxed{V = 3.24 \text{ V}}$$

(d) $V = 45$ mV, $I = 3.5$ mA. Find P.

$$P = VI \qquad P = (45 \times 10^{-3})(3.5 \times 10^{-3})$$

$$= 157.5 \times 10^{-6} \ \text{W} \qquad \boxed{P = 157.5 \ \mu\text{W}}$$

Note that an answer will automatically be in the basic unit if values are substituted in basic units (prefixes replaced with equivalent powers of ten).

Significant Digits   When we multiply two decimal numbers the answer will have as many decimal places as the sum of the places in the two factors; for example,

$$4.\underset{\smile}{26} \times 3.\underset{\smile}{79} = 16.1454$$

$$2 \ + \ \ 2 \ = \qquad 4 \ \text{ decimal places}$$

When we divide one number by a second it is possible to have an answer with a very large number of decimal places. For example, $3.69 \div 2.78 = 1.3273381$ on an eight-digit calculator. Many electrical quantities are measured with instruments which can indicate, at most, four digits. Resistor values are typically specified with only two digits. Therefore, using all of the digits provided in calculator answers, with a few exceptions, is misleading and contrary to good engineering practice. *Henceforth, answers will usually be given with a maximum of four significant digits.*

The terms *significant digit (SD)*, *most significant digit (MSD)*, and *least significant digit (LSD)* are illustrated in the following example:

$$00359.006800$$

$$MSD \longrightarrow \overbrace{\qquad\qquad}^{} \ \ \llcorner LSD$$

$$\underset{9 \ SD}{}$$

"Leading" zeros are not significant digits. Zeros between nonzero digits are significant. Trailing zeros to the right of the decimal point are significant.

To round off a number, a calculator readout for example:
1. Determine the digit position which will be the LSD in the rounded-off number.
2. Look at the digit following the digit that is to be the LSD; if it is

   $<$ 5 leave the LSD unchanged and discard all digits following.

   $\geq$ 5 raise the LSD by 1 and discard all digits following.

Examples: (a) **46.5345 rounded to four significant digits is 46.53**
   (b) **46.5364 rounded to four significant digits is 46.54**

Exercises 2-5 Evaluate an appropriate formula to determine the required quantities in the exercises below. Observe the recommended format (see above) in writing your solutions. Express answers using not more than four significant digits. In Exercises 1–20 values are given for a resistor R and the current I through R. Determine V. ($V = IR$)

1. $2\,\Omega, 6\,A$ 2. $12\,\Omega, 3\,A$
3. $78\,\Omega, 1.5\,A$ 4. $120\,\Omega, 0.65\,A$
5. $4200\,\Omega, 0.024\,A$ 6. $4800\,\Omega, 1.6\,mA$
7. $5.6\,k\Omega, 3.5\,mA$ 8. $1.8\,k\Omega, 27\,mA$
9. $0.68\,k\Omega, 92\,mA$
10. $0.27\,k\Omega, 0.62\,mA$
11. $22\,k\Omega, 250\,\mu A$ 12. $51\,k\Omega, 47\,\mu A$
13. $82\,k\Omega, 87\,\mu A$ 14. $360\,k\Omega, 15\,\mu A$
15. $240\,k\Omega, 52\,\mu A$ 16. $0.91\,M\Omega, 1.9\,\mu A$
17. $1.3\,M\Omega, 2.5\,\mu A$ 18. $6.8\,M\Omega, 0.65\,\mu A$
19. $11\,M\Omega, 0.15\,\mu A$ 20. $0.18\,\Omega, 2.75\,kA$

In Exercises 21–40 values are given for a resistor R, and the voltage V across R. Determine I. ($I = V/R$)

21. $4\,\Omega, 12\,V$ 22. $7\,\Omega, 56\,V$
23. $3\,\Omega, 8.7\,V$ 24. $6.8\,\Omega, 15.8\,V$
25. $180\,\Omega, 275\,V$ 26. $820\,\Omega, 756\,V$
27. $48\,\Omega, 0.86\,V$ 28. $220\,\Omega, 0.26\,V$
29. $3.3\,k\Omega, 9.2\,V$ 30. $5.6\,k\Omega, 16.4\,V$
31. $36\,k\Omega, 250\,mV$ 32. $750\,k\Omega, 675\,mV$
33. $91\,k\Omega, 36\,mV$ 34. $33\,k\Omega, 9.3\,mV$
35. $18\,k\Omega, 1.65\,mV$ 36. $3.6\,k\Omega, 750\,\mu V$
37. $560\,k\Omega, 475\,\mu V$ 38. $0.91\,k\Omega, 56\,\mu V$
39. $0.75\,M\Omega, 2.68\,\mu V$
40. $0.068\,M\Omega, 0.00615\,mV$

In Exercises 41–60 the given values represent the voltage V across a resistor R and the current I through R. Determine R. ($R = V/I$)

41. $14\,V, 3\,A$ 42. $28\,V, 6\,A$
43. $115\,V, 2.7\,A$ 44. $240\,V, 0.8\,A$
45. $2400\,V, 420\,A$ 46. $0.67\,V, 0.23\,A$
47. $0.185\,V, 0.036\,A$ 48. $27\,mV, 6.2\,mA$
49. $33\,kV, 120\,A$ 50. $125\,kV, 480\,mA$
51. $235\,mV, 16\,mA$ 52. $9.3\,V, 127\,mA$
53. $2.6\,V, 2.45\,mA$ 54. $3.6\,mV, 18.5\,\mu A$
55. $1.12\,V, 256\,\mu A$ 56. $1.0\,mV, 0.62\,mA$
57. $0.93\,mV, 0.019\,mA$
58. $0.0105\,mV, 1.18\,mA$
59. $46.5\,\mu V, 2.75\,\mu A$
60. $175\,kV, 427\,\mu A$

In Exercises 61–80 the given values represent the voltage V and current I supplied by an electrical source. Determine P. ($P = IV$)

61. $12\,V, 2\,A$ 62. $26.5\,V, 3.7\,A$
63. $120\,V, 10.5\,A$ 64. $9.1\,V, 15\,mA$
65. $2.75\,V, 8.3\,mA$ 66. $468\,mV, 3.8\,mA$
67. $915\,mV, 56.6\,mA$ 68. $7.2\,kV, 42.5\,A$
69. $33\,kV, 25\,A$ 70. $375\,kV, 95\,A$
71. $1.68\,V, 0.57\,mA$
72. $0.047\,V, 3.6\,mA$
73. $0.67\,mV, 215\,\mu A$
74. $0.089\,mV, 36.7\,\mu A$
75. $0.720\,kV, 27\,A$
76. $4.68\,mV, 0.00175\,mA$
77. $2.75 \times 10^3\,mV, 12\,\mu A$
78. $6.85 \times 10\,V, 74\,\mu A$
79. $3.68 \times 10\,kV, 1.38 \times 10^3\,\mu A$
80. $0.55\,\mu V, 0.0015\,\mu A$

2-6 THE POWER OF A POWER

What is the result when we raise a power of a number to a power? An example will demonstrate the answer.

$$(10^4)^2 = 10^4 \cdot 10^4$$
$$= \underbrace{(10 \cdot 10 \cdot 10 \cdot 10)}_{10^4}\underbrace{(10 \cdot 10 \cdot 10 \cdot 10)}_{10^4} = 10^8$$
$$= 10^{4+4} = 10^8 = 10^{4 \cdot 2}$$
$$(a^2)^3 = a^2 \cdot a^2 \cdot a^2 = a^{2+2+2} = a^6 = a^{3 \cdot 2}$$

We generalize with a rule.

> **To raise a power of a number to a power, multiply the exponents.**
>
> $$(a^u)^v = a^{u \cdot v}$$

Example 15   Simplify the given expressions.

(a) $(10^3)^2$ (a) $(10^3)^2 = 10^{3 \cdot 2} = 10^6$

(b) $(10^{-5})^2$ (b) $(10^{-5})^2 = 10^{-5 \cdot 2} = 10^{-10}$

(c) $(a^x)^3$ (c) $(a^x)^3 = a^{3x}$
                 ($3x$ means "3 times x")

(d) $(10^x)^y$ (d) $(10^x)^y = 10^{xy}$
                 (xy means "$x \cdot y$")

(e) $(x^a)^{4b}$ (e) $(x^a)^{4b} = x^{4ab}$
                 ($4ab = 4 \cdot a \cdot b$)

Product to a Power   We obtain the result of raising a product to a power through application of the definition of the exponent,

$$(2 \times 10)^2 = 2 \cdot 2 \cdot 10 \cdot 10 = 2^2 \cdot 10^2$$
$$= 4 \times 10^2 = 400$$
$$(ab)^3 = (ab) \cdot (ab) \cdot (ab)$$
$$= a \cdot a \cdot a \cdot b \cdot b \cdot b = a^3 b^3$$

Performing this operation also requires utilizing the associative property of multiplication —in the example $(ab)^3$ we grouped the a's together to get a^3 and the b's together to get b^3.

To raise a product to a power, raise each factor to that power:

$$(ab)^u = a^u b^u \qquad (a^u b^v)^x = a^{ux} b^{vx}$$

Example 16   Simplify the given expressions.

(a) $(xy)^2$

(a) $(xy)^2 = x^2 y^2$

(b) $(2 \times 10^3)^2$

(b) $(2 \times 10^3)^2 = 2^2 \times 10^{2 \cdot 3} = 4 \times 10^6$

(c) $(a^2 b^3)^4$

(c) $(a^2 b^3)^4 = a^{2 \cdot 4} b^{3 \cdot 4} = a^8 b^{12}$

(d) $(3.75 \cdot 10^{-4})^2$

(d) $(3.75 \cdot 10^{-4})^2 = 14.0625 \cdot 10^{-8} = 1.406 \cdot 10^{-7}$
(Note: We do not leave a result in *PTN*; we convert either to *SN* or *CN*.)

Application   The power dissipated in a resistance can be calculated using the formula $P =$

VI [see Example 14(d) in Section 2-5]. Two other forms of this formula are popular and are obtained by an operation called *substitution* in algebra. That is, we *substitute* into this formula values for V or I obtained from the Ohm's law formulas $I = V/R$ and $V = IR$.
We substitute for I:

$$P = VI = V \cdot \frac{V}{R} \qquad P = \frac{V^2}{R} \qquad (1)$$

We substitute for V:

$$P = VI = IR \cdot I \qquad P = I^2 R \qquad (2)$$

Formulas (1) and (2) permit us to calculate the power dissipated in a resistance when we know the value of the resistance, and only V [Formula (1)] or only I [Formula (2)]. The unit for power is the watt (W). (See Table 2-1.)

Example 17   Find the power dissipated in a resistance $R = 15$ kΩ when the voltage across it is $V = 175$ mV.

$$P = \frac{V^2}{R}$$
$$P = \frac{(175 \times 10^{-3})^2}{(15 \times 10^3)} = \frac{3.0625 \times 10^{-2}}{15 \times 10^3}$$
$$P = 0.2041666 \times 10^{-2-3} \,\text{W} = \underline{2.042 \; \mu\text{W}}$$

Example 18   Find the power dissipated in a resistance $R = 27$ kΩ when the current through it is $I = 235$ μA.

$$P = I^2 R$$
$$P = (235 \times 10^{-6})^2 \cdot 27 \times 10^3$$
$$= 5.5225 \times 27 \times 10^{-8+3} \,\text{W}$$
$$P = 1.491 \times 10^{-3} \,\text{W} = \underline{1.491 \; \text{mW}}$$

Many calculators incorporate the key labeled (x^2) which allows one to evaluate the square of any number directly: (1) enter the number, and (2) depress the (x^2) key. If your calculator does not have this function you can perform squaring with a little more effort by multiplying the number by itself: (N), $(\times)$, (N), $(=)$, Readout.

Exercises 2-6   In Exercises 1–20 simplify the given expressions by utilizing the rules for the power of a power and the power of a product.

Use your calculator as an aid, when appropriate.

1. $(10^4)^3$ 2. $(10^2)^4$ 3. $(10^{-5})^2$
4. $(10^4)^{-2}$ 5. $(10^{-3})^{-2}$ 6. $(a^3)^2$
7. $(x^a)^b$ 8. $(2^p)^q$ 9. $(x^a)^{4b}$
10. $(2^3)^7$ 11. $(xy)^3$ 12. $(ab)^x$
13. $(a^2 b^6)^3$ 14. $(x^5 y^{-4})^3$
15. $(a \times 10^3)^2$ 16. $(2 \times 10^4)^2$
17. $(3.5 \times 10^{-4})^2$ 18. $(2.6 \times 10^{-6})^2$
19. $(4.75 \times 10^3)^2$ 20. $(5.67 \times 10^{-6})^2$

In Exercises 21–60 evaluate an appropriate formula to determine the desired values. Use the *formula, substitution, solution* format (see Section 2-5) to organize your worksheet. Be sure to include the correct units as part of your solutions.

In Exercises 21–40 the given values represent the current I in a resistor R. Determine P using the formula $P = I^2 R$.

21. 2 A, 4 Ω 22. 8 A, 15 Ω
23. 12.5 A, 68 Ω 24. 75 A, 15 Ω
25. 435 A, 6.8 Ω 26. 0.65 A, 120 Ω
27. 0.024 A, 4200 Ω 28. 1.6 mA, 4800 Ω
29. 3.5 mA, 5.6 kΩ 30. 27 mA, 1.8 kΩ
31. 92 mA, 0.68 kΩ
32. 0.62 mA, 0.27 kΩ
33. 250 μA, 22 kΩ 34. 47 μA, 51 kΩ
35. 87 μA, 82 kΩ 36. 15 μA, 360 kΩ
37. 1.9 μA, 0.91 MΩ 38. 2.5 μA, 1.3 MΩ
39. 0.65 μA, 6.8 MΩ
40. 0.015 A, 11 MΩ

In Exercises 41–60 the given values represent the voltage V across a resistor R. Determine P using the formula $P = V^2/R$.

41. 12 V, 4 Ω 42. 56 V, 7 Ω
43. 8.7 V, 3 Ω 44. 15.8 V, 6.8 Ω
45. 275 V, 180 Ω 46. 756 V, 820 Ω
47. 0.86 V, 48 Ω 48. 0.26 V, 220 Ω
49. 9.2 V, 3.3 kΩ 50. 16.4 V, 5.6 kΩ
51. 250 mV, 36 kΩ 52. 675 mV, 750 kΩ
53. 36 mV, 91 kΩ 54. 33 kV, 15 Ω
55. 7.2 kV, 6.5 Ω 56. 750 μV, 3.6 Ω
57. 475 μV, 560 kΩ 58. 56 μV, 0.91 kΩ
59. 2.68 μV, 0.75 MΩ
60. 0.00615 mV, 0.068 MΩ

2-7 FRACTIONAL EXPONENTS, ROOTS, AND RADICALS

We recall that the symbol $\sqrt{}$, called the *radical sign*, meant "find the square root of" in arithmetic. Thus, $\sqrt{4} = 2$, and in algebra $\sqrt{x}$ means "square root of x." A radical sign with a small index number is used to indicate other roots than the second, for example, $\sqrt[3]{x}$ and $\sqrt[4]{x}$ mean "the cube root of x" and "fourth root of x," respectively.

The square root of a number x is a second number r, which, when multiplied by itself, gives the first number; that is, $r \cdot r = r^2 = x$. Thus, $\sqrt{x} = \sqrt{r^2} = r$ and $\sqrt{x} \cdot \sqrt{x} = x$.

How can we represent $\sqrt{x}$ with an exponent? Let us try a fractional exponent, that is, $\sqrt{x} = x^{1/2}$. To be valid it must conform to the laws for exponents. Thus,

$$\sqrt{x} \cdot \sqrt{x} = x^{1/2} \cdot x^{1/2} = x^{1/2+1/2} = x^1 = x$$

A fractional exponent does, in fact, represent the root of a number:

$x^{1/2}$, read "x to the one-half," means $\sqrt{x}$

$x^{1/3}$, read "x to the one-third," means $\sqrt[3]{x}$

$x^{1/4}$, read "x to the one-fourth," means $\sqrt[4]{x}$ and so on.

The laws for exponents and order of operations apply:

$$(x^{1/3})^3 = x^{1/3} \cdot x^{1/3} \cdot x^{1/3} = x^{3/3} = x$$
$$(x^2)^{1/3} = x^{2/3} = \sqrt[3]{x^2} = (\sqrt[3]{x})^2$$

Most so-called slide rule or scientific calculators incorporate the functions $\boxed{x^2}$, $\boxed{\sqrt{x}}$, and $\boxed{y^x}$ or $\boxed{\sqrt[x]{y}}$ or both. With these functions one can find the square or square root of any number; any number to any power (+ or −, integral or fractional); or any root (+ or −, integral or fractional). Generally, the argument (that is, the y in $\boxed{y^x}$ or in $\boxed{\sqrt[x]{y}}$) is keyed in first, then the desired function key is depressed, and, finally, the exponent or root index is keyed in. Of course, you should study carefully the instruction manual for your particular calculator to know how to perform these operations with it.

Example 19   Utilize the rules for operations with exponents to simplify the given expressions.

(a) $x^{1/2} \cdot x^{3/2}$ (a) $x^{1/2} \cdot x^{3/2} = x^{4/2} = x^2$

(b) $x \cdot x^{1/3}$ (b) $x \cdot x^{1/3} = x^{3/3} \cdot x^{1/3} = x^{4/3}$

(c) $x^{1/3} \cdot x^{4/3}$ (c) $x^{1/3} \cdot x^{4/3} = x^{5/3}$

(d) $(x^{3/2})^2$ (d) $(x^{3/2})^2 = x^{3/2 \cdot 2} = x^3$

(e) $(x^6)^{1/2}$ (e) $(x^6)^{1/2} = x^{6 \cdot 1/2} = x^3$

(f) $(4 \times 10^6)^{1/2}$ (f) $(4 \times 10^6)^{1/2} = 4^{1/2} \times 10^{6 \cdot 1/2} = 2 \times 10^3$

Example 20   Utilize the appropriate functions on your calculator to evaluate the following expressions.

(a) $\sqrt{6.25 \times 10^4}$ (a) $\sqrt{6.25 \times 10^4} = 2.5 \times 10^2 = \underline{250}$

(b) $(6.25 \times 10^6)^{1/2}$ (b) $(6.25 \times 10^6)^{1/2} = 2.5 \times 10^3 = \underline{2500}$

(c) $(6.25 \times 10^{-6})^{1/2}$ (c) $(6.25 \times 10^{-6})^{1/2} = 2.5 \times 10^{-3} = \underline{0.0025}$

(d) $(6.25 \times 10^3)^{1/2}$ (d) $(6.25 \times 10^3)^{1/2} = (62.50 \times 10^2)^{1/2} = (6250)^{1/2} = \underline{79.056942}$

(*Note:* In Examples 20(a), 20(b), and 20(c), the exponent of the ten factor is an even number, and $\frac{1}{2} \times$ an even number is always a whole number. In 20(d) the exponent is 3, an odd number, and $\frac{1}{2} \times$ an odd number is never a whole number. If you enter the complete number (mantissa and power of ten) into a calculator, the $\sqrt{x}$ key will evaluate correctly. If you enter only the mantissa and evaluate the square root of the power of ten mentally you must always convert the notation so that the power of ten is even.)

Application   The formulas $P = V^2/R$ and $P = I^2 R$ provide a means for determining V or I when power and resistance values are known. The algebraic techniques for transforming these formulas will be presented in a later section. For the moment, we present the transformed formulas and demonstrate their application:

$$V = \sqrt{PR} \qquad (3)$$

$$I = \sqrt{\frac{P}{R}} \qquad (4)$$

Example 21   (a) Find the voltage across a 91-kΩ resistor which is dissipating 56.5 μW.

$$V = \sqrt{PR}$$
$$V = \sqrt{56.5 \times 10^{-6} \times 91 \times 10^3}$$
$$= \sqrt{5141.5 \times 10^{-6+3}}$$
$$= \sqrt{5141.5 \times 10^{-3}}$$
$$= \sqrt{5.1415} = 2.2674875 \text{ V}$$

$$\boxed{V = 2.267 \text{ V}}$$

(b) A 0.22-Ω resistor is dissipating 2.5 W. What is the current through it?

$$I = \sqrt{\frac{P}{R}}$$

$$I = \sqrt{\frac{2.5}{0.22}} = \sqrt{11.363636} = 3.3709993 \text{ A}$$

$$\boxed{I = 3.371 \text{ A}}$$

All resistors have a dissipation rating given in watts: $\frac{1}{8}$ W, $\frac{1}{4}$ W, $\frac{1}{2}$ W, 1 W, 2 W, and so forth. In practical situations it is desirable to convert these *rated* wattage values into either "safe current" or "safe voltage" values by applying a "safety factor," commonly 2 or 4, to the power rating, then calculating the equivalent current or voltage corresponding to this reduced power value. Using the safety factor ensures, to some degree, that the resistor chosen for a particular application will not have its power rating exceeded (causing it to be destroyed by burnout and possibly causing a fire) during some unusual circumstance when the design current (or voltage) is exceeded temporarily.

Example 22   The voltage across a 7.5 kΩ resistor is expected to reach a maximum of 25 V. If a resistor rated for $\frac{1}{2}$ W is used, will its safe voltage be exceeded? Use safety factor of 2 ($P_{\text{safe}} = \frac{1}{2} P_{\text{rated}}$).

$$V_{\text{safe}} = \sqrt{\tfrac{1}{2} P_{\text{rated}} \cdot R} = \sqrt{\tfrac{1}{2} \cdot 0.5 \cdot 7500} = 43.30 \text{ V}$$

$$\boxed{V_{\text{safe}} = 43.30 \text{ V}}$$

A voltage of 43.3 V would cause the power dissipated to be $\frac{1}{2}$ of the rated power; therefore, the expected circuit voltage, 25 V, will not exceed the safe voltage.

Example 23 What value of current through a $1.5\text{-}M\Omega$, $\frac{1}{4}\text{-}W$ resistor should not be exceeded to provide the typical safety margin: $P_{safe} = \frac{1}{2}P_{rated}$?

$$I_{safe} = \sqrt{\frac{\frac{1}{2}P_{rated}}{R}} = \sqrt{\frac{\frac{1}{2} \times 0.25}{1.5 \times 10^6}}$$

$$= \sqrt{\frac{0.125}{1.5 \times 10^6}} = \sqrt{0.0833333 \times 10^{-6}}$$

$$= \sqrt{8.33333 \times 10^{-8}}$$

$$= 2.8867508 \times 10^{-4}\ A$$

$$\boxed{I_{safe} = 288.7\ \mu A}$$

Exercises 2-7 In Exercises 1–10 simplify the given expressions by applying appropriate rules for exponents.

1. $(10^6)^{1/2}$ 2. $(10^8)^{1/2}$ 3. $(10^{-6})^{1/2}$
4. $(10^{-4})^{1/2}$ 5. $(4 \times 10^4)^{1/2}$
6. $(9 \times 10^{-6})^{1/2}$ 7. $(40 \times 10^5)^{1/2}$
8. $(90 \times 10^{-7})^{1/2}$ 9. $\sqrt{10^6}$
10. $\sqrt{10^{-4}}$

Study carefully in your calculator instruction manual the explanations for using special keys such as $\sqrt{x}$, $\sqrt[x]{y}$, x^2, and y^x. In Exercises 11–30 use the appropriate calculator functions to evaluate the given expressions.

11. $\sqrt{45.67}$ 12. $\sqrt{789.6}$
13. $\sqrt{56.7 \times 10^4}$ 14. $(3.79 \times 10^6)^{1/2}$
15. $\sqrt{0.03795}$ 16. $(4.159 \times 10^{-4})^{1/2}$
17. $(7.896 \times 10^5)^{1/2}$
18. $(7.896 \times 10^{-3})^{1/2}$
19. $(43.695 \times 10^3)^{1/2}$
20. $(0.1592 \times 10^{-5})^{1/2}$ 21. 7^4
22. $6.5^{1.75}$ 23. $6.93^{2.65}$ 24. $\sqrt[3]{37}$
25. $\sqrt[5]{427}$ 26. $43.7^{0.68}$ 27. $81^{0.25}$
28. $27^{-0.33}$ 29. $49^{-0.5}$ 30. $36^{-0.5}$

In Exercises 31–40 values given are for the power P dissipated in a resistor R. Evaluate an appropriate formula to determine the voltage required to achieve P. Use the *formula, substitution, solution* format (see Section 2-5) to organize your worksheet. Do not forget units.

31. 1 W, 12 kΩ 32. 1 W, 82 kΩ
33. 0.27 W, 150 kΩ 34. 0.056 W, 48 kΩ
35. 2.75 W, 820 Ω 36. 3.97 W, 750 Ω
37. 15.97 W, 250 Ω 38. 0.987 W, 1.5 MΩ
39. 0.25 W, 2.7 MΩ 40. 0.5 W, 750 kΩ

In Exercises 41–50 values given are for P and R. Using an appropriate formula determine the value of I required to produce the given P.

41. 2.5 W, 500 Ω 42. 5 W, 2.5 kΩ
43. 0.047 W, 75 kΩ 44. 0.737 W, 180 kΩ
45. 2.36 mW, 56 kΩ 46. 47.8 mW, 1.2 MΩ
47. 3.76 μW, 2.7 MΩ 48. 41.76 μW, 91 kΩ
49. 375 μW, 68 kΩ 50. 50 kW, 575 Ω

In Exercises 51–60 determine V_{safe} for the given resistors. Power values are for P_{rated}. Use the relationships $P_{safe} = \frac{1}{2}P_{rated}$, $V_{safe} = \sqrt{P_{safe}R}$.

51. 2.2 Ω, 1 W   52. 75 Ω, 0.25 W
53. 2.7 kΩ, 0.5 W   54. 200 Ω, 25 W
55. 10 kΩ, 0.125 W   56. 270 kΩ, 0.5 W
57. 720 kΩ, 0.5 W   58. 0.48 Ω, 2 W
59. 1.2 MΩ, 0.5 W   60. 6.8 MΩ, 1 W

In Exercises 61–70 determine I_{safe} for the given resistors. Power values are for P_{rated}. Use the relationships $P_{safe} = \frac{1}{2}P_{rated}$, $I_{safe} = \sqrt{P_{safe}/R}$.

61. 12 Ω, 10 W   62. 200 Ω, 1 W
63. 200 Ω, 5 W   64. 500 Ω, 100 W
65. 10 kΩ, 0.25 W   66. 270 kΩ, 2 W
67. 56 kΩ, 0.5 W   68. 4.8 kΩ, 0.5 W
69. 0.91 MΩ, 1 W   70. 4.7 MΩ, 0.25 W

71.–80. Solve Exercises 51–60 using a safety factor of 4: $P_{safe} = \frac{1}{4}P_{rated}$.

81.–90. Solve Exercises 61–70 using a safety factor of 4: $P_{safe} = \frac{1}{4}P_{rated}$.

2-8 ESTIMATING SOLUTIONS

Professionals who learned to perform electronics evaluations with the slide rule performed operations called *estimating* as a necessary part of obtaining slide rule answers. Although the hand-held calculator is very convenient and a tremendous time saver, these professionals still prefer to estimate solutions when only quick, approximate solutions are needed. With this skill, one can also check the reasonableness of a calculator answer. The procedure consists of (a) converting the numbers involved in a calculation into scientific notation, (b) rounding off to a single significant digit, and (c) performing the required arithmetic mentally. Utilizing the rules for exponents with powers of ten makes the procedure feasible.

Example 24   (a) Estimate the current that will flow in a 75-kΩ resistor when 29 V is connected across it.

Estimate:

$$I = \frac{V}{R}$$

$$I \simeq \frac{30}{8 \times 10^4} \simeq 4 \times 10^{-4} \text{ A}$$

(30 is almost 32 and 32/8 = 4.)

$$\boxed{I \simeq 400 \ \mu\text{A estimated}}$$

(The symbol $\simeq$ means "is approximately equal to.") Calculator evaluation:

$$I = \frac{29}{7.5 \times 10^4} = 3.867 \times 10^{-4} \text{ A}$$

$$\boxed{I = 386.7 \ \mu\text{A}}$$

(b) Estimate the voltage drop that will appear across a 180-kΩ resistor when 27.5 μA flow through it.

Estimate:

$$V = IR$$
$$V \simeq 30 \times 10^{-6} \times 2 \times 10^5 = 60 \times 10^{-1}$$

$$\boxed{V \simeq 6 \text{ V estimated}}$$

Calculator evaluation:

$$E = 27.5 \times 10^{-6} \times 1.8 \times 10^5 = 4.95 \text{ V}$$

$$\boxed{V = 4.95 \text{ V}}$$

Exercises 2-8   In Exercises 1–16 first estimate the results, then use your calculator to determine results accurate to four significant digits.

1. 380×9100
2. $45{,}690 \times 105{,}682$
3. 279×0.00615
4. 0.000573×0.00182
5. $4100/780$
6. $1060/49{,}500$
7. $0.00775/210$
8. $0.000895/0.00193$
9. $(3900)^2$
10. $(51{,}350)^2$
11. $(0.00078)^2$
12. $(0.00000196)^2$
13. $(413)^{1/2}$
14. $(4130)^{1/2}$ (Remember that power of ten must be even.)
15. $(0.816)^{1/2}$
16. $(0.000896)^{1/2}$

In Exercises 17–24 perform the stated instructions. Utilize an organized format in writing your solutions.

17. Estimate the solution for Exercise 5, Section 2-5.
18. Estimate the solution for Exercise 6, Section 2-5.
19. Estimate the solution for Exercise 31, Section 2-5.
20. Estimate the solution for Exercise 32, Section 2-5.
21. Estimate the solution for Exercise 46, Section 2-5.
22. Estimate the solution for Exercise 47, Section 2-5.
23. Estimate the solution for Exercise 62, Section 2-5.
24. Estimate the solution for Exercise 74, Section 2-5.

Chapter 3
Algebra: Techniques and Operations

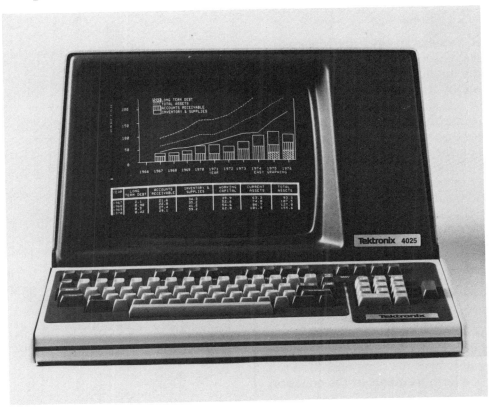

A computer-generated x-y plot illustrates the old saying that "one picture is worth a thousand words." (Courtesy of Tektronix, Inc.)

In this chapter we begin the study of several basic algebraic techniques and operations which provide the essential background for one to understand and use effectively the formulas for Ohm's law, and series, parallel, and combination circuits.

3-1 TRANSFORMING BASIC FORMULAS

An equation (or formula) is a statement that two mathematical expressions are equal. It is said to have three parts, as diagrammed here:

$$2x = 4$$

Left member ⟶ ↑ ↑ ↳ ⟵ Right member
or left side　　　　　　　　or right side

Equal sign

We say that we solve an equation when we mathematically isolate the desired literal quantity, the unknown or variable, so that it is completely by itself on one side (conventionally, the left side) of the equation. The process of achieving this mathematical isolation requires that we perform one or more operations on the equation—we say that we *manipulate* or *transform* the equation. In general, any operation (multiplication, division, addition, subtraction, and so forth) is legitimate—that is, does not destroy the equality of the two members of the equation—so long as we perform the same operation on or to both members of the equation.

Any proper mathematical operation may be performed on an equation without destroying its equality as long as the same operation is performed on both members of the equation.

Equations of the Form $2x = 4$ When an equation is of the form $2x = 4$ we isolate the unknown (x, in this case) by dividing both sides of the equation by the coefficient of the unknown. *A coefficient is a factor (a multiplier) in an indicated product*—2 is the coefficient of x in the product $2x$, a is the coefficient in ay, and so on. Thus, to solve $2x = 4$ we divide both members of the equation by 2,

$$\frac{2x}{2} = \frac{4}{2} \qquad x = 2$$

The equation has been solved. The *solution* or *root* of the equation (the number which makes the statement true) is 2. We check the solution by substituting the root into the original (not an intermediate form) equation:

$$2x = 4 \qquad 2 \cdot 2 = 4 \qquad 4 = 4$$

Example 1   Solve the following equations and check your solutions.

(a) $7x = 14$ 　　(a) $\dfrac{7x}{7} = \dfrac{14}{7}$ 　$\boxed{x = 2}$

Check:　$7 \cdot 2 = 14$
　　　　　　$14 = 14$

(b) $1.5x = -4.5$ 　(b) $\dfrac{1.5x}{1.5} = -\dfrac{4.5}{1.5}$

$\boxed{x = -3}$

Check:　$(1.5)(-3) =$
　　　　　　-4.5
　　　　　　$-4.5 = -4.5$

(c) $-7x = 35$ 　　(c) $\dfrac{-7x}{-7} = \dfrac{35}{-7}$ 　$\boxed{x = -5}$

Check:　$(-7)(-5) = 35$
　　　　　　$35 = 35$

Equations of the Form $x/2 = 4$ In considering the equation $x/2 = 4$ it is important to recognize that

$$\frac{x}{2} \quad \text{means} \quad \frac{x}{1} \cdot \frac{1}{2} \quad \text{or} \quad \frac{1}{2} \cdot \frac{x}{1}$$

Therefore, this equation and all those that are similar are really those of Example 1 and are solved in the fashion demonstrated in Example 1. Of course, the coefficient is a fraction, and when we divide by a fraction we invert and multiply.

Example 2 Solve the following equations and check your solutions.

(a) $\dfrac{x}{2} = 4$

(a) $\dfrac{x}{2} \cdot \dfrac{2}{1} = 4 \cdot \dfrac{2}{1}$

$$\boxed{x = 8}$$

Check: $\dfrac{8}{2} = 4 \qquad 4 = 4$

(b) $\dfrac{x}{1.5} = 3$

(b) $\dfrac{x}{1.5} \cdot \dfrac{1.5}{1} = 3 \cdot \dfrac{1.5}{1}$

$$\boxed{x = 4.5}$$

Check: $\dfrac{4.5}{1.5} = 3$

$3 = 3$

(c) $\dfrac{x}{-3} = 4$

(c) $\dfrac{x}{-3} \cdot \dfrac{-3}{1} = 4 \cdot \left(\dfrac{-3}{1}\right)$

$$\boxed{x = -12}$$

Check: $\dfrac{-12}{-3} = 4$

$4 = 4$

A Shortcut Careful analysis of the transformations demonstrated in Examples 1 and 2 above shows that a quick method of solution is possible. As an aid in remembering this operation we will call it the *checkerboard transposition*. All of our sample equations will fit the checkerboard pattern of Figure 3-1. For any equation which fits this pattern, any *factor* can be transposed *diagonally* across the equal sign. For example,

$\dfrac{\textcircled{2}x}{1} = \dfrac{4}{1}$ then $x = \dfrac{4}{2} = \underline{2}$

$\dfrac{x}{\textcircled{2}} = \dfrac{4}{1}$ then $x = 4 \cdot 2 = \underline{8}$

$\dfrac{\textcircled{3}x}{\textcircled{2}} = \dfrac{6}{1}$ then $x = \dfrac{6 \cdot 2}{3} = \underline{4}$

Check: $\dfrac{3 \cdot 4}{2} = 6 \qquad \dfrac{12}{2} = 6$

$\dfrac{\textcircled{5}x}{\textcircled{3}} = 10$ then $x = \dfrac{10 \cdot 3}{5} = \underline{6}$

Check: $\dfrac{5 \cdot 6}{3} = \dfrac{30}{3} = 10 = 10$

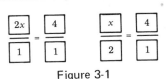

Figure 3-1

Example 3 Use checkerboard transposition to solve the following equations.

(a) $5y = 75$

(a) $\dfrac{5y}{1} = \dfrac{75}{1} \quad y = \dfrac{75}{5} = \underline{15}$

Check: $5 \cdot 15 = 75$

$= 75$

(b) $\dfrac{V}{7} = 2$

(b) $\dfrac{V}{7} = \dfrac{2}{1} \quad V = 7 \cdot 2 = \underline{14}$

Check: $\dfrac{14}{7} = 2 = 2$

(c) $\dfrac{4I}{3} = 12$

(c) $\dfrac{4I}{3} = \dfrac{12}{1} \quad I = \dfrac{3 \cdot 12}{4}$

$I = \underline{9}$

Check: $\dfrac{4 \cdot 9}{3} = \dfrac{36}{3}$

$= 12 = 12$

Literal Equations A literal equation is one which contains two or more variables, as, for example, in the Ohm's law formulas. The solution of a literal equation is one variable expressed in terms of other variables. The operations presented above are applicable to the transformation of literal equations of appropriate form.

Example 4 Solve the following literal equations for the variable indicated in parentheses to the right of the equation.

(a) $ax = b$ (x)

(a) $\dfrac{ax}{1} = \dfrac{b}{1} \quad x = \dfrac{b}{a}$

(b) $\dfrac{y}{a} = b$ (y)

(b) $\dfrac{y}{a} = \dfrac{b}{1} \quad y = \underline{ab}$

(c) $\dfrac{aE}{b} = c$ (E)

(c) $\dfrac{aE}{b} = c \quad E = \dfrac{bc}{a}$

(d) $c = \dfrac{a}{x}$ (x) (d) $\dfrac{c}{1} \searrow \dfrac{a}{x}$ $x = \dfrac{a}{c}$

(e) $I = \dfrac{V}{R}$ (R) (e) $\dfrac{I}{1} \searrow \dfrac{V}{R}$ $R = \dfrac{V}{I}$

Writing Equations   Writing equations to represent word statements is an important skill related to solving equations. Study the following examples carefully.

Example 5

(a) A circuit divides a battery voltage of 9 V into four equal voltages. What is the value of the voltages?

   Let v = the value of each of the four equal voltages; then

$$4v = 9 \qquad v = \dfrac{9}{4} \qquad \boxed{v = 2.25\ \text{V}}$$

   Check: $4 \cdot 2.25 = 9 \qquad 9 = 9$

(b) A circuit divides a battery voltage into three equal parts. One of these parts is 4 V. What is the battery voltage?

   Let V = the battery voltage; then

$$\dfrac{V}{3} = 4 \qquad V = 3 \cdot 4 \qquad \boxed{V = 12\ \text{V}}$$

   Check: $\dfrac{12}{3} = 4 \qquad 4 = 4$

Exercises 3-1   In Exercises 1–28 solve the given equations (a) by performing the appropriate multiplication and/or division operation(s), and (b) by use of checkerboard transposition. Check your solutions. Try to obtain solutions mentally when possible.

1. $2x = 4$ 2. $3I = 9$ 3. $1.5R = 6$
4. $2.5V = 20$ 5. $3x = -12$ 6. $5I = -30$
7. $7.5V = -15$ 8. $3.3I = -66$
9. $-9x = 27$ 10. $-6V = 48$
11. $-7I = -28$ 12. $-8R = -2.4$
13. $\dfrac{x}{5} = 2$ 14. $\dfrac{I}{3} = 1$ 15. $\dfrac{V}{2.5} = 4$
16. $\dfrac{R}{3} = 2.2$ 17. $\dfrac{R}{-4} = 2$ 18. $\dfrac{I}{-5} = 3$
19. $\dfrac{g}{-7} = 4$ 20. $\dfrac{i}{-1} = 3$ 21. $\dfrac{y}{-2} = -4$

22. $\dfrac{C}{-3} = -5$ 23. $\dfrac{v}{-5} = -7$ 24. $\dfrac{v}{-6} = -2$
25. $\dfrac{x}{1.5} = -2$ 26. $\dfrac{i}{-3.7} = 4.2$
27. $\dfrac{v}{-2} = -6.3$ 28. $\dfrac{v}{3.2} = -1.8$

In Exercises 29–40 solve the literal equations for the variable appearing in parentheses to the right of the equation. Show the steps in your transformations.

29. $ax = b$ (x) 30. $\dfrac{y}{a} = b$ (y)

31. $V = IR$ (R) (The two members of an equation may be exchanged without destroying the equality, that is, $V = IR$ may be written $IR = V$.)

32. $I = \dfrac{V}{R}$ (V) 33. $I = \dfrac{V}{R}$ (R)

34. $Q = IR$ (I) 35. $Q = IT$ (T)

36. $I = \dfrac{Q}{T}$ (Q) 37. $I = \dfrac{Q}{T}$ (T)

38. $P = VI$ (V) 39. $P = VI$ (I)

40. $W = PT$ (P)

In Exercises 41–44 find the desired quantities by performing these three steps:
(a) Write an equation of the form $ax = b$.
(b) Solve for the unknown.
(c) Check your result.

41. When a certain quadraphonic amplifier is turned on, the power supply module supplies four equal currents to the sytem. The total of the four currents is 960 mA. What is the value of one current?

42. A certain voltage divider circuit is made up of three equal resistors. The total resistance (the sum) is 16.67 kΩ. What is the value of each resistor?

43. A technician receives $0.15 per mile for driving her own car on the job. One month she received a check for $277.50 for mileage. How many miles did she drive?

44. An oscilloscope display of a signal from a computer clock shows a total time of 16 ns for four pulses. What is the time for one pulse?

In Exercises 45–48 find the desired quantities by performing these three steps:
(a) Write an equation of the form $x/a = b$.
(b) Solve for the unknown.
(c) Check your result.

45. A battery supplies equal currents to each of four circuits. One current is 5 mA. What is I_{total}?

46. The total resistance R_T of a circuit is divided into five equal resistors each with a value of 3.9 kΩ. What is R_T?

47. A technician buys an oscilloscope and pays $105 down which is one-sixth of the total cost. What is the cost of the oscilloscope?

48. The cost of a type 7404 integrated circuit is $0.74 when purchased in lots of 25–99. What is the total cost of the minimum quantity?

3-2 GRAPHS

A *graph* is a kind of mathematical picture of an equation. Like any picture it conveys information more directly than words. We construct a graph on a *set of coordinates* which resembles two number lines drawn at right angles to each other. The horizontal line is called the *x-axis* or *abscissa*. The vertical line is called the *y-axis* or *ordinate*.

Example 6 Construct the graph for the equation $y = 2x$.

We first prepare a table of values obtained by choosing arbitrary values for x, the *independent variable*, and calculating the corresponding values of y, the *dependent variable*. Then we plot these pairs of values as points on the coordinates—each pair of values is one point. Finally we draw a smooth curve through the points.

x	y	Point
0	0	A
1	2	B
2	4	C
3	6	D
4	8	E
5	10	F

The equation $y = 2x$ is said to be a *linear equation* because its graph is a straight line.

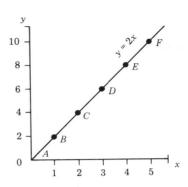

Example 7 Construct the I-versus-V characteristic (graph) for an electrical device which can be represented by the formula $I = 0.5V$. Does the device have a linear or nonlinear characteristic?

V, V	I, A
0	0
1	0.5
2	1
4	2
6	3
8	4
10	5

Note that we plot the independent variable V on the x-axis and the dependent variable I on the y-axis.

When examining the "curve" which represents the device we conclude that the device has a linear I-versus-V characteristic (also sometimes called I–V, or the voltampere characteristic) since the curve is a straight line.

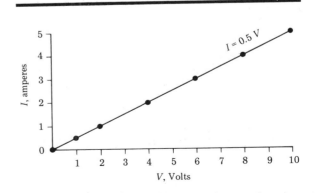

Exercises 3-2
In Exercises 1-12 plot the given equations using the techniques illustrated in Examples 6 and 7. Plot the independent variable on the x-axis.

1. $y = 3x$ 2. $y = 0.25x$ 3. $y = 4x$

4. $y = 10x$ 5. $I = V$ 6. $I = 0.25V$

7. $I = 0.75V$ 8. $I = 2V$ 9. $I = \dfrac{10}{R}$

10. $I = \dfrac{6}{R}$ 11. $I = \dfrac{8}{R}$ 12. $I = \dfrac{24}{R}$

In Exercises 13-18, (a) plot the given sets of values, and (b) describe the curves as linear or nonlinear. Choose divisions along the axes so that points will be neither too crowded nor too spread out.

13.

V, V	I, A
0	0
10	1
20	2
30	3
40	4
50	5

14.

V, V	I, mA
0	0
100	35
200	70
300	105
400	140
500	175

15.

V, mV	I, μA
0	0
15	75
30	150
45	225
60	300
75	375

16.

V, V	I, mA
0	0
5	0
10	5
15	10
20	15
25	20

17.

V, V	I, mA
0	0
5	100
10	150
15	200
20	220
40	300

18.

V, V	I, mA
0	0
5	40
10	48
15	55
20	65
40	110

3-3 ADDITION AND SUBTRACTION OF ALGEBRAIC EXPRESSIONS

Monomials, Binomials, Polynomials An *algebraic expression* is any combination of literal and explicit numbers joined together by mathematical operations. When two or more algebraic expressions representing multiplication only are connected with + or − signs, the expressions are called *terms*. A single term is called a *monomial*, two terms are called a *binomial*, three terms are a *trinomial*. Two or more terms are called a *polynomial*.

Coefficients When a given term represents the product of two or more factors, each factor is the *coefficient* of the other(s). For example, in the term $5a^2b^3c^4$, 5 is the coef-

ficient of $a^2b^3c^4$, b^3 is the coefficient of $5a^2c^4$, and so on. However, the 5 is called the *numerical coefficient* and we often refer to a numerical coefficient as *the coefficient*.

Like Terms Two or more terms are called *like terms* or *similar terms* if they are identical in every respect except the numerical coefficient. In the polynomial $3ab - 4a^2b + 2ab^2 - 5ab + 7ab^2$, $3ab$ and $5ab$ are like terms as well as $2ab^2$ and $7ab^2$, while $4a^2b$ is unlike any other term.

Two or more like terms can be expressed as a single term by combining their coefficients, following the rules for addition and subtraction of signed numbers. This fact permits us to add, algebraically, binomials or trinomials containing like terms, or to simplify polynomials (review Sections 1-3, 1-4, and 1-5). Examples demonstrate the procedure.

Example 8 Perform the indicated operations.
Add:

(a) $5ab, -7ab$
(a) $5ab + (-7ab) = 5ab - 7ab = -2ab$

(b) $2ir, -5ir, 8ir$
(b) $2ir + (-5ir) + 8ir = 2ir - 5ir + 8ir = 5ir$

Subtract:

(c) $-5xy$ from $8xy$
(c) $8xy - (-5xy) = 8xy + 5xy = 13xy$

(d) $4I^2R$ from $-3I^2R$
(d) $-3I^2R - (+4I^2R) = -3I^2R - 4I^2R$
$= -7I^2R$

Add:

(e) $4ax - 7by$ and $-5ax + 3by$

(e) $\begin{array}{r} 4ax - 7by \\ -5ax + 3by \\ \hline -ax - 4by \end{array}$

(f) $2Vg + 5vG$ and $3Vg - 7vG$

(f) $\begin{array}{r} 2Vg + 5vG \\ 3Vg - 7vG \\ \hline 5Vg - 2vG \end{array}$

Subtract:

(g) $-4a^2b + 7cd^2$ from $-7a^2b - 3cd^2$

(g) $\begin{array}{r} -7a^2b - 3cd^2 \\ -4a^2b + 7cd^2 \\ \hline -3a^2b - 10cd^2 \end{array}$

(h) $-3i^2r + 2V^2g$ from $5i^2r - 6V^2g$

(h) $\begin{array}{r} 5i^2r - 6V^2g \\ -3i^2r + 2V^2g \\ \hline 8i^2r - 8V^2g \end{array}$

Simplify:
(i) $3a^2b + 4cd^2 - 5a^2b + 9a^2b - 6cd^2$
(i) $7a^2b - 2cd^2$
(j) $5IR - 7ir + 2ir - 9IR + 6ir$
(j) $ir - 4IR$

Exercises 3-3 In Exercises 1–10 state which are the like terms of the given expressions. (See above under "Like Terms.")

1. $3a + 4b - 2a$
2. $4x^2 - 2xy + 7x^2$
3. $5i - 2v + 7v - 12i$
4. $6I^2R + 5i^2r + 3IR^2 + 2I^2R$
5. $a^2bc^3 + ab^2c^3 - 2a^2bc^3 - 3a^3bc^2$
6. $R_1R_2R_3 + rR_2R_3 + rR_1R_3 + 6R_1rR_3$
7. $3i^3 + 2i^2 - 5i^3 + 7i$
8. $5V_1I_1 + 12I_1V_1 - 6V_1I_2 - 4I_1^2R_1 + 16V_1I_2$
9. $-3V_1G + 4V_1^2G_1 + 3V_1^2G_1$
10. $12I_1^2R_1 - 6R_1^2I_1 + 9I_1^2R_1 - 10R_1^2R_2 + 9R_1^2I_1$

In Exercises 11–14 perform the indicated operations.

11. Add:
(a) $4ax$ and $5ax$
(b) $4R_1R_2R_3$ and $-10R_1R_2R_3$
(c) $-6V^2I$, $-8V^2I$, and $10V^2I$
(d) $5I_1R_1 - 6I_2R_2$ and $-3I_1R_1 + 2I_2R_2$
(e) $5fL_1 - 6fL_2$, $8fL_1 + 2fL_2$, and $-10fL_1 - 3fL_2$

12. Add:
(a) $6x^2yz$ and $-4x^2yz$
(b) $6fLC$ and $-3fLC$
(c) $-4I^2R$, $7I^2R$, and $-5I^2R$
(d) $4V_1G_1 - 8V_2G_2$ and $-6V_1G_1 + 4V_2G_2$
(e) $6IR - 2I'R'$, $4IR + 3I'R'$, and $-2IR + 3I'R'$

13. Subtract:
(a) $-4d^2ef$ from $7d^2ef$
(b) $5IR$ from $3IR$
(c) $6R_1R_2R_3$ from $-10R_1R_2R_3$
(d) $-3I^2R + 3I_1^2R_1$ from $5I^2R - 6I_1^2R_1$
(e) $2V'I' - 10VI$ from $-3V'I' + 4VI$

14. Subtract:
(a) $5st^2$ from $-2st^2$
(b) $-7vg$ from $2vg$
(c) $4h_1iR$ from $-7h_1iR$
(d) $5V_1^2G_1 - 2V_2^2G_2$ from $-3V_1^2G_1 + 5V_2^2G_2$
(e) $4I'R_1 + 3I''R_2$ from $-6I'R_1 + 7I''R_2$

In Exercises 15–20 simplify the given expressions by combining (collecting) like terms.

15. $3x - 4y - 4x + 7y - 5y + 5x$
16. $10i - 12I + 15i - 3i + 7I - 5I + 2i$
17. $2xy - 4xz + 2xy - 5xz + 2yx - 7xz$
18. $5ir - 6IR + 7IR - 2ir + 15IR - 12iR - 5IR$
19. $5VI + 12IV - 6IV - 4I^2R + 16VI$
20. $9I^2R + 2V^2G - 12I^2R + 2I^2R - 3V^2G + 5I^2R$

3-4 MULTIPLICATION OF ALGEBRAIC EXPRESSIONS

Multiplication of Monomials

To find the product of two or more monomials:
1. Find the product of the numerical coefficients by applying the rules for multiplication of signed numbers (Section 1-6).
2. Then multiply by the products of literal factors found by applying rules for exponents (Section 2-3).

Example 9 Find the indicated products.
(a) $3 \cdot 2a$
(b) $4 \cdot 2IR$
(c) $4x \cdot 5y$
(d) $3I \cdot 5R$
(e) $3a \cdot 6ab$
(f) $7I \cdot 5IR$
(g) $-4x \cdot 2xy^2$

(a) $3 \cdot 2a = 6a$
(b) $4 \cdot 2IR = 8IR$
(c) $4x \cdot 5y = 20xy$
(d) $3I \cdot 5R = 15IR$
(e) $3a \cdot 6ab = 18a^2b$
(f) $7I \cdot 5IR = 35I^2R$
(g) $-4x \cdot 2xy^2 = -8x^2y^2$

(h) $-3V \cdot 5GV^2$

(h) $-3V \cdot 5GV^2 =$
$-15GV^3$

(i) $(-2I^2R)(-3IR^2)$

(i) $(-2I^2R)(-3IR^2)$
$= 6I^3R^3$

(j) $(-2ir^2)^2$

(j) $(-2ir^2)^2 = 4i^2r^4$

Multiplication of a Polynomial by a Monomial
In algebra as in arithmetic, when both multiplication and addition (or subtraction) are to be performed, we have the convention (a mutual understanding among all users of mathematics) that multiplication is to be performed before the addition. However, if the addition must be performed first we may use symbols of grouping: parentheses ( ), brackets [ ], and braces { }. Thus, if we need to perform the algebraic addition $4x - 5y - 3z$ before multiplying by $6a$, we write $6a(4x - 5y - 3z)$. Since we cannot actually combine the terms here we make use of another property of mathematics:

$2(3 + 4) = 2 \cdot 3 + 2 \cdot 4$ is the same as $2(7)$

That is,

$2 \cdot 3 + 2 \cdot 4 = 6 + 8 = 14$ and $2 \cdot 7 = 14$

This is known as the *distributive property*. We use it to perform the operation of multiplication of a polynomial by a monomial.

To multiply a polynomial by a monomial, find the product of each term of the polynomial by the monomial, observing the procedure for finding products of monomials. Write the result as a polynomial.

Example 10 Find the indicated products.

(a) $3x(3x - 7y)$

(a) $3x(3x - 7y) = 3x \cdot 3x - 3x \cdot 7y =$
$9x^2 - 21xy$

(b) $2I(R_1 + R_2 + R_3)$

(b) $2I(R_1 + R_2 + R_3) = 2IR_1 + 2IR_2 + 2IR_3$

(c) $3R(I_1 + 2I_2 + 4I_3)$

(c) $3I_1R + 6I_2R + 12I_3R$

(*Note:* In (c) we write the products with the literals in alphabetical order, as is conventional. We are permitted to do this because of the commutative property: $2 \cdot 3 = 3 \cdot 2$.)

(d) $-10i(2ir - Ir - 4IR)$

(d) $-20i^2r + 10iIr + 40iIR$

Observe that when we multiply each term enclosed in parentheses by a factor with a negative sign we change the sign of each of such terms [see Example 10(d)].

Frequently expressions involving multiplication of a polynomial by a monomial include other terms as well. Consider these examples.

Example 11 Find the indicated products and simplify the results by combining like terms.

(a) $3 - 5(a - 2)$

(a) $3 - 5 \cdot a - 5 \cdot (-2) = 3 - 5a + 10 =$
$13 - 5a$

(b) $V - 4(2V + 3v)$

(b) $V - 8V - 12v = -7V - 12v$

(c) $8I^2R - 2I(3IR + 2iR - 4Ir)$

(c) $8I^2R - 6I^2R - 4IiR + 8I^2r$
$= 2I^2R - 4IiR + 8I^2r$

(d) $5I(R_1 - 3R_2) - 3I(2R_2 - 3R_1)$

(d) $5IR_1 - 15IR_2 - 6IR_2 + 9IR_1 =$
$14IR_1 - 21IR_2$

Multiplication of Polynomials We extend the distributive property and determine that

$$(a + b)(c + d)$$

means

$$a(c + d) + b(c + d) = ac + ad + bc + bd$$

To multiply polynomials, find the product of each term of the first expression with each term of the second expression. Express the result as a polynomial after combining like terms.

Example 12 Find the indicated products.

(a) $(a + 2)(a + 3)$

(a) $(a + 2)(a + 3) = a(a + 3) + 2(a + 3) =$

$a^2 + 3a + 2a + 6 = a^2 + 5a + 6$

(b) $(I + 2)(I^2 + 4I + 4)$

(b) $(I + 2)(I^2 + 4I + 4) = I(I^2 + 4I + 4) + 2(I^2 + 4I + 4) = I^3 + 4I^2 + 4I + 2I^2 + 8I + 8 = I^3 + 6I^2 + 12I + 8$

An alternate procedure:

(Product of the second term of the binomial and the trinomial)

$$\begin{array}{r} I^2 + 4I + 4 \\ I + 2 \\ \hline I^3 + 4I^2 + 4I \\ + 2I^2 + 8I + 8 \\ \hline I^3 + 6I^2 + 12I + 8 \end{array}$$ Ans.

(Product of the first term of the binomial and the trinomial)

Exercises 3-4   In Exercises 1–20 find the indicated products.

1. $2 \cdot 6x$ 2. $3 \cdot 4I$ 3. $-4 \cdot 5V$

4. $(-3)(-2v)$ 5. $2x \cdot 3x$ 6. $-3i \cdot 4i^2 r$

7. $4V \cdot (-5I^2R)$ 8. $(-4I^2)(-6V^2I^2)$

9. $(4v)^2$ 10. $(-5i^2)^2$ 11. $(3v^2)^2$

12. $(-2i^2)^2$ 13. $3(x + y)$

14. $-2(2x - 3y)$ 15. $-3(-4I - 5i)$

16. $4I(R_1 + R_2 + R_3 + R_4)$

17. $(G_1 + 2G_2 - 5G_3)5V$

18. $(3a - 4b - 5c)4a$

19. $(R_1 - 2R_2 - 7R_3)(-2I)$

20. $(2IR_1 - 5IR_2 + 7IR_3)(-5I)$

In Exercises 21–40 simplify the given expressions by finding indicated products and combining like terms.

21. $5 + 4(I - 2)$

22. $4I(5R - 2) - 15IR + 6I$

23. $i - 3(i + 2)$ 24. $v + 2(2v - 4V)$

25. $V - 3(V - 2v)$

26. $2(3I - 2i) - 3(3i - 4I)$

27. $(V_1 - 5) + (V_2 + v) - (v - V_3)$

28. $(I_1 + 5) - 2(I_2 - I_3) - (I_1 + I_2 - I_3)$

29. $R(2I - 3i) - I(R + 3r) + 2i(-2R + 4r)$

30. $I(2I - IR) + 2R(I^2 - 3i^2) - (-5I^2 + 4I^2R)$

31. $(x + 1)(x + 2)$ 32. $(x - 2)(x - 3)$

33. $(x + 4)(x - 2)$ 34. $(x + 2)(x + 2)$

35. $(I + i)(R + r)$ 36. $(I + 1)(I + 2)$

37. $(I + i)(R_1 + R_2 + R_3)$ 38. $(V + IR)(V - 2IR)$

39. $3(x + 2)(x - 3)$ 40. $I(R - 2)(R + 4)$

3-5 EXPRESSIONS CONTAINING SYMBOLS OF GROUPING

When we convert the relationships of electronic circuits, or other physical phenomena, into mathematical statements, two or more sets of parentheses or other symbols of grouping may be required. Frequently such expressions may then be simplified by removing the symbols and combining like terms. For example, we simplify the expression $5(I_1 + I_2) - 3(I_1 - I_2) - (I_1 + I_2) + (I_1 - I_2)$ by using the procedures of Section 3-4:

$$5I_1 + 5I_2 - 3I_1 + 3I_2 - I_1 - I_2 + I_1 - I_2$$
$$= 2I_1 + 6I_2$$

We clear the parentheses from the first two terms by multiplying by the numerical factors 5 and -3. Similarly, we consider the factor of the third term to be -1 and of the last term, $+1$.

> To clear an expression of parentheses multiply each term in an enclosure by the factor(s) of the enclosure, taking into account the sign of the factor(s). If an enclosure is preceded by no sign or a + sign only, consider the factor to be +1; if preceded by a − sign, use −1 as the factor.

> ### MEMORY JOGGER
>
> When parentheses preceded by a + sign are removed, the signs of the terms in the grouping remain unchanged; when parentheses preceded by a − sign are removed, the sign of each term is changed to its opposite. (The sign preceding a parenthesis is understood to be + when there is no sign.)

Nested Symbols   For very involved relationships, enclosures within enclosures (nested symbols) may be required. We use the following procedure to remove nested symbols.

To remove nested symbols, remove the inner enclosures first and proceed, step by step, to the outermost. Simplify intermediate expressions by combining like terms.

Example 13 Simplify by removing enclosures and combining like terms.

(a) $3a - (a - b)$

(a) $3a - (a - b) = 3a - a + b = 2a + b$

(b) $3R_1 - [R_1 - 3(R_1 + 2R_2)]$

(b) $3R_1 - [R_1 - 3(R_1 + 2R_2)]$
$= 3R_1 - [R_1 - 3R_1 - 6R_2]$
$= 3R_1 - [-2R_1 - 6R_2]$
$= 3R_1 + 2R_1 + 6R_2$
$= 5R_1 + 6R_2$

(c) $(u + v) -$
$\{x - (u - v) + 3[2x - 4(-u + v)]\}$

(c) First, we remove inner enclosures,

$u + v - \{x - u + v + 3[2x + 4u - 4v]\}$

and again, the innermost,

$u + v - \{x - u + v + 6x + 12u - 12v\}$

Next, we combine like terms,

$u + v - \{11u - 11v + 7x\}$

and finally, we remove the outer enclosures and again combine like terms,

$u + v - 11u + 11v - 7x = -10u + 12v - 7x$

arranging terms with letters in alphabetical order.

Writing Expressions Requiring Symbols of Grouping Sometimes our task of converting the description of a physical situation into a mathematical expression requires the grouping of two or more terms to capture the correct meaning of the relationship. Examples demonstrate the procedure.

Example 14 Using symbols of grouping, convert the following statements to mathematical expressions.

(a) A resistance R_1 added to 4 times the sum of R_1, $3R_2$, and R_3.

(a) $R_1 + 4(R_1 + 3R_2 + R_3)$

(b) A voltage v, less 3 times the sum of $2V$, $3IR$, and -5 times the sum of V' and $-3V''$.

(b) $v - 3[2V + 3IR - 5(V' - 3V'')]$

Inserting Symbols of Grouping Sometimes it is our task to insert symbols of grouping into existing algebraic expressions. It is essential that we do so without changing the actual value of the expression. We follow this procedure.

To insert symbols of grouping:
1. Preserve the sign of each term when the enclosure is preceded by a + sign.
2. Change the sign of each term when the enclosure is preceded by a – sign.

Example 15 Insert parentheses in the following expressions at the points designated with a caret ($\wedge$).

(a) $3a - {}_{\wedge}4b + 5c_{\wedge}$ (a) $3a - (4b - 5c)$

(b) $2I + {}_{\wedge}5i - 6 +$ (b) $2I + (5i - 6 +$
$9V/r_{\wedge}$ $9V/r)$

Exercises 3-5 In Exercises 1–17 remove symbols of grouping and simplify.

1. $(3 + 2)5$ 2. $-3(2 - 4)$ 3. $5(x + 2)$
4. $-a(4 - b)$ 5. $i - 3(i + 2)$
6. $v + 2(2v - 4V)$
7. $(V_1 - 5) + (V_2 + 3) + (V_1 + V_3)$
8. $(I_1 + 5) + (I_2 - I_3) - (I_1 + I_2 - I_3)$
9. $R_1 - 5(R_1 + R_2) + 12(R_2 - 6R_1) + 50R_1$
10. $X_1 - 2(-X_1 + X_2) + 5(X_1 - X_2 + X_3) - 4X_3$
11. $3VI - 2V_1I_1 + 2(VI - V_2I_2)$
12. $4I_1^2R_1 - (2I_1^2R_1 + 6I_2^2R_2)$
13. $R_1 + [R_2 - (5 + R_1) + 7 - (6 - R_2)]$
14. $-V - [V' + (V_1 - V_2) - (3 - V + V_1)] + 7$
15. $\{6 - [X_C + (5 - X_L) - (10 + X_C)] - X_C\} + 7$
16. $7 - \{I_a - [10 - I_b - (I_c - I_a) - (5 - I_b + I_c - I_a)] + 4\}$

子 www.lip   (56.6 kbs)

在 computer 想儲仰首先(按右)边间 new → jond 跟着
打自己的名就打闲自己的 jond, 佗想 儲仰只按 ctrl 跟着
按某人的 jond 傳过自己的 jond (佗想 做什么 都是 w lip.)
佗一张大张 想抹掉某一段 可放老鼠在前或中间按3次
然按圆叉剪按一次 地就 cut 了 过然跟着 paste. (佗想按一次
多放老鼠上前就可以. 想儲两边都可以 看按 jile → page setup
左或右差 (1.2) 空左二位. 空位上或下毛 1. 佗仰段文想重头付会的
地方. 然後按区这个就得或 想 思 按 close 放现两 地失去.
上边的圆按 view → toolbars → formatting 或 standard
www (wold wide web = graphics.

    x   x   x

netcap 打 www. com. 首先打 yahoo
(∧Bookmarks = 记录页码) 佗北寻上材引 按 bookmarks 跟着
lick
add Bookmark
Làm dấu trong trên interenl
Vào bookmark → chon Add bookmark.

17. The time required, in nanoseconds, for a certain computer to add two numbers P and Q, having the number of digits D_p and D_q, respectively, is given by the expression

$$(160 + 40D_q) + 2[160 + 40(D_p - D_q)] - \{160 + 40[D_p - 2(D_q + 7)]\}$$

(a) Simplify the expression.

(b) If P contains ten digits and Q eight digits, calculate the operation time.

In Exercises 18–21 use symbols of grouping in translating the stated relations into algebraic expressions.

18. A number x added to 7 times the sum of twice g and 4 times z.

19. The resistance r added to 4 times the sum of R_1, R_2, and R_3.

20. The sum of $2I$ times the difference R_1 minus R_2, and 3 times the sum of V_1 and V_2.

21. The sum of iR and -3 times the difference V_1 minus $2v$ is to be multiplied by 7, and that product subtracted from $6I_1R_1$.

In Exercises 22–26 insert parentheses at the points designated by a caret ($\wedge$).

22. $5x + {}_\wedge 4y + 3z_\wedge$ 23. $4R - {}_\wedge 6r + 7R'_\wedge$

24. $2i + {}_\wedge 5I_1 - 2I_2 + I_3 v_\wedge - 7I_4$

25. $X_1 - {}_\wedge R_2 + X_2 - X_3{}_\wedge - Z_1$

26. $15 - 4v - {}_\wedge iv - 3I^2R + 6V^2_\wedge - 15IV$

3-6 DIVISION OF ALGEBRAIC EXPRESSIONS

Division of a Monomial by a Monomial

To divide one monomial by another:
1. Find the quotient of the numerical coefficients (cancel common factors whenever possible) in accordance with the rules for signed numbers (see Section 1-6).
2. Combine the literal factors in accordance with rules for exponents (see Section 2-3).

Example 16 Perform the indicated divisions.

(a) $\dfrac{12a}{3}$

(b) $\dfrac{16IR}{4}$

(c) $\dfrac{28a^2b}{7a}$

(d) $\dfrac{18I^2R}{3IR}$

(e) $\dfrac{-21xy^2}{3xy}$

(f) $\dfrac{35V^2G}{-7VG}$

(g) $\dfrac{-28x^2y^2}{-2xy^2}$

(h) $\dfrac{-56i^2r^2}{-8i^2r}$

(a) $\dfrac{12a}{3} = 4a$

(b) $\dfrac{16IR}{4} = 4IR$

(c) $\dfrac{28a^2b}{7a} = 4ab$

(d) $\dfrac{18I^2R}{3IR} = 6I$

(e) $\dfrac{-21xy^2}{3xy} = -7y$

(f) $\dfrac{35V^2G}{-7VG} = -5V$

(g) $\dfrac{-28x^2y^2}{-2xy^2} = 14x$

(h) $\dfrac{-56i^2r^2}{-8i^2r} = 7r$

Division of a Polynomial by a Monomial

To divide a polynomial by a monomial:
1. Divide each term of the polynomial by the monomial, observing the rule for the division of one monomial by another (see above).
2. Write the result as a polynomial, that is, as the algebraic sum of the terms obtained in (1).

Example 17 Perform the indicated divisions.

(a) $\dfrac{3a + 12b}{3}$

(a) $\dfrac{3a + 12b}{3} = \dfrac{3a}{3} + \dfrac{12b}{3} = a + 4b$

(b) $\dfrac{16IR - 24IR'}{4}$

(b) $\dfrac{16IR - 24IR'}{4} = 4IR - 6IR'$

(We may perform the divisions mentally.)

(c) $\dfrac{12IR_1 - 18IR_2 + 27IR_3}{3I}$

(c) $\dfrac{12IR_1 - 18IR_2 + 27IR_3}{3I} =$

$4R_1 - 6R_2 + 9R_3$

(d) $\dfrac{15x^2y - 35xy^2 - 60x^2y^2}{-5xy}$

(d) $\dfrac{15x^2y - 35xy^2 - 60x^2y^2}{-5xy} =$

$\qquad -3x + 7y + 12xy$

(e) $\dfrac{abc + abd + bcd}{ac}$

(e) $b + \dfrac{bd}{c} + \dfrac{bd}{a}$, or $b + bc^{-1}d + a^{-1}bd$

We will now study some important special products and the procedure of factoring before studying the division of one polynomial by another (Section 3-15).

Exercises 3-6 Perform the indicated divisions.

1. $\dfrac{12x}{4}$ 2. $\dfrac{15I}{3}$ 3. $\dfrac{18xy}{-9x}$ 4. $\dfrac{-9IR}{3R}$

5. $\dfrac{-36x^2y^2}{9x}$ 6. $\dfrac{-48I^2R}{12R}$ 7. $\dfrac{-72a^2b^3}{-8a^2b^2}$

8. $\dfrac{-81I^4R^2}{-9I^3R^3}$ 9. $\dfrac{-96V^4G^2}{24V^3G^2}$

10. $\dfrac{-100i^5r^3}{-25i^2r^2}$ 11. $\dfrac{4ab + 2ac}{2}$

12. $\dfrac{15ir + 35IR}{5}$ 13. $\dfrac{18ab + 6bc}{-6}$

14. $\dfrac{3vg - 12vG}{-3}$ 15. $\dfrac{5a + 35ab}{-5a}$

16. $\dfrac{-36I - 27IR}{-9I}$

17. $\dfrac{IR_1 + IR_2 + IR_3 + IR_4}{I}$

18. $\dfrac{R_1R_2R_3 + R_2R_3R_4 + R_3R_4R_5}{R_3R_4}$

19. $\dfrac{4a^2b^4cd^2 + 12b^2cd^3}{2ab^3d}$

20. $\dfrac{15I^4R^2 - 9I^2R}{-3I^2R}$

21. $\dfrac{C_1C_2 + C_2C_3 + C_1C_3}{C_1C_2C_3}$

22. $\dfrac{RC + CL + RL}{RCL}$

23. $\dfrac{12I^2R - 9IR^2 + 3IR^3}{-3IR}$

24. $\dfrac{Vg_1 + Vg_2 + Vg_3 + Vg_4}{V}$

25. $\dfrac{RX_1^2 + RX_2^2 + RX_3^2}{R}$

3-7 FACTORING

Factor A *factor* is one of the quantities in a multiplication operation.

> Factoring an expression consists of (a) recognizing or determining the presence of common factor(s) in products and terms and (b) separating the common factor(s) from the expression in a way that does not destroy the value of the original expression.

For example, $12 = 4 \cdot 3 = 2 \cdot 6 = 2 \cdot 2 \cdot 3$. The factors of 12 are 4 and 3, or 2 and 6, or 2, 2, and 3. Here, 2, 2, and 3 are the *prime factors* of 12 because they do not have any factors (except themselves and 1). What are the factors of $18I^2R$? Do the terms of $4 + 12$ have a common factor? Is there a common factor in the terms of the polynomial, $IR_1 + IR_2 + IR_3$?

$$18I^2R = 2 \cdot 3 \cdot 3 \cdot I \cdot I \cdot R$$

$$4 + 12 = 4(1 + 3)$$

$$IR_1 + IR_2 + IR_3 = I(R_1 + R_2 + R_3)$$

Example 18 Factor the given expressions into their prime factors.

(a) 9   (a) $9 = 3 \cdot 3$

(b) 21   (b) $21 = 3 \cdot 7$

(c) 16   (c) $16 = 2 \cdot 2 \cdot 2 \cdot 2$

(d) 27   (d) $27 = 3 \cdot 3 \cdot 3$

(e) $36x^2yz$ (e) $36x^2yz = 2 \cdot 2 \cdot 3 \cdot 3 \cdot$
$\qquad\qquad\qquad x \cdot x \cdot y \cdot z$

Polynomials with a Common Monomial Factor We know from our work in Section 3-4 that the product of a polynomial and a monomial is a polynomial, every term of which contains the monomial as a factor. For example,

$$3x(4a + 5b + 2c) = 12ax + 15bx + 6cx$$

Therefore, if the terms of a polynomial contain a common factor, we can factor the expression into the product of a monomial and a polynomial. We perform the process by examining each term of the polynomial.

Example 19 Factor the following polynomials.

(a) $ax + ay$
(a) $ax + ay = a(x + y)$
(b) $3a^2 b + 6ax$
(b) $3a^2 b + 6ax = 3a(ab + 2x)$
(c) $8v^2 i^3 g + 14v^4 ig^2$
(c) $2v^2 ig(4i^2 + 7v^2 g)$
(d) $4x^3 + 6x^2 y + 2x$
(d) $2x(2x^2 + 3xy + 1)$

It is helpful to note that in factoring a common factor from the terms of a polynomial we both multiply and divide the expression by the common factor and thus change its form but not its value. Thus, in factoring $IR_1 + IR_2 + IR_3$, the process, if performed in step-by-step detail, proceeds as follows:

$$IR_1 + IR_2 + IR_3 = \frac{I}{I}(IR_1 + IR_2 + IR_3)$$

$$= I \cdot \frac{IR_1 + IR_2 + IR_3}{I}$$

The indicated division

$$\frac{IR_1 + IR_2 + IR_3}{I}$$

is evaluated:

$$\frac{IR_1 + IR_2 + IR_3}{I} = R_1 + R_2 + R_3$$

and thus,

$$I \cdot \frac{IR_1 + IR_2 + IR_3}{I} = I(R_1 + R_2 + R_3)$$

The results of factoring can always be checked by remultiplying the factors,

$$I(R_1 + R_2 + R_3) = IR_1 + IR_2 + IR_3$$

To factor a polynomial whose terms include a common monomial factor:
1. Examine the terms and determine the highest monomial factor common to all terms.
2. Divide each term by the common factor.
3. Write the result as the product of a monomial (the common factor) and a polynomial (the quotient of the original polynomial divided by the common factor).

Example 20 Factor the following polynomials.

(a) $12 + 18$
(a) $12 + 18 = 3 \cdot 4 + 3 \cdot 6 = 3(4 + 6)$
(b) $ax + bx + cx$
(b) $ax + bx + cx = x(a + b + c)$
(c) $I_1 R + I_2 R + I_3 R$
(c) $R(I_1 + I_2 + I_3)$
(d) $x - by + cy$
(d) $x - by + cy = x - y(b - c)$ (Here we group those terms which have a common factor and factor them only.)
(e) $IR_1 - 2IR_3 + 4V - 5IR_2$
(e) $4V + IR_1 - 2IR_3 - 5IR_2 = 4V + I(R_1 - 5R_2 - 2R_3)$

Exercises 3-7 In Exercises 1–8 factor the given expressions into their prime factors.

1. (a) 12 (b) 15 (c) 18 (d) 27
2. (a) 15 (b) 20 (c) 25 (d) 50
3. (a) 14 (b) 21 (c) 28 (d) 42
4. (a) 8 (b) 16 (c) 24 (d) 64   5. $2xyz$
6. $6x^2 y^3 z$ 7. $8I^2 R$ 8. $12V^2 G$

In Exercises 9–30 factor common factors from terms. Check your results by remultiplying the factors.

9. $ax + ay$ 10. $4x^2 + 6x$
11. $5IR_1 + 15IR_2$ 12. $6V - 12V^2 G$
13. $B_0 + RB_0$ 14. $I_c R_L - I_c R_E$
15. $I_L r - I_L R$ 16. $I_b R_e + I_c R_e$
17. $R_1 R_2 + R_1 R_3$
18. $IR_1 + IR_2 + IR_3 + IR_4$

19. $VG_1 + VG_2 + VG_3 + VG_4$

20. $\dfrac{V}{R_1} + \dfrac{V}{R_2} + \dfrac{V}{R_3} + \dfrac{V}{R_4}$

21. $V_1 I + V_2 I + V_3 I + V_4 I$

22. $VI_1 + VI_2 + VI_3 + VI_4$

23. $3 \times 10^3 + 5 \times 10^3 + 2 \times 10^3 + 10 \times 10^3$

24. $0.5 \times 10^{-5} + 2.7 \times 10^{-5} + 3.6 \times 10^{-5} - 7.2 \times 10^{-5}$

25. $am + bm + cm - dm$

26. $pk + qk - rk + sk$

27. $V - IR_1 - IR_2 - IR_3$

28. $I_T - \dfrac{V}{R_1} - \dfrac{V}{R_2}$ 29. $I_T - VG_1 - VG_2$

30. $\dfrac{V^2}{R_T} - \dfrac{V^2}{R_1} - \dfrac{V^2}{R_3}$

3-8 THE SQUARE OF A BINOMIAL

To find the square of a binomial we apply the rule for the product of polynomials,

$(x + 2)^2 = (x + 2)(x + 2) = x(x + 2) + 2(x + 2)$

$= x^2 + 2x + 2x + 4 = x^2 + 4x + 4$

In general,

$$(x + y)^2 = x^2 + 2xy + y^2$$

$$(x - y)^2 = x^2 - 2xy + y^2$$

$$(x \pm y)^2 = x^2 \pm 2xy + y^2$$

The square of a binomial is equal to the square of the first term plus twice the algebraic cross product plus the square of the second term.

Example 21 Find the indicated squares.

(a) $(x + 2)^2$ (a) $(x + 2)^2 = x^2 + 4x + 4$

(b) $(v - 3)^2$ (b) $(v - 3)^2 = v^2 - 6v + 9$

(c) $(3v - 4i)^2$ (c) $9v^2 - 24vi + 16i^2$

Squaring a binomial is sometimes used in arithmetic applications as in the following examples.

Example 22 Find the indicated squares using the rule for the square of a binomial.

(a) 102^2

(a) $102^2 = (100 + 2)^2 = 100^2 + 2(100 \cdot 2) + 2^2 = 10000 + 400 + 4 = 10404$

(b) 0.97^2

(b) $0.97^2 = (1 - 0.03)^2 = 1^2 - 2(0.03 \cdot 1) + 0.03^2 = 1 - 0.06 + 0.0009 = 0.9409$

Exercises 3-8 In Exercises 1–34 square the given binomials by inspection.

1. $(c + d)^2$ 2. $(a + 2)^2$ 3. $(v + i)^2$

4. $(R + r)^2$ 5. $(x - 2)^2$ 6. $(v - 7)^2$

7. $(R - 3)^2$ 8. $(X - 5)^2$ 9. $(2x + y)^2$

10. $(3c + d)^2$ 11. $(4v - i)^2$

12. $(5R - r)^2$ 13. $(i + 5r)^2$

14. $(g + 9r)^2$ 15. $(Z - 8X)^2$

16. $(G - 12B)^2$ 17. $(2a + 3b)^2$

18. $(3x - 4y)^2$ 19. $(5v - 4i)^2$

20. $(6r - 4R)^2$ 21. $(x^2 - 6)^2$

22. $(v^3 + 7)^2$ 23. $(X - 2f^2 L^2)^2$

24. $\left(I^2 R - \dfrac{V^2}{r}\right)^2$ 25. $(12 - v^2 g^2)^2$

26. $(9 - \omega^2 L^2)^2$ 27. $(\tfrac{1}{2}a + \tfrac{2}{3}b)^2$

28. $(x + \tfrac{1}{4})^2$ 29. $(\tfrac{3}{7} - v)^2$

30. $\left(\dfrac{5}{6} - \dfrac{V^2}{I}\right)^2$ 31. $(2v - 0.5)^2$

32. $(3I - 0.2)^2$ 33. $(0.2i - 0.3v)^2$

34. $(0.04R + 0.05r)^2$

In Exercises 35–46 find the value of the indicated squares using the rule for squaring a binomial.

35. 22^2 36. 28^2 37. 32^2 38. 48^2

39. 1.2^2 40. 0.99^2 41. 101^2

42. 99^2 43. 69^2 44. 71^2 45. 83^2

46. 76^2

3-9 FACTORING TRINOMIAL SQUARES

When we encounter a trinomial containing two positive perfect-square terms there is a good chance that it is the square of a binomial. We should suspect as much and test for it: Find the product of the roots of the two perfect-square terms, multiply by 2, and compare with the third term of the trinomial.

For example, in $4v^2 - 12vi + 9i^2$,

$2v$ is the square root of $4v^2$

$-3i$ is the square root of $9i^2$

and

$2(2v)(-3i)$ is $-12vi$

Thus

$$4v^2 - 12vi + 9i^2 = (2v - 3i)^2$$

Example 23 Factor the following trinomial squares.

(a) $x^2 + 4xy + 4y^2$

(a) $x^2 + 4xy + 4y^2 = (x + 2y)^2$

(b) $v^2 - 12v + 36$

(b) $v^2 - 12v + 36 = (v - 6)^2$

(c) $9i^2 - 12ir + 4r^2$

(c) $(9i^2 - 12ir + 4r^2) = (3i - 2r)^2$

(d) $\left(X^2 + XZ + \dfrac{Z^2}{4}\right)$

(d) $\left(X^2 + XZ + \dfrac{Z^2}{4}\right) = \left(X + \dfrac{Z}{2}\right)^2$

Exercises 3-9 In Exercises 1–20 express the trinomials as squares of binomials.

1. $x^2 + 2x + 1$ 2. $a^2 + 6a + 9$

3. $v^2 - 4v + 4$ 4. $i^2 - 6i + 9$

5. $r^2 - 8r + 16$ 6. $g^2 + 10g + 25$

7. $4V^2 + 4V + 1$ 8. $9i^2 + 12ir + 4r^2$

9. $16g^2v^2 + 8giv + i^2$

10. $9R^2 - 6Rr + r^2$ 11. $81 - 90g + 25g^2$

12. $64\pi^2 L^2 + 48\pi LC + 9C^2$

13. $(R^2 + x^2)^2 - 4(R^2 + x^2) + 4$

14. $4i^2 - 40i + 100$ 15. $4v^2 - 24v + 36$

16. $16 - 8v + v^2$ 17. $h^2 i^2 + 4hi + 4$

18. $25r^6 - 30i^2 r^3 + 9i^4$

19. $81s^2 - 198s + 121$

20. $49g^4 v^4 + 84g^2 i^3 v^2 + 36i^6$

In Exercises 21–35 supply the missing term to make each expression a trinomial square.

21. $x^2 + ? + 9$ 22. $v^2 + 4v + ?$

23. $4 - 4i + ?$ 24. $? - 6r + 9$

25. $4v^2 g^2 - ? + 9$ 26. $16 - 24hi + ?$

27. $i^2 r^2 + ? + 9v^2$ 28. $36R^2 + 72Rr + ?$

29. $81R^2 X^2 - 144RXZ + ?$

30. $100\omega^2 L^2 + ? + 4C^2$

31. $64g^2 v^4 + ? + 36$

32. $25h^4 i^4 - 30h^2 i^2 r^3 + ?$

33. $4x^2 + ? + \frac{4}{9}y^2$ 34. $\frac{4}{25}v^2 + \frac{4}{3}iv + ?$

35. $0.25i^2 + ? + 0.16r^2$

3-10 PRODUCT OF THE SUM AND DIFFERENCE OF TWO TERMS

Let us look at the result of finding the indicated product $(a - 3)(a + 3)$.

$$(a - 3)(a + 3) = a(a + 3) - 3(a + 3)$$
$$= a^2 + 3a - 3a - 9$$
$$= a^2 - 9$$

It is a fact that the form of the result we observe here is the same for all cases of special products of this form.

> **The product of the sum and difference of the same two terms is the difference between their squares. The difference of the squares occurs in the same order as the terms.**

Example 24 Determine the indicated special products by inspection.

(a) $(x - 2)(x + 2)$ (a) $x^2 - 4$

(b) $(3v - 7)(3v + 7)$ (b) $9v^2 - 49$

(c) $(4v + 5i)(4v - 5i)$ (c) $16v^2 - 25i^2$

(d) $(8R^2 + 11h^2 i^2) \cdot$ (d) $64R^4 - 121h^4 i^4$
    $(8R^2 - 11h^2 i^2)$

This special product is utilized in a short-cut technique for finding the products of two large numbers when one is conveniently expressed as the sum and the other as the difference of two numbers.

Example 25 Utilizing the technique of the product of the sum and difference of two terms find the indicated products.

(a) $97 \cdot 103$ (a) $(100 - 3)(100 + 3) =$
                          $10{,}000 - 9 = 9991$

(b) $42 \cdot 38$ (b) $(40 + 2)(40 - 2) =$
                         $1600 - 4 = 1596$

Exercises 3-10 In Exercises 1–20 find the indicated products by inspection.

1. $(x - 1)(x + 1)$ 2. $(v - 2)(v + 2)$

3. $(3 + i)(3 - i)$ 4. $(r + 6)(r - 6)$

5. $(hi - 10)(hi + 10)$ 6. $(8 + iR)(8 - iR)$

7. $(3c - 3g)(3c + 3g)$

8. $(iv - I^2R)(iv + I^2R)$

9. $(3v - 5i)(3v + 5i)$

10. $(9gv + 7i)(9gv - 7i)$

11. $(2\beta i + 7\alpha I)(2\beta i - 7\alpha I)$

12. $(11hr + 13h'R)(11hr - 13h'R)$

13. $(\frac{2}{3}i - \frac{3}{4}r)(\frac{2}{3}i + \frac{3}{4}r)$

14. $(\frac{4}{5}R + \frac{3}{7}X)(\frac{4}{5}R - \frac{3}{7}X)$

15. $(0.5\omega L - 0.3\omega C)(0.5\omega L + 0.3\omega C)$

16. $(0.2hi + 0.6h'I)(0.2hi - 0.6h'I)$

17. $(2a^4x^5 + 8b^3y^7)(2a^4x^5 - 8b^3y^7)$

18. $(7v^5g^7 - 17r^3Z^4)(7v^5g^7 + 17r^3Z^4)$

19. $(6n^3p^2 - 1)(6n^3p^2 + 1)$

20. $(\frac{1}{3}i^3v^4 + gr)(\frac{1}{3}i^3v^4 - gr)$

In Exercises 21–30 find the indicated products using the technique of the product of the sum and difference of two terms.

21. $64 \cdot 56$ 22. $17 \cdot 23$ 23. $104 \cdot 96$

24. $205 \cdot 195$ 25. $154 \cdot 146$ 26. $1010 \cdot 990$

27. $394 \cdot 406$ 28. $2020 \cdot 1980$

29. The current through a 52-Ω resistor is 48 A. Find the voltage across the resistor.

30. The voltage across a resistor is 502 mV when the current through it is 498 mA. How much power is it dissipating? (*Note:* millivolts $\times$ milliamperes = microwatts.)

3-11 FACTORING THE DIFFERENCE OF SQUARES

To factor the difference of squares we simply reverse the process we have studied in Section 3-10.

To find the factors of a difference of squares:
1. **Find the square root of each term.**
2. **Write the product of the sum and difference of these roots.**

Example 26   Write the factors of the following, by inspection.

(a) $x^2 - 4$ (a) $(x - 2)(x + 2)$

(b) $16v^2 - 49$ (b) $(4v - 7)(4v + 7)$

(c) $9i^4 - 16r^2$ (c) $(3i^2 + 4r) \cdot$
 $(3i^2 - 4r)$

(d) $\frac{4}{9}h^6i^8 - \frac{25}{49}\beta^4I^{12}$ (d) $(\frac{2}{3}h^3i^4 + \frac{5}{7}\beta^2I^6) \cdot$
 $(\frac{2}{3}h^3i^4 - \frac{5}{7}\beta^2I^6)$

Exercises 3-11   Factor and check the factoring of each of the given expressions.

1. $x^2 - 9$ 2. $v^2 - 1$ 3. $i^2 - 16$

4. $r^2 - 36$ 5. $4R^2 - 49$

6. $81 - 64g^2v^2$ 7. $121i^4r^2 - 100$

8. $144 - 36h^2i^2$ 9. $1 - 49Z^2$

10. $4 - 169\omega^2L^2$ 11. $\frac{9}{25}v^2 - 1$

12. $\frac{4}{49}i^2 - \frac{16}{25}$ 13. $0.25x^2 - 0.16$

14. $0.04r^2 - 0.09$ 15. $I^4R^2 - V^4I^4$

16. $4h^6i^8 - 25g^{10}v^4$ 17. $\dfrac{1}{R^2} - \dfrac{1}{r^2}$

18. $\dfrac{V^4}{R^2} - I^4r^2$

19. $R^2 + 2RX + X^2 - Z^2$ (*Hint:* $R^2 + 2RX + X^2 = (R + X)^2$.)

20. $v^2 - 4iv + 4i^2 - V^2$

3-12 PRODUCTS OF BINOMIALS HAVING SIMILAR TERMS

Let us again review and apply a rule of Section 3-4 to a special situation—the product of binomials in which the respective terms are similar, for example, $(2x - 3y)(x + y)$,

$$\begin{array}{r} 2x - 3y \\ x + y \\ \hline 2x^2 - 3xy \\ + 2xy - 3y^2 \\ \hline 2x^2 - xy - 3y^2 \end{array}$$

With a little practice we can perform such operations mentally, particularly if we analyze the procedure and understand its logic.

1.   The first term of the product is the product of the first terms of the factors:

$$2x^2: \quad (2x - 3y)(x + y) \quad 2x \cdot x = 2x^2$$

2. The middle term of the product is the algebraic sum of the cross products:

$-xy$: $(2x - 3y)(x + y)$ $-3y \cdot x + 2x \cdot y$
$$= -3xy + 2xy = -xy$$

3. The third term is the product of the second terms:

$-3y^2$: $(2x - 3y)(x + y)$ $-3y \cdot y = -3y^2$

4. The overall result is a trinomial made up of the three terms:

$$(2x - 3y)(x + y) = 2x^2 - xy - 3y^2$$

Example 27 Expand, mentally, the indicated products.

(a) $(x + 2)(x - 3)$ (a) $x^2 - x - 6$
(b) $(4v - 2i)(v - 3i)$ (b) $4v^2 - 14iv + 6i^2$
(c) $(2R + 3r)(6R + r)$ (c) $12R^2 + 20Rr + 3r^2$
(d) $\left(3v - \dfrac{V}{4}\right) \cdot \left(4v - \dfrac{V}{3}\right)$ (d) $12v^2 - 2vV + \dfrac{V^2}{12}$

Exercises 3-12 In Exercises 1–20 expand, mentally, the indicated products.

1. $(x + 2)(x + 5)$ 2. $(v + 1)(v + 2)$
3. $(v + 4)(v + 7)$ 4. $(i + 6)(i + 7)$
5. $(r - 2)(r - 3)$ 6. $(r - 8)(r - 2)$
7. $(g - 6)(g - 1)$ 8. $(V - 9)(V - 8)$
9. $(X - 7)(X + 3)$ 10. $(V - 1)(V + 3)$
11. $(hi - 7)(hi + 2)$ 12. $(\omega L + 5)(\omega L - 3)$
13. $(5v - 3V)(2v - 4V)$
14. $(6I - 7i)(8I - 3i)$
15. $(3hr + 2R)(4hr + 8R)$
16. $(9\omega L + 5X)(7\omega L + 9X)$
17. $(v - 3i)(v + 5i)$
18. $(2r + 5R)(r - 12R)$
19. $\left(\dfrac{v}{2} - \dfrac{V}{4}\right)\left(\dfrac{v}{3} - \dfrac{V}{6}\right)$
20. $\left(\dfrac{V}{R} - 2I\right)\left(\dfrac{V}{R} + 4I\right)$

3-13 FACTORING QUADRATIC TRINOMIALS

The products of binomials containing similar terms are called *quadratic trinomials* because such products always contain the squares of the letters. (quad·rat·ic: Involving a quantity or quantities that are squared.) If a trinomial is quadratic it may be factorable and if it is we factor it by determining its two binomial factors in a process which is basically a reversal of that of the preceding section. Let us study the steps required while factoring $6x^2 - 5xy + y^2$.

1. Since, if factorable, the factors will be two binomials with similar terms, we start by writing two sets of parentheses ( )( ).

2. Assuming that the trinomial has been written in standard format, that is, so that the first and third terms are either quadratic, or purely numerical terms, we next factor the first and third terms; for example, for $6x^2 - 5xy + y^2$,

$$6x^2: 6x, x,$$
$$\text{or } 2x, 3x$$
$$y^2: -y, -y$$

(Although the factors of y^2 are also y, y, the − sign of the second term indicates the factors will be $-y, -y$.)

3. We write the possible factors in the parentheses and test for correct results,

(a) $(6x - y)(x - y) = 6x^2 - 7xy + y^2$
(b) $(2x - y)(3x - y) = 6x^2 - 5xy + y^2$

The factors of (b) are obviously the correct ones.

MEMORY JOGGERS

1. If the sign of the third term is +, the signs of its factors will be alike: + if the sign of the middle term is +, − if the middle term is −.

2. If the sign of the third term is −, one of its factors will have a − sign, the other, a + sign.

Example 28   Factor the following quadratic trinomials.

(a) $a^2 + 5a + 6$
Factors of 6: 6, 1 and 3, 2

(a) $(a + 6)(a + 1) = a^2 + 7a + 6$ wrong
$(a + 3)(a + 2) = a^2 + 5a + 6$ Ans.

(b) $v^2 - v - 6$
Factors of -6: $-6, 1$; $6, -1$; $3, -2$; $-3, 2$

(b) $(v - 6)(v + 1) = v^2 - 5v - 6$ wrong
$(v - 1)(v + 6) = v^2 + 5v - 6$ wrong
$(v - 2)(v + 3) = v^2 + v - 6$ wrong
$(v - 3)(v + 2) = v^2 - v - 6$ Ans.

(c) $4i^2 - 14iv + 12v^2$
Factors of $4i^2$: $2i, 2i$; $4i, i$
Factors of $12v^2$: $-6v, -2v$; $-3v, -4v$; $-12v, -v$

(c) $(2i - 6v)(2i - 2v) = 4i^2 - 16iv + 12v^2$ wrong
$(2i - 3v)(2i - 4v) = 4i^2 - 14iv + 12v^2$ Ans.

You may find it necessary to try several combinations of factors before finding the correct set, particularly with terms that have several factors. However, your ability to make quick mental tests of the trial combinations will improve rapidly with practice. And, you will soon be able to recognize the correct combination of factors on the first or second try.

Exercises 3-13   In Exercises 1–32 factor the given expressions.

1. $x^2 + 3x + 2$ 2. $v^2 + 4v + 3$
3. $i^2 + 5i + 6$ 4. $8 + 6r + r^2$
5. $R^2 - 5R + 6$ 6. $X^2 - 7X + 12$
7. $Z^2 - 9Z + 20$ 8. $V^2 - 13V + 12$

9. $I^2 - I - 2$ 10. $v^2 - 2v - 8$
11. $i^2 + 3i - 10$ 12. $r^2 + r - 12$
13. $2s^2 + 5s + 2$ 14. $2g^2 + 5g + 3$
15. $4p^2 + 8p + 3$ 16. $6Q^2 + 17Q + 5$
17. $9f^2 - 18f + 5$ 18. $2v^2 - 3v + 1$
19. $4r^2 - 8r + 3$ 20. $6i^2 - 10i + 4$
21. $8i^2 - 6i - 9$ 22. $5\omega^2 - 11\omega - 12$
23. $7h^2i^2 + 19hi - 6$ 24. $9 - 71P - 8P^2$
25. $x^2 + 13xy + 42y^2$
26. $8v^2 + 22iv + 15i^2$ 27. $3i^2 - 10ir + 7r^2$
28. $4h^2 - 18hr + 14r^2$
29. $5\beta^2 - 23\beta I - 10I^2$
30. $7I^4 + 13I^2R - 2R^2$
31. $6R^4 - 11R^2X^2 - 10X^4$
32. $20\dfrac{V^4}{g^4} - 7\dfrac{V^2}{g^2}I^2 - 6I^4$

3-14 SUMMARY OF FACTORING

Your effectiveness in factoring algebraic expressions will be aided if you remember the generalized form of the various special products we have considered in this chapter and how they are factored and learn to associate the expressions you must factor with their appropriate model. The special products are given in Table 3-1 for your convenience.

An organized approach to a factoring problem can also assist you in achieving effectiveness. The following steps, performed in the order listed, will work well in many instances:

1.   Check for and factor out common monomial factors.
2.   Test a binomial for difference of squares.

Table 3-1

Form of expression	Factors	Reference, section
$xy + xz$	$x(y + z)$	3-4, 3-7
$x^2 + 2xy + y^2$	$(x + y)^2$ $(x \pm y)^2$	3-8, 3-9
$x^2 - 2xy + y^2$	$(x - y)^2$	
$x^2 - y^2$	$(x + y)(x - y)$	3-10, 3-11
$x^2 + (a + b)x + ab$	$(x + a)(x + b)$	3-12, 3-13
$(ac)x^2 + (ad + bc)x + bd$	$(ax + b)(cx + d)$	

3. Test a trinomial as a square of a binomial, then as a product of binomials with similar terms.

4. Carry out factoring until all factors are prime.

5. Check factors by remultiplication.

Exercises 3-14 In Exercises 1–50 find the prime factors of the given expressions.

1. $iR - ir$ 2. $\dfrac{V}{R} + \dfrac{V}{r}$ 3. $4I^2R - 9i^2R$

4. $16hr^2 - 49hR^2$ 5. $4v^2 + 24v + 36$

6. $3i^2 + 30i + 75$ 7. $5v^2 - 40v + 80$

8. $6r^2 - 12r + 6$ 9. $8v^2 + 80v + 128$

10. $6i^2 + 60i + 54$ 11. $10r^2 - 40r + 30$

12. $7R^2 - 49R + 70$ 13. $2v^2 - 2v - 12$

14. $5x^2 + 10x - 75$ 15. $12x^2 + 36x + 27$

16. $24v^2 + 24v + 6$

17. $28X^2 - 140X + 175$

18. $36R^2 - 96R + 64$ 19. $6R^2 + 5R - 6$

20. $10V^2 + 31V - 14$ 21. $20v^2 + 19v + 3$

22. $21G^2 + 38G + 5$ 23. $6v^2 + 13iv + 6i^2$

24. $6i^2 + 23ir + 20r^2$

25. $25v^2 + 20gv - 21g^2$

26. $28h^2 - hi - 45i^2$ 27. $v^4 + 7v^2 + 12$

28. $i^6 + 7i^3 + 10$ 29. $14v^4 + 53i^2v^2 + 45i^4$

30. $10h^4 - 59h^2i^2 - 6i^4$ 31. $4v^3 - 9i^2v$

32. $16i^6v^4 - 36i^2I^4r^2$ 33. $X^3 - 6X^2 + 8X$

34. $t^5 + 14t^4 + 49t^3$

35. $12R^5 - R^4r - 35R^3r^2$

36. $6X^2Z^5 + 7X^3Z^4 - 55X^4Z^3$

37. $x^2 + (y + z)x + yz$

38. $R_1R_2 + R_1X_2 + R_2X_1 + X_1X_2$

39. $vV + Vi + vI + iI$

40. $vV + irv + IRV + iIrR$

41. $ac + bc + ad + bd$

42. $IRv + IRV + iRv + iRV$

43. $x^2 + \frac{4}{3}x + \frac{4}{9}$ 44. $i^2 - \dfrac{iv}{2} + \dfrac{v^2}{16}$

45. $R^2 + \dfrac{R}{6} - \dfrac{1}{3}$ 46. $V^2 + V - \frac{15}{4}$

47. $9I^2 - \dfrac{V^4}{R^2}$ 48. $\dfrac{16v^2}{81Z^2} - 64i^2$

49. $4iv - 3v^2 - \frac{4}{3}i^2$ 50. $hi + \frac{2}{5}h^2 + \frac{5}{8}i^2$

3-15 DIVISION OF ONE POLYNOMIAL BY ANOTHER

To divide one polynomial by another:

1. Factor the polynomials, if factorable (see Sections 3-7 through 3-14), and cancel common factors in dividend and divisor.

2. If polynomials are not obviously factorable, arrange dividend and divisor in descending powers of the literal.

   a. Then, divide the first term of the dividend by the first term of the divisor, and write the result as the first term of the quotient.

   b. Multiply the entire divisor by the first term of the quotient and subtract the product from the dividend.

   c. Divide the first term of the difference by the divisor, write the result as the next term of the quotient, multiply the divisor by this term and subtract from the previous difference.

   d. Repeat step c until the remainder is either zero or a term which is of lower degree than the divisor.

Example 29   Divide $x^2 + 4x + 4$ by $x + 2$.

$$\frac{x^2 + 4x + 4}{x + 2} = \frac{(x + 2)(x + 2)}{x + 2} = x + 2$$

Example 30   Divide $I^2 - 36$ by $I + 6$.

$$\frac{I^2 - 36}{I + 6} = \frac{(I - 6)(I + 6)}{I + 6} = I - 6$$

Example 31   Divide $6R^2 + 5R - 6$ by $3R - 2$

$$\frac{6R^2 + 5R - 6}{3R - 2} = \frac{(3R - 2)(2R + 3)}{3R - 2} = 2R + 3$$

Example 32 Divide $2x^3 + 3x^2 - 4x + 1$ by $2x - 1$.

$$x^2(2x - 1) = \longrightarrow$$
$$2x(2x - 1) = \longrightarrow$$
$$-1(2x - 1) = \longrightarrow$$

$$
\begin{array}{r}
x^2 + 2x - 1 \\
2x - 1\overline{\smash{\big)}\,2x^3 + 3x^2 - 4x + 1} \\
\underline{2x^3 - x^2} \\
4x^2 - 4x \\
\underline{4x^2 - 2x} \\
-2x + 1 \\
\underline{-2x + 1}
\end{array}
$$

$$x^2 + 2x - 1 \quad \text{Ans.}$$

Note: After each subtraction the first term drops out. We continue this process until nothing is left.

Example 33 Perform the indicated division, $\dfrac{8x^3 + 31x - 19x^2 - 3x^4 - 16}{2 - 3x}$.

$$
\begin{array}{r}
x^3 - 2x^2 + 5x - 7 \\
-3x + 2\overline{\smash{\big)}\,-3x^4 + 8x^3 - 19x^2 + 31x - 16} \\
\underline{-3x^4 + 2x^3} \\
6x^3 - 19x^2 \\
\underline{6x^3 - 4x^2} \\
-15x^2 + 31x \\
\underline{-15x^2 + 10x} \\
21x - 16 \\
\underline{21x - 14} \\
-2
\end{array}
$$

Note: It was necessary to rearrange dividend and divisor in descending powers of x.

Remainder

The result may be written

$$x^3 - 2x^2 + 5x - 7 - \frac{2}{-3x + 2}$$

Example 34 Divide $x^4 - y^4$ by $x + y$.

$$
\begin{array}{r}
x^3 - x^2y + xy^2 - y^3 \\
x + y\overline{\smash{\big)}\,x^4 - y^4} \\
\underline{x^4 + x^3y} \\
-y^4 - x^3y \\
\underline{-x^3y - x^2y^2} \\
-y^4 + x^2y^2 \\
\underline{+x^2y^2 + xy^3} \\
-y^4 - xy^3 \\
\underline{-y^4 - xy^3}
\end{array}
$$

Note: We bring the $-y^4$ term down until we can use it.

Exercises 3-15 In Exercises 1–29 perform the indicated divisions.

1. $\dfrac{a^2 + 5a + 6}{a + 2}$

2. $\dfrac{I^2 + 2I - 3}{I - 1}$

3. $\dfrac{R^2 + 2R - 8}{R + 4}$

4. $\dfrac{e^2 - 2e - 8}{e - 4}$

5. $\dfrac{6i^2 + 5i - 4}{3i - 2}$

6. $\dfrac{2h^2 - 7h + 4}{2h - 1}$

7. $\dfrac{14 - 3x^2 - x}{2 - x}$

8. $\dfrac{-11p - 12 - 2p^2}{2p + 3}$

9. $\dfrac{R^3 + 15R^2 + 75R + 125}{R + 5}$

10. $\dfrac{12x + 8 + x^3 + 6x^2}{x + 2}$

11. $\dfrac{4I^5 + 4I^4 - 2I^3 + 2I}{I^2 + I}$

12. $\dfrac{I^4R^2 - 4I^2R + 4}{I^2R - 2}$

13. $\dfrac{V^3I^3 - 27}{VI - 3}$

14. $(16V^4 - 81) \div (2V + 3)$

15. $(\omega^3L^3 + \omega^2L^2 + 3\omega L + 3) \div (\omega^2L^2 + 3)$

16. $[(v - ir)^4 + 3(v - ir)^2 - 10] \div [(v - ir)^2 + 5]$

17. $[36I^6(R + r)^3 + 33I^4(R + r)^2 + 12I^2(R + r) + 4] \div [3I^2(R + r) + 2]$

18. $\dfrac{8L^3 - 36L^2C + 54LC^2 - 27C^3}{2L - 3C}$

19. $\dfrac{e^2 - 2ev + v^2}{e - v}$

20. $\dfrac{R^2 + 4Rr + 4r^2}{R + 2r}$

21. $\dfrac{R^2 - 5RX + 6X^2}{R - 2X}$

22. $\dfrac{8I^3 - 36I^2i + 54Ii^2 - 27i^3}{2I - 3i}$

23. $\dfrac{A^3v^2 - A^2v^2V + A^2vV + V^3}{A^2v - AvV + V^2}$

24. $\dfrac{r^2 - R^2}{r + R}$

25. $\dfrac{R^4 - 4X^4}{R^2 - 2X^2}$

26. $\dfrac{27I^3 - 64i^3}{9I^2 + 12Ii + 16i^2}$

27. $\dfrac{A^3v^3 + 8i^3r^3}{Av + 2ir}$

28. $\dfrac{Rr - jRX_C + jrX_L + X_CX_L}{R + jX_L}$ (*Hint:* Let $j^2 = -1$.)

Chapter 4

Equations and Applications: Series Circuits

A balance scale is the classic analogy for a mathematical equation—the scale will remain balanced if we perform the same operation on both sides. (Marshall Berman)

4-1 EQUATIONS OF THE FORM, $x \pm a = b$

When a quantity is being (1) added to or (2) subtracted from an unknown in an equation, we isolate the unknown by (1) subtracting or (2) adding the quantity on both sides of the equation. Such equations are of the form $x \pm a = b$. Study the following examples carefully to become familiar with the technique.

Example 1 Solve for x and check your solutions.

(a) $x + 3 = 7$

(a) Subtract 3 from both sides

$$\begin{array}{r} x + 3 = 7 \\ -3 \;\; -3 \\ \hline x \quad = 4 \;\; \text{Ans.} \end{array}$$

Check: $4 + 3 = 7$

(b) $x + 5 = 2$

(b) Subtract 5 from both sides

$$\begin{array}{r} x + 5 = \;\; 2 \\ -5 \;\; -5 \\ \hline x \quad = -3 \;\; \text{Ans.} \end{array}$$

Check: $-3 + 5 = 2$

(c) $x - 2 = 3$

(c) Add 2 to both sides

$$\begin{array}{r} x - 2 = \;\; 3 \\ +2 \;\; +2 \\ \hline x \quad = \;\; 5 \;\; \text{Ans.} \end{array}$$

Check: $5 - 2 = 3$

(d) $x - 5 = -6$

(d) Add 5 to both sides

$$\begin{array}{r} x - 5 = -6 \\ +5 = +5 \\ \hline x \quad = -1 \;\; \text{Ans.} \end{array}$$

Check: $-1 - 5 = -6$

Observe that we solve a simple equation of the form $x + a = b$ by subtracting $+a$ from (or adding $-a$ to) both sides of the equation; if the equation is of the form $x - a = b$, we add $+a$ to both sides.

Writing Equations of the Form $x \pm a = b$
Study the following example carefully.

Example 2

A technician is looking for a resistor R_x to make a total of 6.8 kΩ. She already has 3.3 kΩ. What resistor should she look for?

Let R_x = the desired resistor. Then,

$$\begin{array}{r} R_x + 3.3 \text{ k}\Omega = 6.8 \text{ k}\Omega \\ -3.3 \text{ k}\Omega \;\; -3.3 \text{ k}\Omega \\ \hline R_x \qquad\quad = 3.5 \text{ k}\Omega \\ \text{Ans.} \end{array}$$

Check: 3.5 kΩ + 3.3 kΩ = 6.8 kΩ

Exercises 4-1 In Exercises 1–20 solve the given equations by adding or subtracting the appropriate algebraic quantity on both sides of the equation. Check your solutions. Perform the work mentally when possible.

1. $x + 2 = 4$ 2. $y + 3 = 6$
3. $I + 1.5 = 3.7$ 4. $v + 2.7 = 6.2$
5. $x - 2 = 4$ 6. $v - 6 = 3$
7. $I - 1.6 = 5.9$ 8. $g - 2.3 = 4.5$
9. $x + 3 = -7$ 10. $R + 2 = -10$
11. $V + 7.3 = -5.1$ 12. $t + 2.6 = -7.2$
13. $x - 1 = -15$ 14. $V - 2 = -7$
15. $I - 2.3 = 6.1$ 16. $r - 6.8 = -9.1$
17. $x + \frac{1}{2} = \frac{3}{4}$ 18. $I - \frac{1}{3} = \frac{5}{6}$
19. $V + \frac{3}{5} = -\frac{7}{10}$ 20. $R + \frac{1}{9} = -\frac{2}{3}$

In Exercises 21–30 find the desired quantities by performing these three separate steps: (a) write an equation of the form $x \pm a = b$, (b) solve the equation, and (c) check the solution.

21. A circuit requires 12 V for proper operation. How much voltage must be added to a 3-V battery to obtain the desired voltage?

22. When a current I_x is increased by 5 A the total current is 8 A. What is I_x?

23. A technician worked 8 hours overtime one week. His total time for the week was 44 hours. How many hours of regular time did he work?

24. The invoice for an oscilloscope shows a total cost of $875.75 which includes $805 for the oscilloscope plus shipping charges and taxes. What is the amount of the shipping charges and taxes?

25. The total resistance of a circuit is reduced by 15 kΩ to 39 kΩ when a switch is flipped from off to on. What is the resistance of the circuit with the switch off?

26. A technician's weekly take-home pay is $387.50 after deductions of $89.45 have been taken out. What is his gross pay?

27. The terminal voltage of a certain automobile battery drops by 2.35 V to 10.85 V when the battery supplies load current to the starter motor for cranking the engine. What is the no-load terminal voltage of the battery?

28. The base voltage of a certain transistor stage is 4.55 V and is equal to the emitter voltage plus the drop across the base-emitter junction of 0.67 V. What is the emitter voltage?

29. The coaxial lead-in cable for a certain TV antenna installation totals 23 m. There is 18 m of cable from the antenna to a signal splitter. How much cable is there between the splitter and the receiver?

30. A certain commercial power supply can supply 8.7 W to a CB radio transmitter and still have 6.8 W available for some other application. What is the total capacity of the power supply?

4-2 OTHER TECHNIQUES AND MORE DIFFICULT EQUATIONS

Thus far we have studied simple equations which have required only one of the four fundamental operations—addition, subtraction, multiplication, or division—for solution. We now want to learn a few more techniques and apply them to equations that are somewhat more complex.

Transposition To solve an equation of the form $x + a = b$ we have learned to subtract a, the term on the side of the equation with the unknown, from both sides of the equation. However, the net effect of this operation is to move or transfer the a to the opposite side of the equation while reversing its sign,

$$x + a - a = b - a \qquad x = b - a$$

Subtracting a from both sides

is the same as writing a on the other side with sign changed.

We say we *transpose the term.*

```
┌─────────────────────────────────────────────┐
│              TRANSPOSITION                    │
│                                               │
│ Any term may be transferred from one side of  │
│ an equation to the opposite side, without     │
│ destroying the equality, if its sign is       │
│ reversed. If the same term with the same sign │
│ appears on both sides of an equation the term │
│ may be dropped on both sides of the equation. │
└─────────────────────────────────────────────┘
```

Example 3 Solve the given equations using transposition.

(a) $x + 7 = 12$

(a) We transpose the 7:
$x = 12 - 7$
$x = 5$ Ans.
Check: $5 + 7 = 12$

(b) $y - 5 = 2$

(b) $y = 2 + 5$
$y = 7$ Ans.
Check: $7 - 5 = 2$

Changing the Signs of Both Sides of an Equation If we have the equation $5 - x = -3$ we might transpose the 5 obtaining $-x = -3 - 5$; thus $-x = -8$. Then multiplying both sides by -1 we obtain $x = 8$. Or, we could have changed all signs in the original equation: $-5 + x = 3$. Then, transposing, $x = 3 + 5 = 8$. We summarize this conclusion.

```
┌─────────────────────────────────────────────┐
│ All algebraic signs in an equation may be     │
│ changed without destroying the equality.      │
└─────────────────────────────────────────────┘
```

Exchanging the Sides of an Equation It is customary to report the solution of an equation using a form in which the unknown is on the left side of the equation. If we happen to be solving an equation like 5 = x + 2, 5 - 2 = x, 3 = x, we can simply exchange sides completely, that is, x + 2 = 5 is equivalent to 5 = x + 2. Note that since we are exchanging all terms we do not change the signs of the terms as we do in transposition.

The sides of an equation may be exchanged without destroying the equality.

Example 4 Solve and check.

(a) 6 = 4 + x (a) First, we exchange
                          sides,
                          4 + x = 6
                          Then transpose the 4,
                          x = 6 - 4 = 2 Ans.
                          Check: 6 = 4 + 2

(b) 6 = 3x (b) First we exchange sides,
                          3x = 6
                          Then solve in the usual
                          way,
                          $$\frac{\cancel{3}x}{\cancel{3}} = \frac{6}{3}$$
                          x = 2 Ans.
                          Check: 6 = 3 · 2

Many equations require the use of two or more operations for solution. Study carefully the steps illustrated in the following examples.

Example 5 Solve and check: 5x + 9 = 7x - 3

First we transpose the 7x to the left side and +9 to the right side, to get all like terms together,

$$5x - 7x = -3 - 9$$

Then collect like terms,

$$-2x = -12$$

Next we change signs and transpose the coefficient (see "checkerboard transposition,"

Section 3-1),

$$x = \frac{12}{2}$$
$$x = 6 \text{ Ans.}$$

Check: 5 · 6 + 9 = 7 · 6 - 3   (We always
                                         check by substi-
                    39 = 39   tuting the solu-
                                         tion into the
                                         original
                                         equation.)

Example 6 Solve and check:

$$4(x + 2) - 5 = 27$$

First we clear the parentheses and combine like terms,

$$4x + 8 - 5 = 27 \qquad 4x + 3 = 27$$

Next we transpose the +3,

$$4x = 27 - 3 \qquad 4x = 24$$

Finally we transpose the coefficient of the unknown

$$x = \frac{24}{4} \qquad x = 6 \text{ Ans.}$$

Check: 4(6 + 2) - 5 = 27
            32 - 5 = 27

Taking the Reciprocal of Both Sides of an Equation In some instances, to make progress toward solving an equation it is expedient to *invert, or find the reciprocal of,* both sides of the equation—a proper operation. For example, to solve $\frac{4}{x}$ = 2, we divide each side into 1,

$$1 \div \frac{4}{x} = 1 \cdot \frac{x}{4} = \frac{x}{4}$$
$$1 \div 2 = 1 \cdot \frac{1}{2} = \frac{1}{2}$$

Thus, we have

$$\frac{x}{4} = \frac{1}{2} \qquad x = \frac{1}{2} \cdot 4 = 2 \text{ Ans.}$$

Check: $\frac{4}{2}$ = 2

We must be very careful to avoid a common error which is frequently committed when one or both sides of the equation contains more than one term. *We must divide each entire side into 1.*

Example 7 Solve and check: $3/2x = 9/15$

First, we invert both sides of the equation,

$$\frac{2x}{3} = \frac{15}{9}$$

Then multiply by 3 and divide by 2 (multiply by $\frac{3}{2}$),

$$\frac{\cancel{3}}{\cancel{2}} \cdot \frac{\cancel{2}x}{\cancel{3}} = \frac{\cancel{15}^{5}}{\cancel{9}_{\cancel{3}}} \cdot \frac{\cancel{3}}{2}$$

$$x = \frac{5}{2} = 2.5 \quad \text{Ans.}$$

Check: $\dfrac{3}{2 \cdot 2.5} = \dfrac{9}{15}$ $\dfrac{3}{5} = \dfrac{9}{15} = \dfrac{3}{5}$

Example 8 Solve and check: $a/x = b + c$

Again we invert both sides of the equation,

$$\frac{x}{a} = \frac{1}{b + c}$$

but note that we have

$$\frac{1}{b + c}$$

and not

$$\frac{1}{b} + \frac{1}{c}$$

Then we solve by multiplying by a,

$$x = \frac{a}{b + c} \quad \text{Ans.}$$

Check: $\dfrac{a}{\dfrac{a}{b + c}} = a \div \dfrac{a}{b + c} = \cancel{a} \cdot \dfrac{b + c}{\cancel{a}} = b + c$

Therefore,

$$\frac{a}{x} = \frac{a}{\dfrac{a}{b + c}} = b + c$$

> The reciprocal may be taken of both sides of an equation provided all terms of a side are grouped as a single denominator of 1.

The equations of Examples 7 and 8 may also be solved, perhaps more conveniently, by diagonal ("checkerboard") transposition. See Example 9.

Example 9 Solve the given equations using diagonal transposition.

(a) $\dfrac{3}{2x} = \dfrac{9}{15}$ (a) $\dfrac{3}{2\,\textcircled{x}} \diagup \dfrac{\textcircled{9}}{15}$

$$\frac{3}{2 \cdot 9} \diagdown \frac{x}{\textcircled{15}}$$

$$\frac{3 \cdot \cancel{15}^{5}}{2 \cdot \cancel{9}_{\cancel{3}}} = x$$

$$x = \frac{5}{2} = 2.5 \quad \text{Ans.}$$

(b) $\dfrac{a}{x} = b + c$ (b) $\dfrac{a}{\textcircled{x}} \diagup \dfrac{(b + c)}{1}$

$$\frac{a}{(b + c)} = x$$

$$x = \frac{a}{b + c} \quad \text{Ans.}$$

Exercises 4-2 In Exercises 1–30 solve and check the equations utilizing the techniques developed in this section wherever possible.

1. $5 + x = 7$ 2. $4 - i = 2$ 3. $5 = v + 3$
4. $8 = 5 - R$ 5. $2r + 3 = 11$
6. $5 - 3y = 2$ 7. $12 = 7I - 2$
8. $-5 = 7 - 6x$ 9. $2V + 5 = 7V - 15$
10. $12 - 4i = 5i + 30$
11. $4R - 7 - 2R = 8 - 3R$
12. $12I + 6 - 8I = 9I - 4$
13. $3.5 - 4.7R = -8.4R + 7.2$
14. $0.56 + 0.67i = 0.91 - 0.37i$
15. $0.35R + 0.76R = -0.35R - 0.73$

16. $2.95 + 1.63V = 7.52V - 3.67$
17. $3(x - 2) = 9$ 18. $4(x + 3) - 7 = 13$
19. $1.5(x - 2) + 5 = -4.5x + 14$
20. $4.75(x + 3) - 4 = 2.25(x - 1)$
21. $\dfrac{3}{x} = \dfrac{1}{4}$ 22. $\dfrac{5}{R} = -6$
23. $\dfrac{3}{2e} = \dfrac{1}{6}$ 24. $\dfrac{18}{3r} = -3$
25. $\dfrac{1}{x} = c + d$ (solve for x)
26. $R + r = \dfrac{V}{I}$ (solve for I)
27. $\dfrac{4}{x} - 2 = -4$ (*Hint*: Transpose before inverting.)
28. $6 = 3 - \dfrac{15}{x}$ 29. $7 - \dfrac{21}{x} = 4$
30. $7 = 5 + \dfrac{28}{2x}$

4-3 LITERAL EQUATIONS

A literal equation is one which contains two or more variables. The solution of a literal equation, then, is always one variable expressed in terms of one or more other variables.

All of the techniques for solving equations involving one variable and numerals are applicable to literal equations. We assume that all variables, except the desired one, are constants. Examples will illustrate the techniques applied to some familiar electronics formulas.

Example 10 You have measured the voltage and current in a circuit and need to calculate its resistance, but can recall only the formula $V = IR$. Solve for R.

First we exchange the sides of the equation,

$$IR = V$$

and then divide both sides by I, the factor of the desired variable,

$$\frac{\cancel{I}R}{\cancel{I}} = \frac{V}{I}$$

The desired formula is, then,

$$R = \frac{V}{I}$$

Example 11 Manipulate the formula $V_C = V_{CC} - IR_L$ to solve for I.

First, we transpose V_C and IR_L,

$$V_C = V_{CC} - IR_L$$
$$IR_L = V_{CC} - V_C$$

Then, dividing by R_L, the factor of the desired unknown, we have

$$\frac{IR_L}{R_L} = \frac{V_{CC} - V_C}{R_L}$$

$$I = \frac{V_{CC} - V_C}{R_L}$$

Let us recall that any operation may properly be performed on an equation as long as the same operation is performed on both sides of the equation (see Section 3-1). *Raising to a power* and *finding a root* are legitimate operations in the manipulation of formulas.

Example 12 Solve the formula $P = V^2/R$ for V.

$$R \cdot P = \frac{V^2}{R} \cdot R \qquad \text{Multiply by R}$$
$$V^2 = PR \qquad \text{Exchange sides}$$
$$\sqrt{V^2} = \sqrt{PR} \qquad \text{Take the square root of both sides of the equation}$$
$$V = \pm\sqrt{PR}$$

In Example 12 the notation $\pm$, read "plus or minus," means that the square root of the product PR may be either positive or negative since $(+N)^2 = N^2$ and $(-N)^2 = N^2$. Generally, in evaluating this formula, we are interested only in the magnitude of V. Thus, we usually see the formula written $V = \sqrt{PR}$.

Exercises 4-3 In Exercises 1–38 solve the literal equations or formulas for the variable indicated in parentheses after the equation.

1. $x + 2 = 4$ (x) 2. $x + a = b$ (x)
3. $R_1 + R_2 + R_3 = R_T$ (R_2)
4. $L_1 + L_2 + L_3 + L_4 = L_T$ (L_4)
5. $V_C = V_{CC} - IR_L$ (V_{CC})
6. $3y = 12$ (y) 7. $ax = b$ (x)

8. $V = IR$ (I) 9. $V = IR$ (R)

10. $X_L = 2\pi fL$ (f) 11. $X_L = 2\pi fL$ (L)

12. $P = VI$ (I) 13. $P = I^2 R$ (R)

14. $C_i = (1 + A)C_g$ (C_g)

15. $V = (R + r)I$ (I) 16. $\dfrac{x}{4} = 2$ (x)

17. $\dfrac{x}{a} = b$ (x) 18. $I = \dfrac{V}{R}$ (V)

19. $R = \dfrac{V}{I}$ (V) 20. $\text{Gain} = \dfrac{v_o}{v_i}$ (v_o)

21. $\beta = \dfrac{I_c}{I_b}$ (I_c) 22. $\dfrac{3x}{5} = 15$ (x)

23. $\dfrac{14}{x} = 2$ (x) 24. $\dfrac{a}{x} = b$ (x)

25. $I = \dfrac{V}{R}$ (R) 26. $R = \dfrac{V}{I}$ (I)

27. $\beta = \dfrac{I_c}{I_b}$ (I_b) 28. $G = \dfrac{v_o}{v_i}$ (v_i)

29. $2x + 2 = 4$ (x) 30. $ax + b = c$ (x)

31. $V_t = V_g - Ir$ (I)

32. Solve Exercise 31 for r.

33. $V_T = IR_1 + IR_2 + IR_3$ (I)

34. $V_{CC} = V_C + I_C R_L$ (V_C)

35. Solve Exercise 34 for I_C

36. $R_T = R_0 + R_0 \alpha_0 T$ (α_0)

37. $P = I^2 R$ (I)

38. $L = \dfrac{\mu N^2 A}{\ell}$ (N)

4-4 WRITING EQUATIONS

An extremely valuable skill in electronics is that of writing equations to represent the relationships between the variables of real-life situations, such as those of electronic circuits. Since there are so many possible variable relations in life, it is impossible to prescribe a 1-2-3 mechanical routine for writing equations. The skill, of necessity, requires the exercise of the ability to think and analyze. However, it is profitable to have a few suggestions to guide one's thinking. Familiarize yourself with the following before attempting to write equations.

1. Read the statement describing the given situation carefully; or, if you are working with an actual, physical problem, state the situation carefully.

2. Clearly identify, in your mind, the items which are known and those that are to be found. Assign letters to represent the unknown items, letters which will help you relate the symbol to the physical entity, for example, t for time, V for voltage, and so on.

3. Read the statement again, looking for words that can be translated into mathematical operations, for example, "increased by" (+), "diminished by" (−), "sum of" (+), "product of" (×), "times" (×), "quotient of" (÷), and so on. Now translate the word statement into an algebraic statement by connecting the chosen letter symbols with operation symbols (+, −, ×, etc.) obtained from the translation of operation words. Substitute equals for equals wherever possible, to reduce number of variables to one.

4. Validate the final mathematical statement by translating it into words and comparing with the original statement(s).

5. If desired, evaluate the equation obtained.

Example 13 The total resistance of a series circuit is equal to the sum of the individual resistances of the circuit. A given series circuit has three resistances. Write an equation for total resistance.

Let R_T = total resistance of circuit. Let $R_1, R_2,$ and R_3 represent the three resistances. Then,

$$R_T = R_1 + R_2 + R_3 \quad \text{Ans.}$$

Example 14 (1) The total current in a parallel circuit with two or more branches is equal to the sum of the branch currents. (2) In a certain circuit containing two branches, the current in branch a is 3 times the current of branch b. (3) When the current in branch a was increased by 8 A the total current was doubled. Determine the branch currents and the total current (before increase of branch a current).

Let

$$I_b = \text{the current in branch } b$$

Then, from statement (2),

$$3I_b = \text{the current in branch } a$$

and from statement (1),

$$3I_b + I_b = 4I_b = \text{the total current}$$

Using the information in statement (3) we again write an equation for total current,

$$\underbrace{(3I_b + 8)}_{I_a \text{ increased by 8 A}} + I_b = \underbrace{2(4I_b)}_{I_T \text{ is doubled}}$$

To solve, we remove parentheses, transpose, and combine like terms,

$$3I_b + 8 + I_b = 8I_b$$
$$4I_b - 8I_b = -8$$
$$-4I_b = -8$$
$$I_b = 2 \text{ A} \quad \text{Ans.}$$
$$I_a = 3 \cdot 2 = 6 \text{ A} \quad \text{Ans.}$$
$$I_T = 6 + 2 = 8 \text{ A} \quad \text{Ans.}$$

Check: If we increase I_a by 8 A does I_T double?

$$(6 + 8) + 2 = 2 \cdot 8$$
$$16 = 16 \quad \text{Check}$$

Exercises 4-4

1. The formula for four resistors in series is $R_T = R_1 + R_2 + R_3 + R_4$. In a design problem R_T and $R_1, R_2,$ and R_4 are known. How would you find the value required for R_3? Develop an equation.

2. The collector voltage V_C of a transistor in a common-emitter circuit is equal to the collector supply voltage V_{CC} less the product of the collector current I_C and the load resistance R_L (the $I_C R_L$ drop). Write an equation for V_C.

3. The emitter current I_e in a transistor circuit is equal to the sum of the base current I_b and the collector current I_c. Write an equation for the collector current.

4. The terminal voltage V_t of a battery under load is equal to its emf V_s minus the product of its load current I_L and its internal resistance r_i. Write an equation for V_t based on this information.

5. When two resistors R_1 and R_2 are in parallel, their equivalent resistance is equal to their product divided by their sum. Write the formula.

6. The sensitivity of a voltmeter is equal to its resistance R_v in ohms, divided by its full-scale reading V in volts. Write an equation for determining this sensitivity. What are the units of the sensitivity?

7. The amount of voltage v induced in a coil whose inductance is L henries is equal to the rate of change of current, $\Delta i/\Delta t$, through the coil times its inductance. Write a formula for this induced voltage.

8. The total conductance G_T of a parallel circuit is equal to the sum of the conductances of the individual branches. Write the equation for G_T for a circuit containing four different elements.

9. Substitute into your equation for Exercise 8 the equivalent expressions in terms of resistances so as to obtain the formula for resistances in parallel. Recall that resistance is the reciprocal of conductance.

10. The square-wave pulse output voltage of a digital computer "clock," when displayed on an oscilloscope screen, shows 20 pulses on a 10-cm scale. Write a mathematical statement for expressing the pulse rate in pulses/cm. Choose appropriate symbols for representing the quantities. (*Hint:* A sketch might help you to visualize the items involved.)

11. In a certain circuit one resistor is 3 times another resistor. Their sum is 52 Ω. What are the values of the resistors?

12. In a circuit in which one resistor is 3.9 times a second resistor, their sum is 14.7 kΩ. What are the values of the resistors?

13. In the base circuit of a certain transistor one voltage is 4.2 times a second voltage and their sum is 3.64 V. Find the voltages.

14. A technician buys two replacement transistors, one of which costs 5 times the other. Together they cost $4.50, excluding sales tax. Find the cost of each.

15. One resistor is 15 kΩ more than another. Their sum is 81 kΩ. Find their values.

16. In a certain transistor radio, the current in the driver stage is 3.5 mA more than the current in the second IF stage. The total current of the two stages is 4.5 mA. Determine the value of each current.

17. The current in a driver transistor is 25 mA less than the current in an output transistor. The sum of their currents is 31 mA. Find the value of each.

18. The signal level (v_o) at the output of a radio-frequency (rf) amplifier is 12 μV more than the input (v_i). Their sum is 13.5 μV. Find v_i and v_o.

19. The output current of an amplifier stage is 23.5 times the input current. If the output current is 12.8 mA what is the input current?

20. One capacitor is 50 times a second one. If the capacitance of the larger is 1 μF, what is the capacitance of the smaller?

21. The frequency of one radio station is 2.6 times that of another one. If the higher frequency is 1540 kHz what is the lower frequency?

22. The output voltage of a transformer is 12.5 times the input voltage. What is the input voltage when the output voltage is 150 V?

23. In a certain circuit, one resistor is 2.5 times another. Adding 15 Ω to the smaller resistor doubles their sum. What are the values of the resistors?

24. Refer to Exercise 23. Adding 45 Ω to the value of the larger resistor triples their sum (smaller resistor is unchanged). Find the values of the resistors.

25. In a parallel circuit the current in one of two branches is 3.5 times the current in the other branch. If the smaller current is increased by 13.5 A, the total current will be doubled. Find the currents.

26. One inductance has a value $\frac{1}{3}$ of that of a second inductance. If the larger inductance is increased by 0.8 H, the sum will be increased to 1.2 times the original value. Find the values of the inductances.

27. A technician purchases several resistors costing 5 cents each and some capacitors priced at 25 cents each. The total cost of 30 items was $3.50. Determine the number of each component purchased.

28. A TV benchman is paid $45 for repairing color sets and $20 for black-and-white sets. One week he earned $425 while repairing 15 receivers. How many were color? How many were black and white?

29. A technician is constructing a prototype circuit one branch of which requires 5 resistors having a sum of 89 kΩ. The only available resistors have values of 22 kΩ and 15 kΩ. What combination will satisfy the requirements?

30. At the output of an FM ratio detector are two variable voltages whose sum is a constant 2 V. What are the two voltages when the larger is 6 times the smaller?

4-5 APPLICATION—SERIES CIRCUITS

Although electrical theory is the proper concern of theory textbooks, a mathematics textbook can help in the learning of theory by providing the opportunity to practice the development of formulas and their practical application. The purpose of this section is to demonstrate these activities and provide the opportunity for their practice.

Example 15 Basic relationships in a series circuit are stated below. Using these relationships:

(a) Write the literal equation relating the applied voltage V_T to the IR drops in the circuit.

(b) Derive the equation for total resistance of the circuit, R_T, in terms of individual resistances.

Basic Relationships of a Series Circuit (see Figure 4-1)

1. The current I is the same in all elements of the circuit.

2. The applied voltage V_T is equal to the sum of the IR drops. The total or equivalent resistance is defined by the equation $V_T = IR_T$.

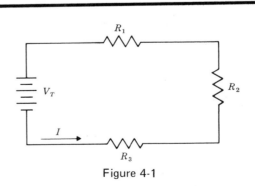

Figure 4-1

Solution

(a) $V_T = V_1 + V_2 + V_3 = IR_1 + IR_2 + IR_3$

(We use subscripted letters—V_T, V_1, V_2, and V_3, for example—to represent the different voltages present in the circuit.

(b) Since $V_T = IR_T$, we substitute IR_T for V_T in the equation of Step (a),

$IR_T = IR_1 + IR_2 + IR_3 + IR_4$

and divide by the common factor I,

$$\frac{IR_T}{I} = \frac{IR_1}{I} + \frac{IR_2}{I} + \frac{IR_3}{I} + \frac{IR_4}{I}$$

Thus,

$$R_T = R_1 + R_2 + R_3 + R_4$$

Example 16   A series circuit containing the resistor values of 560 Ω, 1.8 kΩ, and 3.3 kΩ is to be supplied by a voltage of 35 mV. (a) Draw and label the circuit diagram. (b) Calculate the predicted current. (c) Calculate the predicted voltage drop across each resistor.

(a)

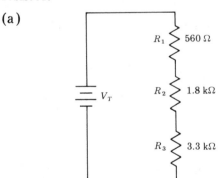

(b) $I = \dfrac{V_T}{R_T}$, so we must find R_T first.

$R_T = R_1 + R_2 + R_3$

To find R_T we put all resistor values in the same units, say kΩ.

$R_T = (0.56 + 1.8 + 3.3)$ kΩ

$R_T = 5.66$ kΩ

Then

$$I = \frac{V_T}{R_T} = \frac{35(10^{-3})}{5.66(10^3)} = 6.184(10^{-6}) \text{ A} =$$

$\boxed{6.184 \ \mu A}$

(c) $V_1 = IR_1 = 6.184(10^{-6}) \times 0.56(10^3)$

$\qquad = 3.463(10^{-3})$ V $= \boxed{3.463 \text{ mV}}$

$V_2 = IR_2 = 6.184(10^{-6}) \times 1.8(10^3)$

$\qquad = 11.1312(10^{-3})$ V $= \boxed{11.13 \text{ mV}}$

$V_3 = IR_3 = 6.184(10^{-6}) \times 3.3(10^3)$

$\qquad = 20.4072(10^{-3})$ V $= \boxed{20.41 \text{ mV}}$

Check the instruction manual for your calculator and learn how you can save steps when you must perform several multiplications with the same factor (multiplication with a constant). We check our values by substituting them into the equation,

$V_T = V_1 + V_2 + V_3$ for series circuits,

$35 = 3.46 + 11.13 + 20.41 = 35.00$

Professionals in the field of electronics report that their use of a few, simple techniques in performing circuit analysis evaluations have contributed significantly to their success. These techniques, demonstrated in the solution of Example 16, are as follows:

1. Construct and label a schematic diagram of the circuit to be analyzed.
2. Following a simple, but orderly, format, (a) write the equation (formula) to be used; (b) rewrite the equation with known quantities (in basic units and powers-of-ten notation) substituted for letters; (c) calculate answer and write the answer obtained from the calculator; (d) rewrite the answer, rounded appropriately, and expressed in a convenient unit; (e) affix the correct unit abbreviation and identify the answer by underlining it or enclosing it in a box.

Exercises 4-5   In writing solutions for Exercises 1–10 refer to the following statements regarding power in a series circuit.

(a) In a series circuit the power P_x dissipated in a resistor R_x is equal to the product of the voltage V_x across the resistor and the current I through it.

(b) The total power P_T dissipated in the circuit is equal to the sum of the power dissipated in the individual resistors. Total power dissipated is equal to total power supplied by the source which is equal to the product of the source voltage V_T and the current being supplied I.

1. Write, in terms of V_x and I_x, the equation for P_x, the power dissipated in any resistor R_x in a series circuit (see statement (a)).

2. Using the relationships of Ohm's law: $V_x = IR_x$ and $I = V_x/R_x$, show the steps required to convert the formula $P_x = IV_x$ into its two other forms, $P_x = V_x^2/R_x$ and $P_x = I^2 R_x$.

3. Refer to statement (b). Using the terms P_1, P_2, P_3, and so on, to represent power dissipated in individual resistors in a series circuit, write the equation for P_T for a circuit with five resistors.

4. Use the equations $P_x = I^2 R_x$, $P_T = I^2 R_T$, and the principle of statement (b) to show the algebraic steps which verify the formula $R_T = R_1 + R_2 + R_3 + \cdots$.

5. The values for P_T, P_1, P_2, and P_3 for a series circuit are known. The circuit includes four resistors. Write an equation for determining P_4 from the known information.

6. You know P_x and the current I through a resistor R_x. Starting with the equation $P_x = I^2 R_x$, show the algebraic steps for obtaining an equation which would enable you to evaluate R_x.

7. You know P_x and R_x. Starting with $P_x = V_x^2/R_x$ show the algebraic steps to an equation for evaluating V_x. (*Hint*: Taking the square root of both sides of an equation is a legitimate solution technique.)

8. You know P_x and R_x. Show the algebraic steps for obtaining an equation for I from the equation $P_x = I^2 R_x$.

9. You know P_1, P_2, and P_3, and R_T for a series circuit with three resistors. Develop an equation for the current I of the circuit in terms of these four quantities (see Exercise 4).

10. In a series circuit containing three resistors R_1, R_2, and R_3, we know the values of the resistors, and P_2, the power being dissipated in R_2. Develop a single equation for determining P_T in terms of these known quantities.

In Exercises 11–20 the V_T- and R-values are for a series circuit. (a) Draw and label the circuit diagram. (b) Calculate the predicted current. (c) Calculate the predicted voltage drop across each resistor. (d) Calculate the predicted power dissipated by each resistor.

11. $V_T = 10$ V, $R_1 = 10\ \Omega$, $R_2 = 20\ \Omega$

12. $V_T = 0.58$ V, $R_1 = 680\ \Omega$, $R_2 = 480\ \Omega$

13. $V_T = 750$ mV, $R_1 = 1.8$ kΩ, $R_2 = 2.7$ kΩ, $R_3 = 910\ \Omega$

14. $V_T = 875$ mV, $R_1 = 22$ kΩ, $R_2 = 33$ kΩ, $R_3 = 41$ kΩ

15. $V_T = 750\ \mu$V, $R_1 = 68\ \Omega$, $R_2 = 56\ \Omega$, $R_3 = 120\ \Omega$

16. $V_T = 37.5\ \mu$V, $R_1 = 910\ \Omega$, $R_2 = 1.2$ kΩ, $R_3 = 2.4$ kΩ

17. $V_T = 12.5$ V, R-values of 8.2 kΩ, 15 kΩ, 82 kΩ, and 39 kΩ

18. $V_T = 2.75$ V, R-values of 620 kΩ, 120 kΩ, 910 kΩ, and 1.2 MΩ

19. $V_T = 1.55$ mV, R-values of 820 Ω, 560 Ω, 470 Ω, 390 Ω, and 1.2 kΩ

20. $V_T = 3.07$ mV, R-values of 12 kΩ, 2.7 kΩ, 8.2 kΩ, 5.6 kΩ, and 18 kΩ

In Exercises 21–30 solve for the indicated quantities.

21. In a series circuit containing three resistors, $P_T = 36.7$ W, $P_1 = 3.9$ W, $P_2 = 26.8$ W. Determine P_3.

22. A certain resistor dissipates 450 mW when it is passing 15 mA. Determine R (see Exercise 6).

23. A 56-kΩ resistor is dissipating 570 μW. Determine the voltage drop across it (see Exercise 7).

24. Determine the current required to dissipate 0.25 W in a 15-kΩ resistor (see Exercise 8).

25. In a series circuit containing three resistors the resistors are dissipating 33.75 mW, 49.5 mW, and 92.25 mW, respectively. The total resistance is 78 kΩ. Determine I for the circuit (see Exercise 9).

26. In a series circuit containing four resistors with values of 24 kΩ, 75 kΩ, 41 kΩ, and 56 kΩ, the 75-kΩ resistor is dissipating 42.19 mW. Determine P_T for the circuit (see Exercise 10).

27. A series circuit contains three resistors. One resistor of 18 kΩ has 4.5 V across it. The applied voltage is 6.85 V. (a) What is I for the circuit? (b) If the other two resistors are identical to each other in value what is that value?

28. In a series circuit with three resistors, R_1, a 15-kΩ resistor, has 5.25 V across it, and $V_T = 7.77$ V. (a) Determine I for the circuit. (b) If R_2 is 0.6 kΩ more than R_3 determine values of R_2 and R_3.

29. In a series circuit with two resistors, one of the resistors, R_2, has overheated and its color code cannot be read. The following is known about the circuit: $V_T = 9$ V, $R_1 = 9.1$ kΩ, and V_1 is normally approximately 5.02 V. What value of resistor would you buy as a replacement for R_2?

30. A friend asks you to help her adapt a 6-V car radio to her new car with a 12-V system. If the radio draws 30 mA when operating normally, and the battery voltage in the 12-V system is actually 13.5 V, (a) what resistor would you recommend as a voltage dropping resistor? (b) What power rating should it have?

Chapter 5

Algebraic Fractions

The concept of fractional parts is an ever-present reality in our daily lives. (Courtesy of Addison-Wesley Publishing Company, Inc.)

5-1 DEFINITIONS AND TERMINOLOGY

Fractions are used to describe or represent a part or "piece" of a whole—$\frac{1}{4}$ of an inch, $\frac{1}{6}$ of a pie, $\frac{2}{3}$ of an apple, and so on. A fraction is also a way of representing a selection of a certain number of equal parts from a group of such parts. For example, if 17 of the 30 students in the class have brown eyes we indicate the selection as $\frac{17}{30}$. *Regardless of how they arise, fractions are a way of indicating division:* $\frac{a}{b} = a \div b = b\overline{|a} = a/b$. *A fraction is an indicated quotient.*

In a fraction such as $\frac{a}{b}$ or a/b (read "a over b") we call the part of the fraction in the position of a the *numerator* and the part in the position of b the *denominator*. Together a and b are called the *terms* of the fraction. If either numerator or denominator, or both, contain a letter the fraction is an *algebraic fraction*.

Since division by zero is undefined, a fraction in which the denominator is zero or a condition under which the denominator may become zero is not permitted. For example, for the fraction $\dfrac{x + 2}{x - 3}$, the value of $x = 3$ is not permitted $\left(\dfrac{3 + 2}{3 - 3} = \dfrac{5}{0}\right)$.

Example 1   Express the following as fractions.

(a) The selection of 5 foreign television sets in a showroom which also includes 6 domestic sets.

(a) $\frac{5}{11}$

(b) The length of the line AB in Figure 5-1.

(b) $\frac{7}{4}$ in.

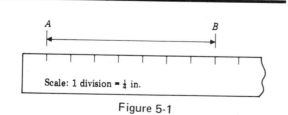

Scale: 1 division = $\frac{1}{4}$ in.

Figure 5-1

(c) Output voltage v_o divided by input voltage v_i.

(c) $\dfrac{v_o}{v_i}$

(d) The root of the equation $V = IR$ when I is the unknown.

(d) $\dfrac{V}{R}$

5-2 EQUIVALENT FRACTIONS

The following box contains the statement of two important properties of fractions used frequently in operations with fractions.

The form of a fraction, but not its value, is changed if:
1. **Both numerator and denominator are multiplied by the same quantity other than zero, or**
2. **Both numerator and denominator are divided by the same quantity other than zero.**
The new fraction is called an equivalent fraction.

Let us note and remember that the preceding statement *does not* say that we may *add* or *subtract* in the numerator and denominator without changing the value of a fraction.

If we increase the value of the numerator and denominator in converting a fraction to its equivalent, we have converted it to *higher terms*. An example of this is the conversion of a set of fractions to a common denominator for performing addition or subtraction operations.

Example 2   Convert the following to equivalent fractions by performing the indicated operation on both numerator and denominator.

(a) $\dfrac{2}{7}$, multiply by 2

(a) $\dfrac{2 \times 2}{7 \times 2} = \dfrac{4}{14}$

(b) $\dfrac{5a}{6b}$, multiply by $4c$

(b) $\dfrac{5a \cdot 4c}{6b \cdot 4c} = \dfrac{20ac}{24bc}$

(c) $\dfrac{v - 2}{v + 3}$, multiply by $v - 2$

(c) $\dfrac{(v - 2) \cdot (v - 2)}{(v + 3) \cdot (v - 2)} = \dfrac{v^2 - 4v + 4}{v^2 + v - 6}$

(d) $\dfrac{12iR}{16ir}$, divide by $4i$

(d) $\dfrac{12iR \div 4i}{16ir \div 4i} = \dfrac{3R}{4r}$

(e) $\dfrac{ri + rI}{Ri + RI}$, divide by $i + I$

(e) $\dfrac{r(i + I) \div (i + I)}{R(I + I) \div (i + I)}$

$= \dfrac{r}{R}$

(To facilitate the division we first factor numerator and denominator.)

When we wish to convert a fraction to its equivalent with a predetermined denominator (or numerator) we must first determine the appropriate multiplier. We do this by dividing the new denominator (or numerator) by the old denominator (or numerator).

Example 3 Convert the following to equivalent fractions with the indicated denominator (or numerator).

(a) $\dfrac{3}{7} = \dfrac{?}{21}$

(a) We find the desired multiplier by dividing the desired denominator by the original denominator: $21 \div 7 = 3$ Hence,

$\dfrac{3}{7} = \dfrac{3 \times 3}{7 \times 3} = \dfrac{9}{21}$ Ans.

(b) $\dfrac{5v}{6g} = \dfrac{20iv}{?}$

(b) $20iv \div 5v = 4i$

$\dfrac{5v}{6g} = \dfrac{5v \cdot 4i}{6g \cdot 4i} = \dfrac{20iv}{24gi}$

(c) $\dfrac{x - 2}{x + 2} = \dfrac{?}{3x^2 + 6x}$

(c) $3x^2 + 6x = 3x(x + 2)$, and $\dfrac{3x(x + 2)}{x + 2} = 3x$

We multiply the numerator as well as the denominator by $3x$:

$\dfrac{x - 2}{x + 2} = \dfrac{(x - 2) \cdot 3x}{(x + 2) \cdot 3x} = \dfrac{3x^2 - 6x}{3x^2 + 6x}$

(d) $\dfrac{r + R}{R - r} = \dfrac{?}{I^2R - I^2r}$

(d) $I^2R - I^2r = I^2(R - r)$, thus $\dfrac{I^2(R - r)}{R - r} = I^2$

$\dfrac{r + R}{R - r} = \dfrac{(r + R) \cdot I^2}{(R - r) \cdot I^2} = \dfrac{I^2r + I^2R}{I^2R - I^2r}$

Exercises 5-2 In Exercises 1–4 write fractions which indicate the units selected from the total (see Example 1).

1. Nine dots from a set of 15 dots.
2. The 21 male students in the electronics class which also includes 15 female students.
3. The 5 color television sets in the service shop which also contains 11 black-and-white sets.
4. The voltage of 4 V across resistor R_1 in the circuit which has a total of 9 V.

In Exercises 5–10 write the indicated quotients as fractions (see Exercise 5).

5. $12 \text{ V} \div 3 \text{ V}$ $\left(\text{Ans.: } 12 \text{ V} \div 3 \text{ V} = \dfrac{12 \text{ V}}{3 \text{ V}}\right)$

6. $I_C \div I_B$ 7. $R_1 \div R_T$

8. $(v + V) \div (r + R)$

9. $(v^2 - 16) \div (v^2 + 4v + 4)$

10. $24V^2g \div (4I^2 + 12I + 9)$

In Exercises 11–16 interpret each fraction as the indicated quotient of the numerator divided by the denominator (see Exercise 11).

11. $\dfrac{5}{11}$ $\left(\dfrac{5}{11} = 5 \div 11\right)$ 12. $\dfrac{V}{R}$ 13. $\dfrac{5iv}{7gR}$

14. $\dfrac{I_C}{I_B}$ 15. $\dfrac{v + V}{i + I}$ 16. $\dfrac{v^2 - i^2}{R^2 + 4R + 4}$

In Exercises 17–24 convert the given fractions to equivalent fractions by performing the indicated operations on both the numerator and the denominator (see Example 2).

17. $\dfrac{3}{5}$, multiply by 4   18. $\dfrac{7Vg}{11IR}$, multiply by $3I$

19. $\dfrac{R + 2}{R + 3}$, multiply by $R - 2$ 20. $\dfrac{v^2 + 6v + 9}{i^2 + 4i + 4}$, multiply by 2

21. $\dfrac{12}{21}$, divide by 3   22. $\dfrac{30I^2R}{36V^2R}$, divide by $6R$

23. $\dfrac{(v + 2)(v + 3)}{(v + 3)(i - 2)}$, divide by $v + 3$ 24. $\dfrac{(2I - 3)(2I + 3)}{(2I + 3)(2I + 3)}$, divide by $2I + 3$

In Exercises 25–36 change the given fractions to equivalent fractions with the indicated denominators (or numerators) (see Example 3).

25. $\dfrac{5}{13} = \dfrac{?}{39}$ 26. $\dfrac{5i}{7v} = \dfrac{?}{35v}$

27. $\dfrac{24v^2}{36i^2} = \dfrac{4v^2}{?}$ 28. $\dfrac{2x - 16}{2x + 16} = \dfrac{?}{x + 8}$

29. $\dfrac{rR}{r + R} = \dfrac{?}{I^2r + I^2R}$ 30. $\dfrac{v - 1}{v + 2} = \dfrac{?}{v^2 + 2v}$

31. $\dfrac{i + 3}{i + 5} = \dfrac{?}{iR + 5R}$

32. $\dfrac{a + b}{2a + 3b} = \dfrac{?}{8a + 12b}$

33. $\dfrac{v + 2}{v - 2} = \dfrac{?}{gv^2 - 2gv}$ 34. $\dfrac{r}{R} = \dfrac{?}{RI - Ri}$

35. $\dfrac{2v + 1}{V - v} = \dfrac{?}{\dfrac{V}{R} - \dfrac{v}{R}}$ 36. $\dfrac{r - X_C}{R + X_L} = \dfrac{?}{rR + rX_L}$

5-3 REDUCING TO LOWEST TERMS

If we reduce or simplify the numerator and denominator in converting a fraction to an equivalent fraction we are *reducing its terms*. When the numerator and denominator are reduced so that they have no common factors except ±1, the fraction is reduced to *lowest terms* or *simplest form*. To reduce a fraction to lowest terms we factor the numerator and denominator into their prime factors and then cancel common factors. Cancellation is equivalent to dividing the numerator and denominator by the same (a common) factor.

Example 4   Reduce the given fractions to lowest terms.

(a) $\dfrac{12}{18}$ (a) $\dfrac{12}{18} = \dfrac{\cancel{2} \cdot 2 \cdot \cancel{3}}{\cancel{2} \cdot \cancel{3} \cdot 3} = \dfrac{2}{3}$

(b) $\dfrac{8a^2 b^3}{12a^3 b^2}$ (b) $\dfrac{8a^2 b^3}{12a^3 b^2} =$

$$\dfrac{\cancel{2} \cdot \cancel{2} \cdot 2 \cdot \cancel{a} \cdot \cancel{a} \cdot \cancel{b} \cdot \cancel{b} \cdot b}{\cancel{2} \cdot \cancel{2} \cdot 3 \cdot \cancel{a} \cdot \cancel{a} \cdot a \cdot \cancel{b} \cdot \cancel{b}}$$

$$= \dfrac{2b}{3a}$$

(c) $\dfrac{v + 2}{4v + 8}$ (c) $\dfrac{\cancel{v + 2}}{4\cancel{(v + 2)}} = \dfrac{1}{4}$

(d) $\dfrac{R_1 R + R_1 r}{IR + Ir}$ (d) $\dfrac{R_1 R + R_1 r}{IR + Ir} = \dfrac{R_1 \cancel{(R + r)}}{I \cancel{(R + r)}}$

$$= \dfrac{R_1}{I}$$

A Common Error   When simplifying fractions containing polynomials it is common to

experience the very great temptation to cancel one or more of the terms of the polynomial (review definition of *terms* in Section 3-3). *Cancelling terms is an incorrect operation*, however. Let us consider an example: Reduce $\dfrac{i(i^2 - 9)}{i + 3}$ to lowest terms.

$$\dfrac{i(i^2 - 9)}{i + 3} = \dfrac{i(i - 3)\cancel{(i + 3)}}{\cancel{i + 3}} = i(i - 3)$$

Let us compare the result to that obtained when we indulge in cancelling terms:

┌─────────────────────────────────────┐
│   INCORRECT PROCEDURE   │
│   │
│ $\dfrac{\cancel{i}(i^2 - 9)}{\cancel{i} + 3} = \dfrac{i^2 - \overset{3}{\cancel{9}}}{\underset{1}{\cancel{3}}} = i^2 - 3$ │
│   │
└─────────────────────────────────────┘

To make it clear that the two answers are not the same, let us choose an arbitrary value for i, say $i = 4$, substitute it into the original expression and the two solutions, and compare the results:

$$\dfrac{i(i^2 - 9)}{i + 3} = \dfrac{4(4^2 - 9)}{7} = \dfrac{4 \cdot 7}{7} = 4$$

$$i(i - 3) = 4(4 - 3) = 4 \cdot 1 = 4$$

$$i^2 - 3 = 4^2 - 3 = 16 - 3 = 13$$

We summarize the procedure for reducing algebraic fractions to lowest terms.

┌──────────────────────────────────────┐
│ **To reduce an algebraic fraction to lowest terms:** │
│ 1. When both the numerator and denominator are monomials, divide each by their highest common factor (or factor into prime factors and cancel all common factors). │
│ 2. When either or both parts of a fraction is a polynomial, factor the parts into prime factors, if factorable, and then cancel common factors, only. │
│   │
│   MEMORY JOGGER   │
│ **Never Cancel the Terms of a Polynomial.** │
└──────────────────────────────────────┘

Exercises 5-3 Reduce the given fractions to lowest terms (see Example 4).

1. $\dfrac{4}{6}$ 2. $\dfrac{14}{21}$ 3. $\dfrac{12iv}{60ir}$ 4. $\dfrac{24gv}{72iv}$

5. $\dfrac{39x^2y}{52xy^2}$ 6. $\dfrac{28i^2Rv^2}{42iR}$ 7. $\dfrac{33p^4y^5}{11p^3y^2}$

8. $\dfrac{25ax^2}{15bx}$ 9. $\dfrac{18a^3b^2}{9a^2b^4}$ 10. $\dfrac{64c^5x^6}{16b^2c^4x^3}$

11. $\dfrac{-15}{35}$ 12. $\dfrac{24}{-8}$ 13. $\dfrac{-28}{-4}$ 14. $\dfrac{-V^2}{-V}$

15. $-\dfrac{i^2R}{r^2R}$ 16. $\dfrac{-h^4i^3}{-h^3i^2}$ 17. $\dfrac{48abc^2x^3y^4}{16a^2bc^3x^2y^5}$

18. $\dfrac{54v^3i^2r^4R^5X^3}{12vi^2r^3R^6X^4}$ 19. $\dfrac{55p^3r^5s^4t^2}{20p^4r^6s^3t}$

20. $\dfrac{6\pi R^2}{2\pi R}$ 21. $\dfrac{x^3(x-2)}{x^2}$ 22. $\dfrac{i^3}{i^2(5-i)}$

23. $\dfrac{5V^2(R-r)}{15V(R-r)}$ 24. $\dfrac{24i^2(R'+R'')^2}{6i(R'+R'')}$

25. $\dfrac{15g(i+I)}{3g^2(i-I)}$ 26. $\dfrac{38I^4(r+R)^4}{19I^3(r+R)^3}$

27. $\dfrac{5x+15y}{3x+9y}$ 28. $\dfrac{16Ir+4ir}{8I+2i}$

29. $\dfrac{3IR+3iR}{6Ir+6ir}$ 30. $\dfrac{RR_1+RR_2}{rR_1+rR_2}$

31. $\dfrac{v+1}{7v+7}$ 32. $\dfrac{I^2R^2-R^4}{I^2-R^2}$

33. $\dfrac{R_1R_2+R_1R_3+R_1R_4}{R_2+R_3+R_4}$

34. $\dfrac{rR+rX_C+rX_L}{IR+IX_C+IX_L}$

5-4 MULTIPLICATION AND DIVISION OF FRACTIONS

To find the product of fractions:
1. **Find the product of their numerators.**
2. **Find the product of their denominators.**
3. **Write:**

$$\frac{\text{product of numerators}}{\text{product of denominators}}$$

$$\frac{a}{b}\cdot\frac{c}{d}=\frac{ac}{bd}$$

Example 5 Multiply the given fractions.

(a) $\dfrac{2}{5}\cdot\dfrac{3}{7}$ (a) $\dfrac{2}{5}\cdot\dfrac{3}{7}=\dfrac{2\cdot3}{5\cdot7}=\dfrac{6}{35}$

(b) $\dfrac{-9}{4}\cdot\dfrac{3}{2}$ (b) $\dfrac{-9}{4}\cdot\dfrac{3}{2}=\dfrac{-9\cdot3}{4\cdot2}=\dfrac{-27}{8}$

(c) $\dfrac{-9}{8}\cdot\dfrac{5}{-3}$ (c) $\dfrac{\overset{3}{\cancel{-9}}}{8}\cdot\dfrac{5}{\underset{1}{\cancel{-3}}}=\dfrac{3\cdot5}{8\cdot1}=\dfrac{15}{8}$

We cancel factors common to the numerator and denominator wherever possible. In (c) we divided both -9 and -3 by -3.

(d) $\dfrac{V}{R}\cdot\dfrac{I}{G}$ (d) $\dfrac{V}{R}\cdot\dfrac{I}{G}=\dfrac{V\cdot I}{R\cdot G}=\dfrac{VI}{RG}$

(e) $\dfrac{R_1}{R_1+R_2}\cdot\dfrac{R_2}{R_1}$

(e) $\dfrac{\cancel{R_1}}{R_1+R_2}\cdot\dfrac{R_2}{\cancel{R_1}}=\dfrac{R_2}{R_1+R_2}$

We cancel factors only, not terms.

A Common Error There is a temptation to cancel the terms of a polynomial when multiplying fractions (see *A Common Error*, Section 5-3). In Example 5(e) the error would be made as follows:

INCORRECT PROCEDURE

$$\frac{\cancel{R_1}}{R_1+\cancel{R_2}}\cdot\frac{\cancel{R_2}}{\cancel{R_1}}=\frac{1}{R_1}$$

We can show that this operation is incorrect by substituting arbitrary values for R_1 and R_2. That is, let $R_1=2$ and $R_2=6$. Then, substituting into the original fractions, we have

$$\frac{\cancel{2}}{2+6}\cdot\frac{6}{\cancel{2}}=\frac{6}{8}=\frac{3}{4}$$

But when we substitute into the result obtained by cancelling a term of the denominator, we have

$$\frac{1}{R_1}=\frac{1}{2}\qquad \frac{1}{2}\neq\frac{3}{4}$$

Multiplication by a Whole Number   In arithmetic we multiply a fraction by a whole number by converting the whole number into a fraction with a denominator of 1. We use the same technique in algebra.

$$\frac{1}{2} \cdot 3 = \frac{1}{2} \cdot \frac{3}{1} = \frac{3}{2} \qquad 4 \cdot \frac{2V}{R} = \frac{4}{1} \cdot \frac{2V}{R} = \frac{8V}{R}$$

Sometimes in algebra it is preferable to leave such a product as an indicated product of a whole number and a fraction. For example, $\frac{R_1}{R_T} \cdot V$ may be written $\frac{R_1}{R_T}V$.

Example 6   Perform the indicated multiplications.

(a) $\dfrac{3}{7} \cdot 4$

(a) $\dfrac{3}{7} \cdot 4 = \dfrac{3}{7} \cdot \dfrac{4}{1} = \dfrac{3 \cdot 4}{7 \cdot 1} = \dfrac{12}{7}$

(b) $\dfrac{x}{y} \cdot 3x$

(b) $\dfrac{x}{y} \cdot 3x = \dfrac{x}{y} \cdot \dfrac{3x}{1} = \dfrac{3x^2}{y}$

(c) $\dfrac{R_1}{R_1 + R_2} \cdot I_T$

(c) $\dfrac{R_1}{R_1 + R_2} \cdot I_T = \dfrac{R_1}{R_1 + R_2} I_T$

(d) two-fifths of 15

(d) $\dfrac{2}{5} \cdot 15 = \dfrac{2}{\cancel{5}} \cdot \dfrac{\overset{3}{\cancel{15}}}{1} = 6$

We interpret "of" following a fraction as "times."

Division of Fractions

To divide one fraction by a second fraction, invert the second fraction (the fraction following the division sign) and multiply:

$$\frac{a}{b} \div \frac{c}{d} = \frac{a}{b} \cdot \frac{d}{c} = \frac{ad}{bc}$$

Example 7   Perform the indicated divisions.

(a) $\dfrac{1}{2} \div \dfrac{3}{4}$

(a) $\dfrac{1}{2} \div \dfrac{3}{4} = \dfrac{1}{\cancel{2}} \cdot \dfrac{\overset{2}{\cancel{4}}}{3} = \dfrac{2}{3}$

(b) $\dfrac{-7}{8} \div \dfrac{5}{9}$

(b) $\dfrac{-7}{8} \div \dfrac{5}{9} = \dfrac{-7}{8} \cdot \dfrac{9}{5} =$

$\dfrac{-7 \cdot 9}{8 \cdot 5} = \dfrac{-63}{40}$

(c) $\dfrac{2V}{3R} \div \dfrac{4}{9R}$

(c) $\dfrac{2V}{3R} \div \dfrac{4}{9R} = \dfrac{\overset{1}{\cancel{2}}V}{\underset{1}{\cancel{3R}}} \cdot \dfrac{\overset{3}{\cancel{9R}}}{\underset{2}{\cancel{4}}} =$

$\dfrac{V \cdot 3}{2} = \dfrac{3V}{2} = \dfrac{3}{2}V$

(d) $\dfrac{R_2}{R_1 + R_2} \div \dfrac{1}{-R_2}$

(d) $\dfrac{R_2}{R_1 + R_2} \div \dfrac{1}{-R_2} = \dfrac{R_2}{R_1 + R_2} \cdot \dfrac{-R_2}{1} =$

$\dfrac{-R_2^2}{R_1 + R_2}$

We cannot cancel the factor R_2 in the numerator with the term R_2 in the denominator.

Division by a Whole Number   Any whole number is considered to be a fraction with a denominator of 1, for example, $3 = \frac{3}{1}$, $x = \frac{x}{1}$, and $a^2 b = a^2 b/1$. To divide a fraction by such quantities we invert them and multiply. This produces fractions with numerators of 1: $\frac{3}{1}$ inverted is $\frac{1}{3}$, $\frac{x}{1}$ inverted is $\frac{1}{x}$, and $a^2 b/1$ inverted is $1/a^2 b$, and so on. *The quantity which is the result of inverting a number is called the reciprocal of the number:* the reciprocal of 3 is $\frac{1}{3}$, the reciprocal of x is $\frac{1}{x}$, the reciprocal of R is $\frac{1}{R}$.

The reciprocal of the reciprocal of a number is the number: the reciprocal of $\frac{1}{3}$ is $\frac{1}{3}$ inverted or 3, the reciprocal of $\frac{1}{R}$ is R, and so on.

The product of any number and its reciprocal is 1:

$$\frac{3}{1} \cdot \frac{1}{3} = 1 \qquad \frac{R}{1} \cdot \frac{1}{R} = 1$$

In general, division by any quantity is the same as multiplication by its reciprocal and can be performed in any order. For example,

$$\frac{3}{5} \div 3 = \frac{\cancel{3}}{5} \cdot \frac{1}{\cancel{3}} = \frac{1}{\cancel{3}} \cdot \frac{\cancel{3}}{5} = \frac{1}{5}$$

$$\frac{R_1 + R_2}{R_2} \div 2 = \frac{R_1 + R_2}{R_2} \cdot \frac{1}{2} = \frac{1}{2} \cdot \frac{R_1 + R_2}{R_2}$$

$$= \frac{R_1 + R_2}{2R_2}$$

Slide rule calculators incorporate the facility for converting any quantity to its reciprocal directly. The key is usually labeled $(1/x)$. The operation is as follows: Find the reciprocal of 5.786. Enter 5.786, $(1/x)$, Readout = 0.1728309 ($1/5.786 = 0.1728309$).

Example 8 Perform the indicated divisions.

(a) $\dfrac{3}{7} \div 6$

(a) $\dfrac{3}{7} \div 6 = \dfrac{3}{7} \cdot \dfrac{1}{6} =$

$\dfrac{1}{14}$

(b) $6 \div \dfrac{3}{7}$

(b) $6 \div \dfrac{3}{7} = \dfrac{6}{1} \cdot \dfrac{7}{3} =$

14

(c) $\dfrac{4x^2}{5y} \div 2x$

(c) $\dfrac{4x^2}{5y} \div 2x =$

$\dfrac{4x^2}{5y} \cdot \dfrac{1}{2x} = \dfrac{2x}{5y}$

(d) $\dfrac{R_1}{R_1 + R_2} \div 2R_1$

(d) $\dfrac{R_1}{R_1 + R_2} \div 2R_1 =$

$\dfrac{R_1}{R_1 + R_2} \cdot \dfrac{1}{2R_1} =$

$\dfrac{1}{2(R_1 + R_2)}$

(e) Find the reciprocal of 97.356 on a calculator.

(e) 97.356, $(1/x)$, Readout: 0.0102715

Exercises 5-4 In Exercises 1–36 perform the indicated operations. Cancel common factors. Express results in lowest terms.

1. $\dfrac{1}{2} \cdot \dfrac{2}{3}$ 2. $\dfrac{3}{5} \cdot \dfrac{7}{9}$ 3. $\dfrac{2}{3} \cdot \dfrac{4}{7} \cdot \dfrac{2}{5}$

4. $\dfrac{37}{53} \cdot 75$ 5. $\dfrac{-5}{9} \cdot (-45)$ 6. $\dfrac{-4}{15} \cdot \dfrac{-27}{-16}$

7. $\dfrac{x}{y} \cdot \dfrac{a}{b}$ 8. $\dfrac{ir}{V} \cdot \dfrac{gV}{iR}$ 9. $\dfrac{R_1 R_2}{R_3 R_4} \cdot \dfrac{R_3 R_5}{R_1 R_6}$

10. $\dfrac{R^2 X}{4} \cdot \dfrac{-8}{RX^2}$ 11. $\dfrac{-V}{iv} \cdot \dfrac{3v}{-4V^2}$

12. $\dfrac{I^2 R}{P_2} \cdot P^2$

13. $\dfrac{R}{r + R} \cdot \dfrac{r}{3}$ (Remember: Do not cancel terms.)

14. $\dfrac{V}{i + I} \cdot \dfrac{4}{IV}$ 15. $\dfrac{x + 2}{a - 3} \cdot \dfrac{(a - 3)(a + 3)}{x + 2}$

16. $\dfrac{5V + 15}{5} \cdot \dfrac{16R}{4V + 12}$

17. $\dfrac{R(v - V)}{i} \cdot \dfrac{i^2(R + r)}{R^2 + Rr}$

18. $\dfrac{ir + Ir}{gV} \cdot \dfrac{gr}{Ir^2 + ir^2}$

19. $(R_1 + R_2) \dfrac{V + v}{(R_1 + R_2)(R_1 - R_2)}$

20. $\dfrac{(i + I)R}{IV} \cdot \dfrac{rV}{IrR + irR}$ 21. $\dfrac{1}{2} \div \dfrac{2}{3}$

22. $\dfrac{3}{5} \div \dfrac{7}{-9}$ 23. $\dfrac{-2}{3} \div 2$ 24. $\dfrac{-5}{9} \div (-45)$

25. $16 \div \dfrac{4}{5}$ 26. $\dfrac{5x}{7} \div \dfrac{15x}{21y}$ 27. $\dfrac{7}{3v} \div \dfrac{-35}{9vr}$

28. $\dfrac{R_1}{R_2 R_3} \div \dfrac{R_2}{R_1 R_3}$ 29. $R \div \dfrac{R}{V}$

30. $R \div \dfrac{R}{R + r}$ 31. $\dfrac{R_3 + R_4}{R_3 R_4} \div (R_1 + R_2)$

32. $\dfrac{iR}{Ir} \div \dfrac{R}{r}$ 33. $\dfrac{V}{r} \div \dfrac{R + r}{Rr}$

34. $\dfrac{r - X}{R + x} \div (r - X)$ 35. $\dfrac{Rr}{R + r} \div R$

36. $\dfrac{-gv}{(r + 5)(r + 4)} \div \dfrac{1}{r + 5}$

5-5 ADDITION AND SUBTRACTION OF FRACTIONS

We recall from arithmetic that we can add $\frac{1}{6}$ and $\frac{4}{6}$ directly, but not $\frac{1}{2}$ and $\frac{2}{3}$:

$$\tfrac{1}{6} + \tfrac{4}{6} = \tfrac{5}{6} \qquad \tfrac{1}{2} + \tfrac{2}{3} = ?$$

The fractions $\frac{1}{6}$ and $\frac{4}{6}$ are *like fractions* because they have the same denominator.

To add like fractions, add their numerators and keep the denominator.

To add unlike fractions we must convert them to equivalent fractions with the same denominator, a *common denominator*. In general,

$$\frac{a}{b} + \frac{c}{d} = \frac{ad}{bd} + \frac{bc}{bd} = \frac{ad + bc}{bd}$$

A common denominator is an expression divisible without remainder by all the denominators of the fractions to be added. If these denominators contain common factors a common denominator smaller than their product is possible. The smallest of these (a product of the minimum number of common factors) is called the *lowest common denominator* or *LCD*. Although it is never wrong to find a common denominator simply by taking the product of the denominators, in many cases it is more expedient to find and convert to an *LCD*.

To find the LCD of two or more fractions, factor each denominator into its prime factors. Obtain a product (the lowest common multiple, LCM) which includes each prime factor as many times as it appears most frequently in any one denominator.
   To convert fractions to an LCD, divide the LCD by each original denominator to obtain the conversion factor, CF. Multiply each numerator by its respective CF.

Example 9   Perform the indicated operations.

(a) $\dfrac{1}{2} + \dfrac{2}{3}$

(a) $\dfrac{1}{2} + \dfrac{2}{3} = \dfrac{1 \cdot 3}{2 \cdot 3} + \dfrac{2 \cdot 2}{3 \cdot 2} = \dfrac{3 + 4}{6} = \dfrac{7}{6}$

(b) $\dfrac{2a}{3b} + \dfrac{3c}{4d}$

(b) $\dfrac{2a}{3b} + \dfrac{3c}{4d} = \dfrac{2a \cdot 4d}{3b \cdot 4d} + \dfrac{3c \cdot 3b}{4d \cdot 3b} = \dfrac{8ad + 9bc}{12bd}$

(c) $\dfrac{3b}{12x^2y} + \dfrac{4c}{18xy^3}$

(c) We factor the denominators,
$12x^2y = 2 \cdot 2 \cdot 3 \cdot x \cdot x \cdot y$
$18xy^3 = 2 \cdot 3 \cdot 3 \cdot x \cdot y \cdot y \cdot y$
Then
LCD $= 2 \cdot 2 \cdot 3 \cdot 3 \cdot x^2y^3 = 36x^2y^3$
Dividing LCD by the first denominator,
$$\frac{36x^2y^3}{12x^2y} = 3y^2$$
and by the second,
$$\frac{36x^2y^3}{18xy^3} = 2x$$
Then
$$\frac{3b}{12x^2y} + \frac{4c}{18xy^3} = \frac{3b \cdot 3y^2}{12x^2y \cdot 3y^2} +$$
$$\frac{4c \cdot 2x}{18xy^3 \cdot 2x} = \frac{9by^2 + 8cx}{36x^2y^3}$$

(d) $\dfrac{3a + b}{a^2 - b^2} + \dfrac{2}{a + b} - \dfrac{1}{a - b}$

(d) By inspection the LCD is $a^2 - b^2$. Thus we have

$$\frac{3a + b + 2(a - b) - (a + b)}{a^2 - b^2} = \frac{4a - 2b}{a^2 - b^2}$$

We observe in Example 7(d) that, to subtract a fraction, we convert it to an LCD and subtract its numerator. It is important to recall here that the fraction line or *vinculum* is a symbol of grouping. A negative sign before a fraction applies to a numerator the same as it would if the numerator was in parentheses—the signs of all terms in the numerator must be changed before they are combined with other terms.

Exercises 5-5   In Exercises 1–32 perform the indicated operations. Express results in lowest terms.

1. $\dfrac{2}{3} + \dfrac{3}{5}$ 2. $\dfrac{3}{7} - \dfrac{1}{3}$ 3. $\dfrac{1}{4} + \dfrac{4}{5} - \dfrac{3}{10}$

4. $\dfrac{5}{6} - \dfrac{3}{4} + \dfrac{7}{12}$ 5. $\dfrac{9}{5} - \dfrac{2}{3} + \dfrac{11}{6} - \dfrac{7}{10}$

6. $\dfrac{5}{7} + \dfrac{12}{5} - \dfrac{7}{10} + \dfrac{11}{15}$

7. $2 + \dfrac{3}{4} - \dfrac{5}{8} + \dfrac{5}{6} - \dfrac{5}{12}$ $\left(\textit{Hint: } \text{Treat 2 as the fraction } \dfrac{2}{1}.\right)$

8. $\dfrac{15}{9} - \dfrac{2}{3} + \dfrac{7}{12} - \dfrac{7}{4} + 3$ 9. $\dfrac{4}{a} - \dfrac{3}{b}$

10. $\dfrac{V}{4} + \dfrac{v}{7}$ 11. $\dfrac{2a}{8b} - \dfrac{3a^2}{6b^2}$

12. $\dfrac{4}{3IR} - \dfrac{3}{6I^2R}$ 13. $\dfrac{5}{ac} + \dfrac{4}{bc} - \dfrac{3}{ab}$

14. $\dfrac{5}{R_1 R_2} + \dfrac{7}{R_1 R_3} - \dfrac{8}{R_2 R_3}$ 15. $\dfrac{1}{R_1} + \dfrac{1}{R_2}$

16. $\dfrac{1}{R_1} + \dfrac{1}{R_2} + \dfrac{1}{R_3}$ 17. $3 + \dfrac{4x}{yz} + \dfrac{5y}{xz} + \dfrac{3}{xy}$

18. $\dfrac{R}{rX} + \dfrac{r}{RX} + \dfrac{X}{rR} + V$ 19. $\dfrac{4}{iR + IR} + \dfrac{2}{R}$

20. $\dfrac{V}{I + i} - \dfrac{v}{ir + Ir}$ 21. $\dfrac{4}{2v - 6} + \dfrac{2}{3v - 9} - \dfrac{1}{6}$

22. $\dfrac{1}{3} + \dfrac{1}{4R + 12X} - \dfrac{2}{2R + 6X}$

23. $\dfrac{R}{R + X} + \dfrac{X}{R - X}$ 24. $\dfrac{I}{i + I} - \dfrac{i}{i - I} + 1$

25. $\dfrac{a}{a + b} - \dfrac{2ab}{a^2 - b^2} + \dfrac{b}{a - b}$ $\begin{bmatrix}\textit{Hint: } a^2 - b^2 = \\ (a - b)(a + b).\end{bmatrix}$

26. $\dfrac{r}{r - x} - \dfrac{x}{r + x} + \dfrac{2rx^3}{r^2 - x^2}$

27. $\dfrac{-x}{x - y} - \dfrac{2xy - 2x^2}{x^2 - 2xy + y^2}$ $\begin{bmatrix}x^2 - 2xy + y^2 = \\ (x - y)(x - y)\end{bmatrix}$

28. $\dfrac{R^2}{(R + X)(R + X)} - \dfrac{1}{R + X}$

29. $\dfrac{4ab}{(a + b)(a - 3b)} - \dfrac{b}{a + b} + \dfrac{a}{a - 3b}$

30. $\dfrac{2R}{v + 4} + \dfrac{3R}{(v + 4)(v - 3)} - \dfrac{5}{v - 3}$

31. $\dfrac{2}{i^2 - 1} + \dfrac{3}{(i - 1)(i - 1)}$

32. $\dfrac{V}{R + 2} - \dfrac{2}{R^2 + 4R + 4} + \dfrac{V}{R^2 - 4}$
    $[R^2 + 4R + 4 = (R + 2)(R + 2)]$

5-6 SIGNS OF FRACTIONS

Let us remember that a fraction is a number (the quotient) obtained as a result of dividing one number (the numerator) by a second number (the denominator). Therefore, it is necessary to recognize that *a fraction has three signs: the sign of the numerator, the sign of the denominator, and the sign in front of the fraction called the sign of the fraction.* For example, in the fraction $-\dfrac{+9}{-3}$, the sign of the numerator is +, and the sign of the fraction and of the denominator is −.

Since $(-1) \cdot (-1) = 1$ and $\dfrac{-1}{-1} = 1$ it is possible to change any two signs of a fraction without changing its value. Consider the following:

$$-\dfrac{+9}{-3} = -\dfrac{+9(-1)}{-3}(-1) = +\dfrac{-9}{-3} = +3 \quad (1)$$

$$-\dfrac{+9}{-3} = -\dfrac{+9(-1)}{-3(-1)} = -\dfrac{-9}{+3} = +3 \quad (2)$$

$$-\dfrac{+9}{-3} = -\dfrac{+9}{-3(-1)}(-1) = +\dfrac{+9}{+3} = +3 \quad (3)$$

Note that in equation (1) we changed the signs of the fraction and the numerator; in (2) we changed the signs of the numerator and denominator; and in (3) we changed the signs of the fraction and the denominator. In each case, however, the answer is +3. Hence, all three forms of the fraction are equivalent. *Any two signs of a fraction may be changed without changing its value.*

Example 10 Change appropriate pairs of signs and write the three equivalent forms of each of the following.

(a) $+\dfrac{+a}{+b}$ (a) $\dfrac{a}{b} = \dfrac{-a}{-b} = -\dfrac{-a}{b} = -\dfrac{a}{-b}$

(b) $\dfrac{x - y}{a - b}$ (b) $\dfrac{x - y}{a - b} = \dfrac{y - x}{b - a} = -\dfrac{y - x}{a - b} =$

$\qquad\qquad -\dfrac{x - y}{b - a}$

Note: $(-1)(x - y) = -x + y = y - x$

Frequently the binomial terms of fractions differ only by their signs. The correct change of signs in many of these situations allows further simplification.

Example 11 Perform the indicated operations and simplify.

(a) $\dfrac{a}{a - b} - \dfrac{b}{b - a}$

(a) The denominators are alike except for their signs. Thus, if we change the signs of the second fraction and its denominator, we have

$$\frac{a}{a-b} + \frac{b}{a-b} = \frac{a+b}{a-b}$$

(b) $\dfrac{V-v}{r-R} \cdot \dfrac{r^2-R^2}{v^2-V^2}$

(b) $-\dfrac{\cancel{v-V}}{\cancel{r-R}} \cdot \dfrac{\cancel{(r-R)}(r+R)}{\cancel{(v-V)}(v+V)} = -\dfrac{r+R}{v+V}$

Exercises 5-6 In Exercises 1–4 express each fraction as an equivalent fraction with the sign of the numerator changed.

1. $\dfrac{-5}{7}$ 2. $\dfrac{3a}{4b}$ 3. $\dfrac{x-y}{a-b}$ 4. $\dfrac{v^2-i^2}{R^2-r^2}$

In Exercises 5–8 express the given fractions as equivalent fractions with the sign of the fraction +.

5. $-\dfrac{5V^2}{-4R^2}$ 6. $-\dfrac{R+X}{R-X}$ 7. $-\dfrac{V^2+v^2}{R-r}$

8. $-\dfrac{-(R+r)}{X-R}$

In Exercises 9–12 reduce the given fractions to lowest terms.

9. $\dfrac{V-v}{v-V}$ 10. $-\dfrac{R-X}{X^2-R^2}$

11. $\dfrac{(i-I)(i-I)}{(I-i)(I-i)}$ 12. $\dfrac{(V+3IR)(V-2IR)}{(3IR-V)(2IR-V)}$

In Exercises 13–20 perform the indicated operations and simplify.

13. $\dfrac{3V}{R-r} - \dfrac{4V}{r-R}$ 14. $\dfrac{I}{g-G} + \dfrac{IG}{G^2-g^2}$

15. $\dfrac{5x+6}{(x-2)(x-2)} - \dfrac{2}{2-x}$

16. $\dfrac{L+1}{(L+3)(L-2)} + \dfrac{3}{2-L}$

17. $\dfrac{V-2}{R+3} \cdot \dfrac{R-4}{2-V}$ 18. $\dfrac{I-4}{G+6} \div \dfrac{4-I}{6+G}$

19. $\dfrac{6V^2}{R_1^2-R_2^2} \cdot \dfrac{R_2-R_1}{3V}$

20. $\dfrac{V^2-9}{R-4} \div \dfrac{(V-3)(V-3)}{(R+1)(4-R)}$

5-7 MIXED EXPRESSIONS

In arithmetic we learned that a fraction such as $2\frac{3}{4}$, which contains both an integer and a fraction, is called a *mixed fraction*. Recall that when we write the mixed fraction we mean $2\frac{3}{4} = 2 + \frac{3}{4}$. In algebra, on the other hand, we imply multiplication when we write two expressions adjacent to each other. For example $2\frac{x}{y}$ means $2 \cdot \frac{x}{y}$. We do have mixed algebraic expressions however: $x + \frac{x}{y}$, $2 + V + \frac{V}{R}$, and so on.

In arithmetic we convert a mixed fraction to an *uncommon fraction* by multiplying the integer by the denominator and adding the result to the numerator:

$$2\frac{3}{4} = \frac{2}{1} + \frac{3}{4} = \frac{2\cdot4}{1\cdot4} + \frac{3}{4} = \frac{2\cdot4+3}{4} = \frac{8+3}{4} = \frac{11}{4}$$

We convert an uncommon fraction to a mixed fraction by performing the indicated division and expressing the remainder as a fraction:

$$\frac{18}{7} = 2 + \frac{4}{7} = 2\frac{4}{7}$$

The corresponding procedures in algebra are similar.

Example 12 Express the following as single fractions.

(a) $3\dfrac{4}{5}$

(a) $3\dfrac{4}{5} = \dfrac{3\cdot5+4}{5} = \dfrac{19}{5}$

(b) $3 + \dfrac{x}{y}$

(b) $3 + \dfrac{x}{y} = \dfrac{3\cdot y+x}{y} = \dfrac{3y+x}{y}$

(c) $V + IR - \dfrac{4V+VR}{R+4}$

(c) $\dfrac{V(R+4)+IR(R+4)-4V-VR}{R+4}$

$= \dfrac{VR+4V+IR^2+4IR-4V-VR}{R+4}$

$= \dfrac{IR^2+4IR}{R+4} = \dfrac{IR(R+4)}{R+4} = IR$

Example 13 Perform the indicated operations.

(a) $\left(V - \dfrac{v}{R}\right) \cdot \dfrac{R}{v-V}$

(a) Since we must perform operations inside parentheses first, we convert $V - \dfrac{v}{R}$ to a single fraction:

$$\left(V - \frac{v}{R}\right)\frac{R}{v - V} = \frac{RV - v}{R} \cdot \frac{R}{v - V} = \frac{RV - v}{v - V}$$

(b) $\dfrac{V - 2}{R^2 - 1} \div \left(4 - \dfrac{3R}{R + 1}\right)$

(b) $4 - \dfrac{3R}{R + 1} = \dfrac{4(R + 1) - 3R}{R + 1}$

$$= \frac{4R + 4 - 3R}{R + 1} = \frac{R + 4}{R + 1}$$

$$\frac{V - 2}{R^2 - 1} \div \frac{R + 4}{R + 1} = \frac{V - 2}{(R - 1)(R+1)} \cdot \frac{R+1}{R + 4}$$

$$= \frac{V - 2}{R^2 + 3R - 4}$$

Example 14 Express the following fractions as mixed expressions.

(a) $\dfrac{23}{15}$

(b) $\dfrac{15V + 4}{5V}$

(c) $\dfrac{x^3 - 3x^2 + 5}{x}$

(a) $\dfrac{23}{15} = 1\dfrac{8}{15}$

(b) $\dfrac{15V}{5V} + \dfrac{4}{5V} = 3 + \dfrac{4}{5V}$

(c) $\dfrac{x^3}{x} - \dfrac{3x^2}{x} + \dfrac{5}{x} =$

$x^2 - 3x + \dfrac{5}{x}$

Exercises 5-7 In Exercises 1–8 express the given expressions as single fractions (see Example 12).

1. $3\dfrac{7}{8}$ 2. $11\dfrac{2}{3}$ 3. $2 + \dfrac{x}{2}$ 4. $I + \dfrac{V}{R}$

5. $I - 3 + \dfrac{4R_1}{R_1 + R_2}$ 6. $15 - \left(\dfrac{V}{R}\right)^2$

7. $a - \dfrac{b}{2} + \dfrac{3a^2 + 4b^2}{a + b}$ 8. $I - \dfrac{4V - 5R}{R - 2} + 4$

In Exercises 9–14 perform the indicated operations and simplify (see Example 13).

9. $\left(x - \dfrac{3}{4}\right)\dfrac{8}{8x - 6}$ 10. $\dfrac{v - r}{v + r}\left(\dfrac{v}{r} + 1\right)$

11. $\left(1 - \dfrac{v^2}{r}\right) \div \dfrac{(v + 1)^2}{r}$

12. $\left(2 + i - \dfrac{i - 2i^2}{i}\right)\dfrac{i^2}{6i^2 + 2i}$

13. $\left(V - \dfrac{V^2}{R}\right) \div \left(1 - \dfrac{V}{R}\right)$

14. $\left(\dfrac{1}{R_1 + R_2} - \dfrac{1}{R_1}\right) \div \left(\dfrac{1}{R_1} + \dfrac{1}{R_2}\right)$

In Exercises 15–24 express the given fractions as mixed expressions (see Example 14).

15. $\dfrac{17}{5}$ 16. $\dfrac{5 + x}{5}$ 17. $\dfrac{V - IR}{R}$

18. $\dfrac{V_{CC} - IR_L}{I}$ 19. $\dfrac{h_{12}I_b + I_b R_L + 30V_C}{I_b}$

20. $\dfrac{R_1 + R_2 + R_1 R_2}{R_1 + R_2}$ 21. $\dfrac{V^2 + v^2}{V^2}$

22. $\dfrac{I^2 - 4I + 4}{I}$ 23. $\dfrac{v^2 - 6v + 7}{v}$

24. $\dfrac{4v^2 + 6vV + 15V^2}{2v + 3V}$

In Exercises 25–36 state whether the given equations are TRUE or FALSE. Correct any that are false.

25. $\dfrac{R + 3}{3} = R$ 26. $\dfrac{R - X}{R} = 1 - \dfrac{X}{R}$

27. $\dfrac{I^2 - 4}{I^2 - 8} = \dfrac{1}{2}$ 28. $\dfrac{i(R + 2)}{i^2 + 2} = \dfrac{R + 2}{i + 2}$

29. $\dfrac{V + v}{R} \cdot \dfrac{R}{V + v} = 0$ 30. $\dfrac{V}{R} + \dfrac{v}{r} = \dfrac{V + v}{R + r}$

31. $1 \div \left(\dfrac{1}{R_1} + \dfrac{1}{R_2}\right) = R_1 + R_2$

32. $\dfrac{V^2}{I^2} = \dfrac{V}{I}$ 33. $\dfrac{IR + V}{R} = I + V$

34. $\dfrac{-V^2}{R - r} = \dfrac{V^2}{r - R}$ 35. $\dfrac{V}{R} \cdot \dfrac{1}{2} = \dfrac{V}{2R}$

36. $\dfrac{V}{R} \cdot \dfrac{R}{V + v} = \dfrac{1}{v}$

5-8 COMPLEX FRACTIONS

A fraction in which the numerator or denominator or both contains a fraction is called a *complex fraction*. Complex fractions often arise when the relationships of electronics phenomena are stated using mathematical expressions.

Complex fractions can be transformed to a simpler or at least to a more useful form by performing the division operation which the fraction represents. For example, the complex fraction

$$\frac{\dfrac{x+y}{x}}{\dfrac{x^2-y^2}{xy}}$$

really means "divide the fraction $\dfrac{x+y}{x}$ by the fraction $\dfrac{x^2-y^2}{xy}$." Thus,

$$\frac{x+y}{x}\div\frac{x^2-y^2}{xy}=\frac{\cancel{x+y}}{\cancel{x}}\cdot\frac{\cancel{x}y}{(x-y)\cancel{(x+y)}}=\frac{y}{x-y}$$

In general, $\dfrac{\frac{a}{b}}{\frac{c}{d}}$ simplifies as follows:

$$\frac{\frac{a}{b}}{\frac{c}{d}}=\frac{a}{b}\div\frac{c}{d}=\frac{a}{b}\cdot\frac{d}{c}=\frac{ad}{bc}$$

From this we observe that during simplification: (1) the numerators of the fractions of both numerator and denominator (a and c in this example) remain in their respective positions, and (2) the denominators of the numerator and denominator fractions (b and d above) move to the other term of the fraction (the denominator of the numerator fraction moves to the denominator of the main fraction, the denominator of the denominator fraction moves to the numerator of the main fraction). Also note, then, that common factors in numerators can be cancelled, and common factors in denominators can be cancelled.

Algebraic sums must be converted to single fractions before simplification.

Example 15 Simplify the given expressions.

(a) $\dfrac{\frac{2}{15}}{\frac{2}{5}}$

(a) $\dfrac{\frac{2}{15}}{\frac{2}{5}}=\dfrac{\cancel{2}}{\cancel{15}3}\cdot\dfrac{\cancel{5}}{\cancel{2}}=\dfrac{1}{3}$

or

$\dfrac{\cancel{2}\cdot\cancel{5}}{\underset{3}{\cancel{15}}\cdot\cancel{2}}=\dfrac{1}{3}$

(b) $\dfrac{\frac{ax}{by}}{\frac{ac}{bd}}$

(b) $\dfrac{\frac{\cancel{a}x}{\cancel{b}y}}{\frac{\cancel{a}c}{\cancel{b}d}}=\dfrac{dx}{cy}$ or

$\dfrac{\cancel{a}x\cdot\cancel{b}d}{\cancel{b}y\cdot\cancel{a}c}=\dfrac{dx}{cy}$

(c) $\dfrac{2+\frac{2}{3}}{1+\frac{7}{9}}$

(c) $\dfrac{2+\frac{2}{3}}{1+\frac{7}{9}}=\dfrac{\frac{6+2}{3}}{\frac{9+7}{9}}=\dfrac{\cancel{8}\cdot\overset{3}{\cancel{9}}}{\cancel{3}\cdot\underset{2}{\cancel{16}}}$

$=\dfrac{3}{2}$

(d) $\dfrac{i-\frac{v^2}{4i}}{i+v+\frac{v^2}{4i}}$

(d) $\dfrac{i-\frac{v^2}{4i}}{i+v+\frac{v^2}{4i}}=$

$\dfrac{\frac{4i^2-v^2}{4\cancel{i}}}{\frac{4i^2+4iv+v^2}{4\cancel{i}}}=$

$\dfrac{(2i-v)\cancel{(2i+v)}}{(2i+v)\cancel{(2i+v)}}=\dfrac{2i-v}{2i+v}$

Exercises 5-8 In Exercises 1–26 simplify the given expressions.

1. $\dfrac{\frac{3}{4}}{\frac{9}{8}}$

2. $\dfrac{\frac{15}{7}}{\frac{5}{28}}$

3. $\dfrac{\frac{x^2y^2}{ab}}{\frac{xy}{bd}}$

4. $\dfrac{\frac{R_1R_2}{R}}{\frac{R_2R_3}{R}}$

5. $\dfrac{\frac{1}{4}-\frac{5}{3}}{\frac{5}{12}+\frac{7}{8}}$

6. $\dfrac{\frac{2}{7}-\frac{7}{3}}{\frac{1}{2}+\frac{5}{7}}$

7. $\dfrac{3-\frac{9}{4}}{1+\frac{11}{8}}$

8. $\dfrac{2+\frac{1}{3}-\frac{1}{2}}{\frac{1}{5}-\frac{11}{6}+\frac{5}{2}}$

9. $\dfrac{x+\frac{a}{b}}{b+\frac{a}{x}}$

10. $\dfrac{\frac{1}{R}+\frac{1}{r}}{\frac{1}{R}-\frac{1}{r}}$

11. $\dfrac{V}{\frac{1}{R}+\frac{1}{r}}$

12. $\dfrac{1}{\frac{1}{R}+\frac{1}{r}}$

13. $\dfrac{i^2-\frac{V^2}{R^2}}{R+\frac{V}{i}}$

14. $\dfrac{1-\frac{2r}{R+r}}{2-\frac{R+r}{R}}$

15. $\dfrac{V+\frac{v-V}{2}}{1+\frac{V}{v}}$

16. $\dfrac{4+i+\frac{4}{i}}{i-\frac{4}{i}}$

17. $\dfrac{3-\frac{v-i}{v+i}}{2+\frac{v}{i}}$

18. $\dfrac{R_1+\frac{R_2R_3}{R_2+R_3}}{R_4+\frac{R_2R_3}{R_2+R_3}}$

19. $\dfrac{\frac{R_pR_m}{R_p+R_m}}{\frac{R_pR_m}{R_p+R_m}+R_s}$

20. $$\dfrac{\dfrac{R_p}{R_p + R_s}}{\dfrac{R_p R_m}{R_p + R_m}} \bigg/ \dfrac{\dfrac{R_p R_m}{R_p + R_m} + R_s}$$

21. $$\dfrac{1}{r_p + \dfrac{r_c R_L}{r_c + R_L}}$$

22. $$\dfrac{\dfrac{rv}{r + R}}{\dfrac{rR}{r + R}}$$

23. $$\dfrac{V}{R} + \dfrac{\dfrac{v}{r}}{1 + \dfrac{r}{R}}$$

24. $$\dfrac{\dfrac{v}{r} + \dfrac{V}{r}}{1 - \dfrac{v}{r} \cdot \dfrac{V}{r}}$$

25. $$\dfrac{\dfrac{v^2}{R^2} - \dfrac{2v}{R} + 1}{1 - \dfrac{v^2}{R^2}}$$

26. $$\dfrac{\dfrac{1}{Z^2 - X^2} - \dfrac{1}{Z + X}}{\dfrac{1}{X} + \dfrac{1}{Z}}$$

Chapter 6

Equations with Fractions, and Applications — Parallel and Combination Circuits: Equations with Radicals

A computer microprocessor chip contains many parallel circuits. (Courtesy of INTEL Corporation)

6-1 CLEARING EQUATIONS OF FRACTIONS

When an equation contains fractions, solving it may become very tedious. We simplify the process if we first simplify the equation by performing the procedure called *clearing of fractions.*

We learned in Chapter 3 that if we performed the same operation on both members of an equation we do not destroy its equality but simply transform it to an equivalent equation. Hence, if we multiply both members of an equation containing fractions by its LCD we transform it into an equivalent equation without fractions. We solve the transformed equation using the most appropriate techniques (see Chapters 3 and 4).

Example 1 Clear the given equations of fractions and solve.

(a) $\dfrac{x}{2} - \dfrac{2x}{3} + \dfrac{5x}{9} = \dfrac{7}{9}$

(a) The LCD is 18. We multiply each term by 18:

$$\overset{9}{\cancel{18}}\cdot\frac{x}{\cancel{2}} - \overset{6}{\cancel{18}}\cdot\frac{2x}{\cancel{3}} + \overset{2}{\cancel{18}}\cdot\frac{5x}{\cancel{9}} = \overset{2}{\cancel{18}}\cdot\frac{7}{\cancel{9}}$$

$$9x - 12x + 10x = 14$$

Combine like terms,

$$7x = 14$$

and divide by the coefficient of x,

$$x = 2$$

Check root in original equation.

$$\frac{2}{2} - \frac{2\cdot2}{3} + \frac{5\cdot2}{9} = \frac{7}{9}$$

$$\frac{9 - 12 + 10}{9} = \frac{7}{9}$$

$$\frac{7}{9} = \frac{7}{9}$$

(b) $\dfrac{x+2}{3} - \dfrac{x-4}{5} = \dfrac{8}{15}$

(b) The LCD is 15. We multiply each term by 15,

$$\overset{5}{\cancel{15}}\cdot\frac{x+2}{\cancel{3}} - \overset{3}{\cancel{15}}\cdot\frac{x-4}{\cancel{5}} = \frac{8}{\cancel{15}}\cdot\cancel{15}$$

$$5(x+2) - 3(x-4) = 8$$

$$5x + 10 - 3x + 12 = 8$$

$$2x = 8 - 22 = -14$$

$$x = -7$$

Check root in original equation.

$$\frac{-7+2}{3} - \frac{-7-4}{5} = \frac{8}{15}$$

$$\frac{-25 + 33}{15} = \frac{8}{15}$$

$$\frac{8}{15} = \frac{8}{15}$$

Exercises 6-1 In Exercises 1–16 solve and check the given equations.

1. $\dfrac{x}{2} + \dfrac{3x}{5} = \dfrac{11}{10}$

2. $\dfrac{2V}{3} - \dfrac{6V}{7} = -\dfrac{16}{21}$

3. $\dfrac{2x}{3} - \dfrac{1}{2} = \dfrac{x}{6}$

4. $\dfrac{R}{2} + \dfrac{R}{5} = \dfrac{7}{5}$

5. $\dfrac{I}{3} - \dfrac{3I}{4} = -5$

6. $\dfrac{1}{3} + \dfrac{3}{5} = \dfrac{R}{15}$

7. $V - \dfrac{5}{6} = \dfrac{2V}{3}$

8. $\dfrac{V}{4} - \dfrac{2}{3} = \dfrac{5}{6} - \dfrac{V}{8}$

9. $\dfrac{I-15}{4} = -\dfrac{9}{2}$

10. $\dfrac{V-12}{6} + \dfrac{V-3}{3} + \dfrac{V}{4} = 0$

11. $\dfrac{3V+1}{2} = \dfrac{40}{5}$

12. $\dfrac{I-5}{2} = \dfrac{29-I}{4}$

13. $\dfrac{2R-4}{7} - \dfrac{6-6R}{5} = \dfrac{5R+2}{10}$

14. $\dfrac{V}{3} - 1 = \dfrac{V+1}{4}$

15. $\dfrac{3}{5}(x+1) - \dfrac{5x+2}{9} = \dfrac{7}{15}(3-x)$

16. $\dfrac{7R-9}{25} - \dfrac{2R+3}{5} = \dfrac{1}{3}(R+2) - \dfrac{4R+9}{15}$

6-2 EXTRANEOUS ROOTS

When the unknown appears in a denominator of an equation we clear fractions in the usual way. However, some equations of this form yield solutions which will not satisfy the original equation, even though all operations leading to the solution have been performed correctly.

Example 2 Solve and check the equation $\dfrac{x}{x-2} - 2 = \dfrac{3}{x-3}$.

Multiply each term by the *LCD* to clear of fractions,

$$(x - 3)\left(\frac{x}{x - 3}\right) - (x - 3) \cdot 2 = \frac{3}{x - 3} \cdot (x - 3)$$

$$x - 2(x - 3) = 3$$

Transpose and collect like terms,

$$x - 2x + 6 = 3$$
$$-x = -3$$
$$x = 3$$

Check in the original equation.

$$\frac{3}{3 - 3} - 2 = \frac{3}{3 - 3} \qquad \frac{3}{0} - 2 = \frac{3}{0}$$

The solution $x = 3$ is a true solution of the equation obtained after clearing fractions. It is an extraneous root of the original equation.

Checking solutions of equations by substituting roots into the original equation is important for two reasons: (1) to determine whether our work is correct, and (2) to ferret out extraneous roots.

<u>Exercises 6-2</u>   In Exercises 1–10, solve and check the given equations.

1. $\dfrac{2}{3R} + \dfrac{1}{6} = \dfrac{5}{6R}$ 2. $\dfrac{x}{x - 5} + 3 = \dfrac{5}{x - 5}$

3. $\dfrac{2I + 5}{12I} - \dfrac{7}{36}(I - 2) = \dfrac{5 - I}{8} - \dfrac{5I + 15}{72}$

4. $\dfrac{1}{R} + \dfrac{3}{2R} - \dfrac{5}{3R} = \dfrac{5}{6}$

5. $\dfrac{3V + 1}{V - 3} - \dfrac{5}{V} = 3$ [Hint: LCD = $V(V - 3)$]

6. $\dfrac{5}{7 - i} + \dfrac{3}{i} - \dfrac{5i}{i^2 - 7i} = \dfrac{-4}{i - 7}$
   [LCD = $i(i - 7)$]

7. $\dfrac{3R + 1}{R - 3} + \dfrac{5}{R} = 3$ 8. $\dfrac{1}{v} + \dfrac{3}{2v} - \dfrac{5}{3v} = \dfrac{5}{6}$

9. $b = \dfrac{a - i}{i - a}$ (Solve for i)

10. $\dfrac{3v - 4}{2v} - \dfrac{2v - 7}{7v} + \dfrac{7v - 3}{14} = \dfrac{4v + 7}{8}$

6-3 LITERAL EQUATIONS CONTAINING FRACTIONS

We recall from Section 4-3 that literal equations are those in which quantities other than the unknown are represented with letters. Having learned a number of new skills concerning operations with algebraic expressions

we are now prepared to learn to manipulate literal equations and formulas of greater complexity. This is one of the most important skills we can learn in mathematics for application in electronics.

As we saw in Section 4-3 we solve literal equations using the same techniques applied to numerical equations. When the literal equation contains fractions we start by clearing it of fractions.

Example 3

(a) Solve for x: $\dfrac{x}{a} + \dfrac{2x}{b} = c$

Multiply both members by ab, the LCD,

$$\not{ab} \cdot \frac{x}{\not{a}} + a\not{b} \cdot \frac{2x}{\not{b}} = ab \cdot c$$

Collect like terms,

$$(2a + b)x = abc$$

Divide both members by $2a + b$, the coefficient of x,

$$x = \frac{abc}{2a + b}$$

Check the solution by substituting $\dfrac{abc}{2a + b}$ for x in the original equation,

$$\frac{\not{a}bc}{\not{a}(2a + b)} + \frac{2a\not{b}c}{\not{b}(2a + b)} = c$$

$$\frac{2ac + bc}{2a + b} = c \qquad \frac{c(2a + b)}{2a + b} = c$$

$$c = c$$

(b) Solve for r_p: $A = \dfrac{\mu R_L}{r_p + R_L}$

Multiply both members by $r_p + R_L$,

$$A(r_p + R_L) = \frac{\mu R_L}{\not{r_p + R_L}} \cdot \frac{\not{r_p + R_L}}{1}$$

$$Ar_p + AR_L = \mu R_L$$

Transpose AR_L,

$$Ar_p = \mu R_L - AR_L$$

Factor R_L from the right member and divide both members by A, the coefficient of r_p,

$$r_p = R_L \frac{(\mu - A)}{A} = R_L \left(\frac{\mu}{A} - 1\right)$$

Check by substituting the result for r_p in the original formula,

$$A = \frac{\mu R_L}{R_L\left(\dfrac{\mu}{A} - 1\right) + R_L}$$

Divide the numerator and denominator by R_L which is common to all terms of the fraction

$$A = \frac{\mu}{\dfrac{\mu}{A} - 1 + 1} = \frac{\cancel{\mu}}{\dfrac{\cancel{\mu}}{A}} = A$$

Exercises 6-3 In Exercises 1–25 solve the given equation or formula for the indicated letter(s).

1. $I = \dfrac{V}{R}$, for V, R

2. $I = \dfrac{V}{R + r}$, for V, R, r

3. $R = \dfrac{V}{I}$, for I, V

4. $T = \dfrac{1}{f}$, for f

5. $f = \dfrac{1}{T}$, for T

6. $\lambda = \dfrac{c}{f}$, for f

7. $X_C = \dfrac{1}{2\pi f C}$, for f, C

8. $V_1 = \dfrac{R_1}{R_T} V_T$, for R_1, R_T, V_T

9. $I_2 = \dfrac{R_1}{R_2} I_1$, for R_1, R_2, I_1

10. $I_1 = \dfrac{R_2}{R_1 + R_2} I_T$, for R_1, R_2, I_T

11. $R_T = \dfrac{R_1 R_2}{R_1 + R_2}$, for R_1, R_2

12. $R_2 = \dfrac{V_T}{I} - R_1$, for V_T, R_1, I

13. $\dfrac{1}{R_T} = \dfrac{1}{R_1} + \dfrac{1}{R_2}$, for R_T, R_1

14. $\dfrac{1}{C_T} = \dfrac{1}{C_1} + \dfrac{1}{C_2}$, for C_T, C_2

15. $\beta = \dfrac{\alpha}{1 - \alpha}$, for α

16. $A' = \dfrac{A}{1 - A\beta}$, for A, β

17. $r_i = \dfrac{V_{NL} - V_L}{I_L}$, for V_{NL}, I_L

18. $VR = \dfrac{V_{NL} - V_L}{V_L}$, for V_{NL}, V_L

19. $A = \dfrac{\mu R_L}{r_p + R_L}$, for r_p, R_L

20. $A_i = \dfrac{h_{21}}{h_{22} R_L + 1}$, for h_{21}, h_{22}

21. $t = \dfrac{\theta}{360} \cdot \dfrac{1}{f}$, for θ, f

22. $T_C = \dfrac{5}{9}(T_F - 32°)$, for T_F

23. $Q = \dfrac{1}{2\pi f C R}$, for C

24. $d = \dfrac{\omega - \omega_0}{\omega_0}$, for ω, ω_0

25. $\omega_1 = \omega_0 - \dfrac{\omega_0}{2Q}$, for Q, ω_0

6-4 PARALLEL CIRCUITS

The performance of a parallel circuit (see Figure 6-1) can be outlined with two statements of basic principles:

1. In a parallel circuit the voltage is the same across all branches*

$$V_T = V_1 = V_2 = V_3 = \text{and so on}$$

2. The total current is the sum of the branch currents

$$I_T = I_1 + I_2 + I_3 + \cdots$$

Using these two basic principles and Ohm's law we can develop algebraically all of the well-known formulas for parallel circuits.

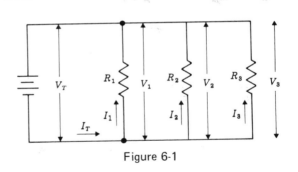

Figure 6-1

Example 4 Derive the reciprocal formula for resistances in parallel.

We start with the equation for the total current in a parallel circuit,

$$I_T = I_1 + I_2 + I_3 + \cdots$$

Then, substituting the Ohm's law equivalent for each current, we have

$$\frac{V_T}{R_T} = \frac{V_1}{R_1} + \frac{V_2}{R_2} + \frac{V_3}{R_3} + \cdots$$

*In a parallel circuit a *branch* is any circuit path from one of the common sides of the circuit to the other common side.

But since

$$V_T = V_1 = V_2 = V_3 = \cdots$$

we substitute V_T for V_1, V_2, V_3, and so on,

$$\frac{V_T}{R_T} = \frac{V_T}{R_1} + \frac{V_T}{R_2} + \frac{V_T}{R_3} + \cdots$$

and divide by the common factor V_T,

$$\frac{1}{R_T} = \frac{1}{R_1} + \frac{1}{R_2} + \frac{1}{R_3} + \cdots$$

Example 5   The circuit of Figure 6-1 has the following values: $V_T = 12$ V, $R_1 = 27$ kΩ, $R_2 = 39$ kΩ, and $R_3 = 68$ kΩ. (a) Calculate the predicted branch currents, I_1, I_2, and I_3 and I_T. (b) Calculate the power that would be dissipated by each resistor.

(a) $I_1 = \dfrac{V_T}{R_1} = \dfrac{12}{27 \times 10^3} = 4.44 \times 10^{-4}$ A

$I_1 = 444.4\ \mu A$

$I_2 = \dfrac{V_T}{R_2} = \dfrac{12}{39 \times 10^3} = 3.077 \times 10^{-4}$ A

$I_2 = 307.7\ \mu A$

$I_3 = \dfrac{V_T}{R_3} = \dfrac{12}{68 \times 10^3} = 1.765 \times 10^{-4}$ A

$I_3 = 176.5\ \mu A$

$I_T = I_1 + I_2 + I_3 = 444.4 + 307.7$
$\qquad + 176.5 = 928.6\ \mu A$

(b) $P_1 = \dfrac{V_T}{R_1} = \dfrac{(12)^2}{27 \times 10^3} = 5.333 \times 10^{-3}$ W

$P_1 = 5.333\ mW$

$P_2 = \dfrac{V_T}{R_2} = \dfrac{144}{39 \times 10^3} = 3.692 \times 10^{-3}$ W

$P_2 = 3.692\ mW$

$P_3 = \dfrac{V_T}{R_3} = \dfrac{144}{68 \times 10^3} = 2.118 \times 10^{-3}$ W

$P_3 = 2.118\ mW$

Example 6   Show the algebraic steps required to develop the formula

$$R_T = \frac{1}{\dfrac{1}{R_1} + \dfrac{1}{R_2} + \dfrac{1}{R_3} + \cdots}$$

given that

$$R_T = \frac{V_T}{I_T} \text{ and } I_T = \frac{V_T}{R_1} + \frac{V_T}{R_2} + \frac{V_T}{R_3} + \cdots$$

We start with $R_T = \dfrac{V_T}{I_T}$ and substitute the equivalent of I_T; thus,

$$R_T = \frac{V_T}{\dfrac{V_T}{R_1} + \dfrac{V_T}{R_2} + \dfrac{V_T}{R_3} + \cdots}$$

V_T is a common factor of the terms of the denominator,

$$R_T = \frac{\cancel{V_T}}{\cancel{V_T}\left(\dfrac{1}{R_1} + \dfrac{1}{R_2} + \dfrac{1}{R_3} + \cdots\right)}$$

Therefore,

$$R_T = \frac{1}{\dfrac{1}{R_1} + \dfrac{1}{R_2} + \dfrac{1}{R_3} + \cdots}$$

Example 7   Use the formula of Example 6 and a calculator to determine R_T for a circuit with the following resistors in parallel: $R_1 = 1.2$ kΩ, $R_2 = 820\ \Omega$, $R_3 = 1.8$ kΩ, $R_4 = 2.7$ kΩ. We follow the format, formula, substitution, solution:

$$R_T = \frac{1}{\dfrac{1}{R_1} + \dfrac{1}{R_2} + \dfrac{1}{R_3} + \dfrac{1}{R_4}}$$

$$= \frac{1}{\dfrac{1}{1.2 \times 10^3} + \dfrac{1}{820} + \dfrac{1}{1.8 \times 10^3} + \dfrac{1}{2.7 \times 10^3}}$$

For some calculators which have the [$1/x$] function available, a most effective and rapid method of evaluating a problem of this type, a very common electronics problem, is sometimes possible as follows: (Consult the instruction manual for your calculator.)

Enter: 1.2, EE, 3, $1/x$, +, 820, $1/x$, +, 1.8, EE, 3, $1/x$, +, 2.7, EE, 3, = (Readout = 2.97877 $\times 10^{-3}$), $1/x$, Readout: 335.70889

$$R_T = \underline{335.7\ \Omega}$$

Study the calculator operations of Example 7 very carefully and note especially the following:

1. Depressing the [$1/x$] key immediately calculates the reciprocal of a number and allows this reciprocal value to be used in other operations (addition, in this case) in chained calculations.

2. The [=] key provides the "sum of the reciprocals" in this example.

3. Since R_T is the "reciprocal of the 'sum of the reciprocals' of the individual resistors" we only need to use the [1/x] key again (on the readout of the sum of the reciprocals) to obtain R_T.

A Valuable Shortcut for Calculations The value of R_T for a parallel circuit will be in the same unit as the values of the individual resistors if they are all in the same unit, that is, all Ω's or all in MΩ's. Therefore, fewer calculator entries will be required if all resistor values are converted to the same unit and only the significant digits (the mantissa) of the unit values used. Example 6 would then be set up as follows:

$$R_T = \frac{1}{\frac{1}{1.2} + \frac{1}{0.82} + \frac{1}{1.8} + \frac{1}{2.7}} \text{ k}\Omega$$

Calculator: 1.2, 1/x, +, 0.82, 1/x, +, 1.8, 1/x, +, 2.7, 1/x, =, 1/x, Readout = 0.3357

$$R_T = 0.3357 \text{ k}\Omega = \underline{335.7 \ \Omega}$$

Exercises 6-4 In Exercises 1–12 use Ohm's law and the basic principles of a parallel circuit, whenever appropriate, to obtain the required derivations.

1. Show the algebraic steps required to obtain the formula for two resistors in parallel, $\frac{R_1 R_2}{R_1 + R_2}$, from the "reciprocal" formula,
$$\frac{1}{R_T} = \frac{1}{R_1} + \frac{1}{R_2} + \frac{1}{R_3} + \cdots$$

2. Starting with the equation $I_T = I_1 + I_2 + I_3 + \cdots$ show the steps for developing the equation which supports the statement, "the total power in a parallel circuit is equal to the sum of the powers of the branches."

3. Derive the equation for the total resistance of n equal resistances, of value R, in parallel:
$$R_T = \frac{R}{n}$$

4. Show the algebraic steps for developing the equation

$$\frac{I_1}{I_2} = \frac{R_2}{R_1}$$

for a parallel circuit.

5. In calculating R_T for a parallel circuit some professionals prefer to use the equation
$$R_T = \frac{R_1}{1 + \frac{R_1}{R_2} + \frac{R_1}{R_3} + \cdots}$$

Show how to derive this formula from the basic reciprocal formula (see Exercise 1).

6. If you knew the total current and three of the four branch currents of a four-branch parallel circuit, how would you calculate the fourth current? Derive the equation.

7. In a two-branch parallel circuit you know I_1, R_1, and I_T. Develop equations for calculating V_T, R_2, and I_2.

8. For a parallel circuit you know V_T and I_T and need to increase I_T by 5 mA without changing V_T or any part of the original circuit. How would you do it? Develop any equations you would use.

9. Conductance is defined as the reciprocal of resistance: $G = \frac{1}{R}$. Show the steps for obtaining the equation $G_T = G_1 + G_2 + G_3 + \cdots$ for a parallel circuit.

10. Demonstrate algebraically that for a series circuit,
$$\frac{1}{G_T} = \frac{1}{G_1} + \frac{1}{G_2} + \frac{1}{G_3} + \cdots$$
(See Exercise 9.)

11. Derive the "current divider" formula for a two-branch parallel circuit,
$$I_1 = \frac{R_2}{R_1 + R_2} I_T$$

12. A resistor R_x is to be connected in parallel with R to achieve a desired R_T. Derive the equation
$$R_x = \frac{R \cdot R_T}{R - R_T}$$

In Exercises 13–22 the V_T- and R-values are for a parallel circuit. (a) Draw and label the circuit diagram. (b) Calculate the predicted branch currents and the total current. (c) Calculate the power that would be dissipated by each resistor. (Use the memory function of your calculator wherever possible to save keystrokes.)

13. $V_T = 10$ V, $R_1 = 10$ Ω, $R_2 = 20$ Ω

14. $V_T = 0.58$ V, $R_1 = 680$ Ω, $R_2 = 480$ Ω

15. $V_T = 750$ mV, $R_1 = 1.8$ kΩ, $R_2 = 2.7$ kΩ, $R_3 = 910$ Ω

16. $V_T = 875$ mV, $R_1 = 22$ kΩ, $R_2 = 33$ kΩ, $R_3 = 41$ kΩ

17. $V_T = 750$ μV, $R_1 = 68$ Ω, $R_2 = 56$ Ω, $R_3 = 120$ Ω

18. $V_T = 37.5$ μV, $R_1 = 910$ Ω, $R_2 = 1.2$ kΩ, $R_3 = 2.4$ kΩ

19. $V_T = 12.5$ V, R-values of 8.2 kΩ, 15 kΩ, 82 kΩ, and 39 kΩ

20. $V_T = 27.5$ V, R-values of 620 kΩ, 120 kΩ, 910 kΩ, and 1.2 MΩ

21. $V_T = 1.55$ mV, R-values of 820 Ω, 560 Ω, 470 Ω, 390 Ω, and 1.2 kΩ

22. $V_T = 3.07$ mV, R-values of 12 kΩ, 2.7 kΩ, 8.2 kΩ, 5.6 kΩ, and 18 kΩ

In Exercises 23–30 calculate R_T for a parallel circuit containing the given R-values. (See method of Example 7.)

23. 47 Ω, 27 Ω, 91 Ω

24. 180 Ω, 430 Ω, 510 Ω 820 Ω

25. 9.1 kΩ, 12 kΩ, 33 kΩ, 20 kΩ

26. 560 kΩ, 1.0 MΩ, 1.8 MΩ, 910 kΩ

27. 820 Ω, 15 kΩ, 470 kΩ, 1.2 MΩ

28. 680 Ω, 470 Ω, 390 Ω, 910 Ω, 1.2 kΩ

29. 390 kΩ, 560 kΩ, 750 kΩ, 910 kΩ, 2.2 MΩ

30. 9.1 kΩ, 8.2 kΩ, 5.6 kΩ, 4.7 kΩ, 3.3 kΩ, 2.4 kΩ

In Exercises 31–40 solve the given problems involving parallel circuits.

31. In a parallel circuit containing four resistors, $I_T = 27.5$ mA, $I_2 = 2.3$ mA, $I_3 = 5.6$ mA, and $I_4 = 11.3$ mA. Determine I_1.

32. If $P_T = 75$ mW, $P_1 = 12$ mW, $P_3 = 27$ mW, and $P_4 = 21$ mW in a four branch parallel circuit, what is P_2?

33. A 22-kΩ resistor is dissipating 3.68 mW as a branch of a parallel circuit. What is V_T for the circuit?

34. If $I_2 = 6.86$ mA $= 5.67 I_1$ and $R_1 = 6.8$ kΩ, in a parallel circuit, what is the value of R_2? of V_T?

35. If $R_1 = 39$ kΩ, $I_1 = 37.5$ μA, $R_2 = 18$ kΩ, and $R_3 = 68$ kΩ in a three-branch parallel circuit, determine: (a) V_T; (b) I_2, I_3, and I_T; (c) R_T.

36. Repeat Exercise 27 for $R_1 = 1.2$ MΩ, $I_1 = 350$ nA, $R_2 = 910$ kΩ, $R_3 = 3.3$ MΩ.

37. A parallel circuit with $R_T = 4.56$ kΩ and $I_T = 2.75$ mA is to be altered so that $I_T = 4.594$ mA. However, neither V_T nor any of the original branches are to be changed. (a) How would you accomplish the desired result? Give equations and calculate values. (b) What is R_T after the change?

38. A power supply set for 5.8 V is supplying a total of 37.65 mA to several transistor stages connected in parallel. (a) What R_T is the overall circuit equivalent to? (b) A technician unsolders a one-resistor branch of the circuit and the current drops to 36.42 mA. What was the value of the resistor unsoldered?

39. A television receiver power supply is protected by a 0.5-A fuse. The fuse blows and a test of the three parallel branches fed by the fuse reveals that their normal currents are 45 mA, 150 mA, and 185 mA. What is the minimum overload current (the current in excess of normal current) that could have caused the fuse to blow?

40. An isolation transformer in a TV repair shop is rated at 1.5 kW. Already connected (in parallel) to the transformer are two color TV receivers and a black-and-white receiver which draw 450 W, 375 W, and 175 W, respectively. Is it safe to connect a video tape recorder, which consumes 575 W, to the transformer without unplugging any of the TV sets? Support your conclusion with relevant equation(s) and numbers.

6-5   SERIES-PARALLEL CIRCUITS

Most actual circuits of the real world of electronics are neither purely series circuits nor parallel circuits. They are combinations of these two circuit forms. And, since there is virtually an unlimited number of possible circuit combinations, it is impossible to state a simple 1-2-3-4 routine for analyzing all such circuits. Gaining a thorough understanding of the worked-out analyses of a few common configurations is a most effective means of learning the general approach to the analysis of compound circuits. Study the following examples carefully.

Example 8 Find R_T for the circuit of Figure 6-2(a); that is, find the total (or equivalent) resistance seen "looking into" the circuit at terminals a and b.

The circuit is a three-branch parallel circuit. Each branch is a series circuit. Our approach is to convert each branch to a single resistance equivalent to the series circuit of its branch. This converts the circuit into a simple parallel circuit [see Figure 6-2(b)] which can be analyzed with the techniques of Section 6-4. Let us call the total resistance of the left branch R_{TL}. Then

$$R_{TL} = R_1 + R_2 = 4 + 6 = 10 \ \Omega$$

Similarly, for the center branch,

$$R_{TC} = R_3 + R_4 + R_5 = 2 + 4 + 6 = 12 \ \Omega$$

and, for the right branch,

$$R_{TR} = R_6 + R_7 = 8 + 6 = 14 \ \Omega$$

Then, for the equivalent parallel circuit,

$$\frac{1}{R_T} = \frac{1}{R_{TL}} + \frac{1}{R_{TC}} + \frac{1}{R_{TR}}$$

Substituting known values for letters,

$$\frac{1}{R_T} = \frac{1}{10} + \frac{1}{12} + \frac{1}{14}$$

(See Example 7 for a method of calculator evaluation of this type of problem.) Then

$$R_T = \frac{1}{0.2547618} = \underline{3.925 \ \Omega}$$

Example 9 Find R_T for the circuit of Figure 6-3(a)—the resistance looking in at terminals a, b.

Here the circuit consists of three parallel circuits, called *banks*, connected in series. We convert each bank into an equivalent single resistance which produces a series circuit of three resistances [see Figure 6-3(b)]. We treat the series circuit with the techniques of Section 4-5.

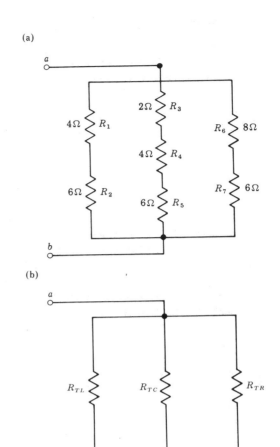

(a)

(b)

Figure 6-2

(a)

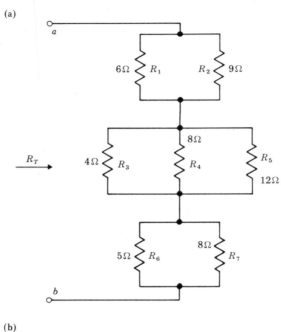

(b)

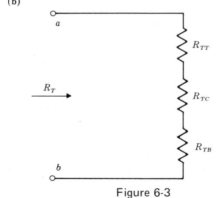

Figure 6-3

Let us call the equivalent resistance for the top bank R_{TT}. Then

$$R_{TT} = \frac{R_1 R_2}{R_1 + R_2} = \frac{6 \cdot 9}{6 + 9} = 3.60 \ \Omega$$

Similarly, for the center bank,

$$\frac{1}{R_{TC}} = \frac{1}{R_3} + \frac{1}{R_4} + \frac{1}{R_5} = \frac{1}{4} + \frac{1}{8} + \frac{1}{12}$$
$$R_{TC} = 2.1818 \ \Omega$$

For the bottom bank,

$$R_{TB} = \frac{R_6 R_7}{R_6 + R_7} = \frac{5 \cdot 8}{5 + 8} = 3.0769 \ \Omega$$

Analyzing the equivalent series circuit of Figure 6-3(b),

$$R_T = R_{TT} + R_{TC} + R_{TB}$$
$$= 3.60 + 2.1818 + 3.0769$$
$$R_T = \underline{8.859 \ \Omega} \quad \text{(rounded to four significant digits)}$$

Example 10   Find the resistance seen looking in at terminals a, b of the circuit of Figure 6-4(a).

The circuit has one resistance R_1 in series with a parallel bank, $R_2 \parallel R_3$. (The symbol $\parallel$ means "in parallel with.") The first step is to convert the parallel bank to a single equivalent resistance which we will arbitrarily designate R_{23},

$$R_{23} = R_2 \parallel R_3 = \frac{R_2 R_3}{R_2 + R_3} = \frac{10 \cdot 15}{10 + 15} = 6 \ \Omega$$

Then, analyzing the series circuit of Figure 6-4(b),

$$R_T = R_1 + R_{23} = 6 + 6 = 12 \ \Omega \quad R_T = \underline{12 \ \Omega}$$

Example 11   Find the R_T for the circuit of Figure 6-5(a), as seen at input terminals a, b.

This circuit is a very common one in electronics—a "π" circuit (R_1, R_2, and R_3) with load R_L. Redrawing the circuit [see Figure 6-5(b)], we see that it is a two-branch parallel circuit with R_1 as one branch and R_2 in series with the parallel bank $R_3 \parallel R_L$ as the second branch. First, we find the equivalent for $R_3 \parallel R_L$, R_{3L},

$$R_{3L} = R_3 \parallel R_L$$
$$= \frac{R_3 R_L}{R_3 + R_L} = \frac{30 \cdot 40}{30 + 40} = 17.143 \ \Omega$$

Next, combine R_2 and R_{3L},

$$R_{23L} = R_2 + R_{3L} = 10 + 17.143 = 27.143 \ \Omega$$

Finally, analyze $R_1 \parallel R_{23L}$,

$$R_T = R_1 \parallel R_{23L} = \frac{24 \cdot 27.143}{24 + 27.143} = 12.737 \ \Omega$$

$$R_T = \underline{12.74 \ \Omega}$$

Example 12 Find R_T for the circuit of Figure 6-6(a) as seen at input terminals a, b.

Like the circuit in Figure 6-5, the circuit can be made simpler to analyze if we redraw it, as in Figure 6-6(b). We see that it is basically a two-branch parallel circuit with R_1 as one branch in parallel with a branch consisting of R_2 in series with two parallel banks. We proceed by simplifying the two banks. For the lower bank,

$$R_{TL} = \frac{R_6(R_7 + R_8)}{R_6 + R_7 + R_8} = \frac{16(12 + 10)}{16 + 12 + 10}$$
$$= 9.263 \ \Omega$$

and for the upper bank,

$$R_{TU} = \frac{R_5(R_3 + R_4)}{R_5 + R_3 + R_4} = \frac{10(8 + 14)}{10 + 8 + 14}$$
$$= 6.875 \ \Omega$$

The right branch of the circuit is now a series circuit consisting of R_2, R_{TU}, and R_{TL} [see Figure 6-6(c)]. The total resistance of the right branch, R_{RB}, is

$$R_{RB} = R_2 + R_{TU} + R_{TL} = 5 + 6.875 + 9.263$$
$$= 21.138 \ \Omega$$

Then

$$R_T = \frac{R_1 R_{RB}}{R_1 + R_{RB}} = \frac{20 \cdot 21.138}{20 + 21.138} = 10.2766 \ \Omega$$

$$R_T = \underline{10.28 \ \Omega}$$

Frequently in electronics finding the total (equivalent) resistance of some complex compound circuit such as that of Figure 6-6 is only a step toward finding the current through just one particular resistor in the circuit. The next example illustrates the procedure. Study it carefully until the reasons for the various steps are clear.

Example 13 Find the current through resistor R_7 of Figure 6-6 when 6.4 V is applied between terminals a and b.

Referring to Figure 6-6(c) we see that the 6.4 V will be applied across the right branch

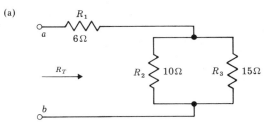

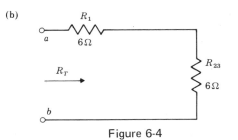

Figure 6-4

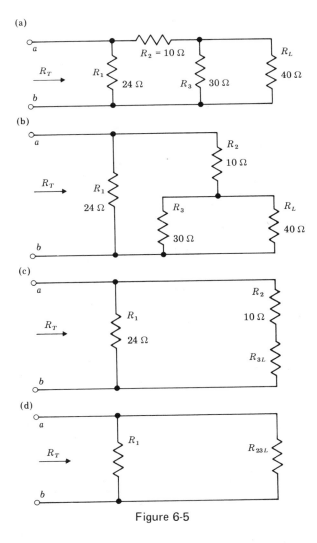

Figure 6-5

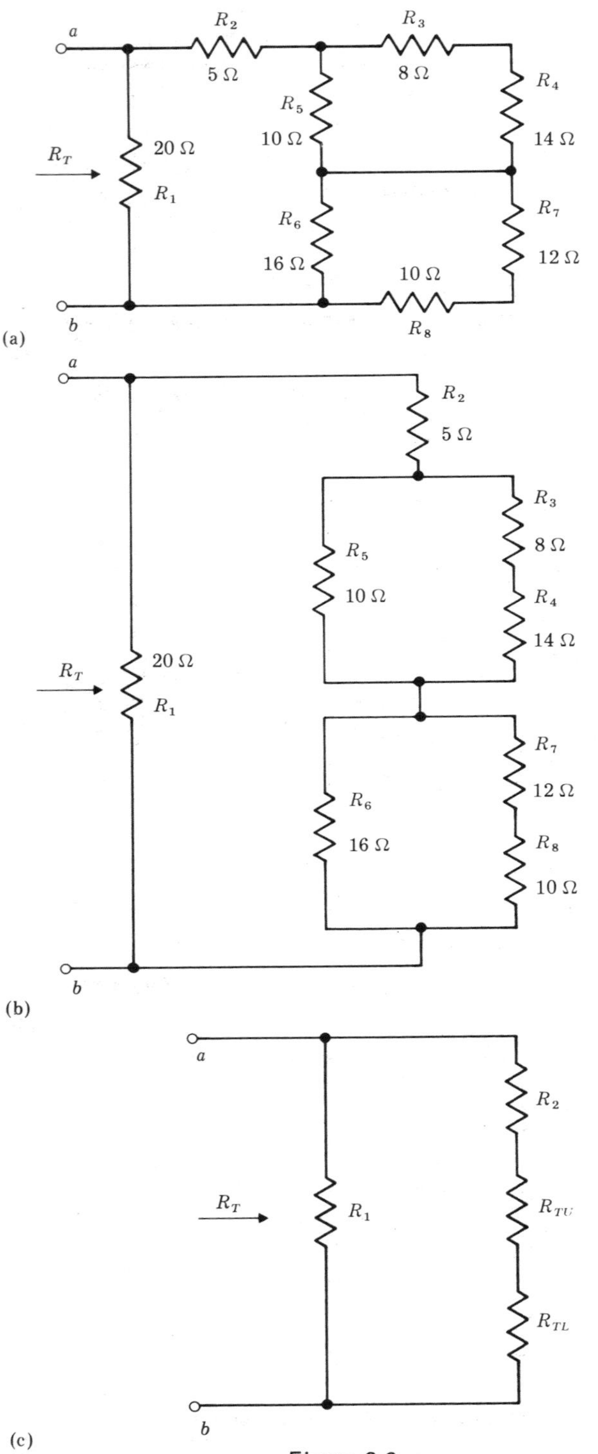

(a)

(b)

(c)

Figure 6-6

consisting of R_2, R_{TU}, and R_{TL} in series. The current through that branch will then be

$$I_{RB} = \frac{V_T}{R_2 + R_{TU} + R_{TL}} = \frac{6.4}{21.138} = 0.3028 \text{ A}$$

The voltage across the lower bank is equal to the product of the right branch current, I_{RB}, and the total resistance of the lower bank, R_{TL},

$$V_{LB} = I_{RB} \cdot R_{TL} = 0.3028 \cdot 9.263 = 2.805 \text{ V}$$

The current through R_7 and R_8, which are in series, is then

$$I_7 = I_8 = \frac{V_{LB}}{R_7 + R_8} = \frac{2.805}{12 + 10}$$

$$I_7 = \underline{0.1275 \text{ A}}$$

Exercises 6-5

1. Determine R_T at terminals a, b for the circuit of Figure 6-7, given the following resistor values: $R_1 = 24 \ \Omega$, $R_2 = 20 \ \Omega$, $R_3 = 28 \ \Omega$, $R_4 = 60 \ \Omega$, $R_5 = 40 \ \Omega$, $R_6 = 24 \ \Omega$.

2. What is R_T looking in at terminals a and b of the circuit of Figure 6-7 if $R_1 = 1.2 \text{ k}\Omega$, $R_2 = 1.8 \text{ k}\Omega$, $R_3 = 2.4 \text{ k}\Omega$, $R_4 = 4.8 \text{ k}\Omega$, $R_5 = 2.7 \text{ k}\Omega$, and $R_6 = 3.9 \text{ k}\Omega$?

3. Determine the power that will be dissipated in resistor R_6 of the circuit of Exercise 1 if a source $V_T = 18.6$ V is connected between terminals a and b.

4. Predict the current flow through resistor R_2 if a voltage source $V_s = 27.5$ mV is applied to the terminals a, b of the circuit of Exercise 2.

5. Determine R_T at the terminals X_1, X_2 for the circuit of Figure 6-8 if $R_1 = 12 \ \Omega$, $R_2 = 15 \ \Omega$, $R_3 = R_4 = 24 \ \Omega$, $R_5 = 18 \ \Omega$, $R_6 = 9 \ \Omega$, $R_7 = 7 \ \Omega$, $R_8 = 16 \ \Omega$.

6. Repeat Exercise 5 for a circuit (Figure 6-8) with the values $R_1 = 22 \text{ k}\Omega$, $R_2 = 33 \text{ k}\Omega$, $R_3 = R_4 = R_5 = 39 \text{ k}\Omega$, $R_6 = 12 \text{ k}\Omega$, $R_7 = 24 \text{ k}\Omega$, $R_8 = 36 \text{ k}\Omega$.

7. What voltage will appear across R_6 of the circuit of Exercise 5 if a source of $V_s = 750 \ \mu$V is applied between terminals X_1, X_2?

8. Determine the voltage across R_7 of the circuit of Exercise 6 if R_6 becomes

shorted while a source voltage of 65 mV is being applied between input terminals X_1, X_2.

9. Determine I_3 (current through R_3) for the circuit of Figure 6-9.

10. What is I_3 for the circuit of Exercise 9 if R_2 opens?

11. Determine voltage across R_3 of the circuit of Exercise 9 if a resistor $R_4 = 10 \text{ k}\Omega$ is connected in parallel with R_2, R_3.

12. Determine power dissipated in R_1 of the circuit of Exercise 9, (a) when circuit values are as given, and (b) with R_2 shorted.

13. Determine P_L, the power that will be dissipated in the resistor R_L of the circuit in Figure 6-10.

14. What is P_L for the circuit of Exercise 13 if R_3 opens?

15. Assume that in the circuit of Exercise 13 we desire to make R_T, the resistance seen by the 1.55-V source, equal to approximately 5.1 kΩ. We will change R_1 but leave the rest of the circuit unchanged. What value should we use for R_1?

16. Refer to Figure 6-10. What value of resistor would we have to use as R_2 to increase the voltage across R_L to 1.0 V?

17. The voltage (V_{out}) across R_8 in the circuit of Figure 6-11 is 1.8 V. Determine V_s.

18. What total current would the source V_s of Exercise 17 supply?

19. Refer to Exercise 17. What total current would V_s supply if the output terminals O, O' were shorted? (Use the value of V_s found in Exercise 17.)

20. Let V_s for the circuit of Figure 6-11 be as determined in Exercise 17. What change will occur in the current through R_2 if R_7 becomes shorted? Calculate I_2 for the two conditions: R_7 normal and R_7 shorted.

6-6 EQUATIONS CONTAINING RADICALS

A number of the formulas we use in electronics contain the radical sign $\sqrt{}$; for example,

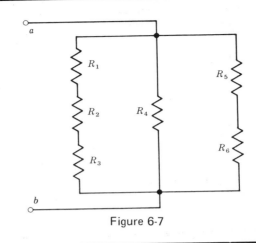

Figure 6-7

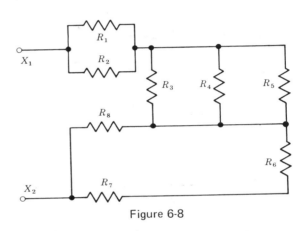

Figure 6-8

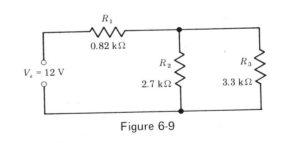

Figure 6-9

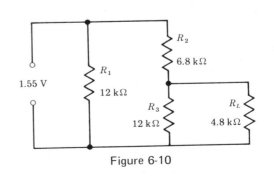

Figure 6-10

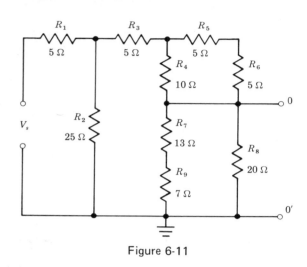

Figure 6-11

$$f_r = \frac{1}{2\pi \sqrt{LC}},\qquad \text{formula for resonant frequency}$$

$$M = k\sqrt{L_1 L_2},\qquad \text{formula for mutual inductance}$$

When our task is to solve such formulas for one of the quantities under the radical sign, an effective procedure is as follows:

1. Isolate the radical sign with its contents to one side of the equation.
2. Square both members of the equation.
3. Solve the equation.

Example 14   Solve $f_r = \dfrac{1}{2\pi \sqrt{LC}}$ for L.

We multiply both members by $\sqrt{LC}$ and divide by f_r $\left(\text{multiply by } \dfrac{\sqrt{LC}}{f_r}\right)$.

$$\cancel{f_r} \cdot \frac{\sqrt{LC}}{\cancel{f_r}} = \frac{1}{2\pi \sqrt{\cancel{LC}}} \cdot \frac{\sqrt{\cancel{LC}}}{f_r}$$

Square both members,

$$LC = \frac{1}{4\pi^2 f_r^2}$$

Divide by C.

$$L = \frac{1}{4\pi^2 f_r^2 C}$$

Exercises 6-6   In Exercises 1–14 solve the given equations for the indicated letter(s).

1. $f_r = \dfrac{1}{2\pi \sqrt{LC}}$, for C

2. $M = k\sqrt{L_1 L_2}$, for L_1

3. $N_1 = \sqrt{\dfrac{Z_1}{Z_2}}\, N_2$, for Z_1

4. $Z = \sqrt{R^2 + X^2}$, for R^2

5. $Q_s = \dfrac{\sqrt{L/C}}{R}$, for L

6. $Q_p = \dfrac{R}{\sqrt{L/C}}$, for C

7. $n = \dfrac{\sqrt{L(9r + 10l)}}{r^2}$, for L, l

8. $Z_0 = \sqrt{Z_1 Z_2}$, for Z_1

9. $i = \sqrt{\dfrac{p}{r + R}}$, for p, r

10. $n = \sqrt{1 - \dfrac{k}{f^2}}$, for k, f^2

11. $Z_0 = \sqrt{\dfrac{L}{C}}$, for C 12. $d = \sqrt{2h}$, for h

13. $V_z = \sqrt{\dfrac{2eV_A}{m}}$, for e, m

14. $u = \sqrt{V_0^2 + \dfrac{2ev}{m}}$, for e, m

Chapter 7
Ratios, Variation, and Percent

A baseball player's batting average is the ratio of "hits" to "times at bat." (Kent Reno/Jeroboam, Inc.)

7-1 RATIOS

A ratio is one number divided by another number. (This is also the definition of a fraction. In fact, a fraction in which the numerator and denominator are integers is called a *rational* number.) Ratios are very common in electronics. A few examples are: amplifier gain (ratio of output to input); transformer winding ratio (secondary turns divided by primary turns); decibels (10 times the logarithm of the power ratio); h_{fe} (ratio of collector current to base current in a transistor); g_m (ratio of an output current to an input voltage).

A ratio is used to make comparisons. For example, we may hear it said, "amplifier X has a voltage gain of 500." This means that if we compare the output voltage of the amplifier to its input voltage, by division, the result is 500. That is, $v_o \div v_i = 500$. The quantities being compared in a ratio may be similar quantities, for example, voltages, currents, resistances, turns of transformer windings, and so on. When quantities are similar they must be expressed in identical units. And, the units cancel—the ratio is a pure number.

Example 1 Express the ratios in simplest form.

(a) 2 V to 40 mV

(a) $\dfrac{2\,\cancel{V}}{0.04\,\cancel{V}} = 50$

(b) 10 mA to 50 μA

(b) $\dfrac{10,000\,\cancel{\mu A}}{50\,\cancel{\mu A}} = 200$

(c) 15 kΩ to 45,000 Ω

(c) $\dfrac{15\,\cancel{k\Omega}}{45\,\cancel{k\Omega}} = \dfrac{1}{3}$ or 0.3333

(Note: The units cancel in these ratios leaving a pure number as the result.)

Ratios may be written to compare quantities that are dissimilar. In these instances the ratio will require units, usually of the form: miles per gallon (mi/gal), ohms per volt (Ω/V), and so on.

Example 2 Express the ratios in simplest form and affix appropriate units.

(a) 200 miles on 10 gallons of gasoline

(a) $\dfrac{200\text{ mi}}{10\text{ gal}} =$ 20 mi/gal

(b) 10,000 Ω for 0.5 V

(b) $\dfrac{10,000\ \Omega}{0.5\text{ V}} =$ 20,000 Ω/V

(c) 10 mA for 0.4 V

(c) $\dfrac{10\text{ mA}}{0.4\text{ V}} =$ 25 mA/V or 25 mS

(d) 10 V for 2 mA

(d) $\dfrac{10\text{ V}}{2\text{ mA}} =$ 5000 V/A or 5000Ω

These ratios must have units since the quantities being compared are dissimilar (units do not cancel).

Exercises 7-1 In Exercises 1–20 express the ratios as decimal numbers (see Example 1).

1. 50 V to 2 V   2. 75 mV to 0.15 V
3. 400 turns (transformer winding) to 50 turns
4. 120 turns to 720 turns
5. 7.5 A to 1.5 A   6. 800 mA to 4000 μA
7. 48 kΩ to 1.2 kΩ
8. 560,000 Ω to 1.5 MΩ
9. 750 μS to 50 μS   10. 27 mS to 0.75 S
11. −1.8 V to 15 mV
12. 450 μV to −15 mV
13. 7.5 mA to −50 μA
14. −500 μA to −0.75 mA
15. 27 kΩ to 1200 Ω
16. 720 Ω to 12 kΩ 17. $150 to $39
18. $56 to $165   19. 260 pF to 65 pF
20. 108 MHz to 88 MHz

In Exercises 21–30 express the ratios in simplest form. Affix appropriate units when quantities are dissimilar.

21. 300 mi on 20 gal   22. 45 mi on 2 gal

23. 240 students to 8 instructors
24. 560 students to 17 instructors
25. 15 mA for 0.25 V
26. 0.8 mA for 25 mV
27. 15 mA to 250 μA   28. 12 V to 9Ω
29. 100,000 Ω to 0.75 V
30. 3000 revolutions in 0.8 minute

In Exercises 31–50 find the required ratios.

31. The voltage gain (G_V) of an electronic amplifier is defined as the ratio of output voltage (v_o) to input voltage (v_i). Find G_V if:
    (a) v_o = 50 V, v_i = 2 V
    (b) v_o = 1.8 V, v_i = 4 mV
    (c) v_o = 200 V, v_i = 50 μV
    (d) v_o = 10 mV, v_i = 0.5 V

32. Find the voltage gain of an amplifier (see Exercise 31) if:
    (a) v_o = 1.5 V, v_i = 0.03 V
    (b) v_o = 75 mV, v_i = 40 μV
    (c) v_o = 45 mV, v_i = 225 mV
    (d) 15 mV, 5 μV

33. The power gain G_P of an electronic amplifier is defined as the ratio of output power (P_o) to input power (P_i). Find G_P if:
    (a) P_o = 100 mW, P_i = 40 μW
    (b) P_o = 200 W, P_i = 8 mW
    (c) P_o = 475 μW, P_i = 0.5 mW
    (d) P_o = 0.75 W, P_i = 0.30 mW

34. Find G_P for amplifiers which have the following P_o, P_i values, respectively:
    (a) 75 mW, 0.35 mW
    (b) 50 kW, 450 W
    (c) 870 mW, 350 μW
    (d) 18 mW, 65 mW

35. An important characteristic of bipolar junction transistors (BJTs) is the ratio of collector current (I_C) to base current (I_B). This comparison is the current transfer ratio and is designated β (β = I_C/I_B). Find β for the following sets of values (I_C is the first value given):
    (a) 15 mA, 50 μA
    (b) 1.5 mA, 15 μA
    (c) 200 mA, 500 μA
    (d) 2.0 A, 25 mA

36. Find the β (see Exercise 35) of transistors which have the following I_C and I_B values values, respectively:
    (a) 2.5 mA, 45 μA
    (b) 45 mA, 650 μA
    (c) 175 mA, 1.05 mA
    (d) 1.8 A, 250 mA

37. An important characteristic of so-called *voltage-controlled* electronic devices, such as vacuum tube triodes and field-effect transistors (FETs), is called mutual conductance g_m and is defined as "the ratio of the change in output current Δi_o to the change in input voltage Δv_i which produces it." (The symbol Δ is the Greek letter delta and is used in mathematics to mean "a small change in." Thus Δi_o is read "delta i_o" and means "a small change in i_o.") Find g_m for the following Δi_o, Δv_i values:
    (a) 10 mA, 1.2 V
    (b) 25 mA, 4.5 V
    (c) 150 mA, 40 V
    (d) 750 μA, 35 mV

38. Find the g_m of devices which have the following Δi_o, Δe_i values (see Exercise 37):
    (a) 250 mA, 30 V
    (b) 700 μA, 2.5 V
    (c) 12.5 mA, 0.75 V
    (d) 375 μA, 15 mV

39. An important characteristic of transformers is the turns ratio, N_2/N_1, where N_2 is the number of turns of conductor on the secondary or output side of the transformer and N_1 is the number of turns on the primary or input side of the transformer. Calculate the turns ratios of transformers which have the following primary and secondary turns respectively:
    (a) 100, 200   (b) 300, 15
    (c) 500, 10,000   (d) 7000, 1200

40. A transformer is called a step-up transformer if $N_2/N_1 > 1$. Define a step-down transformer and indicate for each of the pairs of values in Exercise 39 whether the transformer is a step-up or step-down unit.

41. Recall the Ohm's law formula for current. Complete the sentence, "Current is

the ratio of _____ to _____ ."
And "amperes equal _____ per
_____ ."

42. Recall the Ohm's law formula for resistance and complete the sentences, "Resistance is the ratio of _____ to _____ . Ohms equal _____ per _____ ."

43. Complete the sentences, "Conductance is the ratio of _____ to _____ . Siemens (see Chapter 3) equal _____ per _____ ."

44. In a series circuit, the voltage V_x across a resistance R_x is equal to the ratio of that resistance R_x to the total resistance R_T in the circuit times the total voltage V_T. Write this statement as an equation.

45. In a parallel circuit with two resistances R_1 and R_2 the current I_1 through the first resistance is equal to the ratio of the other resistance R_2 to the sum of the resistances $R_1 + R_2$, times the total current I_T. Write this statement as an equation.

46. In a transformer the ratio of the secondary voltage V_2 to the primary voltage V_1 is equal to the turns ratio, secondary to primary (see Exercise 39). Write the equation.

47. The tuning ratio of a radio receiver is equal to the ratio of the highest frequency f_h it can receive to the lowest frequency f_l receivable. What is the tuning ratio for: (a) a broadcast AM receiver: f_h = 1.6 MHz, f_l = 535 kHz (b) a broadcast FM receiver: f_h = 108 MHz, f_l = 88 MHz

48. A computer printer can print 18,000 lines of data in 30 minutes. Describe its capability in lines per minute (lpm).

49. A card reader (computer input device) can read, and transfer to computer memory, 2000 punched-hole cards in 8 minutes. Describe the capability of the machine using an appropriately descriptive unit based on a ratio (cards per minute).

50. The tape transport mechanism of a video recorder (VTR) moves 2250 ft of tape past the recording head while recording a 30-minute program. What is the tape speed? $1\frac{7}{8}$ IPS (inches per second), $3\frac{3}{4}$ IPS, $7\frac{1}{2}$ IPS, or 15 IPS?

7-2 PROPORTION

A proportion is a statement of the equality of two ratios. For example, in a transformer the voltage ratio is equal to the turns ratio. This can be written simply, $V_2/V_1 = N_2/N_1$, although proportions in mathematics texts are sometimes written in the form $V_2 : V_1 = N_2 : N_1$ or even $V_2 : V_1 :: N_2 : N_1$. These are read, "V_2 is to V_1 as N_2 is to N_1."

As equations, proportions can be manipulated by any of the operations applicable to equations.

Example 3 Solve the proportions for the unknown quantities.

(a) $\dfrac{V_1}{100\ \text{V}} = \dfrac{200\,T}{500\,T}$

(a) We multiply both members by 100 V,

$$\cancel{100\ \text{V}} \cdot \frac{V_1}{\cancel{100\ \text{V}}} = \frac{200\,\cancel{T}}{500\,\cancel{T}} \cdot 100\ \text{V}$$

$$V_1 = 40\ \text{V}$$

(b) In a series circuit

$$\frac{V_x}{V_T} = \frac{R_x}{R_T}$$

Solve for V_x.

(b) We multiply both members by V_T,

$$\frac{V_x}{V_T} \cdot V_T = \frac{R_x}{R_T} \cdot V_T$$

$$V_x = \frac{R_x}{R_T} \cdot V_T$$

(c) In a parallel circuit

$$\frac{I_1}{I_2} = \frac{R_2}{R_1}.$$

Solve for I_2.

(c) First we "cross multiply," that is, multiply both members by the LCD, $I_2 R_1$,

$$I_1 R_1 = I_2 R_2$$

Now we divide both members by R_2, to isolate I_2, and exchange sides of the equals sign,

$$\frac{I_1 R_1}{R_2} = I_2 \qquad I_2 = \frac{R_1}{R_2} I_1$$

Exercises 7-2 In Exercises 1–10 solve the given proportions for the unknown quantity.

1. $\dfrac{R_1}{12} = \dfrac{15}{30}$ 2. $\dfrac{8}{V} = \dfrac{10}{30}$ 3. $\dfrac{40}{10} = \dfrac{N_1}{200}$

4. $\dfrac{50}{2} = \dfrac{180}{R_T}$ 5. $\dfrac{3.9}{I_T} = \dfrac{4.7}{6.9}$

6. $\dfrac{6.3}{15.9} = \dfrac{X_C}{46.5}$ 7. $\dfrac{R_1}{R_2} = \dfrac{R_x}{R_3}$ for R_x

8. $\dfrac{V_1}{V_2} = \dfrac{I_2}{I_1}$ for V_2 9. $\dfrac{I_1}{I_2} = \dfrac{R_2}{R_1}$ for R_1

10. $\dfrac{V_x}{V_T} = \dfrac{R_x}{R_T}$ for V_x

A Wheatstone bridge is a circuit of the type shown in Figure 7-1. When the bridge is balanced, $I_m = 0$, and $I_1 = I_1'$ and $I_2 = I_2'$.

In Exercises 11–16 refer to the circuit of Figure 7-1 and assume that the bridge is balanced.

11. One of the characteristics of a balanced Wheatstone bridge is that $I_1'R_1 = I_2'R_3$ and $I_1 R_2 = I_2 R_4$ (see Figure 7-1). Using this information, show the algebraic steps necessary to derive the basic equation of the bridge: $R_1/R_2 = R_3/R_4$.

12. Derive an expression for R_3 in terms of R_1, R_2, and R_4 (see Exercise 11).

13. If in a balanced Wheatstone bridge $R_1/R_2 = 3/27$ what is the ratio of R_4 to R_3?

14. The "arms" of a balanced Wheatstone bridge have these values: $R_1 = 2\ \Omega$, $R_2 = 200\ \Omega$, and $R_3 = 400\ \Omega$. What is the value of R_4?

15. The control knob settings on a commercial model of a Wheatstone bridge are as follows: $R_1/R_2 = 1011$, $R_4 = 0.558\ \Omega$. Determine R_3.

16. Does the value of V_s in the Wheatstone bridge affect the accuracy with which an unknown resistor R_3 can be measured? Why or why not? Use the mathematical relationships of the bridge to explain your answer.

7-3 VARIATION

When we take a fixed (nonvariable) resistance R and connect it in a circuit where the voltage V across it can be varied, as in Figure 7-2(a),

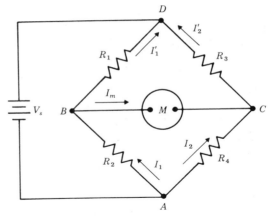

Figure 7-1

we find that if the current is I_1 when $V = V_1$ and I_2 when $V = V_2$, the ratio of the currents equals the ratio of the voltages: $I_2/I_1 = V_2/V_1$. Thus, for example, if $V_2/V_1 = 2$ or $V_2 = 2\,V_1$, then $I_2 = 2I_1$, and so on. Similarly, if V is held fixed and R is varied, as in Figure 7-2(b), we find that $I_2/I_1 = R_1/R_2$. For example, if R is increased so that $R_2 = 2R_1$, I will decrease and $I_2 = \frac{1}{2}I_1$. It is from these proportions that a statement found in most textbooks on electricity is derived: *The current in a circuit is directly proportional to the applied voltage*

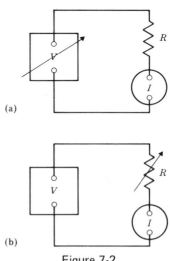

(a)

(b)

Figure 7-2

and inversely proportional to the resistance of the circuit.

The phrases "directly proportional to" and "inversely proportional to" are used in many applications in science and technology. The symbol $\propto$ is used to represent "proportional to." Using this symbol, we can write the statements of variation in an electric circuit as, $I \propto V$ and $I \propto 1/R$. It is helpful to be able to convert statements of proportionality into equations so that they can be dealt with using a wide variety of operations. In general, the sign of proportionality can be replaced with the equals sign and a constant of proportionality. For example,

$$I \propto V \quad \text{becomes} \quad I = kV$$

and

$$I \propto \frac{1}{R} \quad \text{becomes} \quad I = k\frac{1}{R}$$

The constant of proportionality can be evaluated in terms of the known values of a relationship.

Example 4 In a test circuit like that of Figure 7-2(a) I is directly proportional to V and $I_1 = 2$ A when $V_1 = 20$ V. Derive an equation for the circuit.

Since $I \propto V$, then $I = kV$. Therefore,

$$2 = k20$$

$$k = \frac{2}{20} = \frac{1}{10}$$

And the equation of this particular circuit is

$$I = \frac{V}{10}$$

Note the similarity of this equation to one form of Ohm's law: $I = V/R$. Thus, when R is not variable in a circuit $1/R$ is simply the constant of proportionality between I and V for the circuit.

In the physical world one quantity may vary directly and/or inversely with a variety of other quantities and in a variety of ways.

Example 5 Write an equation for the given variations.

(a) The power in an electric   (a) $P = kI^2R$
    circuit varies directly as

the product of the square of the current and the resistance.

(b) The operating frequency   (b) $f = k\dfrac{1}{\sqrt{C}}$
    of the radio station
    "tuned in" by a radio
    receiver is inversely
    proportional to the
    square root of its tuning
    capacitor C.

(c) The power dissipated in   (c) $P = k\dfrac{V^2}{R}$
    a resistance is directly
    proportional to the
    square of the voltage
    across the resistance
    and inversely propor-
    tional to the resistance.

We can evaluate a constant of proportionality if we know one set of values for the variables in the statement of variation. Having determined the constant we can then use it to solve for an unknown variable in situations where all but one variable is known.

Example 6 In Example 5(b) we stated the relationship between received frequency f for a radio receiver and the capacitance C of its tuning circuit. If $C = 65$ pF when $f = 1650$ kHz, determine C when $f = 535$ kHz. From Example 5(b)

$$f = k\frac{1}{\sqrt{C}}$$

We substitute known values for f and C and solve for k,

$$1650 = k\frac{1}{\sqrt{65}}$$
$$k = 1650 \cdot \sqrt{65} = 13{,}303$$

To find C when $f = 535$ kHz we again substitute known values into the equation,

$$535 = \frac{13{,}303}{\sqrt{C}}$$
$$\sqrt{C} = \frac{13{,}303}{535} = 24.87$$
$$C = (24.87)^2 = 618.5 \text{ pF}$$

Exercises 7-3 In Exercises 1–12 write equations for the stated variations using a constant of proportionality.

1. In an electric circuit current varies directly as the product of the applied voltage and the conductance G.

2. The resistance of a conductor varies directly with its length l and inversely with its cross-sectional area A.

3. The collector current I_C of a transistor varies directly as the product of its current ratio β and its base current I_B.

4. The antenna current I of a radio transmitter is directly proportional to the square root of the radiated power P.

5. The "length" λ of a radio wave is inversely proportional to the frequency f of the wave.

6. The period T (time for one cycle) of an electronic signal is inversely proportional to its frequency f.

7. The conductance G of an electric circuit is inversely proportional to its resistance.

8. Inductive reactance X_L is directly proportional to the product of frequency f and inductance L but capacitive reactance X_C is inversely proportional to the product of frequency and capacitance C.

9. The capacitance of a parallel-plate capacitor is directly proportional to the cross-sectional area A of the plates and inversely proportional to the separation d between the plates.

10. The mutual inductance L_M of two coils L_1 and L_2 with mutual coupling is directly proportional to the square root of the product of L_1 and L_2.

11. The inductance L of a coil is directly proportional to the product of the cross-sectional area A of the coil and the square of the number of turns N, and inversely proportional to the length l of the coil.

12. The amount of self-inducted voltage v_L produced in a coil because of a change of current di/dt is directly proportional to the product of inductance L and di/dt.

In Exercises 13–24 determine the unknown variable after evaluating the constant of proportionality from the given set of values.

13. Refer to Exercise 1. Given: $I = 10$ mA, $V = 20$ V, $G = 500$ μS. Find I when $V = 60$ mV.

14. A conductor with a cross-sectional area of of 10.55 mm^2 has a resistance of 1.634 Ω/km. Determine the resistance of 1 km of a conductor made of the same metal with a cross section of 42.41 mm^2 (see Exercise 2).

15. A transistor has a collector current of 15 mA when its base current is 300 μA. What is I_C when $I_B = 125$ μA? (See Exercise 3.)

16. Refer to Exercise 4. Antenna current is 10 A when $P = 5$ kW. When the transmitter is modulated 100%, P increases to 7.5 kW. Calculate I_{ant} when modulation is 100%.

17. The relationship between wavelength and frequency is stated in Exercise 5. If $\lambda = 300$ m when $f = 1$ MHz, determine λ when $f = 640$ kHz.

18. The period of a signal with $f = 455$ kHz is 2.198 μs. Determine T when $f = 15{,}750$ Hz (see Exercise 6).

19. A resistance of 15 kΩ has a conductance of 66.67 μS. What is the conductance if $R = 1.8$ MΩ? (See Exercise 7.)

20. Inductive reactance and capacitive reactance are each 150 Ω when $f = 100$ kHz. What are X_L and X_C for the same L and C when $f = 75$ kHz? 125 kHz? (See Exercise 8.)

21. The capacitance of a parallel-plate capacitor is 100 pF when $d = 1.5$ mm. What will C be for the same plates if d is reduced to 0.90 mm? (See Exercise 9.)

22. Refer to Exercise 10. The constant of proportionality in the relationship for mutual inductance is also called the coefficient of coupling. What coefficient of coupling is required to make $L_M = 2.5$ H if $L_1 = 9$ H and $L_2 = 4$ H?

23. A coil has an inductance of 500 mH when $N = 750$ turns. What is L if $N = 400$ turns and l and A remain unchanged? (See Exercise 11.)

24. Refer to Exercise 12. If 120 V is induced in a certain coil when $di/dt = 450$ mA/s, what is v_L if $di/dt = 675$ mA/s.

7-4 RATIOS IN ANALYSIS OF SERIES CIRCUITS

A series circuit is a voltage divider. The ratio of voltages (across portions of the circuit) in a

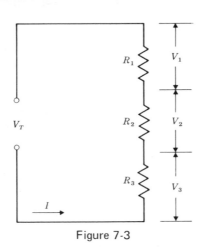

Figure 7-3

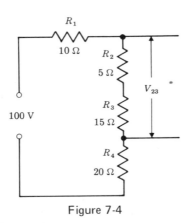

Figure 7-4

series circuit is equal to the ratio of the resistances across which they exist. For example, for the circuit of Figure 7-3,

$$\frac{V_1}{V_2} = \frac{R_1}{R_2} \qquad (1)$$

$$\frac{V_1}{V_T} = \frac{R_1}{R_1 + R_2 + R_3} \qquad (2)$$

and so forth.

The validity of this relationship can be shown if we remember that the current I is the same in all elements of a series circuit. We multiply the numerators and denominators of the right members of equations (1) and (2) by I,

$$\frac{V_1}{V_2} = \frac{IR_1}{IR_2}$$

$$\frac{V_1}{V_T} = \frac{IR_1}{I(R_1 + R_2 + R_3)}$$

Since V_1 is in fact equal to IR_1, $V_2 = IR_2$, and $V_T = I(R_1 + R_2 + R_3)$ the original proportions are validated.

An extremely powerful, useful, and popular circuit analysis tool is obtained when we solve equation (2) for V_1,

$$V_1 = \frac{R_1}{R_1 + R_2 + R_3} V_T \qquad (3)$$

We generalize equation (3) into the more familiar form,

$$V_x = \frac{R_x}{R_T} V_T \qquad (4)$$

Equation (4), converted to words, states:

> **The voltage V_x across any resistance R_x in a series circuit is equal to the product of the ratio, of that resistance to the total resistance, and the total voltage V_T across the series circuit.**

Example 7

(a) Using the ratio method for series circuits (the voltage-divider formula) find the voltage V_{23} across resistors R_2, R_3 in the circuit of Figure 7-4.

$$V_{23} = \frac{R_2 + R_3}{R_1 + R_2 + R_3 + R_4} \cdot V_T$$

$$= \frac{5 + 15}{10 + 5 + 15 + 20} \cdot 100 = \frac{\overset{2}{\cancel{20}}}{\cancel{50}} \cdot \cancel{100}$$

$$V_{23} = 40 \text{ V}$$

(b) Using the voltage-divider formula, determine V_o for the circuit of Figure 7-5.

$$V_o = \frac{R_7}{R_5 + R_6 + R_7} \cdot 15$$

$$= \frac{0.82}{1.8 + 1.5 + 0.82} \cdot 15$$

$$= \frac{0.82}{4.12} \cdot 15$$

$$V_o = 2.985 \text{ V}$$

Exercises 7-4   In Exercises 1–14 use the ratio method (voltage-divider formula) for series circuits in obtaining solutions.

1. Determine V_4 the voltage across R_4 in the circuit of Figure 7-6 if $V_T = 25$ V, $R_1 = 10\ \Omega$, $R_2 = 3\ \Omega$, $R_3 = 7\ \Omega$ and $R_4 = 5\ \Omega$.

2. Refer to Exercise 1. Determine V_1, the voltage across R_1.

3. Determine V_{34}, the voltage across R_3 and R_4 together, in the circuit of Exercise 1.

4. Predict what the voltage across R_4 would be in the circuit of Exercise 1 if a short circuit was connected across R_2 and R_3.

5. Determine V_4, the voltage across R_4 in the circuit of Figure 7-6 if $V_T = 1.55$ mV, $R_1 = 150\ \Omega$, $R_2 = 15$ kΩ, $R_3 = 720\ \Omega$, and $R_4 = 6.8$ kΩ.

6. Refer to Exercise 5. Determine V_1, the voltage across R_1.

7. Determine V_{23}, the voltage across R_2 and R_3 together, in the circuit of Exercise 5.

8. Predict the voltage across R_4 of the circuit of Exercise 5 for the situation in which R_1 and R_2 become shorted.

9. Refer to Figure 7-7. Given: $V_2 = 11.5$ V, $R_3 = 1.5$ kΩ, $R_4 = 5.6$ kΩ, $R_5 = 9.1$ kΩ. Determine V_o.

10. Refer to Exercise 9. A 7.2-kΩ load is connected to terminals A, B. This causes V_2 to drop to 9.25 V. Resistors R_3, R_4, and R_5 are unchanged. What is V_o?

11. In the circuit of Figure 7-7 $R_3 = 10$ kΩ, $R_4 = 1.8$ kΩ, $R_5 = 4.7$ kΩ, and $V_o = 45$ mV. Find V_2. (*Hint:* Solve the equation $V_x = (R_x/R_T)\ V_T$ for V_T.)

12. Refer to Exercise 11. When an 8.2-kΩ load is connected to terminals A and B, V_o drops to 25.8 mV. Determine V_2 for this condition.

13. Refer to Figure 7-8. Given: $V_{CC} = 6$ V, $R_1 = 15$ kΩ, $R_2 = 4.7$ kΩ. What is V_A if the connection between A and B is opened?

14. Refer to Exercise 13 and assume that the R_{eq} of the circuit from A to B to C (through the base-emitter junction) is 10 kΩ. Determine V_A for this condition.

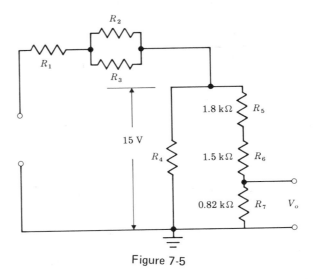

Figure 7-5

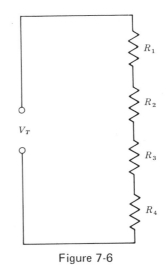

Figure 7-6

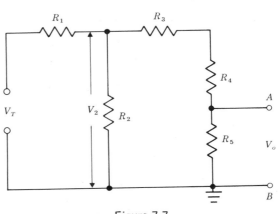

Figure 7-7

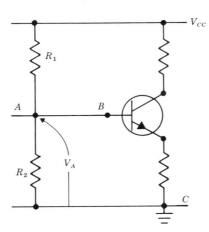

Figure 7-8

7-5 THE RATIO METHOD APPLIED TO PARALLEL CIRCUITS

Parallel circuits are current dividers. In parallel circuits the voltage is the same across all branches. Therefore, in a two-branch circuit (see Figure 7-9),

$$I_1 R_1 = I_2 R_2 \qquad I_1 R_1 = I_T R_T \qquad I_2 R_2 = I_T R_T$$

From the first of these equations we derive

$$I_1 = \frac{R_2}{R_1} I_2 \quad \text{and} \quad I_2 = \frac{R_1}{R_2} I_1$$

From the second equation we obtain

$$I_1 = \frac{R_T}{R_1} I_T$$

but

$$R_T = \frac{R_1 R_2}{R_1 + R_2}$$

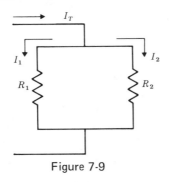

Figure 7-9

Substituting this R_T equivalent, we obtain

$$I_1 = \frac{R_1 R_2}{R_1 (R_1 + R_2)} I_T = \frac{R_2}{R_1 + R_2} I_T$$

Similarly,

$$I_2 = \frac{R_1}{R_1 + R_2} I_T$$

We list these ratio, or current-divider formulas, for a two-branch parallel circuit, for easy reference:

$$I_1 = \frac{R_2}{R_1} I_2 \qquad (1)$$

$$I_2 = \frac{R_1}{R_2} I_1 \qquad (2)$$

$$I_1 = \frac{R_2}{R_1 + R_2} I_T \qquad (3)$$

$$I_2 = \frac{R_1}{R_1 + R_2} I_T \qquad (4)$$

It is helpful to remember the following summary of equations (3) and (4):

> **A two-branch parallel circuit always divides the total current so that the current through one branch is equal to the product of the total current and the ratio of the resistance of the *other* branch to the sum of the resistances of the two branches.**

Example 8   Using the current-divider formulas (3) or (4), find the current I_2 for the circuit of Figure 7-9. Given $I_T = 15$ mA, $R_1 = 6.8$ kΩ, $R_2 = 3.3$ kΩ.

$$I_2 = \frac{R_1}{R_1 + R_2} I_T = \frac{6.8}{6.8 + 3.3} \cdot 15 = \underline{10.1 \text{ mA}}$$

Example 9 In the circuit of Figure 7-9, $R_1 = 56$ kΩ, $I_1 = 2.37$ mA, $I_2 = 7.37$ mA. Use a ratio formula to determine R_2.

First we transpose equation (1),

$$\frac{R_1}{I_2} \cdot I_1 = \frac{R_2}{\cancel{R_1}}\cancel{I_2} \cdot \frac{\cancel{R_1}}{\cancel{I_2}} \qquad R_2 = \frac{I_1}{I_2} R_1$$

Then substitute known values,

$$R_2 = \frac{2.37}{7.37} \cdot 56 = \underline{18 \text{ k}\Omega}$$

(*Note:* In both solutions above the units of the ratios cancel, leaving a pure number. Therefore, we simply do not show the units, or equivalent powers of ten, in the substitution step.)

Exercises 7-5 In Exercises 1–10 use the current-divider formulas for two-branch parallel circuits in obtaining solutions.

1. In a two-branch parallel circuit $R_1 = 2.4$ kΩ, $R_2 = 3.9$ kΩ, $I_1 = 5.8$ mA. Determine I_2.

2. In a two-branch parallel circuit $R_1 = 0.82$ kΩ, $R_2 = 4.1$ kΩ, $I_2 = 450$ μA. Find I_1.

3. Resistors R_1 and R_2 are in parallel; $R_1 = 720$ kΩ, $I_1 = 2.85$ μA, $I_2 = 15.6$ μA. Find R_2.

4. Resistors R_a and R_b are in parallel; $R_b = 1.8$ MΩ, $I_a = 59$ nA, $I_b = 0.125$ μA. Find R_a.

5. Resistors r and R are in parallel; $r = 15$ Ω, $R = 27$ Ω, $I_T = 0.47$ A. Find the two currents.

6. In a two-branch parallel circuit $R_1 = 56$ kΩ, $R_2 = 100$ kΩ, and $I_T = 75$ μA. Find I_1 and I_2.

7. If $R_p = 270$ kΩ, $R_q = 680$ kΩ, and $I_T = 1.85$ μA in a two-branch parallel circuit, find I_p and I_q.

8. When $R_1 = 1.8$ kΩ is in parallel with $R_2 = 4.3$ kΩ and $I_T = 29.5$ mA, what are I_1 and I_2?

9. Figure 7-10 shows the circuit of a dc ammeter provided with a shunt R_{sh} to extend its range. Derive an expression for R_{sh} in terms of I_m, I_T, and R_m. (*Hint:* Let $I_{sh} = I_T - I_m$.)

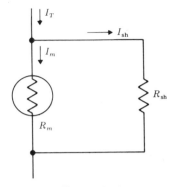

Figure 7-10

10. In the circuit of Figure 7-11, I_T, I_1, and R_1 are known or measurable. Derive an expression for R_{in}, the input resistance, of the electronic system.

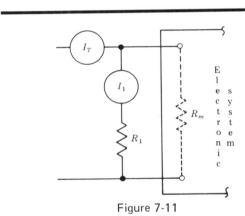

Figure 7-11

7-6 PERCENT

Three elements, the *base*, the *rate*, and the *percentage*, can be identified in every relationship in which the percent notation is used. For example, electronics components such as resistors, capacitors, semiconductors, and so on, are commonly offered for sale with discounts (reductions in price) for quantity purchases: a particular firm might advertise a 10% discount for quantities of 25–49, $12\frac{1}{2}$% for 50–99, 15% for 100–199, and so on. The expressions 10%, $12\frac{1}{2}$%, and 15% are called the *rate of the discount*—they are the *rate* element in the relationship. The advertised basic prices (the list prices) of the particular types of components are examples of the element which is called the *base* in all percent problems. The *amounts* of the discounts, in dollars and cents, are examples of the *percentage* element of percent problems.

Let us use B to represent base, P to represent percentage, and R to represent rate. The mathematical relationships of these three elements in *all* percentage problems are expressed in the following formulas:

$$R = \frac{P}{B} \times 100\% \qquad (1)$$

$$P = \frac{R}{100} \times B \qquad (2)$$

$$B = \frac{P}{R/100} \qquad (3)$$

Example 10   Tricity Electronic Corp. offers an 18% discount on its transistors on quantities of 100–199. What will be the amount of the discount per transistor when 150 2N2475's are purchased at a list price of $1.41 each?

$$P = \frac{R}{100}B \quad P = \frac{18}{100} \times 1.41 \quad P = \$0.25$$

Example 11   Libdisc Electronics Co. will sell 1000 or more 2N3252 transistors for $1.62 each. The list price is $3.00. What is the rate of discount?

First, we must calculate the amount of the discount.

Amount of discount = List price − net price
$$= 3.00 - 1.62$$
$$= \$1.38$$

Then,

$$R = \frac{P}{B} \times 100\% = \frac{1.38}{3.00} \times 100\%$$

$$R \text{ (rate of discount)} = 46\%$$

Example 12   Saveless Electronics Sales states that it gives 8% discount, "a big 10¢ off" when you buy 100 or more of its 2N3014 transistors. What is their list price for this transistor?

$$B = \frac{P}{R/100} = \frac{0.10}{0.08} = 1.25$$

$$B \text{ (list price)} = \$1.25$$

Example 13   A certain metal film resistor has a temperature coefficient α of ±100 PPM/°C in the temperature range −55° to 165°C. (PPM/°C means "parts per million per degree Celsius.") (a) What is the percent change equivalent for 100 PPM? (b) If the temperature of a 110-kΩ resistor increases from 22°C to 72°C during a warm-up period, what is the amount of the change in the ohmic value?

(a)   100 PPM = 100 ÷ 1,000,000 = 0.0001
      0.0001 × 100% = 0.01%

(b) $\Delta T = 72 - 22 = 50°$
      $\Delta R = R \cdot \Delta T \cdot \alpha = 110(10^3) \cdot 50 \cdot 0.0001$
      = 0.55 kΩ

It is important to note that the rate element is always expressed with the percent symbol— 5%, $12\frac{1}{2}\%$, 50%, and so on. It is converted to its decimal equivalent by dividing by 100: 5% = 5/100 = 0.05, $12\frac{1}{2}\%$ = 12.5/100 = 0.125. *A rate quantity must always be converted to its decimal equivalent before it is used in a mathematical evaluation.*

In a similar way, when we divide a percentage element by an appropriate base element (as in the formula $R = P/B$) we first obtain a simple ratio. *To express this ratio, or any number, as a percent we multiply by 100.* Thus, the formula to obtain a percent rate is usually written

$$R = \frac{P}{B} \times 100\%$$

Exercises 7-6   In Exercises 1–8 express the given numbers in percent notation.

1. 0.25, 0.33, 0.50   Ans.: 25%, 33%, 50%
2. $0.05\frac{1}{2}$, 0.0625, $0.07\frac{1}{4}$
3. 0.15, 0.30, 0.90   4. 0.1, 0.75, 1.25
5. 2, 5, 10   6. $1.33\frac{1}{3}$, 2.5, 6.75
7. 0.0015, 0.00025, 0.019   8. 15, 50, 70

In Exercises 9–12 perform the indicated divisions and express the answers as percents.

9. $\dfrac{1}{2}, \dfrac{3}{4}, \dfrac{1}{3}$ 10. $\dfrac{1}{10}, \dfrac{1}{5}, \dfrac{3}{5}$

11. $\dfrac{25}{15}, \dfrac{150}{20}, \dfrac{495}{36}$

12. $\dfrac{2950}{357}, \dfrac{12}{59,356}, \dfrac{3.3}{2,568,589}$

In Exercises 13–16 convert the percent quantities to their decimal-number equivalents.

13. 25%, 33%, 50%   14. $5\frac{1}{2}$%, $12\frac{1}{2}$%, $37\frac{1}{2}$%

15. 450%, 1000%, 2500%

16. 0.1%, $\frac{1}{2}$%, $\frac{1}{2}$ of 1%

In Exercises 17–32 solve for the base, rate, or percentage, as required.

17. What is the amount of discount on an order for $89.00 if the discount rate is $8\frac{1}{2}$%?

18. You can get a $22\frac{1}{2}$% discount on a $47.95 calculator at E Z BY SALES stores. What is the amount of the price reduction?

19. You have deposited $1500 in a savings account which earns interest an an annual rate of 8.03%. How much interest does the deposit earn in one year?

20. You will receive an automatic $6\frac{1}{2}$% pay increase at Data Electronics at the end of your first six months with the company. If your starting rate is $4.25 per hour, what will be the amount of the increase?

21. A digital multimeter has a specified accuracy of "±0.5% of reading." If the reading is 15.64 mA, what is the amount of possible error?

22. An audio frequency generator has a specified accuracy of "±2% of frequency setting." What is the possible error in Hz if the unit is set for 8750 Hz?

23. Quantities of 100–999 of 2N3444 transistors sell at $2.50 each (list price $4.46) at Disco Sales. What is the discount rate?

24. The CLC Company has announced a new calculator with a list price of $36.95. Big Cut Sales advertises the calculator for $22.95. Determine the percent discount.

25. Your deposit of $600 earns $48.50 in interest in one year. What is the interest rate?

26. Your hourly rate increased from $4.30 per hour to $4.70 per hour after six months with Data Electronics. What was the percent of pay increase?

27. A voltmeter indicates 1.544 V when measuring a precisely known 1.560 V. What is the amount of error? the percent error?

28. The operating frequency of a broadcast transmitter is found to be 546.739 kHz when checked with a precision frequency meter. The frequency assigned to the station by the Federal Communications Commission (FCC) is 546.000 kHz. What is the percent of frequency deviation?

29. A voltmeter reads 0.05 V low while being checked for accuracy. It is in error by 1.5%. What was the voltage being measured?

30. Your savings account earns 8.15% interest. Last year you received $65.20 in interest from the account. What was the amount in the account at the first of the year?

31. A radio station's transmitter was monitored by the FCC and its carrier frequency was found to be off by 0.14715 MHz, a deviation error of 0.15%. What is the station's operating frequency?

32. You saved $8.25 when you received a discount of 22% on the purchase of a calculator at Big Deal Sales. What was the list price of the calculator?

In Exercises 33–36, α = temperature coefficient, ΔT = change in temperature. (a) What is α in percent? (b) What is the change in R for the given ΔT? (See Example 13.)

33. $R = 1.3$ MΩ, $\alpha = 50$ PPM/°C, $\Delta T = 78$°C
34. $R = 75$ kΩ, $\alpha = 400$ PPM/°C, $\Delta T = 70$°C
35. $R = 820$ kΩ, $\alpha = 700$ PPM/°C, $\Delta T = 30$°C
36. $R = 39$ kΩ, $\alpha = 1200$ PPM/°C, $\Delta T = 50$°C

7-7 MORE APPLICATIONS OF PERCENT

When we purchase an item at discount we are usually interested in determining the *net price*, not just the *amount of the discount*. Similarly, we want to know what our new salary will be, not just the amount of the raise; we want to know what our saving account balance is at the end of the year, not just the amount of interest earned, and so on.

All of these problems have something in common: we must combine the *percentage* with the *base*, in either addition or subtraction, to obtain a *new base*. For example, in discount-type problems we have

Net price = List price − discount

In interest-type problems,

New balance = Old balance + interest

All such problems lend themselves to an application of algebra. Let us call the original base B_o. The desired end result, then, is the new base, B_n. Thus, calculating discounts, we have

$$B_n = B_o - \frac{R}{100}B_o$$

and for interest-type problems,

$$B_n = B_o + \frac{R}{100}B_o$$

In general, then, many percentage problems can be evaluated using the formula

$$B_n = B_o \pm \frac{R}{100}B_o$$

which we now simplify by factoring B_o,

$$B_n = B_o \left(1 \pm \frac{R}{100}\right)$$

Example 14 You will receive a $17\frac{1}{2}\%$ discount on the purchase of a $44.00 calculator at Hi Disc Sales. What is your net price?

$$B_n = B_o \left(1 - \frac{R}{100}\right) = 44(1 - 0.175)$$
$$= 44(0.825) = 36.30$$
Net price = $36.30

Example 15 A 27-kΩ, ±5% (gold fourth band in color code) resistor is used in the base-bias network of a transistor amplifier. What are the maximum and minimum values the resistor may have and still be in tolerance?

$$R_{max} = R_{nominal}\left(1 + \frac{Tol}{100}\right) = 27 \times 10^3(1.05)$$
$$= 28.35 \text{ k}\Omega$$

$$R_{min} = R_{nominal}\left(1 - \frac{Tol}{100}\right) = 27 \times 10^3(0.95)$$
$$= 25.65 \text{ k}\Omega$$

Example 16 The output voltage of a regulated dc power supply is measured to be 8.802 V. This is 2.2% below its rated voltage. What is the rated voltage?

$$B_n = B_o\left(1 - \frac{R}{100}\right) \qquad 8.802 = B_o(1 - 0.022)$$
$$0.978B_o = \underline{8.802} \qquad B_o = \frac{8.802}{0.978} = \underline{9.0 \text{ V}}$$

Example 17 A customer service engineer has his service VOM (voltohmmeter) checked against a precision, laboratory-standard meter. When his meter is reading 6.15 V, the reading is $2\frac{1}{2}\%$ (of the correct voltage) high. (a) What is the correct voltage? (b) What percent correction should he apply to readings which are approximately 6 V, taken with his meter?

(a) $B_n = B_o\left(1 + \frac{R}{100}\right)$

$\qquad 6.15 = B_o(1 + 0.025)$

$\qquad 1.025B_o = 6.15 \qquad B_o = \frac{6.15}{1.025} = \underline{6.0 \text{ V}}$

The lab meter reading was 6.0 V.

(b) The error of the field meter was

$\qquad 6.15 - 6.0 = \underline{0.15 \text{ V}}$

We use the field meter reading as the base to calculate the percent error,

$\qquad R = \frac{P}{B} \times 100\% = \frac{0.15}{6.15} \times 100 = \underline{2.439\%}$

The field engineer should apply a correction of -2.439% to the readings of his meter when he is measuring voltages approximating 6 V.

Voltage Regulation The terminal voltage of a generator, battery, or electronic power supply drops when an electrical load drawing a load current is connected. The change in terminal voltage, designated ΔV, is undesirable in most instances. Voltage regulation is a term which is used to identify a quantitative measure of the ability of a source to supply a load with minimum voltage change. It is defined by the formula

$$VR = \frac{\Delta V}{V_{FL}} \cdot 100\%$$

$$\Delta V = V_{NL} - V_{FL}$$

V_{NL} = terminal voltage when load current is zero

V_{FL} = terminal voltage when source is supplying rated load current

Voltage regulation is related to the internal resistance R_i of the source and the load resistance R_L by the formula

$$VR = \frac{R_i}{R_L} \cdot 100\%$$

Example 18 The terminal voltage of an unregulated dc power supply changes from 9.00 V to 8.67 V when the load current changes from $I_L = 0$ to $I_L = 500$ mA (rated). What is VR? (An unregulated power supply is one which does not employ any electronic circuitry to overcome the voltage drop caused by load current.)

$$VR = \frac{\Delta V}{V_{FL}} \cdot 100\% = \frac{9.00 - 8.67}{8.67} \cdot 100\%$$
$$VR = 3.8\%$$

Example 19 An automobile battery has an internal resistance of 0.0025 Ω. What is the voltage regulation of the battery if its full load current of 75 A is equivalent to $R_L = 0.1813$ Ω?

$$VR = \frac{0.0025}{0.1813} \cdot 100\% = 1.38\%$$

Example 20 A variable-voltage power supply is specified to have a voltage regulation of 0.25%. What will be its output voltage when supplying rated current if set for 12.5 V, no load?

We must solve the voltage regulation formula for V_{FL},

$$VR = \frac{V_{NL} - V_{FL}}{V_{FL}} \cdot 100\%$$
$$0.01VR = \frac{V_{NL} - V_{FL}}{V_{FL}}$$
$$0.01VR \cdot V_{FL} = V_{NL} - V_{FL}$$
$$V_{FL}(1 + 0.01VR) = V_{NL}$$
$$V_{FL} = \frac{V_{NL}}{(1 + 0.01VR)} = \frac{12.5}{(1.0025)}$$
$$V_{FL} = 12.47 \text{ V}$$

Efficiency The efficiency η (eta) of any energy conversion system is defined by the formula

$$\eta = \frac{\text{output energy}}{\text{input energy}} \times 100\%$$

Since power P is energy per unit of time, we often see the efficiency formula expressed as

$$\eta = \frac{P_{out}}{P_{in}} \times 100\%$$

Example 21 A mobile radio transmitter has a radiated output of 75 W for a dc input of 125 W. What is its efficiency?

$$\eta = \frac{P_{rf}}{P_{dc}} \times 100 = \frac{75}{125} \times 100 = 60\%$$

Example 22 A CB transmitter operates at an efficiency of 55%. What is the rf output when the dc input is 5 W?

Solving efficiency formula for P_{rf},
$$P_{rf} = 0.01\eta P_{dc} = 0.01 \cdot 55 \cdot 5 = 2.75 \text{ W}$$

Meter Loading A voltmeter will load (draw current from and thus change the characteristics of) a circuit to which it is connected unless its internal resistance R_m is very much larger than that of the resistance R_p of the portion of the circuit across which it is connected (see Figure 7-12). When meter loading occurs, a meter reading V_r may be corrected by the addition of the following term,

$$\text{Meter correction} = \frac{V_r R_p}{R_m \left(1 + \dfrac{R_p}{R_s}\right)}$$

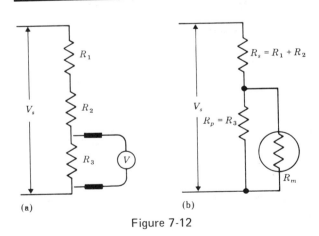

Figure 7-12

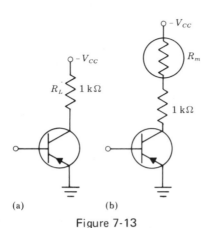

Figure 7-13

(See Figure 7-12 for significance of R_m, R_p, and R_s.)   Thus, a corrected voltage measurement V_c could be obtained by using the equation

$$V_c = V_r + \frac{V_r R_p}{R_m \left(1 + \dfrac{R_p}{R_s}\right)}$$

Example 23   In the circuit of Figure 7-12(a), $R_1 = 270$ kΩ, $R_2 = 330$ kΩ, $R_3 = 560$ kΩ, $R_m = 150$ kΩ, and $V_r = 1.980$ V.   Determine (a) the correction factor, (b) corrected voltage, and (c) correction as a percent of corrected voltage.

(a) $\text{CF} = \dfrac{V_r R_p}{R_m \left(1 + \dfrac{R_p}{R_s}\right)} = \dfrac{1.980 \cdot 560}{150 \left(1 + \dfrac{560}{270 + 330}\right)}$

$= \dfrac{1.980 \cdot 560}{150(1 + 0.933)}$

$\text{CF} = \underline{3.823 \text{ V}}$

(b) $V_c = V_r + \text{CF} = 1.980 + 3.823 = \underline{5.803 \text{ V}}$

(c)   Percent correction $= \dfrac{\text{CF}}{V_c} \times 100\% =$

$\dfrac{3.823}{5.803} \times 100\% = \underline{65.88\%}$

A current meter (ammeter) will also "load" a circuit.   Its internal resistance R_m, if greater than about $\frac{1}{10}$ of the resistance of the circuit in which it is inserted, will increase the resistance of the circuit and thus reduce significantly the current we are trying to measure. A correction factor is available as follows:

True current = Current meter reading
+ correction factor

$$I_t = I_r + \frac{R_m}{R_T} I_r$$

where
I_r = Meter reading
R_m = Resistance of current meter
R_T = Total resistance of circuit without the meter

Example 24   A current meter is inserted in the collector circuit of a transistor stage (see Figure 7-13) to measure the collector current. If $R_m = 100$ Ω, $R_L = 1$ kΩ, the estimated resistance of the transistor is 250 Ω, and $I_r = 6.675$ mA, determine: (a) the corrected meter reading, (b) correction as a percent of actual current.

(a) $I_t = I_r + \dfrac{R_m}{R_T} I_r = 6.675 + \dfrac{100}{1250} \cdot 6.675$

$= 6.675 + 0.534 = \underline{7.209 \text{ mA}}$

(b) $\dfrac{\text{CF}}{I_t} \times 100\% = \dfrac{0.534}{7.209} = \underline{7.41\%}$

Exercises 7-7   In Exercises 1–10 utilize the equation $B_n = B_o \left(1 \pm \dfrac{R}{100}\right)$ to solve the problems.

1. You are considering purchasing a $49.50 calculator because a discount of 24% has been offered.   What net price would you pay?

2. The net price of a calculator is $34.65. State sales tax rate is 6%.   What would this calculator cost including sales tax?

3. What are the maximum and minimum values a 470-kΩ resistor may have and still be within tolerance if its fourth color band is silver (±10% tolerance)?

4. The fourth color band of an 820-Ω resistor is black (±20% tolerance).   What maximum and minimum values would be permitted by the tolerance specification?

5. A 36-kΩ resistor with gold fourth color band (±5% tolerance) measures 33.96 kΩ on a DMM.   Is the resistor value within permitted manufacturing tolerance? What is the permitted minimum value?

6. A manufacturer's schematic specifies a

15-kΩ, $\pm$10% resistor in the bias network of a transistor. While repairing equipment containing this circuit you find this resistor open. In preparing to replace it you find you do not have a 15-kΩ resistor but you do have an 18-kΩ resistor which measures 16.45 kΩ. Would the equipment be likely to work properly with this resistor as a replacement?

7. The line voltage on a biomedical equipment installation is measured at 111 V which is 7.5% (of rated value) low. What is the rated voltage for the equipment?

8. You can obtain a calculator with desired features for $26.95, a net price after a discount of 15% is applied to the list price. How does the list price of this calculator compare with that of Brand X, list price $30.00? Give amount of difference.

9. A digital multimeter costs $314.48, including a 6% sales tax. What is the selling price (before tax is added)?

10. A company advertises that its entry-level technician's pay, which includes a $22\frac{1}{2}$% (of base salary) fringe benefit package, is equivalent to a monthly salary of $1010.63. What is the base salary?

11. The voltage of a power supply drops from 35 V to 33.85 V when rated load is connected. What is percent voltage regulation?

12. The output of an audio generator changes from 15 mV on open circuit to 14.05 mV when a 600-Ω load (rated load) is connected. Determine percent VR.

13. A dry-cell battery has an internal resistance of 0.1 Ω. Rated load resistance is 4560 Ω. Determine percent VR.

14. A solar cell array has an internal resistance of 197 Ω. Rated load resistance is 4560 Ω. Determine percent VR.

15. A computer power supply has a voltage regulation of 0.75%. If output voltage is 5.955 V under normal load, what is V_{NL}?

16. A camera mechanism "energizer" cell operates in an application where its voltage regulation is 4.5%. If V_{NL} is 1.35 V, what is the cell voltage when supplying normal load?

17. An FM mobile transmitter produces 20 W of rf power when the dc input power is 42.7 W. What is the efficiency of the transmitter?

18. A CB transmitter produces 2.35 W of rf power when the dc input power is 5.0 W. What is the efficiency?

19. When a regulated dc power supply provides a rated 10 A at 30 V (dc) the input from the ac line is 350 W. What is the efficiency of the unit?

20. When the output of a transformer is 12.5 A at 6.3 V the input is 0.68 A at 120 V. What is the efficiency of the transformer?

21. The efficiency of the output stage of a transmitter is 61%. If the dc input to the stage is 50 kW, what is the rf output?

22. Refer to Exercise 21. How much dc power is dissipated in the output stage itself?

In Exercises 23–26 the given values refer to the circuit of Figure 7-12(a). For each set of values determine the following:
(a) the correction factor, CF,
(b) the corrected voltage reading, and
(c) the correction as a percent of corrected voltage (see Example 23).

23. R_1 = 36 kΩ, R_2 = 0 Ω, R_3 = 56 kΩ, R_m = 100 kΩ, V_r = 0.35 V

24. R_1 = 2.4 kΩ, R_2 = 1.2 kΩ, R_3 = 8.2 kΩ, R_m = 15 kΩ, V_r = 0.27 V

25. R_1 = 560 kΩ, R_2 = 470 kΩ, R_3 = 2.4 MΩ, R_m = 10 MΩ, V_r = 35 V

26. R_1 = 1 MΩ, R_2 = 0 Ω, R_3 = 100 MΩ, R_m = 10 MΩ, V_r = 0.45 V

In Exercises 27–30 R_T refers to resistance of a circuit before a current meter is inserted, R_m is the internal resistance of the current meter, and I_r is the reading of the meter. In each exercise determine:
(a) the correction factor, CF = $(R_m/R_T)I_r$,
(b) the corrected meter reading, and
(c) the correction as a percent of corrected reading (see Example 24).

27. R_m = 100 Ω, R_T = 950 Ω, I_r = 1.654 mA
28. R_m = 1 Ω, R_T = 85 Ω, I_r = 146.7 mA
29. R_m = 1000 Ω, R_T = 6000 Ω, I_r = 97.54 μA
30. R_m = 10 Ω, R_T = 750 Ω, I_r = 18.05 mA

Chapter 8
Circuit Analysis Methods I

The German physicist Georg Simon Ohm (1789–1854) found experimentally that there is a consistent relationship between the current in a circuit element and the voltage across it. The description of this relationship is called Ohm's Law in his honor. (The Bettman Archive)

8-1 BASIC CIRCUIT LAWS

All equations written to represent the physical relationships in electric circuits must satisfy one or more of the following three basic physical principles:

1. *Ohm's Law* The current in any circuit element is equal to the voltage across that element divided by its impedance. Expressing this in mathematical symbols, for a resistive circuit we have

$$I = \frac{V}{R}, \qquad V = IR, \qquad R = \frac{V}{I}$$

2. *Kirchhoff's Current Law* (KCL) The algebraic sum of the currents at any point in a circuit must be zero.

$$\text{At any point: } \Sigma I = 0$$

(The symbol Σ is the Greek letter sigma and is used in mathematics to mean "the algebraic sum of.") The practical effect of this law is often stated:

$$\Sigma \text{ currents toward a point} =$$
$$\Sigma \text{ currents away from that point}$$

3. *Kirchhoff's Voltage Law* (KVL) The algebraic sum of the voltages around a closed path is zero.

$$\Sigma V = 0$$

In the past some professionals have used E or e to represent the voltages produced by sources, batteries, generators, and power supplies, and V or v to represent the voltages, usually called *voltage drops*, produced by the current in passive elements, such as resistors and reactors. However, the latest international standard, SI (see Section 2-2), calls for V or v to be used exclusively for all voltages. The summation formula may be expressed

$$\Sigma V_{sources} + \Sigma V_{drops} = 0$$

Or, to represent practical reality more directly

$$\Sigma \text{ source voltages} = \Sigma \text{ voltage drops}$$

The current and voltage terms that we use in the algebraic summations of the Kirchhoff circuit laws must have algebraic signs as well as magnitude. We consider algebraic signs in electric circuits in the next section. (Review Section 1-5 on *algebraic summation*.)

8-2 CURRENT AND VOLTAGE NOTATION

In the analysis of all but the most simple of electric circuits it is essential that we consider and identify the *direction* of electric currents and the *polarity* of voltages. Algebraic signs, + and –, are used in this process. The process involves (a) the identification of directions and polarities on circuit diagrams, and (b) the assignment of algebraic signs to current or voltage values when they are used in mathematical operations. There are several conventions (widely accepted methods) in use. We describe the conventions which we use throughout this text.

Current

1. *Circuit marking—single-subscript and arrow method.* Identify a circuit current with an arrow and the letter I and a single letter or numeral subscript; for example,

The arrow indicates the direction

of a current. "Current" may mean "conventional current" or "electron flow." ["Conventional current" corresponds to the movement of *positive current carriers;* hence, it is the opposite of electron flow. The conventional-current concept was advanced by Benjamin Franklin before the development of our modern *electron theory* of electricity. Conventional current remains the official convention of the Institute of Electrical and Electronics Engineers (IEEE).] The direction indicated by the arrow may be known (as a result of measurement or by analysis) or assumed (because unknown) and therefore arbitrary. Unless indicated otherwise we use electron flow in this text—the arrow indicates direction of movement of electrons which is away from a point of negative charge toward one of less negative (more positive) charge. If a current, say I_1 of 5 A, is actually in the direction indicated on a diagram we say it is a positive current—$I_1 = 5$ A. If this current is really in a direction opposite of that indicated, it has a negative value—$I_1 = -5$ A.

2. *Double-subscript notation.* Sometimes we find it convenient to use a double-subscript notation, for example,

Here I_{ab} is defined as the current from a to b. If current is really from b to a it is $-I_{ab}$ or I_{ba}.

Voltage

1. *Circuit marking—single-subscript and polarity-marks method.* Identify each voltage *source* (battery, generator, power supply, and so on) with its known electrical polarity, that is, + on one terminal, – on the other, and an appropriate symbol such as V_s with single subscript, for example,

Identify the voltage across each *passive element* with single subscripted V and electrical polarity marks + or – as follows:

Electron current flow. Place – polarity on the terminal that current (known or assumed) enters, + polarity on the terminal that current leaves, for example,

Conventional current. Place + polarity mark on the terminal that current enters, – polarity on the terminal that current leaves.

2. *Double-subscripted voltages.* In some applications where the terminals of devices have standardized labels, or in circuits which we have intentionally labeled circuit points with letters or numerals to aid discussion, it is meaningful to use a double-subscript voltage notation. For example,

Here V_{BA} is defined as "the voltage of B with respect to A." If B is more negative than A by 0.3 v, then $V_{BA} = -0.3$ V. [If a digital multimeter (DMM), with automatic polarity indication, is connected to measure the

voltage V_{HC} so that the "common" test lead is on point C and the "hot" lead is on point H, the meter will indicate + if H is more positive than C (an analog voltohmmeter VOM will read "upscale"), – if H is less positive than C (a VOM will read "downscale").] Another way to look at the circuit is to consider the voltage of A with respect to B, V_{AB}. Then $V_{AB} = +0.3$ V. Reversing the order of the subscripts reverses the sign of the term. Let us consider some examples of the use of these conventions.

Example 1   Construct and label sample circuit diagrams to illustrate the given currents.

(a) $I_1 = 4$ A

(a)

(b) $I_a = -2$ A

(b)

(←——Actual current direction)

(c) $I_{pq} = -I_1$

(c)

(d) The equivalent of 3 A of electron charge moving from right to left through a resistor.

(d)

(e) 5 A of electron flow from point c to d.

(e)

Example 2   Label the given circuit diagrams with single-subscript voltage notation and polarity marks. Indicated current direction is for electron flow.

(a)

(a)

(b)

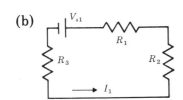

(b)

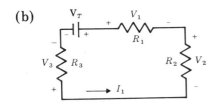

(c)

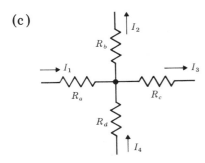

(c)

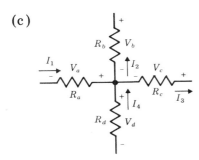

(d) Label with double-subscript voltage notation so that all values are positive.

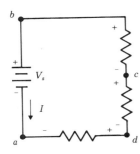

(d)

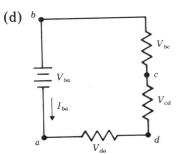

Example 3 Construct diagrams to illustrate the given voltage values.

(a) $V_1 = 3$ V, negative terminal on left.

(a)
$$\underset{\text{3 V}}{\overset{V_1}{-\!\!\wedge\!\!\wedge\!\!\wedge\!\!+}}$$

(b) $V_{pk} = 12$ V, negative terminal at lower end.

(b)

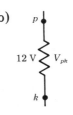

(c) Redraw and relabel polarity to make voltage value positive.

(c)
$$\underset{R_a}{\overset{V_a = 2\text{ V}}{+\!\!\wedge\!\!\wedge\!\!\wedge\!\!-}}$$

(d) $V_{12} = -6$ V (a source)

(d)

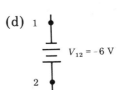

(e) Voltage at output terminals O, O' is $V_{OO'} = -3.5$ V. Indicate voltage polarity on the diagram with + and − symbols.

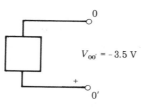

$V_{oo'} = -3.5$ V

Exercises 8-2   In Exercises 1–10 construct and label sample circuit diagrams to represent the given currents. (See Example 1.) All currents represent electron flow.

1. $I_1 = 5$ mA   2. $I_b = -10$ mA
3. $I_{ab} = 8$ mA   4. $I_{cd} = -27$ mA
5. $I_{de} = -I_1$ 6. $I_a = -I_{12}$
7. 50 mA flowing from left to right through a resistor.
8. 35 mA flowing from node c to node d.
9. I_m is the opposite of 40 mA flowing from right to left.
10. I_{ef} is actually flowing from f to e.

In Exercises 11–20 redraw the given circuit diagrams and label with single-subscript voltage notation and polarity marks.

11.

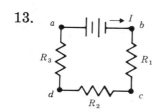

12.

13.

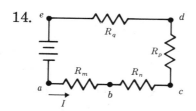

14.

15.

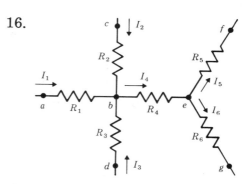

16.

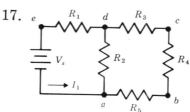

17.

18.

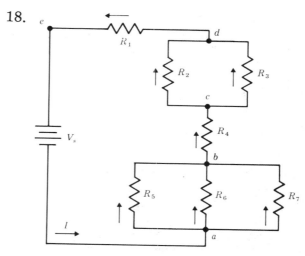

19.

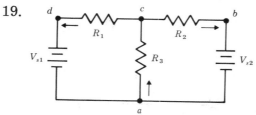

20.

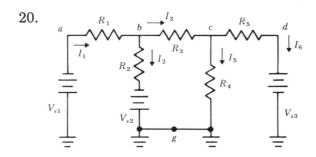

In Exercises 21–30 redraw the circuits of Exercises 11–20 and label with double-subscript voltage notation.

Some professionals prefer to use still another type of circuit notation for voltage, the *single-subscript and arrow notation:* Draw an arrow on diagram with head adjacent to a + polarity, tail adjacent to a – polarity. Identify arrow with a single-subscripted V. For example,

In Exercises 31–40 redraw the circuits of Exercises 11–20 and label with single-subscript and arrow notation.

In Exercises 41–50 construct and label model circuits to illustrate the given voltage values. Use notation of your choice or as requested by your instructor.

41. $V_1 = 9$ V (source)
42. $V_{ab} = -3$ V (source)
43. $V_a = -15$ V (source)
44. $V_{mn} = 6$ V (source)
45. $V_1 = 15$ mV (drop)
46. $V_a = -27$ mV (drop)
47. $V_{ab} = 16$ mV (drop)
48. $V_{12} = -15\ \mu$V (drop)
49. $V_1 = -V_{cd} = -24$ mV (drop)
50. $V_{mn} = -V_x = 15$ V (drop)

8-3 APPLYING KIRCHHOFF'S LAWS

The first step in applying the circuit laws in any formal analysis procedure is to label the circuit as described in Section 8-2. The next step is to write an equation summing either (a) the currents at a junction point, or (b) the voltages around a closed circuit path. In a KCL equation each *term* represents one of the currents at the summation junction point.

Each term must be assigned an algebraic sign to reflect the direction (known or assumed) of that particular current. Similarly, in a KVL equation, each *term* represents one of the voltages around the closed loop being considered. Each term must be assigned an algebraic sign taking into consideration (a) polarity and (b) the direction being considered. We summarize the procedures for assigning signs to current and voltage terms that will be used in this textbook.

In writing current law (KCL) equations:
1. Consider currents whose indicated direction is toward the summation point as positive.
2. Consider currents with directions away from that point as negative.

In writing voltage law (KVL) equations:
1. Trace a loop in direction of your choice, either clockwise (CW) or counterclockwise (CCW).
2. As you approach each circuit element, look at the polarity, + or –, of the first terminal you see. Write that sign with the term representing the voltage of the element.

Study the following examples carefully until you understand these procedures.

Example 4   Figure 8-1 represents just one node (junction point) of a complex network. Currents flow in the branches (current paths)

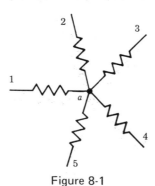

Figure 8-1

which connect at node a as follows: 1, toward a; 2, toward a; 3, away from a; 4, toward a; 5, away from a.

(a) Redraw the diagram and label with current and voltage notations.

(a)

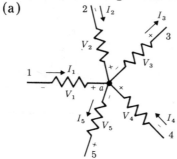

(b) Write a KCL equation for node a.
(b) KCL: $\Sigma I_a = 0$ $I_1 + I_2 - I_3 + I_4 - I_5 = 0$

Example 5   Determine the value of I_2 in Example 4 if the other currents have the following values: $I_1 = 5$ A, $I_3 = -3$ A, $I_4 = 7$ A, $I_5 = 6$ A.

We start with the KCL equation,

$$I_1 + I_2 - I_3 + I_4 - I_5 = 0$$
$$5 + I_2 - (-3) + 7 - 6 = 0$$
$$I_2 + 15 - 6 = 0$$
$$I_2 = -9 \text{ A}$$

I_2 is 9 A and flows away from node a (the negative sign indicates its actual direction is opposite to that stated originally).

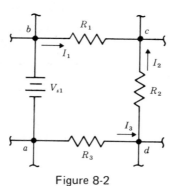

Figure 8-2

Example 6   The circuit in Figure 8-2 is just one loop (closed circuit path) of a complex network which has many loops. It has currents flowing as shown.

(a) Redraw the circuit and label with voltage notation.
(b) Write a KVL equation for the loop in terms of $V_s, I_1 R_1, I_2 R_2$, and $I_3 R_3$.
(c) Determine the magnitude and polarity of the voltage across R_3 if $V_s = 12$ V, $I_1 = 2$ A, $R_1 = 1.5\ \Omega$, $I_2 = 3.5$ A, and $R_2 = 2\ \Omega$.

(a)

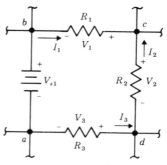

Figure 8-3

(b) KVL: Σ sources + Σ drops = 0
By arbitrary choice we start at node a and go around the loop CW:

$$-V_{s1} - I_1 R_1 + I_2 R_2 + I_3 R_3 = 0$$

(c) We substitute values into the KVL:

$$-12 - 2 \cdot 1.5 + 3.5 \cdot 2 + V_3 = 0$$
$$-15 + 7 + V_3 = 0, \qquad V_3 = 8 \text{ V}$$

The value of V_3 is positive and therefore its polarity as marked in Figure 8-3 is correct.

Exercises 8-3   In Exercises 1–8 arrows indicate direction of (electron) current flow in the branches. (a) Redraw the circuits and label with current and voltage notations (see Section 8-2). (b) Write KCL equations for the nodes depicted. (c) Solve for the unknown currents.

1.

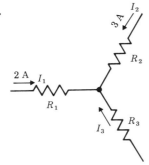

2.

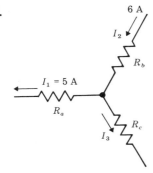

3.

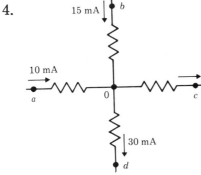

4.

5.

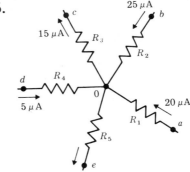

6.

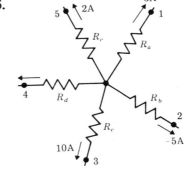

7.

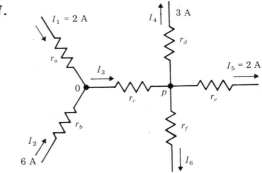

8.

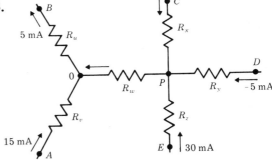

In Exercises 9–16 the circuits represent single loops from complex networks. (a) Redraw the circuits and label with current and voltage notations (see above). (b) Write KVL equations for the loops tracing CW. (c) Solve for the unknown voltages. (Assume that the indicated current directions are for actual currents.)

9.

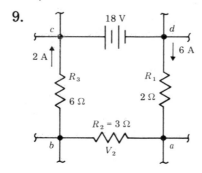

10.

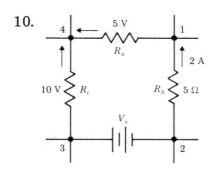

11.

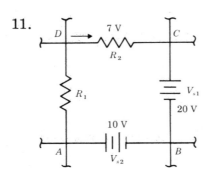

12.

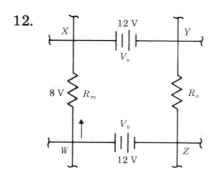

13.

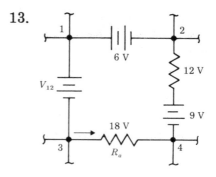

14.

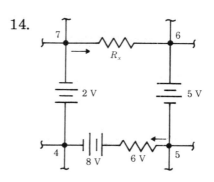

15.

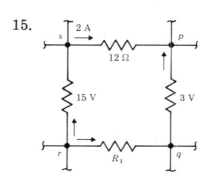

16.

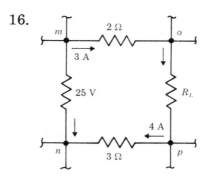

8-4 CIRCUIT VOLTAGE ANALYSIS AND VOLTAGE DIVIDER CIRCUITS

The circuit of Figure 8-4 is that of a transistor voltage amplifier and is drawn in a style typical of manufacturers' schematics. The diagram is labeled with the voltages of the collector (C), base (B), and emitter (E) with respect to ground—-4.5 V, -1.75 V, and -1.15 V, respectively. A typical everyday problem in the life of an electronics technician is to determine the voltage V_{BE}, the voltage of the base with respect to the emitter. How can Kirchhoff's voltage law help us with this very practical problem? Must we have a thorough knowledge of transistors to determine V_{BE}? The answer is that we can apply the voltage law to determine the voltage between two points in a circuit and need to know only the voltage sources and voltage drops in the circuit or the voltages of the two points with respect to a third point, commonly ground.

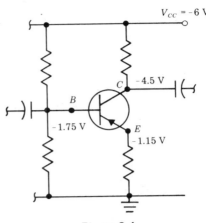

Figure 8-4

To determine the voltage (V_{XY}) of any point X in a circuit with respect to any other point Y in that circuit, find the algebraic sum of the voltages of a circuit path between X and Y. Obtain the algebraic sum by starting at point X and tracing the circuit to Y. As you approach each circuit element (source or passive device) write down the first polarity sign (+ or -) you see and the value (letter or numeral of its voltage).

We can find the voltage of B with respect to G by measuring with a DMM:

1. The voltage from B to E (hot lead to B, common lead to E) and obtaining $V_s = +9$ V.
2. The voltage from E to G (hot lead to E, common lead to G) and obtaining $I_e R_e = -3$ V.
3. Adding the two values algebraically, $V_{BG} = 9 - 3 = +6$ V.

Example 8

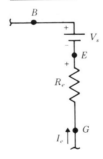

(a) $V_{BG} = V_{BE} + V_{EG} = \underline{+V_s + I_e R_e}$
(b) $V_{BG} = 9 + 3 = \underline{+12 \text{ V}}$

The circuits shown in the following examples are part of complete networks. For each find (a) V_{BG} in terms of V_s and $I_e R_e$, and (b) the magnitude and polarity of V_{BG} if $V_s = 9$ V and $I_e R_e = 3$ V.

Example 7

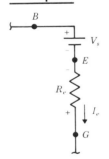

(a) $V_{BG} = V_{BE} + V_{EG} = \underline{+V_s - I_e R_e}$
(b) $V_{BG} = 9 - 3 = \underline{+6 \text{ V}}$

Let us relate the mathematics to the shop.

Example 9

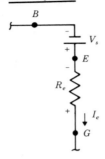

(a) $V_{BG} = V_{BE} + V_{EG} = \underline{-V_s - I_e R_e}$
(b) $V_{BG} = -9 - 3 = \underline{-12\ V}$

Example 10

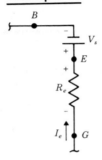

(a) $V_{BG} = V_{BE} + V_{EG} = \underline{-V_s + I_e R_e}$
(b) $V_{BG} = -9 + 3 = \underline{-6\ V}$

Exercises 8-4A   In Exercises 1–12 find the required voltages. (See Examples 7, 8, 9, and 10.)

1. Find V_{AC}.

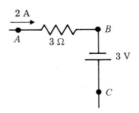

2. Find V_{CA} for the circuit of Exercise 1.
3. Find V_{CA}.

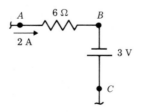

4. Find V_{AC} for the circuit of Exercise 3.
5. Find V_{CA}.

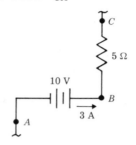

6. Find V_{AC} for the circuit of Exercise 5.
7. Find V_{AB}.

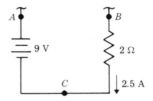

8. Find V_{BA} for the circuit of Exercise 7.
9. Find V_{AD}, V_{AE}, and V_{DA}.

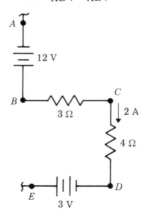

10. For the circuit of Exercise 9 find V_{BE}, V_{CA}, and V_{CE}.
11. For the circuit of Exercise 9 assume that node C is connected to ground and there is no other change in the circuit. (a) Redraw the diagram with ground connection shown. (b) Find V_A (point A to ground), V_B (B to ground), V_D, and V_E.
12. For the circuit of Exercise 9 assume that node D is connected to ground without any other change in the circuit. (a) Redraw the diagram with ground connection shown. (b) Find V_A (that is, A to ground), V_B, V_C, and V_E.

Let us apply the procedure to more complicated circuits.

Example 11

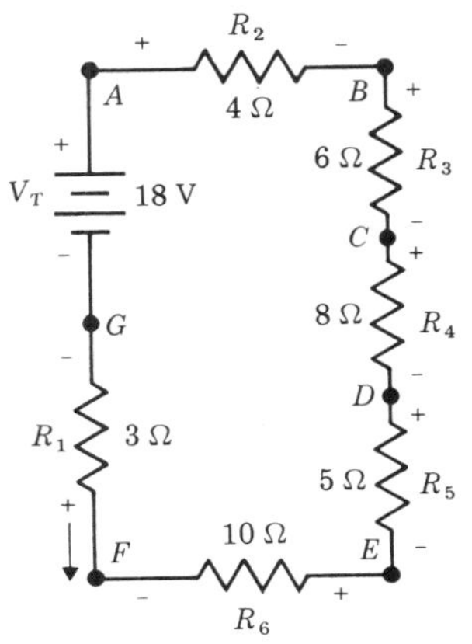

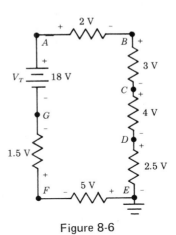

Figure 8-5

Example 12

Figure 8-6

For the circuit shown in Figure 8-5 find the following voltages: V_1, V_2, V_3, V_4, V_5, V_6, V_{AD}, V_{BF}, V_{GD}.

Sample calculation for V_1, V_2, V_3, and so on: We use the ratio formula of Section 7-4,

$$V_1 = \frac{R_1}{R_T} V_T = \frac{3}{36} \cdot 18 = 1.5 \text{ V}$$

$$V_1 = \underline{1.5 \text{ V}}, \qquad V_2 = \underline{2 \text{ V}}, \qquad V_3 = \underline{3 \text{ V}},$$

$$V_4 = \underline{4 \text{ V}}, \qquad V_5 = \underline{2.5 \text{ V}}, \qquad V_6 = \underline{5 \text{ V}}$$

The polarities of these voltages as shown on the diagram are determined by the direction of the current. To determine V_{AD} we start at A and trace, either CW or CCW, to D adding the voltages along the way:

CW: $V_{AD} = V_2 + V_3 + V_4 = 2 + 3 + 4 = \underline{+9 \text{ V}}$

CCW: $V_{AD} = V_T - V_1 - V_6 - V_5$
$= 18 - 1.5 - 5 - 2.5 = \underline{+9 \text{ V}}$

We find V_{BF} and V_{DG} in a similar fashion:

V_{BF} (CCW) $= -V_2 + V_T - V_1 = -2 + 18 - 1.5$
$= \underline{14.5 \text{ V}}$

V_{GD} (CW) $= -V_1 - V_6 - V_5$
$= -1.5 - 5 - 2.5 = \underline{-9 \text{ V}}$

We ground the circuit of Example 11 at node E (see Figure 8-6). Let V_A, V_B, and so forth, represent the voltages of the nodes (A, B, C, etc.) with respect to ground. Find V_A, V_B, V_C, V_D, V_F, and V_G.

V_A (tracing A, G, F, to E) $= 18 - 1.5 - 5$
$= \underline{+11.5 \text{ V}}$

V_B (tracing B, C, D, to E) $= 3 + 4 + 2.5$
$= \underline{+9.5 \text{ V}}$

$V_C = 4 + 2.5 = \underline{+6.5 \text{ V}}$

$V_D = \underline{+2.5 \text{ V}}$

$V_F = -5 \text{ V}$

$V_G = -1.5 - 5 = \underline{-6.5 \text{ V}}$

Example 13   Refer to the circuit of Figure 8-7. (a) Calculate the voltage across each resistor: V_1, V_2, V_3, and so on. (b) Using the KVL technique demonstrated in Examples 10, 11, and 12, determine the following voltages: V_{BD}, V_{VE}, V_{BG}, V_{BF}, V_{BH}, V_{CD}, V_{CE}, V_{CG}, V_{CF}, V_{CH}, V_{DF}, V_{DH}, V_{DG}, V_{EF}, V_{EH}, V_{EG}, V_{FG}, V_{HG}. Remember: V_{BE} means "the voltage of B with respect to E" ("hot" lead of a meter to B, COM lead to E). The sign of V_{BE} is + if B is more positive than E.

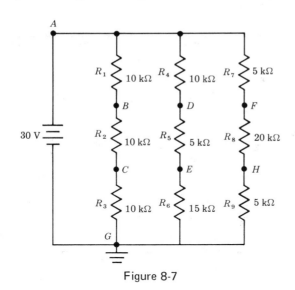

Figure 8-7

(a) Sample solution for resistor voltages:

$$V_5 = \frac{R_5}{R_4 + R_5 + R_6} V_T$$

$$= \frac{5k}{10k + 5k + 15k} \cdot 30 = \underline{5\text{ V}}$$

$V_1 = \underline{10\text{ V}}, \qquad V_2 = \underline{10\text{ V}}, \qquad V_3 = \underline{10\text{ V}},$
$V_4 = \underline{10\text{ V}}, \qquad V_5 = \underline{5\text{ V}}, \qquad V_6 = \underline{15\text{ V}},$
$V_7 = \underline{5\text{ V}}, \qquad V_8 = \underline{20\text{ V}}, \qquad V_9 = \underline{5\text{ V}}$

(b) V_{BD} (tracing $B - A - D$) $= -10 + 10 = \underline{0\text{ V}}$

$V_{BE} = -10 + 10 + 5 = \underline{+5\text{ V}}$

V_{BG} ($B - C - G$) $= +10 + 10 = \underline{+20\text{ V}}$

$V_{BF} = -10 + 5 = \underline{-5\text{ V}}$

$V_{BH} = -10 + 5 + 20 = \underline{+15\text{ V}}$

V_{CD} ($C - B - A - D$) $= -10 - 10 + 10$
$\qquad = \underline{-10\text{ V}}$

V_{CE} ($C - G - E$) $= +10 - 15 = \underline{-5\text{ V}}$

$V_{CG} = \underline{+10\text{ V}}$

$V_{CF} = -10 - 10 + 5 = \underline{-15\text{ V}}$

$V_{CH} = +10 - 5 = \underline{5\text{ V}}$

$V_{DF} = -10 + 5 = \underline{-5\text{ V}}$

$V_{DH} = -10 + 5 + 20 = \underline{+15\text{ V}}$

$V_{DG} = +5 + 15 = \underline{+20\text{ V}}$

$V_{EF} = +15 - 5 - 20 = \underline{-10\text{ V}}$

$V_{EH} = +15 - 5 = \underline{+10\text{ V}}$

$V_{EG} = \underline{+15\text{ V}}$

$V_{FG} = +20 + 5 = \underline{+25\text{ V}}$

$V_{HG} = \underline{+5\text{ V}}$

A careful analysis of the voltages for the circuit of Example 13 will enable us to develop the technique for finding the voltage between two points when only their voltages with respect to a common third point are known. Let us consider points D and H in Figure 8-7. We found that the voltage of D with respect to ground was +20 V (V_{DG} = +20 V), V_{HG} = +5 V and V_{DH} = 15 V. Note that we could find V_{DH} even if we knew only V_{DG} and V_{HG}:

$$V_{DH} = V_{DG} - V_{HG} = 20 - 5 = +15\text{ V}$$

That is, when the voltages of two circuit points X and Y with respect to ground (V_{XG} and V_{YG}) are known, the voltage of point X with respect to point Y is found using the relation

$$V_{XY} = V_{XG} - V_{YG}$$

Let us see if this idea works with other points in the circuit of Figure 8-7. Consider V_{CF}:

$$V_{CF} = V_{CG} - V_{FG} = 10 - 25 = -15\text{ V}$$

and this agrees with our previous finding in Example 13.

Example 14

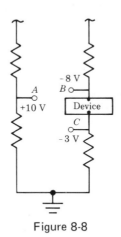

Figure 8-8

The circuit of Figure 8-8 is part of a complex network. Voltages shown are the voltages of the points with respect to ground. Determine V_{CA}, V_{CB}, and V_{BA}.

$$V_{CA} = -3 - 10 = \underline{-13\text{ V}}$$
$$V_{CB} = -3 - (-8) = \underline{+5\text{ V}}$$
$$V_{BA} = -8 - 10 = \underline{-18\text{ V}}$$

We are now ready to find the voltage of the base with respect to the emitter of the circuit of Figure 8-4:

$$V_{BE} = V_B - V_E = -1.75 - (-1.15) = \underline{-0.60 \text{ V}}$$

We find the voltage of collector with respect to base and emitter,

$$V_{CB} = V_C - V_B = -4.5 - (-1.75) = \underline{-2.75 \text{ V}}$$
$$V_{CE} = V_C - V_E = -4.5 - (-1.15) = \underline{-3.35 \text{ V}}$$

Example 15

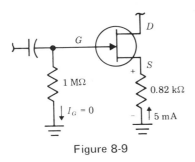

Figure 8-9

Determine V_{GS}, the gate-to-source voltage of the self-biased FET stage (see Figure 8-9).

V_{GS} (tracing G-Ground-S)

$$= V_G - V_S$$
$$= 0 \times 1 \times 10^6 - (5 \times 10^{-3})(0.82 \times 10^3)$$
$$= \underline{-4.1 \text{ V}}$$

Example 16

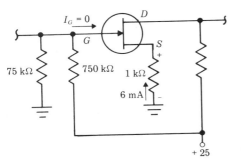

Figure 8-10

Determine the gate bias (V_{GS}) for the FET in the circuit of Figure 8-10.

V_G (gate to ground)

$$= \frac{75k}{75k + 750k} \cdot 25$$
$$= \underline{2.273 \text{ V}}$$
$$V_S = (6 \times 10^{-3})(1 \times 10^3) = \underline{6 \text{ V}}$$
$$V_{GS} = V_G - V_S = 2.273 - 6.0$$
$$= \underline{-3.727 \text{ V}}$$

Exercises 8-4B In Exercises 13–22 find the required voltages.

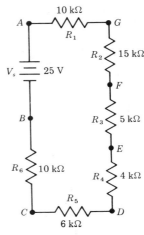

Figure 8-11

13. For the circuit of Figure 8-11 find: (a) voltages across all resistors, and (b) V_{AC}, V_{AE}, V_{EB}, V_{EA}, V_{GC} (see Example 11).

14. For the circuit of Figure 8-11 assume that V_s is changed to 15 V and its direction reversed. (a) Redraw the circuit. Find (b) voltages across all resistors, and (c) V_{BF}, V_{BG}, V_{GB}, V_{GD}, V_{GA}.

15. Refer to Exercise 13. Assume that a ground is connected to node G. (a) Redraw the circuit. (b) Find the voltages of the nodes with respect to ground: V_A, V_B, V_C, V_D, V_E, V_F (see Example 12).

16. Refer to Exercise 14. Assume that a ground is connected to node C. (a) Re-

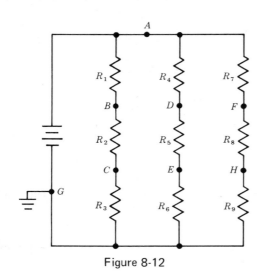

Figure 8-12

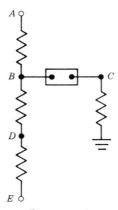

Figure 8-13

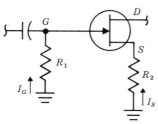

Figure 8-14

draw the circuit. (b) Find the voltages of the nodes with respect to ground: V_A, V_B, V_D, V_E, V_F, V_G.

17. Refer to the circuit of Figure 8-12. Given: $V_s = 80$ V, $R_1 = 20$ kΩ, $R_2 = 40$ kΩ, $R_3 = 20$ kΩ, $R_4 = 50$ kΩ, $R_5 = 25$ kΩ, $R_6 = 5$ kΩ, $R_7 = 8$ kΩ, $R_8 = 22$ kΩ, $R_9 = 50$ kΩ (see Example 13). Find (a) the voltages across all resistors, (b) V_{BD}, V_{BE}, V_{BF}, V_{BH}, V_{DF}, V_{DH}, V_{EF}, V_{EH}, and (c) the voltages of nodes to ground, V_B, V_C, V_D, V_E, V_F, V_H.

18. Refer to Exercise 17. Given: $V_s = -40$ V (polarity reversed) and R-values as in Exercise 17. Repeat parts (a), (b), and (c) as in Exercise 17.

19. Refer to the circuit of Figure 8-13. The voltages of the labeled circuit points with respect to ground are $V_A = +12$ V, $V_B = +2.5$ V, $V_C = +1.9$ V, $V_D = -2$ V, $V_E = -6$ V. Find V_{BC}, V_{CB}, V_{DC}, V_{EC}, and V_{EB} (see Example 14).

20. Refer to the circuit of Figure 8-13. Given the node voltages with respect to ground: $V_A = -9$ V, $V_B = -2.5$ V, $V_C = -2.2$ V, $V_D = +2$ V, $V_E = +3$ V. Find V_{BC}, V_{CB}, V_{DC}, V_{EC}, and V_{EB}.

21. Refer to Figure 8-13. Given: $V_{AB} = +4$ V, $V_{BC} = +0.6$ V, $V_C = +1.4$ V, $V_{DB} = -2$ V, $V_{ED} = -4$ V. Find (a) the voltages of the nodes with respect to ground, and (b) V_{DC}, V_{CE}, V_{BE}.

22. Refer to Figure 8-13. Given: $V_{AB} = -5$ V, $V_{BC} = -0.3$ V, $V_C = -1.7$ V, $V_{DB} = +2$ V, $V_{DE} = -2$ V. Find (a) the voltages of the nodes with respect to ground, and (b) V_{DC}, V_{CE}, V_{BE}.

23. Refer to the circuit of Figure 8-14. Given: $R_1 = 910$ kΩ, $R_2 = 0.91$ kΩ, $I_G = 0$, $I_S = 4$ mA. Find V_{GS}, the gate-to-source voltage (see Example 15).

24. Repeat Exercise 23. Given: $R_1 = 1.2$ MΩ, $R_2 = 0.82$ kΩ, $I_G = 0$, $I_S = 6$ mA.

25. Refer to the circuit of Figure 8-15. Given: $R_1 = 100$ kΩ, $R_2 = 910$ kΩ, $R_3 = 1$ kΩ, $V_{DD} = 20$ V, $I_G = 0$, and $I_S = 4.5$ mA. Find V_G and V_{GS} (see Example 16).

26. Repeat Exercise 25 for the following conditions: $R_1 = 120$ kΩ, $R_2 = 1.0$ MΩ, $R_3 = 910$ Ω, $V_{DD} = 12$ V, $I_G = 0$, and $I_S = 5.5$ mA.

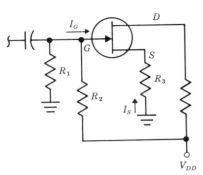

Figure 8-15

Circuits with More than One Source Many multisource networks require the more advanced techniques found in subsequent sections. However, some can be analyzed with the tools we have already used.

Example 17

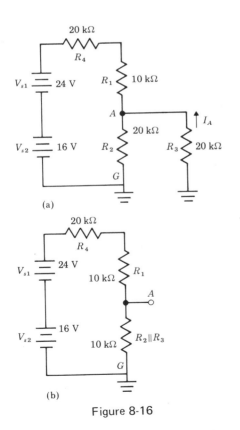

(a)

(b)

Figure 8-16

For the circuit of Figure 8-16(a), determine V_A (voltage of node A to ground), and I_A.

First, let us note that R_2 is in parallel with R_3. If we replace R_2 and R_3 with their equivalent, $R_2 \parallel R_3$, the circuit is a simple series circuit with two sources aiding and a ground at G [see Figure 8-16(b)].

We find the equivalent $R_2 \parallel R_3$,

$$R_2 \parallel R_3 = \frac{20 \text{ k}\Omega}{2} = 10 \text{ k}\Omega$$

then

$$V_A = V_{2\parallel3} = \frac{R_2 \parallel R_3}{R_T} \; V_T = \frac{10 \text{ k}\Omega}{40 \text{ k}\Omega} \cdot (24 + 16)$$

$$V_A = \underline{+10 \text{ V}}$$

$$I_A = \frac{V_A}{R_3} = \frac{10}{20 \text{ k}\Omega} = \underline{0.5 \text{ mA}}$$

Example 18

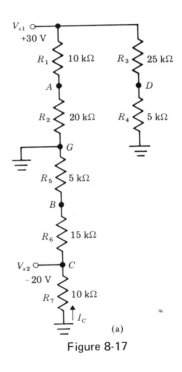

(a)

Figure 8-17

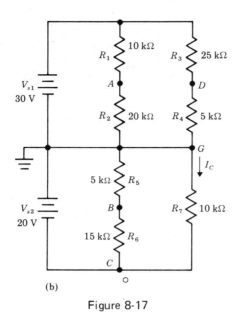

(b)

Figure 8-17

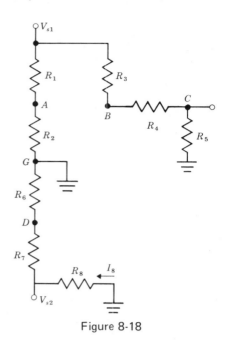

Figure 8-18

Analyze the circuit of Figure 8-17 to determine V_A, V_B, V_C, V_D (voltages of nodes to ground), V_{AD}, V_{DB}, V_{BA}, and I_C.

This circuit is really two independent parallel circuits with separate sources but a common ground point [see Figure 8-17(b)].

$$V_A = \frac{R_2}{R_1 + R_2} V_{s1} = \frac{20\ k\Omega}{30\ k\Omega} \cdot 30 = \underline{+20\ V}$$

$$V_D = \frac{R_4}{R_3 + R_4} V_{s1} = \frac{5\ k\Omega}{30\ k\Omega} \cdot 30 = \underline{+5\ V}$$

$$V_B = \frac{R_5}{R_5 + R_6} V_{s2} = \frac{5\ k\Omega}{20\ k\Omega} (-20) = \underline{-5\ V}$$

$$V_C = V_{s2} = \underline{-20\ V}$$

$$V_{AD} = V_A - V_D = 20 - 5 = \underline{+15\ V}$$

$$V_{DB} = V_D - V_B = 5 - (-5) = \underline{+10\ V}$$

$$V_{BA} = V_B - V_A = -5 - 20 = \underline{-25\ V}$$

$$I_C = \frac{V_C}{R_7} = \frac{-20}{10\ k\Omega} = \underline{-2\ mA}$$
(I_C will flow from C to ground.)

Exercises 8-4C In Exercises 27 and 28 find the required voltages.

27. Refer to Figure 8-18. Given: $V_{s1} = +18$ V, $V_{s2} = -12$ V, $R_1 = 10$ kΩ, $R_2 = 8$ kΩ, $R_3 = 3$ kΩ, $R_4 = 10$ kΩ, $R_5 = 5$ kΩ, $R_6 = 18$ kΩ, $R_7 = 6$ kΩ,

$R_8 = 4$ kΩ. Determine V_A, V_B, V_C, V_D, V_{CA}, V_{CD}, V_{AB}, and I_8 (see Example 18).

28. If the resistor values of the circuit in exercise 27 remain the same, but $V_{s1} = -9$ V and $V_{s2} = 6$ V, what do the quantities called for in that exercise become?

8-5 THE SUPERPOSITION THEOREM

Many electronics networks which contain two or more sources cannot be analyzed by the techniques presented in the preceding sections. An important principle, the *superposition theorem*, however, provides a method for analyzing these more complex networks— a method requiring only the algebraic techniques of preceding chapters.

Superposition comes from the verb *superpose* which means "to lay or place on, over, or above something else." And that is just what we do in applying the method. That is, first we analyze a network using only one source at a time, we determine the voltages or currents that each source acting alone will produce. Then we "lay" these analyses on top of each other and combine them algebraically to determine the "net" values. While we are analyzing with one source we must reduce the effect of the other sources to zero. We demonstrate the application of the principle with several examples.

Example 19 Use the superposition theorem to find V_A, the voltage of point A with respect to ground, in the voltage divider circuit of Figure 8-19(a).

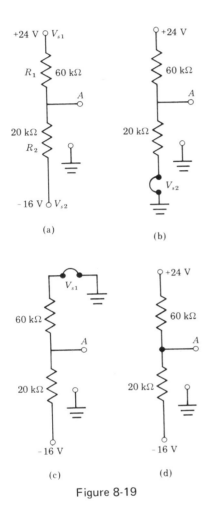

(a)

(b)

(c)

(d)

Figure 8-19

We first redraw the circuit as in Figure 8-19(b), reducing the effect of V_{s2} to zero by shorting it out. (In a real circuit we would remove the battery and connect a short piece of wire in its place.) Considering V_{s1} only, which we will call the first source, we have

$$V_{A1} = \frac{R_2}{R_1 + R_2} V_{s1} = \frac{20(10^3)}{60(10^3) + 20(10^3)} \times 24$$

$$V_{A1} = 6 \text{ V}$$

Next we short out V_{s2} and consider the effect of the second source V_{s2} acting alone [see Figure 8-19(c)]:

$$V_{A2} = \frac{R_1}{R_1 + R_2} V_{s2}$$

$$= \frac{60(10^3)}{60(10^3) + 20(10^3)} \times (-16)$$

$$V_{A2} = -12 \text{ V}$$

Finally, we find V_A by combining the effects of the two sources algebraically [see Figure 8-19(d)],

$$V_A = V_{A1} + V_{A2} = 6 + (-12)$$

$$V_A = \underline{-6 \text{ V}}$$

Example 20 Use the superposition method to find the currents in R_1, R_2, and R_3 (I_1, I_2, and I_3) in the circuit of Figure 8-20(a).

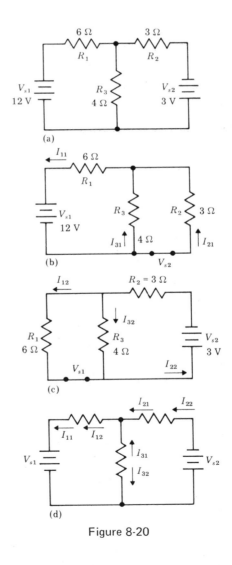

(a)

(b)

(c)

(d)

Figure 8-20

We start the solution by redrawing the network. We reduce V_{s2} to zero by shorting across its terminals [see Figure 8-20(b)]. Next we find R_{T1}, the total resistance for this first-source network (the network used with the first source considered in the analysis);

$$R_{T1} = R_1 + R_2 \parallel R_3 = 6 + \frac{3 \times 4}{3 + 4} = 7.714 \ \Omega$$

Next we determine the first-source currents, I_{11}, I_{21}, and I_{31}, the currents in the resistors due to V_{s1} acting alone:

$$I_{11} = \frac{V_{s1}}{R_{T1}} = \frac{12}{7.714} = 1.556 \ \text{A}$$

We use the current-divider formula to find I_{21} and I_{31} (review Section 7-5):

$$I_{21} = \frac{R_3}{R_2 + R_3} I_{11} = \frac{4}{7} \times 1.556 = 0.8891 \ \text{A}$$

$$I_{31} = \frac{R_2}{R_2 + R_3} I_{11} = \frac{3}{7} \times 1.556 = 0.6669 \ \text{A}$$

We carefully label the first-source network [see Figure 8-20(b)] with its currents, including their direction.

Next we analyze the second-source network [see Figure 8-20(c)] in the same way:

$$R_{T2} = R_2 + R_1 \parallel R_3 = 3 + \frac{6 \times 4}{6 + 4} = 5.4 \ \Omega$$

$$I_{T2} = I_{22} = \frac{V_{s2}}{R_{T2}} = \frac{3}{5.4} = 0.5556 \ \text{A}$$

$$I_{12} = \frac{R_3}{R_1 + R_3} I_{T2} = \frac{4}{6 + 4} \times 0.5556$$
$$= 0.2222 \ \text{A}$$

$$I_{32} = \frac{R_1}{R_1 + R_3} I_{T2} = \frac{6}{6 + 4} \times 0.5556$$
$$= 0.3334 \ \text{A}$$

We label the circuit [see Figure 8-20(c)].

Finally, we transfer the current values to the final network drawing [Figure 8-20(d)], which again shows both sources, and combine the currents algebraically:

$$I_1 = I_{11} + I_{12} = 1.556 + 0.2222 = \underline{1.778 \ \text{A}}$$

Since both I_{11} and I_{12} are in the same direction they have the same sign:

$$I_2 = I_{21} + I_{22} = 0.8891 + 0.5556 = \underline{1.445 \ \text{A}}$$

$$I_3 = I_{31} + I_{32} = 0.6669 + (-0.3334) = \underline{0.3335 \ \text{A}}$$

In R_3 the two sources force currents in opposite directions. The actual current is the algebraic sum of the two separate currents and is

in the direction of the larger, the first-source current.

Example 21   Determine $I_1, I_2, I_3, I_4, I_5, V_A$, and V_B for the circuit of Figure 8-21(a).

We proceed as in Example 20:

1.   Find first-source currents and then second-source currents.

2.   Superpose the currents in the original network and combine them algebraically.

Analyzing with V_{s1} acting alone, V_{s2} replaced with its internal resistance, we first find R_{T1}. Studying the circuit of Figure 8-21(c) we see that R_3 and R_4 are in parallel, $R_3 \parallel R_4$. R_5 is in series with that combination, $R_5 + R_3 \parallel R_4$. That combination is, in turn, in parallel with R_2, $R_2 \parallel (R_5 + R_3 \parallel R_4)$. Finally, R_1 is in series with all of that,

$$R_{T1} = R_1 + R_2 \parallel (R_5 + R_3 \parallel R_4)$$

$$R_{T1} = R_1 + \frac{R_2 \left(R_5 + \dfrac{R_3 R_4}{R_3 + R_4} \right)}{R_2 + R_5 + \dfrac{R_3 R_4}{R_3 + R_4}}$$

$$= 5 + \frac{15 \left(50 + \dfrac{25 \times 15}{25 + 15} \right)}{15 + 50 + \dfrac{25 \times 15}{25 + 15}}$$

$$= 5 + \frac{15(59.375)}{74.375} = 16.97 \ \text{k}\Omega$$

(Since all resistor values are in kilohms, for convenience we use these values in our calculations and do not change to ohms. We must remember to use the correct unit with our answers.)

$$I_{T1} = I_{11} = \frac{V_{s1}}{R_{T1}} = \frac{40}{16.97} = 2.356 \ \text{mA}$$

$$I_{51} = \frac{R_2}{R_2 + R_5 + R_3 \parallel R_4} I_{T1} = \frac{15}{74.375} \times 2.356$$
$$= 0.2017 \times 2.356 = 0.4752 \ \text{mA}$$

$$I_{21} = \frac{R_5 + R_3 \parallel R_4}{R_2 + R_5 + R_3 \parallel R_4} I_{T1} = \frac{59.375}{74.375} \times 2.356$$
$$= 0.7983 \times 2.356 = 1.881 \ \text{mA}$$

$$I_{31} = \frac{R_4}{R_3 + R_4} I_{51} = \frac{15}{25 + 15} \times 0.4752$$
$$= 0.1782 \ \text{mA}$$

$$I_{41} = \frac{R_3}{R_3 + R_4} I_{51} = \frac{25}{25 + 15} \times 0.4752$$
$$= 0.2970 \ \text{mA}$$

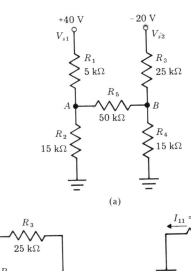

(a)

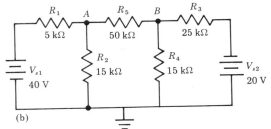

(b)

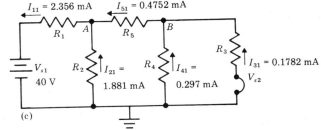

(c)

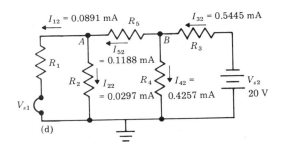

(d)

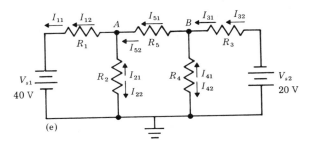

(e)

Figure 8-21

We label the first-source network with these currents [see Figure 8-21(c)] and then proceed to find the second-source currents [see Figure 8-21(d)].

$$R_{T2} = R_3 + R_4 \parallel (R_5 + R_1 \parallel R_2)$$

$$= 25 + \frac{15\left(50 + \dfrac{5 \times 15}{5 + 15}\right)}{15 + 50 + \dfrac{5 \times 15}{5 + 15}}$$

$$= 25 + \frac{15(53.75)}{68.75} = 36.73 \text{ k}\Omega$$

$$I_{T2} = I_{32} = \frac{V_{s2}}{R_{T2}} = \frac{20}{36.73} = 0.5445 \text{ mA}$$

$$I_{42} = \frac{R_5 + R_1 \parallel R_2}{R_4 + R_5 + R_1 \parallel R_2} I_{T2}$$

$$= \frac{53.75}{68.75} \times 0.5445 = 0.4257 \text{ mA}$$

$$I_{52} = \frac{R_4}{R_4 + R_5 + R_1 \parallel R_2} I_{T2}$$

$$= \frac{15}{68.75} \times 0.5445 = 0.1188 \text{ mA}$$

$$I_{12} = \frac{R_2}{R_1 + R_2} I_{52} = \frac{15}{5 + 15} \times 0.1188$$

$$= 0.0891 \text{ mA}$$

$$I_{22} = \frac{R_1}{R_1 + R_2} I_{52} = \frac{5}{5 + 15} \times 0.1188$$

$$= 0.0297 \text{ mA}$$

We label the second-source circuit with these currents [see Figure 8-21(d)]. Finally, we superpose the currents in the network with both sources [see Figure 8-21(e)] and combine them algebraically to determine the net current for each component:

$I_1 = I_{11} + I_{12} = 2.356 + 0.0891 = \underline{2.445 \text{ mA}}$

$I_2 = I_{21} + I_{22} = 1.881 + (-0.0297) = \underline{1.851 \text{ mA}}$

(The current through R_2 is in the direction of the current caused by V_{s1} acting alone since I_{21} is the larger current.)

$I_3 = I_{31} + I_{32} = 0.1782 + 0.5445 = \underline{0.7227 \text{ mA}}$

$I_4 = I_{41} + I_{42} = 0.2970 + (-0.4257)$

$\quad = \underline{-0.1287 \text{ mA}}$

(The − sign means simply that the current is in the direction that I_4 would flow if V_{s2} acted alone.)

$I_5 = -I_{AB} = I_{51} + I_{52} = 0.4752 + 0.1188$

$\quad = \underline{0.5940 \text{ mA}}$

We may check our solutions by applying Kirchhoff's current law, say at nodes A and B.

Node A: $-I_1 + I_2 + I_5 = -2.445 + 1.851$
$\quad\quad\quad\quad + 0.5940 = 0$

Node B: $I_3 - I_4 - I_5 = 0.7227 - 0.1287$
$\quad\quad\quad\quad - 0.5940 = 0$

Finally, we determine V_A and V_B:

$V_A = I_2 R_2 = (1.851)(15) = \underline{27.77 \text{ V}}$

$V_B = I_4 R_4 = (-0.1287)(15) = \underline{-1.931 \text{ V}}$

(Node B is negative with respect to ground.)

We state the principle that we have used in these solutions.

SUPERPOSITION THEOREM

If the resistances and sources of a network are (1) linear, and (2) bilateral, the currents and voltages which will develop in the network in response to two or more sources will be equal to the algebraic sum of the respective currents and voltages considering each source separately.

The condition "linear" means that the current in an element is directly proportional to the voltage across it (review "Variation," Section 7-3), no matter what that voltage may be (within practical limits, of course). It means that an I-versus-V graph for the element is a straight line (review "Graphs," Section 3-2). *Bilateral* means that current will flow equally well in either direction in the element. Most passive components—resistors, capacitors, and inductors—are linear and bilateral. Most active components—electron tubes, semiconductor diodes, and transistors—are neither linear nor bilateral. The principle applies whether the sources produce direct or alternating current (dc or ac). In the case of ac, the passive components are referred to as *impedances*.

Exercises 8-5 In Exercises 1–15 use the superposition theorem in your solutions.

1. In the circuit of Figure 8-22(a), $R_1 = 30 \ \Omega$, $R_2 = 50 \ \Omega$, $V_{s1} = 10$ V, and $V_{s2} = 18$ V. Find V_A (voltage of A with respect to ground).

2. Refer to Exercise 1. We alter the circuit by changing V_{s1} to -15 V and V_{s2} to -24 V (change amplitude and polarity of both V_{s1} and V_{s2}). Find V_A.

3. In the circuit of Figure 8-22(b), $R_1 = 20 \ \text{k}\Omega$, $R_2 = 30 \ \text{k}\Omega$, $R_3 = 10 \ \text{k}\Omega$, $V_{s1} = 50$ V, and $V_{s2} = 10$ V with polarities as marked. Find V_A and V_B.

4. Refer to Exercise 3. Find V_A and V_B if $V_{s1} = -40$ V, $V_{s2} = -20$ V, and all resistor values are unchanged. (*Note:* The signs for V_{s1} and V_{s2} indicate polarities opposite to those marked.)

5. Refer to Figure 8-22(c). Given $V_{s1} = +12$ V, $V_{s2} = -6$ V (signs indicate polarities of terminals with respect to ground), $R_1 = 4 \ \Omega$, $R_2 = 3 \ \Omega$, $R_3 = 6 \ \Omega$. Find V_A and the magnitude and direction of current in R_3.

6. Refer to Exercise 5. What is the current supplied by V_{s1}?

7. What current will V_{s2} supply in the circuit of Exercise 5?

8. Refer to Figure 8-22(c). Given: $V_{s1} = +240$ V, $V_{s2} = -180$ V, $R_1 = 80 \ \text{k}\Omega$, $R_2 = 100 \ \text{k}\Omega$, $R_3 = 40 \ \text{k}\Omega$. Find V_A and the magnitude and direction of current in R_3.

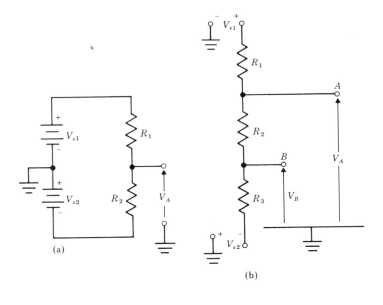

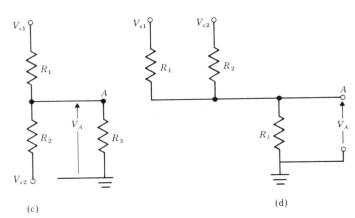

Figure 8-22

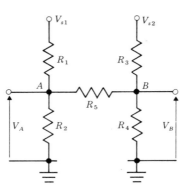

Figure 8-23

9. The circuit elements in Figure 8-22(d) have the following values: $V_{s1} = +15$ V, $V_{s2} = +10$ V, $R_1 = 10$ Ω, $R_2 = 5$ Ω, $R_L = 5$ Ω. Find V_A and I_L.

10. Refer to Figure 8-22(d). Given: $V_{s1} = +90$ V, $V_{s2} = +60$ V, $R_1 = 6$ kΩ, $R_2 = 10$ kΩ, and $R_L = 4$ kΩ. Find V_A and I_L.

11. Refer to Exercise 10. Find I_L and V_A if $R_L = 0$.

12. Refer to Figure 8-22(d). Given: $V_{s1} = -12$ V, $V_{s2} = -9$ V, $R_1 = 15$ kΩ, $R_2 = 12$ kΩ, $R_L = 4.8$ kΩ. Find V_A and I_L.

13. Refer to Figure 8-23. Given: $V_{s1} = +30$ V $V_{s2} = -20$ V, $R_1 = 5$ Ω, $R_2 = 10$ Ω, $R_3 = 6$ Ω, $R_4 = 4$ Ω, $R_5 = 10$ Ω. Find V_A, V_B, and I_{AB}.

14. For the circuit of Figure 8-23, $V_{s1} = +100$ V, $V_{s2} = +80$ V, $R_1 = 70$ Ω, $R_2 = 30$ Ω, $R_3 = 20$ Ω, $R_4 = 80$ Ω, $R_5 = 50$ Ω. Find V_A, V_B, and I_{AB}.

15. For the circuit of Figure 8-23, $V_{s1} = -12$ V, $V_{s2} = -9$ V, $R_1 = 6.8$ kΩ, $R_2 = 18$ kΩ, $R_3 = 4.7$ kΩ, $R_4 = 5.6$ kΩ, and $R_5 = 2.4$ kΩ. Find V_A, V_B, and I_{AB}.

Chapter 9

Circuit Analysis Methods II: Techniques for Network Simplification

Circuit analysis is an essential task in both circuit design and trouble diagnosis. The signature analyzer is an electronic instrument which automates the analysis of digital electronic circuits by comparing their actual "signatures" (patterns of signals during operation) with their known correct signatures. (Courtesy of Hewlett-Packard Company)

9-1 EQUIVALENCY PRINCIPLE

The analysis of circuits containing resistors in series, parallel, or series-parallel combinations generally involves reducing such circuits to the simplicity of a single resistance—a resistance equivalent to the original circuit. If we connect the same voltage, say V_s, to a complex network of resistors, and then to a single resistor R, and obtain the same current I in each case, then the single resistor is *equivalent* to the network.

Two two-terminal circuits are equivalent if they have the same voltage-current characteristic.

Networks containing sources and resistances, as well as resistances only, can also be reduced to equivalent circuits of great simplicity. The techniques of these simplifications may take various forms. *Thévenin's theorem, Norton's theorem,* and *Millman's theorem* are three very popular techniques for such simplifications. We present them in this chapter.

9-2 THÉVENIN'S THEOREM

With Thévenin's theorem we are able to simplify a complex network of resistances and sources to a simple series circuit with only one source and one resistance. We state the principle and process in three parts:

1. *For the purpose of finding the electrical effect on a circuit connected to any two points of a network, that network, no matter how complex and containing both sources and resistances, can be represented*

by a simple series circuit. *This equivalent circuit consists of a resistance, hereafter designated R_{th}, in series with an ideal voltage source, V_{th} (see Figure 9-1).*

2. *The ideal voltage source V_{th} is equal to the voltage at the two points where the "source" network connects to the "load" device (or circuit, or network) with the load network disconnected. That is,*

$$V_{th} = V_{oc}$$

3. *The resistance R_{th} is equal to the resistance "seen looking back into" the source network from the load connection points, and with the sources reduced to zero.*

Although a statement of the theorem is heavy, its application is not, if the words are not allowed to turn us away before we try it.

Let us study Figure 9-1 and note that if a current-measuring meter with zero resistance, instead of R_L, were connected to either source network, a short-circuit current I_{sc} would be measured. But looking at the Thévenin equivalent circuit it would have to be true that $R_{th} = V_{th}/I_{sc}$. And, since $V_{th} = V_{oc}$, we could find R_{th} using the relation

$$R_{th} = \frac{V_{oc}}{I_{sc}} \qquad (1)$$

In the field or laboratory we may "thevenize" (find the Thévenin equivalent of) an actual circuit by measuring V_{oc} and I_{sc} and using these values in equation (1). At the desk we calculate V_{oc} and R_{th} using appropriate analysis techniques which may include calculating I_{sc} then R_{th} from equation (1).

Example 1 Using Thévenin's theorem determine the current through the load resistor R_L of Figure 9-2a. Let R_L have values of 5 Ω, 10 Ω, and 25 Ω.

First, we note that the circuit may be drawn as in Figure 9-2b. We calculate V_{oc},

$$V_{oc} = \frac{R_2}{R_1 + R_2} V_s = \frac{12}{27} \times 9 = 4 \text{ V}$$

Then we set V_s to zero by replacing it with a short conductor [see Figure 9-2d] and calculate R_{th},

$$R_{th} = R_1 \| R_2 = \frac{12 \cdot 15}{12 + 15} = 6.667 \ \Omega$$

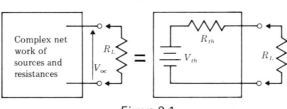

Figure 9-1

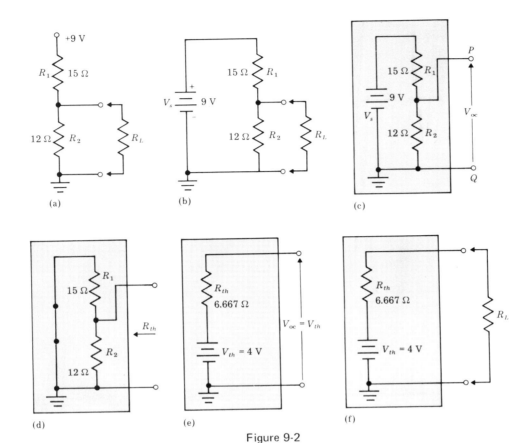

Figure 9-2

The Thévenin equivalent circuit is shown in Figure 9-2e. Next we calculate the various load currents [see Figure 9-2f].
For $R_L = 5 \; \Omega$:

$$I_L = \frac{V_{\text{th}}}{R_{\text{th}} + R_L} = \frac{4}{6.667 + 5} = \underline{342.8 \text{ mA}}$$

For $R_L = 10 \; \Omega$:

$$I_L = \frac{4}{6.667 + 10} = \underline{240.0 \text{ mA}}$$

For $R_L = 25 \; \Omega$:

$$I_L = \frac{4}{6.667 + 25} = \underline{126.3 \text{ mA}}$$

To demonstrate the alternative procedure, let us calculate R_{th} by calculating I_{sc} and using the formula $R_{\text{th}} = V_{\text{oc}}/I_{\text{sc}}$. If we short terminals P and Q in Figure 9-2c we have

$$I_{\text{sc}} = \frac{V_s}{R_1} = \frac{9}{15} = 0.6 \text{ A}$$

Then

$$R_{\text{th}} = \frac{V_{\text{oc}}}{I_{\text{sc}}} = \frac{4}{0.6} = 6.667 \; \Omega$$

For this circuit, then, calculating I_{sc} and V_{oc}, and finding R_{th} from $R_{\text{th}} = V_{\text{oc}}/I_{\text{sc}}$ may be faster than evaluating R_{th} by setting sources to zero. In any case, since we may use either method, we learn to choose the method which saves time and labor.

Example 2 Determine the Thévenin equivalent circuit for the voltage divider circuit of Figure 9-3a. Find I_L for $R_L = 1 \text{ k}\Omega$, $2 \text{ k}\Omega$, $4 \text{ k}\Omega$. Find $V_L = V_A$ for each of the values of R_L.

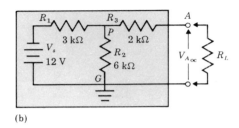

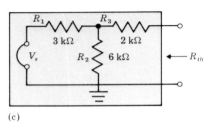
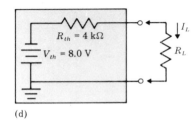

Figure 9-3

We first find $V_{A_{oc}}$ with R_L disconnected [see Figure 9-3b],

$$V_{th} = V_{A_{oc}} = V_{PG} = \frac{R_2}{R_1 + R_2} V_s$$

$$= \frac{6(10^3)}{3(10^3) + 6(10^3)} (-12)$$

$$V_{th} = \underline{-8.0 \text{ V}}$$

$$R_{th} = R_3 + R_1 \| R_2 = 2(10^3) + \frac{3(10^3)6(10^3)}{9(10^3)}$$

$$= \underline{4 \text{ k}\Omega}$$

Having determined the Thévenin equivalent circuit [see Figure 9-3d] we can easily evaluate I_L and V_L for various values of load resistors R_L.

For $R_L = 1$ kΩ:

$$V_L = \frac{R_L}{R_L + R_{th}} V_{th} = \frac{1(10^3)}{1(10^3) + 4(10^3)} \times (-8)$$

$$= \underline{-1.6 \text{ V}}$$

$$I_L = \frac{V_L}{R_L} = \frac{1.6}{1(10^3)} = \underline{1.6 \text{ mA}}$$

For $R_L = 2$ kΩ:

$$V_L = \frac{2(10^3)}{2(10^3) + 4(10^3)} (-8) = \underline{-2.667 \text{ V}}$$

$$I_L = \frac{2.667}{2(10^3)} = \underline{1.333 \text{ mA}}$$

For $R_L = 4$ kΩ:

$$V_L = \frac{4(10^3)}{4(10^3) + 4(10^3)} (-8) = \underline{-4.0 \text{ V}}$$

$$I_L = \frac{4.0}{4(10^3)} = \underline{1 \text{ mA}}$$

Exercises 9-2A   In Exercises 1–14 (a) convert the given circuits to their Thévenin equivalents, and (b) determine the stated values using the equivalent circuit.

1. For the circuit of Figure 9-4(a), determine I_L and V_L when R_L has values of 5 Ω, 10 Ω, and 25 Ω. Given: $V_s = 20$ V, $R_1 = 10$ Ω, and $R_2 = 20$ Ω.

2. Refer to Figure 9-4(a). Given: $V_s = 12$ V, $R_1 = 4$ Ω, and $R_2 = 12$ Ω. What are I_L and V_L when R_L is 2 Ω, 6 Ω, and 12 Ω?

3. In the circuit of Figure 9-4(b) let $V_s = 15$ V, $R_1 = 6$ kΩ, $R_2 = 12$ kΩ, and $R_3 =$

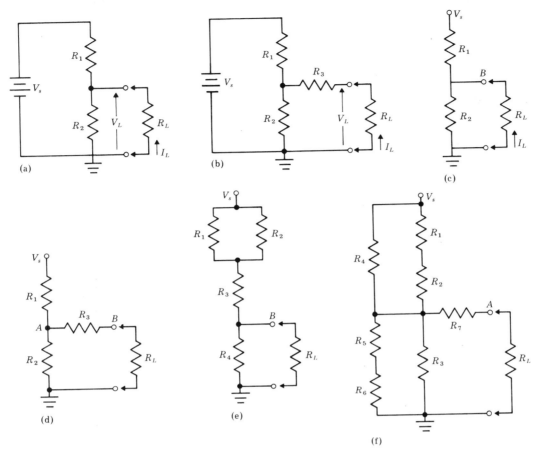

Figure 9-4

3 kΩ. Determine I_L and V_L when R_L is 0 Ω, 3 kΩ, and 10 kΩ.

4. Refer to Figure 9-4(b). Given: $V_s = 24$ V, $R_1 = 150$ kΩ, $R_2 = 50$ kΩ, and $R_3 = 25$ kΩ. Find I_L and V_L when R_L is 20 kΩ, 50 kΩ, and 100 kΩ.

5. In the circuit of Figure 9-4(c) V_s is -12 V, $R_1 = 15$ kΩ, $R_2 = 5$ kΩ. Find V_B (voltage of B to ground) and I_L when R_L is 0 Ω, 10 kΩ, and 100 kΩ.

6. For the circuit of Figure 9-4(c) $V_s = -9$ V, $R_1 = 18$ kΩ, and $R_2 = 9$ kΩ. Find V_B when $R_L = R_{th}$. What is P_L for this condition?

7. For circuit of Figure 9-4(d) $V_s = -100$ V, $R_1 = 60$ kΩ, $R_2 = 20$ kΩ, and $R_3 = 10$ kΩ. Find V_B and P_L when $R_L = 2R_{th}$.

8. For Figure 9-4(d) $V_s = 200$ V, $R_1 = 80$ kΩ, $R_2 = 40$ kΩ, and $R_3 = 20$ kΩ.

Find V_A and V_B when $R_L = R_{th}/2$. (*Hint:* Find V_B and I_L, then V_3.)

9. For Figure 9-4(e) $V_s = 40$ V, $R_1 = 40$ Ω, $R_2 = 60$ Ω, $R_3 = 12$ Ω, and $R_4 = 36$ Ω. Find I_L and P_L when R_L is equal to $R_{th}/2$, and $2R_{th}$.

10. For the circuit of Figure 9-4(e) $V_s = 500$ mV, $R_1 = 70$ kΩ, $R_2 = 30$ kΩ, $R_3 = 9$ kΩ, and $R_4 = 30$ kΩ. Find I_L and P_L when R_L is equal to $R_{th}/2$, and $2R_{th}$.

11. Refer to Figure 9-4(f). Given: $V_s = 200$ mV, $R_1 = 4$ Ω, $R_2 = 6$ Ω, $R_3 = 100$ Ω, $R_4 = 100$ Ω, $R_5 = 3$ Ω, $R_6 = 7$ Ω, $R_7 = 5$ Ω. Find I_L when R_L is equal to R_{th} and $2R_{th}$.

12. Consider the circuit of Exercise 11. Since R_3 and R_4 are very much larger than the circuits they are in parallel with, what would be the effect of neglecting (not using) them in your cal-

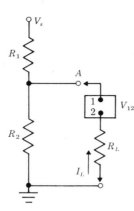

Figure 9-5

culations? To answer this question re-work Exercise 11 but do not use R_3 and R_4 in your calculations.

13. Refer to Figure 9-5. Given: $V_s = -6$ V, $R_1 = 15$ kΩ, $R_2 = 5$ kΩ. Find V_A and I_L when $V_{12} = -1$ V and $R_L = 2$ kΩ.

14. Refer to Figure 9-5. Given: $V_s = +6$ V, $R_1 = 16$ kΩ, $R_2 = 4$ kΩ. Find V_A and I_L when $V_{12} = +0.6$ V and $R_L = 1$ kΩ.

Thévenin's Theorem and Multisource Networks   Thévenin's theorem is applicable to networks with more than one source. Examples demonstrate the application.

Example 3   Thevenize the network of Figure 9-6(a) letting terminals A-B be the load terminals. Calculate V_L and I_L for $R_L = 3$ Ω, 4 Ω, and 10 Ω.

This circuit is also seen in the form shown in Figure 9-6(b). We calculate V_{oc} using the fact that V_{s1} and V_{s2} are connected series aiding [see Figure 9-6(c)],

$$V_2 = \frac{R_2}{R_1 + R_2}(V_{s1} + V_{s2}) = \frac{3}{6+3} \times 15 = 5.0 \text{ V}$$

$$V_{th} = V_{oc} = V_{AB} = V_2 - V_{s2} = +5.0 - 3$$
$$= +2.0 \text{ V}$$

and from Figure 9-6(d),

$$R_{th} = \frac{R_1 R_2}{R_1 + R_2} = \frac{6 \times 3}{6+3} = 2.0 \text{ }\Omega$$

The Thévenin equivalent circuit is shown in Figure 9-6(e). We calculate the required load voltages and currents using this circuit. For $R_L = 3$ Ω:

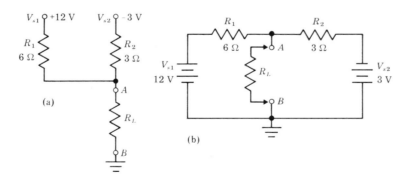

(a)

(b)

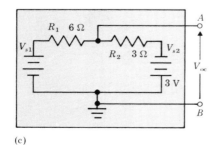

(c)

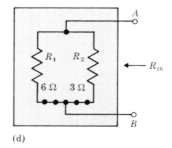

(d)

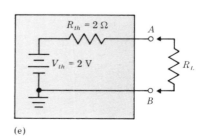

(e)

Figure 9-6

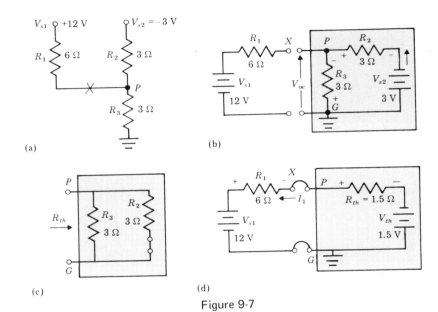

Figure 9-7

$$V_L = \frac{R_L}{R_L + R_{th}} V_{th} = \frac{3}{3+2} \times 2 = \underline{1.2 \text{ V}}$$

$$I_L = \frac{V_L}{R_L} = \frac{1.2}{3} = \underline{400 \text{ mA}}$$

For $R_L = 4 \; \Omega$:

$$V_L = \frac{4}{4+2} \times 2 = \underline{1.333 \text{ V}}$$

$$I_L = \frac{1.333}{4} = \underline{333.3 \text{ mA}}$$

For $R_L = 10 \; \Omega$:

$$V_L = \frac{10}{10+2} \times 2 = \underline{1.667 \text{ V}}$$

$$I_L = \frac{1.667}{10} = \underline{166.7 \text{ mA}}$$

(*Note*: $I_L = I_{BA}$ as determined by the polarity of V_{th}.)

Example 4 Use Thévenin's theorem to determine the current in R_1 of the circuit of Figure 9-7(a). (Observe that this is the circuit of Example 3 also.)

We temporarily open the circuit to the left of node P [at X in Figure 9-7(a)] and consider V_{s1} and R_1 the load network [see equivalent circuit in Figure 9-7(b)],

$$V_{oc} = \frac{R_3}{R_2 + R_3} V_{s2} = \frac{3}{3+3} (-3) = -1.5 \text{ V}$$

$$R_{th} = \frac{R_2 R_3}{R_2 + R_3} = \frac{3 \times 3}{3+3} = 1.5 \; \Omega$$

[see Figure 9-7(c)]

We write a KVL equation for the loop of Figure 9-7(d) (review Section 8-3),

$$+1.5 - 1.5I_1 - 6I_1 + 12 = 0$$

$$7.5I_1 = 13.5$$

$$I_1 = \frac{13.5}{7.5} = 1.8 \text{ A}$$

Example 5 Use Thévenin's theorem to determine I_2, the current in R_2, in the circuit of Figure 9-8(a).

We now open the circuit of Examples 3 and 4 to the right of node P [see Figure 9-8(a)],

$$V_{oc} = \frac{R_3}{R_1 + R_3} V_{s1} = \frac{3}{3+6} \cdot 12 = 4 \text{ V}$$

$$R_{th} = \frac{R_1 R_3}{R_1 + R_3} = \frac{6 \cdot 3}{6+3} = 2 \; \Omega$$

[see Figure 9-8(b)]

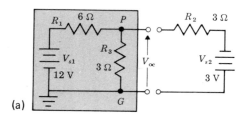

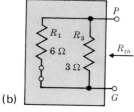

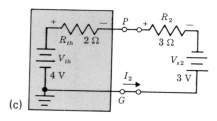

Figure 9-8

We write a KVL equation for the loop of Figure 9-8(c),

$$-4 + 2I_2 + 3I_2 - 3 = 0$$
$$5I_2 = 7$$
$$I_2 = \frac{7}{5} = \underline{1.4 \text{ A}}$$

We use Kirchhoff's current law (review Section 8-3) to check solutions from Examples 2, 3, and 4 (see Figure 9-9).

$$\Sigma I \text{ (at node } P\text{)} = 0$$

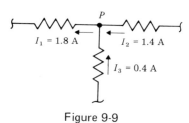

Figure 9-9

$$-I_1 + I_2 + I_3 = 0$$
$$-1.8 + 1.4 + 0.4 = 0$$
$$0 = 0$$

We apply the procedure to other common circuits.

Example 6 Use Thévenin's theorem to determine I_{AB}, V_{AC}, and V_{BC} in the unbalanced bridge circuit of Figure 9-10(a).

Our strategy here will be to consider the network to consist of two parts: (a) The "source" circuit as seen looking into terminals A-C, and (b) the "load" circuit (which we thevenize, as well as the source circuit) consists of R_m in series with the equivalent cir-

cuit seen looking into terminals B-C [see Figure 9-10(b)]. (*Note:* Since V_s has no internal resistance, the current it supplies to the R_1, R_2 branch will have no effect on the current it is called on to supply to the R_3, R_4 branch.) It is as if the two branches are being supplied by separate sources. We proceed with the conversion to the equivalent circuits,

$$V_{\text{th}A} = V_{\text{oc}A} = \frac{R_2}{R_1 + R_2} V_s = \frac{4}{2 + 4} \cdot 3 = 2 \text{ V}$$

$$R_{\text{th}A} = \frac{R_1 R_2}{R_1 + R_2} = \frac{2 \cdot 4}{2 + 4} = 1.333 \ \Omega$$

$$V_{\text{th}B} = V_{\text{oc}B} = \frac{R_4}{R_3 + R_4} V_s = \frac{2}{8 + 2} \cdot 3 = 0.6 \text{ V}$$

$$R_{\text{th}B} = \frac{R_3 R_4}{R_3 + R_4} = \frac{2 \cdot 8}{2 + 8} = 1.6 \ \Omega$$

We write a KVL equation to find I [see Figure 9-10c],

$$-V_{\text{th}A} + IR_{\text{th}A} + IR_m + IR_{\text{th}B} + V_{\text{th}B} = 0$$
$$-2 + 1.333I + 6I + 1.6I + 0.6 = 0$$
$$8.933I = 1.4 \qquad I = \frac{1.4}{8.933} = 156.7 \text{ mA}$$

but I flows from B to A; therefore, $I_{AB} = -I = -156.7$ mA

$$V_{AC} = -IR_{\text{th}A} + V_{\text{th}A} = -(0.1567)(1.333) + 2$$
$$= +1.791 \text{ V}$$
$$V_{BC} = -IR_m + V_{AC} = -(0.1567)(6) + 1.791$$
$$= +0.8508 \text{ V}$$

If we are interested only in finding I_{AB} or V_{AB} we can reduce the network further to a single source and single resistance [see Figure 9-10(d)]. From circuit of Figure 9-10(c),

$$V_{\text{th}} = V_{\text{oc}AB} = V_{\text{th}A} - V_{\text{th}B} = 2 - 0.6 = +1.4 \text{ V}$$
$$R_{\text{th}} = R_{\text{th}A} + R_{\text{th}B} = 1.333 + 1.6 = 2.933 \ \Omega$$

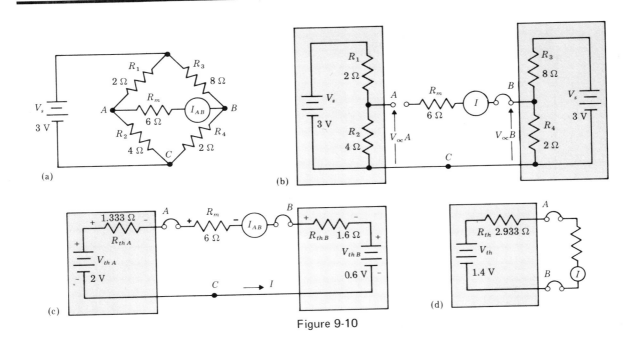

Figure 9-10

Then, for the circuit of Figure 9-10(d),

$$I_{AB} = \frac{-V_{th}}{R_{th} + R_m} = \frac{-1.4}{2.933 + 6}$$

$$I_{AB} = -156.7 \text{ mA}$$

Example 7 Determine I_{AB}, V_A, and V_B for the circuit of Figure 9-11(a).

Let us note that this circuit may appear as in Figure 9-11(b) and (c). We consider V_{s1}, R_1, and R_2 as the source network and R_5, R_4, R_3, and V_{s2} as the load network,

$$V_{A\,oc} = \frac{R_2}{R_1 + R_2} V_{s1} = \frac{15 \text{ k}\Omega}{5 \text{ k}\Omega + 15 \text{ k}\Omega} \cdot 40$$

$$= 30 \text{ V}$$

$$R_{thA} = \frac{R_1 R_2}{R_1 + R_2} = \frac{5 \text{ k}\Omega \cdot 15 \text{ k}\Omega}{5 \text{ k}\Omega + 15 \text{ k}\Omega} = 3.75 \text{ k}\Omega$$

$$V_{B\,oc} = \frac{R_4}{R_3 + R_4} \cdot V_{s2} = \frac{15 \text{ k}\Omega}{25 \text{ k}\Omega + 15 \text{ k}\Omega} \cdot (-20)$$

$$= -7.5 \text{ V}$$

$$R_{thB} = \frac{R_3 R_4}{R_3 + R_4} = \frac{15 \text{ k}\Omega \cdot 25 \text{ k}\Omega}{15 \text{ k}\Omega + 25 \text{ k}\Omega} = 9.375 \text{ k}\Omega$$

We write a KVL for the circuit of Figure 9-11(f),

$$-V_{thA} + IR_{thA} + IR_5 + IR_{thB} - V_{thB} = 0$$

$$-30 + 3.75(10^3)I + 50(10^3)I + 9.375(10^3)I$$

$$- 7.5 = 0$$

$$63.125(10^3)I = 37.5 \text{ V}$$

$$I_{AB} = -I = -\frac{37.5}{63.125(10^3)}$$

$$= \underline{-594.1 \ \mu\text{A}}$$

$$V_A = -IR_{thA} + V_{thA}$$

$$= -(3.75)(10^3)(594.1)(10^{-6})$$

$$+ 30 = -2.228 + 30$$

$$V_A = \underline{+27.77 \text{ V}}$$

$$V_B = IR_{thB} - V_{thB}$$

$$= 9.375(10^3)(594.1)(10^{-6}) - 7.5$$

$$= 5.570 - 7.5$$

$$V_B = \underline{-1.93 \text{ V}}$$

Exercises 9-2B In Exercises 15–18 use Figure 9-12(a) and the following circuit values: $V_{s1} = 6$ V, $R_1 = 3 \ \Omega$, $R_2 = 6 \ \Omega$, $R_3 = 4 \ \Omega$, and $V_{s2} = 2$ V. Use Thévenin's theorem to determine the stated quantities (see Example 3).

15. Find V_B and the current through R_3.

16. Find the Thévenin equivalent circuit seen by V_{s1}. What current will V_{s1} supply?

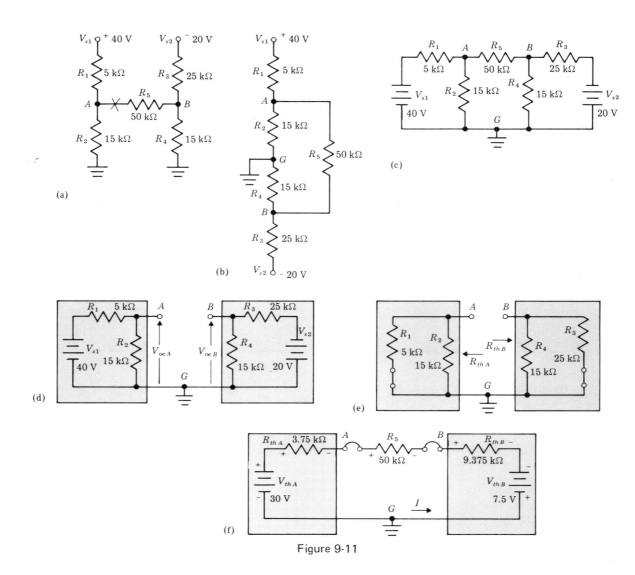

Figure 9-11

17. Repeat Exercise 16 for V_{s2}.

18. If V_{s2} becomes shorted what current will flow through R_2? R_1?

For Exercises 19–22 use the circuit of Figure 9-12(b) [or Figure 9-12(c) which is identical, electrically, to (b)] and the following values: V_{s1} = 200 V, V_{s2} = – 100 V, R_1 = 60 kΩ, R_2 = 40 kΩ, and R_3 = 16 kΩ. Use Thévenin's theorem in performing the required analyses (see Example 3).

19. Find the Thévenin equivalent circuit, as seen by R_L, for the circuit of Figure 9-12(b).

20. Find V_B and I_L for the circuit of Figure 9-12(b) when R_L = 0.8R_{th}.

21. Find V_B and I_L for the circuit of Figure 9-12(b) when R_L = 0.1R_{th}.

22. Find what V_B and I_L would be for the circuit of Figure 9-12(b) if R_2 opens. Let R_L = 18 kΩ.

In Exercises 23–26 use Figure 9-12(d) and the following values: V_s = 10 V, R_1 = 4 kΩ, R_2 = 6 kΩ, R_3 = 7 kΩ, and R_4 = 3 kΩ. Utilize Thévenin's theorem in performing the required analyses (see Example 6).

23. For the circuit of Figure 9-12(d) find I_{AB}, V_{AC}, and V_{BC} if R_5 = 5 kΩ.

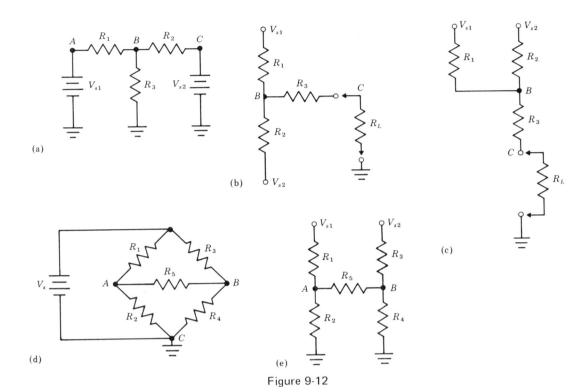

Figure 9-12

24. For the circuit of Figure 9-12(d) find I_{AB}, V_{AC}, and V_{BC} if $R_5 = 3$ kΩ and R_4 is changed to 0 Ω.

25. For the circuit of Figure 9-12(d) find I_{AB}, V_{AC}, and V_{BC} if $R_5 = 4$ kΩ and R_4 is changed to infinite ohms.

26. For the circuit of Figure 9-12(d) find I_{AB}, V_{AC}, and V_{BC} if $R_5 = 5$ kΩ and R_3 is changed to 2 kΩ.

In Exercises 27–30 use Figure 9-12(e) and the following values: $V_{s1} = +180$ V, $V_{s2} = -100$ V, $R_1 = 36$ kΩ, $R_2 = 64$ kΩ, $R_3 = 72$ kΩ, $R_4 = 28$ kΩ. Utilize Thévenin's theorem in performing the required analyses.

27. Find I_{AB}, V_A, and V_B for the circuit of Figure 9-12(e) if $R_5 = 12$ kΩ.

28. Find I_{AB}, V_A, and V_B for the circuit of Figure 9-12(e) if $R_5 = 3$ kΩ.

29. Refer to the circuit of Figure 9-12(e). Assume that a battery $V_{s3} = +45$ V is inserted between R_4 and ground (negative to ground). Find I_{AB} if $R_5 = 6$ kΩ.

30. Refer to the circuit of Figure 9-12(e). Assume that a battery $V_{s3} = +60$ V is in-

serted in series with R_5 and node B (negative to B). Find I_{AB} if $R_5 = 12$ kΩ.

31. Refer to Figure 9-13(a). Given: $V_{s1} = +100$ V, $V_{s2} = -40$ V, $V_{s3} = +160$ V, $V_{s4} = -60$ V, $R_1 = 20$ kΩ, $R_2 = 18$ kΩ, $R_3 = 20$ kΩ, $R_4 = 14$ kΩ, $R_5 = 6$ kΩ, and $R_6 = 20$ kΩ. Determine the following: V_A, V_B, V_C, V_D, V_E, V_F, V_{FC}, V_{ED}, V_{AB}, V_{AC}, V_{BD}, V_{AE}, V_{BF}, V_{BE}, V_{AF}, V_{AD}, V_{FE}, I_{AB}, I_{AC}, I_{BD}, I_{AE}, I_{BF}, and I_{FE}. (Hint: Consider R_5 the load for a Thévenin equivalent circuit. Find V_A and V_B under loaded conditions, then other voltages and currents.)

32. Refer to Figure 9-13(a). Given: $V_{s1} = +60$ V, $V_{s2} = -10$ V, $V_{s3} = +40$ V, $V_{s4} = -20$ V, $R_1 = 3$ kΩ, $R_2 = 2$ kΩ, $R_3 = 4$ kΩ, $R_4 = 6$ kΩ, $R_5 = 1$ kΩ, and $R_6 = 4$ kΩ. Perform the analysis specified in Exercise 31.

33. Refer to Figure 9-13(b). Given: $V_{s1} = +100$ V, $V_{s2} = -120$ V, $V_{s3} = +160$ V, $V_{s4} = -160$ V, $R_1 = 60$ kΩ, $R_2 = 20$ kΩ, $R_3 = 40$ kΩ, $R_4 = 60$ kΩ, $R_5 = 20$ kΩ, and $R_6 = 40$ kΩ. Find the following: V_A,

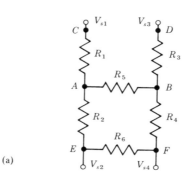

(a)

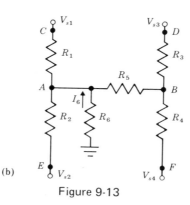

(b)

Figure 9-13

V_B, V_C, V_D, V_E, V_F, V_{AB}, V_{AC}, V_{AE}, V_{BD}, V_{BF}, V_{CE}, V_{CF}, V_{CD}, V_{ED}, V_{EF}, I_{AB}, I_6, I_{CA}, I_{EA}, I_{DB}, and I_{FB}.

34. Refer to Figure 9-13(b). Given: V_{s1} = +20 V, V_{s2} = -30 V, V_{s3} = +10 V, V_{s4} = -20 V, R_1 = 10 kΩ, R_2 = 10 kΩ, R_3 = 8 kΩ, R_4 = 14 kΩ, R_5 = 5 kΩ, and R_6 = 8 kΩ. Perform the analysis specified in Exercise 33.

35. Refer to Figure 7-12, Chapter 7. Use Thévenin's theorem to develop equations for the voltages across R_3: (a) with, and (b) without, the voltmeter connected.

9-3 VOLTAGE AND CURRENT SOURCES

An ideal voltage source is one which has no internal resistance. The schematic circuit symbols for an ideal battery and generator are shown in Figures 9-14(a) and (b), respectively. An ideal voltage source will supply a constant voltage V_s no matter what load device is connected to it. Practical (nonideal) voltage sources have inherent, distributed,

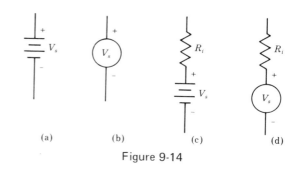

Figure 9-14

internal resistance. For most analyses such resistance can be represented by discrete resistors often designated R_i. The schematic symbols of practical voltage sources include these series resistors as in Figures 9-14(c) and (d). Practical voltage sources supply the constant voltage V_s diminished by a voltage drop $I_L R_i$ caused by load current in the internal resistance. The terminal voltage of a practical source is then $V_t = V_s - I_L R_i$, or for any source network it is, using Thévenin's theorem, $V_t = V_{th} - I_L R_{th}$.

An ideal current source is one which supplies a constant current I no matter what is connected to its terminals. It has infinite internal resistance. The schematic circuit symbol of such a source is shown in Figure 9-15(a). The arrow indicates the direction of current flow. A practical current source is represented schematically as an ideal source in parallel with a resistor R_i as in Figure 9-15(b). An actual load current for a load resistance connected to the source may be less than I because the load shares current with R_i [see Figure 9-15(c)].

We now describe the relationship between equivalent voltage and current sources.

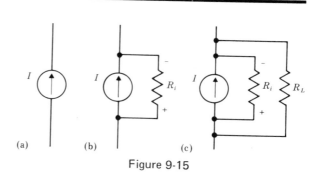

Figure 9-15

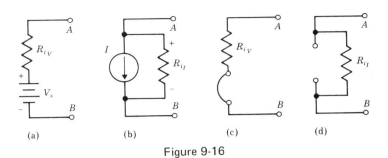

Figure 9-16

Refer to Figure 9-16 and assure yourself of the truth of the following observations:

1. An ammeter with zero internal resistance will measure:

(a) $I_{V_{sc}} = \dfrac{V_s}{R_{i_V}}$ when connected to terminals A-B in (a).

(b) I when connected to terminals A-B in b. (R_{i_I} will get no current when a short is connected in parallel with it.)

2. A voltmeter with infinite internal resistance (requires no current) will measure across terminals A-B:

(a) $V_{oc_V} = V_s$ in Figure 9-16(a),

(b) $V_{oc_I} = IR_{i_I}$ in (b).

3. The resistance seen looking into terminals A-B with V_s and I reduced to zero will be:

(a) R_{i_V} for the voltage source circuit [V_s is reduced to zero by a short as in Figure 9-16(c)].

(b) R_{i_I} for the current equivalent circuit [I is reduced to zero by opening its circuit, as in Figure 9-16(d)].

4. A voltage source and a current source are equivalent (produce same current in a load resistor) when the following are true:

(a) $V_{oc_V}(=V_s) = V_{oc_I}$.

(b) $R_{i_V} = R_{i_I} = R_i$.

(c) $I_{V_{sc}} = I_{I_{sc}}(=I)$.

Can we thevenize circuits containing current sources? Yes, but we must remember that reducing a current source to zero means opening I. Consider these examples.

Example 8 Thevenize the circuit of Figure 9-17(a) and find the current in an $R_L = 10\ \Omega$.

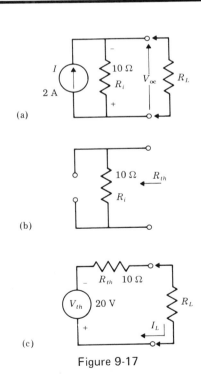

Figure 9-17

We find V_{oc} using Figure 9-17a,

$$V_{th} = V_{oc} = IR_i = (-2)(10) = -20 \text{ V}$$

Then we open I to reduce source to zero [see Figure 9-17(b)], and calculate R_{th}, $R_{th} = R_i = 10\ \Omega$.

The Thévenin equivalent circuit is shown in Figure 9-17(c).

Then

$$I_L = \frac{V_{th}}{R_L + R_{th}} = \frac{20}{10 + 10} \qquad I_L = \underline{1.0 \text{ A}}$$

Example 9   Thevenize the circuit of Figure 9-18(a) and find the current in an $R_L = 10 \text{ k}\Omega$.

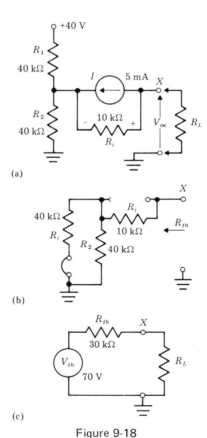

(a)

(b)

(c)

Figure 9-18

We find V_{oc} in two steps. First, we find V_2 the voltage across R_2 with R_L disconnected [see Figure 9-18(a)],

$$V_2 = \frac{R_2}{R_1 + R_2} V_s = \frac{40(10^3)}{40(10^3) + 40(10^3)} \times (+40)$$
$$= +20 \text{ V}$$

The voltage V_I across the current source (with R_L disconnected) is

$$V_I = IR_i = (5 \times 10^{-3})(10 \times 10^3) = 50 \text{ V}$$

The polarity of V_I is as shown on the diagram of Figure 9-18(a) and is determined by the direction of I in its internal resistance.

We find V_{oc} by summing voltages from load terminal X to ground,

$$V_{oc} = V_I + V_2 = 50 + 20 = 70 \text{ V}$$

Thus,

$$V_{th} = \underline{70 \text{ V}}$$

We find R_{th} by looking in at the load terminals [see Figure 9-18(b)]. Note that to reduce the sources to zero V_s is shorted and I is opened.

$$R_{th} = R_i + R_1 \| R_2 = 10 \text{ k}\Omega + 20 \text{ k}\Omega = \underline{30 \text{ k}\Omega}$$

The Thévenin equivalent circuit is shown in Figure 9-18(c). We use this circuit to find I_L,

$$I_L = \frac{V_{th}}{R_{th} + R_L} = \frac{70}{30 \times 10^3 + 10 \times 10^3}$$
$$= 1.75 \times 10^{-3} \text{ A}$$
$$I_L = \underline{1.75 \text{ mA}}$$

Exercises 9-3   In Exercises 1–10 thevenize the specified circuits and then calculate the load currents for the given values of R_L.

1. In the circuit of Figure 9-19(a), $I = 2$ A, $R_i = 8\ \Omega$. Find I_L when $R_L = 4\ \Omega$.

2. Find I_L for the circuit of Figure 9-19(a) when $R_L = 10 \text{ k}\Omega$, if $I = 10$ mA and $R_i = 3 \text{ k}\Omega$.

3. Refer to the circuit of Figure 9-19(b). Given: $I = 50$ mA, $R_i = 400\ \Omega$, and $R_1 = 600\ \Omega$. Find I_L when $R_L = 5 \text{ k}\Omega$.

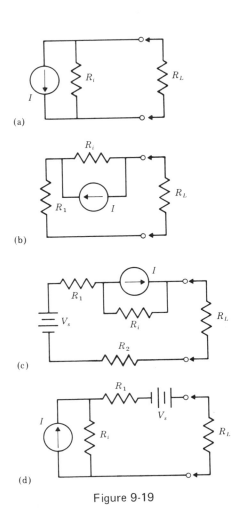

(a)

(b)

(c)

(d)

Figure 9-19

8. Find I_L for $R_L = R_{th}$ and $2R_{th}$ for the circuit of Figure 9-19(d). Given: $I = 3$ A, $R_i = 2\ \Omega, R_1 = 10\ \Omega$, and $V_s = 20$ V.

9. In the circuit of Figure 9-19(d), $I = 20$ mA, $R_i = 4$ kΩ, $R_1 = 6$ kΩ, and $V_s = 100$ V. Find I_L when $R_L = 20$ kΩ.

10. Refer to Exercise 9. Assume that the polarity of V_s is reversed. Calculate I_L for $R_L = 20$ kΩ.

9-4 NORTON'S THEOREM

The principle by which a complex network of sources and resistances can be reduced to a simple series circuit equivalent (see Section 9-2, Thévenin's theorem) can be applied in a slightly different way to yield a simple parallel circuit or current-source equivalent. We describe the procedure called *Norton's theorem.*

NORTON'S THEOREM

The current produced in a load device or network connected to any two points of a complex source network is the same as that produced by a simple parallel equivalent circuit. This equivalent circuit consists of an ideal current source of magnitude I_N in parallel with the Thévenin resistance R_{th} of the source network. The current I_N is the current produced at the load terminals when the load network is replaced by a short circuit. Symbolically,

$$I_N = \frac{V_{oc}}{R_{th}}$$

Study the following examples carefully, observing the technique for finding and utilizing the Norton equivalent circuits for various networks.

Example 10 Find the Norton equivalent circuit for the network of Figure 9-20(a). (*Note:* This is the circuit of Example 3, Section 9-2.) Find I_{AB} when R_L is 3 Ω, 4 Ω, and 10 Ω.

4. In the circuit of Figure 9-19(b) $I = 200\ \mu$A, $R_i = 100$ kΩ, and $R_1 = 300$ kΩ. Find I_L if $R_L = 500$ kΩ.

5. For the circuit of Figure 9-19(c) $V_s = 60$ V, $R_1 = 20$ kΩ, $R_2 = 40$ kΩ, $I = 1.0$ mA, $R_i = 20$ kΩ. Find I_L when $R_L = 0.5\ R_{th}$ and $5\ R_{th}$.

6. Refer to Exercise 5. Assume that the polarity of V_s is reversed from that shown on the diagram. Calculate I_L for $R_L = 0.1 R_{th}$ and $10 R_{th}$.

7. If the direction of I is reversed in the circuit of Exercise 5 what is the value of I_L for the given values of R_L?

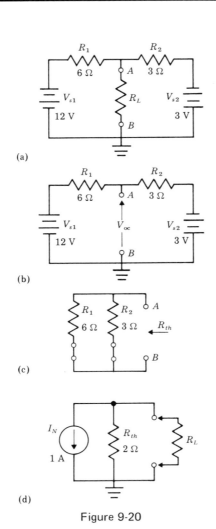

(a)

(b)

(c)

(d)

Figure 9-20

We find V_{oc} and R_{th} exactly as in Example 3; hence,

$$V_{oc} = 2.0 \text{ V}$$
$$R_{th} = 2.0 \text{ } \Omega$$

and, then,

$$I_N = \frac{V_{oc}}{R_{th}} = \frac{2.0}{2.0} = 1 \text{ A}$$

We use the current-divider formula to find I_L for various values of R_L [see Figure 9-20(d)]:
    For $R_L = 3 \text{ } \Omega$:

$$I_L = \frac{R_{th}}{R_L + R_{th}} I_N = \frac{2.0}{3.0 + 2.0} \times 1.0 = \underline{400 \text{ mA}}$$

For $R_L = 4 \text{ } \Omega$:

$$I_L = \frac{2.0}{4.0 + 2.0} \times 1.0 = \underline{333.3 \text{ mA}}$$

For $R_L = 10 \text{ } \Omega$:

$$I_L = \frac{2.0}{10.0 + 2.0} \times 1.0 = \underline{166.7 \text{ mA}}$$

The current values check with those found using Thevenin's theorem in Example 3, Section 9-2. Note that we may also find R_{th} using the relationship

$$R_{th} = \frac{V_{oc}}{I_{sc}}$$

We apply the principle of superposition (review Section 8-5) to determine $I_{sc_{AB}}$.
    Let V_{s1} be the first source; then

$$I_{AB1} (AB \text{ shorted}) = -\frac{V_{s1}}{R_1 + 0} = -\frac{12}{6} = -2 \text{ A}$$

Using V_{s2} as the second source,

$$I_{AB2} = \frac{V_{s2}}{R_2 + 0} = \frac{3}{3} = 1.0 \text{ A}$$

Thus,

$$I_{AB_{sc}} = I_{AB1} + I_{AB2} = -2 + 1.0 = -1.0 \text{ A}$$

and

$$R_{th} = \frac{V_{oc}}{I_{sc}} = \frac{2.0}{1.0} = 2.0 \text{ } \Omega$$

This is, as it should be, equal to the value for R_{th} found by looking into the terminals A-B with sources reduced to zero.

Source Network Containing a Current Source
Can a network containing a current source be converted to a Norton equivalent circuit? Yes, and the procedure is demonstrated in the following example.

Example 11   Find the Norton equivalent circuit for the network of Figure 9-21(a). Find I_L when $R_L = 3 \text{ } \Omega$, 4 Ω, and 10 Ω.

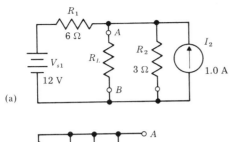

(a)

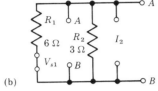

(b)

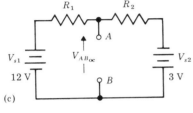

(c)

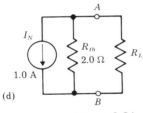

(d)

Figure 9-21

We determine R_{th} easily with the aid of Figure 9-21(b),

$$R_{th} = \frac{R_1 R_2}{R_1 + R_2} = \frac{6 \times 3}{6 + 3} = 2.0 \ \Omega$$

Note that I_2 is opened, V_{s1} is shorted.

To find I_N we will first find V_{th} ($V_{oc_{AB}}$). This requires that we convert I_2 to an equivalent voltage source (see Section 9-3):

$$R_{V_2} = R_{I_2} = 3.0$$
$$V_{s2} = I_2 R_{I_2} = 1.0 \times 3.0 = 3.0 \ V$$

The circuit, redrawn with V_{s2} substituted for

I_2, is as shown in Figure 9-21(c). We find $V_{AB_{oc}}$:

$$V_2 = \frac{R_2}{R_1 + R_2}(V_{s1} + V_{s2}) = \frac{3}{6 + 3} \times 15 = 5 \ V$$
$$V_{AB_{oc}} = +5 - 3 = +2 \ V$$

And, since

$$I_N = \frac{V_{oc}}{R_{th}}, \qquad I_N = \frac{2.0}{2.0} = 1.0 \ A$$

The Norton equivalent circuit is shown in Figure 9-21(d). By now you have no doubt discovered that this network is equivalent to that of Examples 3 and 10. Therefore, we can write immediately:

For $R_L = 3 \ \Omega$:

$$I_L = \frac{R_{th}}{R_L + R_{th}} I_N = \frac{2.0}{3.0 + 2.0} \times 1.0 = \underline{400 \ mA}$$

For $R_L = 4 \ \Omega$:

$$I_L = \underline{333.3 \ mA}$$

and for $R_L = 10 \ \Omega$:

$$I_L = \underline{166.7 \ mA}$$

Relationship of Thévenin and Norton Equivalent Circuits   The preceding examples demonstrate that the Thévenin and Norton equivalent circuits are closely related:

$$I_N = \frac{V_{th}}{R_{th}} = V_{th} G_{th} \qquad (1)$$

$$V_{th} = I_N R_{th} = \frac{I_N}{G_{th}} \qquad (2)$$

$$R_{th} = R_{th} \qquad (3)$$

$$G_{th} = \frac{1}{R_{th}} \qquad (4)$$

A given network may be easier to reduce to a Norton equivalent than the Thévenin even though the Thévenin equivalent is desired. If so, we find the Norton circuit first and then easily change it to the Thévenin equivalent using equation (2). Or, if thevenizing is easier when the Norton equivalent is desired, we thevenize and then use equation (1).

<u>Exercises 9-4</u> In Exercises 1–18 (a) convert the given circuits to their Norton equivalents, and (b) find the values called for using these equivalent circuits. The circuits to be analyzed are described in the exercises of Sections 9-2 and 9-3, as indicated.

1. Exercise 2, Section 9-2
2. Exercise 3, Section 9-2
3. Exercise 2, Section 9-3
4. Exercise 3, Section 9-3
5. Exercise 4, Section 9-2
6. Exercise 5, Section 9-2
7. Exercise 8, Section 9-2
8. Exercise 7, Section 9-2
9. Exercise 10, Section 9-2
10. Exercise 9, Section 9-2
11. Exercise 4, Section 9-3
12. Exercise 5, Section 9-3
13. Exercise 6, Section 9-3
14. Exercise 11, Section 9-2
15. Exercise 14, Section 9-2
16. Exercise 13, Section 9-2
17. Exercise 8, Section 9-3
18. Exercise 9, Section 9-3

9-5 MILLMAN'S THEOREM

A common problem in electronic circuit analyses is to find the voltage between two nodes which have two or more sources between them [see Figure 9-22(a)]. For example, in Figure 9-22(a) we would have to find $V_{AA'}$ if we wanted to find V_{oc} for the Thévenin or Norton equivalent of the network as seen at these nodes. Or, if we could find $V_{AA'}$ easily, we could then calculate I_3 simply by applying Ohm's law. Millman's theorem provides a time-saving method for finding such node voltages.

We state without proof the formula for Millman's theorem as applied to the generalized network of Figure 9-22(b):

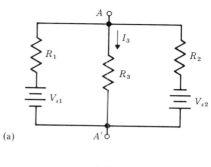

(a)

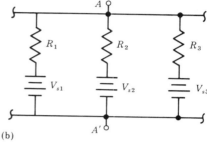

(b)

Figure 9-22

$$V_{AA'} = \frac{\dfrac{V_{s1}}{R_1} + \dfrac{V_{s2}}{R_2} + \dfrac{V_{s3}}{R_3} + \cdots}{\dfrac{1}{R_1} + \dfrac{1}{R_2} + \dfrac{1}{R_3} + \cdots} \qquad (1)$$

The letter symbols in the formula have these meanings:

$V_{AA'}$ = the voltage between two adjacent nodes.

$V_{s1}, V_{s2}, V_{s3}, \ldots$ = the total voltage-source equivalent voltages of the branches between nodes A and A'. (The algebraic signs of these voltage terms is that of the terminals nearest A because we are finding voltage of A with respect to A'.)

$R_1, R_2, R_3, \ldots$ = the equivalent resistances of the branches between nodes A and A'. When there is more than one resistance in a branch these are combined to form an equivalent branch resistance.

The terms in the numerator of equation (1), V_{s1}/R_1, V_{s2}/R_2, V_{s3}/R_3, ... represent the values I_1, I_2, I_3, ... of current-source (Norton) equivalents of each branch. The terms of the denominator of (1) represent the conductances G_1 ($=1/R_1$), G_2, G_3, ... of the current sources [see Figure 9-22(b)]. Rewriting (1) we have

$$V_{AA'} = \frac{I_1 + I_2 + I_3 + \cdots}{G_1 + G_2 + G_3 + \cdots} = \frac{I_T}{G_T} = I_T R_T$$

Examples will help demonstrate how we apply this theorem.

Example 12 For the network shown in Figure 9-23(a) find (a) the current I_3 in R_3, and (b) the expected current I_L for an additional load resistance $R_L = 8\ \Omega$ connected to nodes $A - A'$.

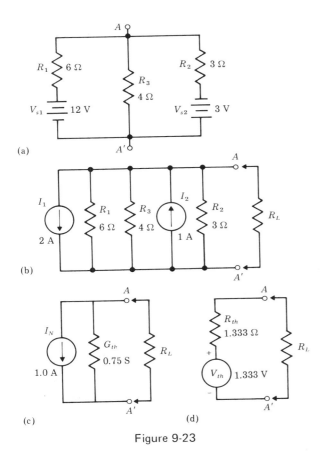

(a)

(b)

(c)   (d)

Figure 9-23

Our approach will be to find the voltage $V_{AA'}$ using Millman's theorem. This will enable us to find I_3 from $I_3 = V_{AA'}/R_3$. Also, knowing $V_{AA'}$, we can easily find the Thévenin or Norton equivalent circuits with which to evaluate I_L.

(a) $V_{AA'} = \dfrac{V_{s1}/R_1 + V_{s2}/R_2 + 0/R_3}{1/R_1 + 1/R_2 + 1/R_3}$

$\qquad = \dfrac{12/6 + (-3)/3 + 0/4}{1/6 + 1/3 + 1/4} = \dfrac{2 - 1 + 0}{9/12}$

$\qquad = \underline{1.333\ \text{V}}$

Refer to the Millman's equivalent network of Figure 9-23(b) and note that we show the direction of the current sources on the diagram and take these directions into account in the evaluation for $V_{AA'}$. The net current $I_T = 1$ A is in the direction of I_1 [see Figure 9-23(c)] making $V_{AA'}$ positive.

We find I_3 applying Ohm's law to the original network [Figure 9-23(a)],

$$I_3 = \frac{V_{AA'}}{R_3} = \frac{1.333}{4} = \underline{0.3333\ \text{A}}$$

(Note that this current checks with that found for the identical network of Figure 9-6, Example 3, Section 9-2.)

The equation for $V_{AA'}$ really defines a Norton equivalent circuit: (a) the numerator is I_N, and (b) the denominator is $G_{\text{th}} = 1/R_{\text{th}}$ [see Figure 9-23(c)]. Or, $V_{AA'}$ and R_T are the values for Thévenin equivalent circuit of the network as seen from nodes $A - A'$, since $V_{AA'} = V_{\text{oc}}$ and $R_T = R_{\text{th}}$ [see Figure 9-23(d)]. We may use either to calculate I_L for $R_L = 8\ \Omega$ as requested in part (b) of this example.

(b) $I_L = \dfrac{R_{\text{th}}}{R_L + R_{\text{th}}} I_N = \dfrac{1.333}{8 + 1.333} \times 1.0$

$\qquad I_L = \underline{0.1428\ \text{A}}$

Or, using the Thévenin equivalent circuit,

$$I_L = \frac{V_{\text{th}}}{R_L + R_{\text{th}}} = \frac{1.333}{8 + 1.333}$$

$$I_L = \underline{0.1428\ \text{A}}$$

Example 13 An energy conservationist has rigged a wind charger and a solar-cell array to charge an automobile battery as part of a home electrical system [see Figure 9-24(a)]. (a) What will be the charging current I_c for the conditions shown? (b) What current will the house load, R_L, draw?

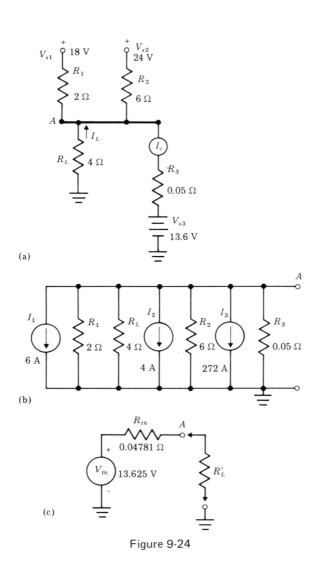

(a)

(b)

(c)

Figure 9-24

(a) If we determine the voltage of the bus (common connection point A) with respect to ground we will then be able to find I_c using a KVL equation, and I_L using Ohm's law.

$$V_A = \frac{\dfrac{V_{s1}}{R_1} + \dfrac{0}{R_L} + \dfrac{V_{s2}}{R_2} + \dfrac{V_{s3}}{R_3}}{\dfrac{1}{R_1} + \dfrac{1}{R_L} + \dfrac{1}{R_2} + \dfrac{1}{R_3}}$$

$$= \frac{\dfrac{18}{2} + \dfrac{0}{4} + \dfrac{24}{6} + \dfrac{13.6}{0.05}}{\dfrac{1}{2} + \dfrac{1}{4} + \dfrac{1}{6} + \dfrac{1}{0.05}}$$

$$= \frac{9 + 0 + 4 + 272}{20.917} = \frac{285}{20.917}$$

$$= \underline{13.625 \text{ V}}$$

We write a KVL equation to include the branch containing the battery,

$$I_c R_3 + V_{s3} = V_A$$
$$0.05 I_c + 13.6 = 13.625$$
$$0.05 I_c = 0.025$$
$$I_c = \underline{0.50 \text{ A}}$$

Applying Ohm's law to the R_L branch,

$$I_L = \frac{V_A}{R_L} = \frac{13.625}{4}$$
$$I_L = \underline{3.406 \text{ A}}$$

Example 14 If an additional load, $R_L' = 8 \ \Omega$, is connected between node A and ground in Figure 9-24(a) will the battery continue to be charged?

To answer this question we note that I_c will be zero if V_A drops to 13.6 V as a result of the additional load connected to node A. We can evaluate V_A using the Thévenin equivalent of the network before R_L' is connected [see Figure 9-24(c)],

$$V_{\text{th}} = V_A \text{ without } R_L' \text{ connected} = 13.625 \text{ V}$$

$$R_{\text{th}} = \frac{1}{G_T} = \frac{1}{20.917} = 0.04781 \ \Omega$$

We find V_A', the voltage at node A with R_L' connected, by applying the voltage divider formula to the circuit of Figure 9-24(c),

$$V_A' = \frac{R_L'}{R_L' + R_{\text{th}}} V_A = \frac{8}{8.04781} \times 13.625$$
$$= \underline{13.54 \text{ V}}$$

Since the voltage of node A is less than that of the "chemical" voltage (13.6 V) of the battery, the battery will now be supplying energy rather than receiving energy for charging.

Exercises 9-5   In Exercises 1–20 use Millman's theorem in your solutions.

1. In Figure 9-25(a), V_{s1} = 10 V, V_{s2} = 5 V, R_1 = 5 Ω, R_2 = 10 Ω, and R_3 = 15 Ω. Find $V_{AA'}$ and I_3.

2. Use the following values for the circuit of Figure 9-25(a): V_{s1} = 12 V, V_{s2} = 9 V, R_1 = 6 kΩ, R_2 = 12 kΩ, and R_3 = 8 kΩ. Find $V_{AA'}$ and I_3.

3. Refer to the circuit of Figure 9-25(b). Let V_{s1} = 10 V, V_{s2} = 20 V, V_{s3} = 15 V, R_1 = 20 Ω, R_2 = 40 Ω, and R_3 = 30 Ω. Find V_A (voltage of A with respect to ground).

4. (a) Thevenize the circuit of Exercise 1. (b) Predict the load current for an R_L = 10 Ω connected to terminals A-A'.

5. (a) Reduce the circuit of Exercise 1 to its Norton equivalent, and (b) predict the load current for an R_L = 5 Ω connected to terminals A-A'.

6. (a) Find the Norton equivalent circuit for the circuit of Exercise 3, and (b) predict the load current for an R_L = 30 Ω.

7. (a) Thevenize the circuit of Exercise 3, and (b) predict the load current for an R_L = 40 Ω.

8. For the circuit of Figure 9-25(c) let

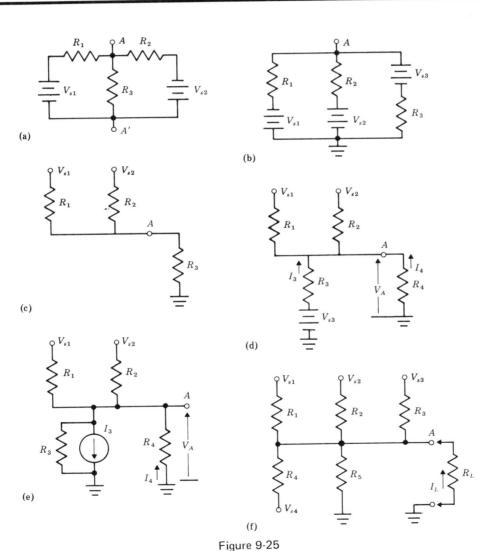

(a)

(b)

(c)

(d)

(e)

(f)

Figure 9-25

V_{s1} = 100 V, V_{s2} = 120 V, R_1 = 60 kΩ, R_2 = 40 kΩ, and R_3 = 50 kΩ. Find V_A and I_3.

9. For the circuit of Figure 9-25(c) let V_{s1} = 24 V, V_{s2} = 12 V, R_1 = 18 kΩ, R_2 = 6.8 kΩ, and R_3 = 10 kΩ. Find V_A and I_3.

10. For the circuit of Figure 9-25(c) let V_{s1} = -5 V, V_{s2} = -6 V, R_1 = 9.1 kΩ, R_2 = 8.2 kΩ, and R_3 = 5 kΩ. Find V_A and I_3.

11. For the circuit of Figure 9-25(d) let V_{s1} = 18 V, V_{s2} = 24 V, V_{s3} = 12 V, R_1 = 12 Ω, R_2 = 18 Ω, R_3 = 0.2 Ω, and R_4 = 2 Ω. Find (a) V_A, (b) I_3, and (c) I_4 (see Example 13).

12. Refer to Figure 9-25(d). Given: V_{s1} = 16 V, V_{s2} = 28 V, V_{s3} = 13.6 V, R_1 = 2 Ω, R_2 = 24 Ω, R_3 = 0.02 Ω, and R_4 = 1 Ω. Find (a) V_A, (b) I_3, and (c) I_4.

13. Refer to Exercise 12. Change R_4 to 20 Ω and repeat the analysis.

14. Refer to Exercise 11. Change R_4 to 0.5 Ω and repeat the analysis.

15. Assume that, for the circuit of Figure 9-25(e), V_{s1} = 12 V, V_{s2} = 5 V, I_3 = 10 mA, R_1 = 24 kΩ, R_2 = 10 kΩ, R_3 = 0.8 kΩ, and R_4 = 12 kΩ. Find V_A and I_4.

16. Refer to Exercise 15. (a) Determine the Thévenin equivalent of the network. (b) Find I_L for R_L = 4 kΩ connected between terminal A and ground.

17. Refer to Exercise 15. (a) Determine the Norton equivalent of the network. (b) Find I_L for R_L = 8 kΩ connected between terminal A and ground.

18. Refer to Figure 9-25(f). Thevenize the circuit with R_L disconnected. Find I_L and V_A for R_L = R_{th}. Given: V_{s1} = 40 V, V_{s2} = -20 V, V_{s3} = 60 V, V_{s4} = 30 V, R_1 = 40 kΩ, R_2 = 80 kΩ, R_3 = 50 kΩ, R_4 = 60 kΩ, and R_5 = 80 kΩ.

19. Refer to Figure 9-25(f). Given: V_{s1} = 100 V, V_{s2} = 180 V, V_{s3} = -160 V, V_{s4} = -200 V, R_1 = 60 kΩ, R_2 = 80 kΩ, R_3 = 70 kΩ, R_4 = 100 kΩ, and R_5 = 60 kΩ. Find the Norton equivalent of the circuit with R_L disconnected. Find I_L and V_A for R_L = R_{th}.

20. A space vehicle that carries a 120-V chemical battery (r_i = 42 Ω) to supply its electronic instrumentation-communica-

tion package also has a solar-cell array arranged for charging the chemical battery. While the vehicle is on the launching pad, the battery is charged by a battery charger operated from the 120-V ac-service line. The electronics package represents an equivalent load of 78 Ω. The internal resistance of the solar-cell array is 100 Ω, and that of the charger on the launching gantry, 70 Ω. The emf generated by the ground-based charger is 122 V.

(a) Draw the circuit diagram of the three sources and the load.

(b) While the vehicle is being prepared for launch, the sun cells at times receive bright sunlight. If the emf thus generated by the solar-cell array is 124 V, what is the no-load voltage available to the electronics package?

(c) What will be the load voltage if the package is in normal operation?

(d) What power is being supplied to the package?

(e) If the load is disconnected, how much charging current is being supplied to the chemical battery when the solar-cell array is excited?

9-6 π AND T NETWORK CONVERSIONS

The analyses of some networks are possible only when the arrangements of the components are converted to other, equivalent connection patterns. Others are simplified greatly by conversions. Such conversions involve networks referred to as π (pi) or Δ (delta) networks and T (tee) or Y (wye) networks. The descriptive titles are based on the shapes of the networks. In Figures 9-26(a) and 9-26(b) are shown the electrically identical π and Δ networks; Figure 9-27 includes examples of T and Y networks. The titles

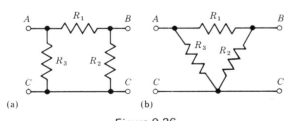

Figure 9-26

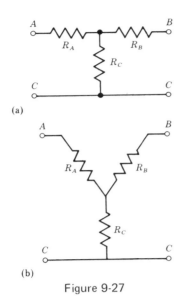

(a)

(b)

Figure 9-27

"π" and "T" are generally used in the fields of electronics and telecommunications; Δ and Y (and sometimes "star") are used in the electric power industry.

We convert a network of one type into a network of the opposite type by substituting component values into sets of standardized formulas. The resulting network is electrically equivalent to the original network. The resistances seen looking into the networks are the same for any corresponding pairs of terminals.

π-to-T (Δ-to-Y) Conversions   The π-to-T conversion is the most frequently used conversion. Figure 9-28 shows a T network drawn inside a π network to illustrate more directly

the relationship between the components of the equivalent network. The following formulas, utilizing the component designations of Figure 9-28, are used to make the π-to-T conversion.

π-to-T (Δ-to-Y) Conversions

$$R_A = \frac{R_1 R_3}{R_1 + R_2 + R_3}$$

$$R_B = \frac{R_1 R_2}{R_1 + R_2 + R_3}$$

$$R_C = \frac{R_2 R_3}{R_1 + R_2 + R_3}$$

$$R_{Tee} = \frac{\text{Product of two adjacent resistors of } \pi}{\text{Sum of } \pi \text{ resistors}}$$

T-to-π (Y-to-Δ) Conversions   The formulas for this conversion also utilize the component designations of Figure 9-28.

T-to-π (Y-to-Δ) Conversions

$$R_1 = \frac{R_A R_B + R_A R_C + R_B R_C}{R_C}$$

$$R_2 = \frac{R_A R_B + R_A R_C + R_B R_C}{R_A}$$

$$R_3 = \frac{R_A R_B + R_A R_C + R_B R_C}{R_B}$$

$$R_\pi = \frac{\text{Sum of all possible } T\text{-pair products}}{\text{Opposite } T \text{ resistor}}$$

Example 15   Convert the π network shown in Figure 9-29(a) to the equivalent T network.

We start by drawing and labeling a T network with standard notation (see Figure 9-29). Next, we substitute the values of the π network into the π-to-T formulas,

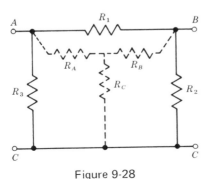

Figure 9-28

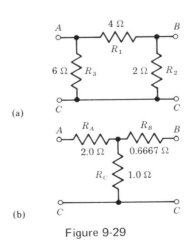

(a)

(b)

Figure 9-29

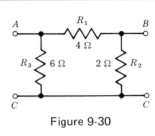

Figure 9-30

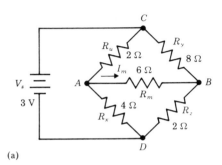

(a)

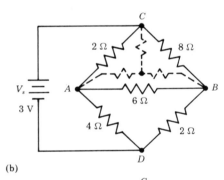

(b)

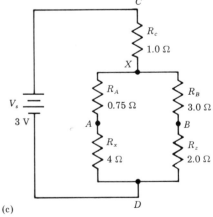

(c)

Figure 9-31

$$R_A = \frac{R_1 R_3}{R_1 + R_2 + R_3} = \frac{4 \times 6}{4 + 6 + 2} = \underline{2.0 \ \Omega}$$

$$R_B = \frac{R_1 R_2}{12} = \frac{4 \times 2}{12} = \underline{0.6667 \ \Omega}$$

$$R_C = \frac{R_2 R_3}{12} = \frac{2 \times 6}{12} = \underline{1.0 \ \Omega}$$

The network in Figure 9-29(b) has been labeled with these values.

Example 16   Convert the T network shown in Figure 9-29(b) to its equivalent π network.

We again start by drawing the equivalent π network and labeling with standard notation (see Figure 9-30). Then we use the T-to-π formulas,

$$R_1 = \frac{R_A R_B + R_A R_C + R_B R_C}{R_C}$$

$$= \frac{(2.0)(0.6667) + (2.0)(1.0) + (0.6667)(1.0)}{1.0}$$

$$R_1 = \frac{4.0}{1.0} = \underline{4 \ \Omega}$$

$$R_2 = \frac{4.0}{R_A} = \frac{4.0}{2.0} = \underline{2.0 \ \Omega}$$

$$R_3 = \frac{4.0}{R_B} = \frac{4.0}{2/3} = \underline{6.0 \ \Omega}$$

The network in Figure 9-30 has been labeled with the values obtained from the formulas. As should be expected, the network is identical to that of Example 15.

Example 17   Use a π-to-T conversion to simplify the bridge circuit of Figure 9-31(a) and find V_{AB} and I_{AB}.

The resistors in the loop ABC form a π or Δ network. We will convert this π to an equivalent T. We redraw the network to show how the T corresponds to the π [see Figure 9-31(b)]. Next we use the π-to-T formulas,

$$R_{Tee} = \frac{\text{Product of two adjacent } \pi \text{ resistors}}{\text{Sum of } \pi \text{ resistors}}$$

$$R_A = \frac{2 \times 6}{2 + 6 + 8} = \frac{12}{16} = 0.75 \ \Omega$$

$$R_B = \frac{6 \times 8}{16} = 3.0 \ \Omega$$

$$R_C = \frac{2 \times 8}{16} = 1 \ \Omega$$

The circuit is again redrawn for clarification [see Figure 9-31(c)] and labeled with the T-network values. We will find V_{AB} by first finding V_{AD} and V_{BD}. Before we can find these we must find R_{XD} and V_{XD},

$$R_{XD} = (R_A + R_X)\|(R_B + R_Z)$$
$$= \frac{(0.75 + 4)(3 + 2)}{0.75 + 4 + 3 + 2} = 2.436\ \Omega$$

We use the ratio formula to find V_{XD},

$$V_{XD} = \frac{R_{XD}}{R_C + R_{XD}} V_s = \frac{2.436}{1.0 + 2.436} \times 3$$
$$= 2.127\ V$$

Similarly,

$$V_{AD} = \frac{R_X}{R_A + R_X} V_{XD} = \frac{4}{4.75} \times 2.127$$
$$= 1.791\ V$$

$$V_{BD} = \frac{2}{5} \times 2.127 = 0.8508\ V$$

Thus,

$$V_{AB} = V_{AD} - V_{BD} = 1.791 - 0.8508 = 0.9402\ V$$
$$I_{AB} = -\frac{V_{AB}}{R_m} = -\frac{0.9402}{6}$$
$$I_{AB} = -0.1567\ A$$

The bridge network here is identical to that used for Example 6 in Section 9-2. The solution checks with the solution in that example obtained by using Thévenin's theorem.

Exercises 9-6

1. For Figure 9-26(a) let $R_1 = 1\ \Omega$, $R_2 = 2\ \Omega$, and $R_3 = 3\ \Omega$. Find the T-equivalent network.

2. For Figure 9-26(a) let $R_1 = 5\ \Omega$, $R_2 = 10\ \Omega$, $R_3 = 15\ \Omega$. Find the T-equivalent network.

3. In Figure 9-26(a) if $R_1 = 50\ k\Omega$, $R_2 = 30\ k\Omega$, and $R_3 = 50\ k\Omega$, find the values of the elements of an equivalent T network.

4. Refer to Figure 9-26(a). Let $R_1 = 6.8\ k\Omega$, $R_2 = 8.2\ k\Omega$, and $R_3 = 4.8\ k\Omega$. Find the values for the elements of an equivalent T network.

5. Refer to Figure 9-27(a). Let $R_A = R_B = 15\ \Omega$ and $R_C = 25\ \Omega$. Find the equivalent π network.

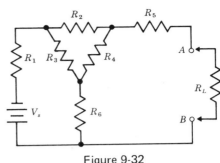

Figure 9-32

6. Let $R_A = 4\ \Omega$, $R_B = 6\ \Omega$, and $R_C = 8\ \Omega$ for the T network of Figure 9-27(a). Find the equivalent π network.

7. For Figure 9-27(a) let $R_A = 2.2\ k\Omega$, $R_B = 3.3\ k\Omega$, and $R_C = 4.7\ k\Omega$. Find the equivalent π network.

8. For Figure 9-27(a) let $R_A = 40\ k\Omega$, $R_B = 60\ k\Omega$, and $R_C = 40\ k\Omega$. Find the equivalent π network.

9. Refer to the bridge circuit of Figure 9-31(a). Let $V_s = 3\ V$, $R_u = 4\ \Omega$, $R_x = 3\ \Omega$, $R_y = 6\ \Omega$, $R_z = 8\ \Omega$, and $R_m = 6\ \Omega$. Utilizing a π-to-T conversion find I_m and V_{AB}.

10. Refer to the bridge circuit of Figure 9-31(a). Let $V_s = 6\ V$, $R_u = 10\ k\Omega$, $R_x = 20\ k\Omega$, $R_y = 20\ k\Omega$, $R_z = 15\ k\Omega$, and $R_m = 10\ k\Omega$. Utilizing a π-to-T conversion find I_m and V_{AB}.

11. For the circuit of Figure 9-32 let $V_s =$

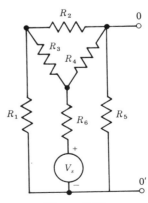

Figure 9-33

10 V, $R_1 = 5\ \Omega$, $R_2 = 5\ \Omega$, $R_3 = 20\ \Omega$, $R_4 = 20\ \Omega$, $R_5 = 5\ \Omega$, and $R_6 = 15\ \Omega$. Thevenize the circuit for the load terminals A-B. Find I_L and V_L for $R_L = R_{th}$. Use a π-to-T conversion in your solution.

12. For the circuit of Figure 9-32 let $R_1 = 20\ k\Omega$, $R_2 = 30\ k\Omega$, $R_3 = 50\ k\Omega$, $R_4 = 50\ k\Omega$, $R_5 = 10\ k\Omega$, $R_6 = 20\ k\Omega$, and $V_s = 100$ V. Use a π-to-T conversion and find the Norton equivalent circuit for the network as seen from terminals A-B. Find I_L and V_L for $R_L = R_{th}$.

13. For the circuit of Figure 9-33 let $V_s = 10$ V, $R_1 = 15\ \Omega$, $R_2 = 20\ \Omega$, $R_3 = 12\ \Omega$, $R_4 = 12\ \Omega$, $R_5 = 25\ \Omega$, and $R_6 = 5\ \Omega$. Use a π-to-T conversion and find the Thevenin equivalent circuit for the network, as seen from terminals 0-$0'$.

14. Refer to Figure 9-33. Using a π-to-T conversion, find the Norton equivalent circuit of the network, as seen from terminals 0-$0'$, if the components have the following values: $V_s = 100$ V, $R_1 = 10\ k\Omega$, $R_2 = 5\ k\Omega$, $R_3 = 10\ k\Omega$, $R_4 = 10\ k\Omega$, $R_5 = 25\ k\Omega$, and $R_6 = 5\ k\Omega$.

Chapter 10

Other Network Analysis Techniques: Kirchhoff's Methods

The German physicist Gustav Robert Kirchhoff (1824–1887)
was a 23-year-old student when he first published a description
of two important basic principles that he had discovered
experimentally regarding voltages and currents in networks.
The statements of these principles are now called Kirchhoff's
Voltage and Current Laws. (The Bettman Archive)

10-1 INTRODUCTION

Consider the network of Figure 10-1, a network which we have previously analyzed using Thévenin's and the superposition theorems. Other techniques for analyzing such networks with two or more sources include the direct application of Kirchhoff's laws. For example, in the loop-current technique (presented in detail in Section 10-6) we assume that there are two *loop currents* in the network, as shown in Figure 10-1(b). Having assumed these currents we polarize the resistors and write Kirchhoff voltage laws for the loops (review Section 8-3).

For loop *abda*,

$$I_1 R_1 + (I_1 - I_2)R_3 - V_{s1} = 0 \qquad (1)$$

and for loop *bcdb*,

$$I_2 R_2 + (I_2 - I_1)R_3 - V_{s2} = 0 \qquad (2)$$

Substituting known values and transposing the *V*-terms, we have

$$6I_1 + 4I_1 - 4I_2 = 12 \qquad (1')$$

$$3I_2 + 4I_2 - 4I_1 = 3 \qquad (2')$$

Combining like terms, and aligning unknowns, we have

$$10I_1 - 4I_2 = 12 \qquad (1'')$$

$$-4I_1 + 7I_2 = 3 \qquad (2'')$$

Each of the equations (1″) and (2″) contain two unknowns, I_1 and I_2. It is not possible to get an explicit solution to a single equation if it contains more than one literal. For example, if we solve (1″) for I_1 we have I_1 in terms of I_2,

$$I_1 = \frac{12 + 4I_2}{10} \qquad (3)$$

And if we do not know the value of I_2 there is no way to determine I_1. However, we do have another equation (2″) which provides more information about I_1 and I_2. Such sets of equations involving two or more unknown quantities are called *simultaneous equations*.

Note from equation (3) that if we knew a value for I_2 we could calculate I_1. For each value of I_2, even if there is an infinite number of such values, there is a unique value of I_1. When unknowns are related, as expressed by sets of simultaneous equations, however, there is *only one set of values* which will satisfy *all* of the equations. (There must be as many independent equations as there are unknowns.) Solving the equations simultaneously we find this unique set of values. There are various methods by which simultaneous solutions may be obtained. In this chapter we present several of these techniques and then develop Kirchhoff's methods of network analysis which require such solutions.

10-2 GRAPHIC SOLUTION OF SIMULTANEOUS EQUATIONS

Simultaneous equations with two unknowns can be solved by graphing the equations (review Section 3-2). This technique is utilized in an important electronics analysis method called *load-line analysis*. To find the solution in this method we (1) graph the equations, and (2) locate the point of intersection of the graphs. The point of intersection is the solution.

Example 1 Solve the following simultaneous equations using the graphic technique.

$$2x - y = 3 \qquad (1)$$

$$x + 2y = 4 \qquad (2)$$

To plot these equations we take advantage of a property of linear equations: A straight line is defined by two points. (Linear equations

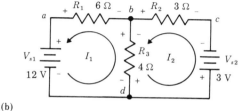

(a)

(b)

Figure 10-1

are equations whose graph is a straight line.
Equations will be linear if the unknowns appear only to the first power.) To plot equation (1) we solve it for two pairs of values.

For $x = 0$:

$$-y = 3 \quad \text{or} \quad y = -3$$

which gives the point $(0, -3)$

For $y = 0$:

$$2x = 3 \quad x = \tfrac{3}{2}$$

and we define the point $(\tfrac{3}{2}, 0)$

The graph of equation (1) is obtained by plotting the points $(0, -3)$ and $(\tfrac{3}{2}, 0)$ (see Figure 10-2). Similarly, we find two points for equation (2).

For $x = 0$:

$$2y = 4 \quad y = 2$$

One point is, then, $(0, 2)$. The second point is found by letting $y = 0$, $x + 0 = 4$. The point is $(4, 0)$. The graph of the equation $x + 2y = 4$ is also shown in Figure 10-2.

The two lines intersect at the point $(2, 1)$. This is the solution of the equations. We check it by substituting into the original equations. Substituting $x = 2$, $y = 1$ into equation (1),

$$2x - y = 3, \quad 2 \cdot 2 - 1 = 3, \quad 4 - 1 = 3$$

and into equation (2),

$$x + 2y = 4, \quad 2 + 2 \cdot 1 = 4, \quad 2 + 2 = 4$$

Load-Line Analysis The graphic technique for obtaining simultaneous solutions for certain types of circuits is a frequently used approach in electronics.

Example 2 Using the graphic technique determine (a) the current I_b, (b) the voltage V_b, and (c) the voltage $V_{bb} - V_b$ for the circuit of Figure 10-3(a). Let $V_{bb} = 10$ V, $r_p = 4$ Ω, and $R_L = 6$ Ω.
We write an Ohm's law equation for r_p,

$$V_b = I_b r_p \quad \text{or} \quad V_b - I_b r_p = 0 \qquad (3)$$

and we write a KVL equation for the loop,

$$V_{bb} - V_b - I_b R_L = 0 \qquad (4)$$

Substituting known values into equations (3) and (4) we obtain a pair of simultaneous equations.

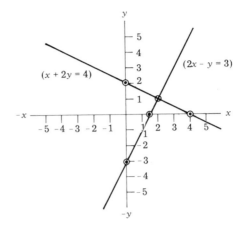

Figure 10-2

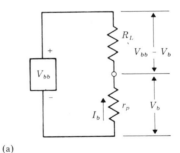

(a)

(b)

Figure 10-3

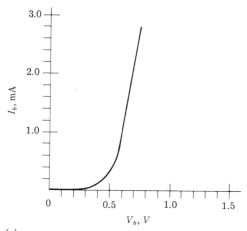

(a)

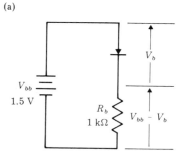

(b)

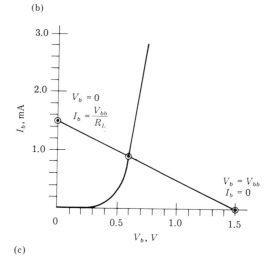

(c)

Figure 10-4

From (3)

$$V_b - 4I_b = 0$$

and from (4)

$$10 - V_b - 6I_b = 0$$

or

$$V_b + 6I_b = 10$$

Thus,

$$V_b - 4I_b = 0 \qquad (5)$$
$$V_b + 6I_b = 10 \qquad (6)$$

Using the technique of Example 1 we obtain points for plotting the equations.
For equation (5):

$$\text{if } I_b = 0, \qquad V_b = 0$$
$$\text{if } I_b = 1, \qquad V_b = 4$$

These values define the points $(0, 0)$ and $(4\text{ V}, 1\text{ A})$.
For equation (6):

$$\text{when } \quad I_b = 0, \qquad V_b = 10\text{ V}$$
$$\text{when } \quad V_b = 0 \qquad I_b = \frac{10}{6} = 1.667\text{ A}$$

Thus we have the points $(10\text{ V}, 0)$ and $(0, 1.667\text{ A})$. The equations are shown plotted in Figure 10-3(b). The point of intersection of the two lines is $I_b = 1$ A and $V_b = 4$ V. Therefore, the required answers are:

(a) $\underline{I_b = 1\text{ A}}$ (b) $\underline{V_b = 4\text{ V}}$
(c) $\underline{V_{bb} - V_b = 10 - 4 = 6\text{ V}}$

The load-line analysis technique is more typically used when one circuit element, such as a semiconductor diode or transistor, is nonlinear. In such cases the I-V (current-versus-voltage) characteristic of the nonlinear device is plotted from measured data and the equation of the linear device is plotted on the same graph.

Example 3   The I-V characteristic of a certain diode is shown in Figure 10-4(a). The diode is to be used in the circuit of Figure 10-4(b). Using the load-line method, determine the operating point (I_b and V_b) of the circuit.
The graph for the diode is provided; we determine the equation for R_L and plot it on the same graph.

$$V_{bb} - V_b - I_b R_L = 0 \qquad \text{(KVL equation for the loop)}$$

or

$$1.5 - V_b - 1000I_b = 0$$
$$V_b + 1000I_b = 1.5 \qquad (7)$$

To plot equation (7) we determine two points.

At $I_b = 0$:
$$V_b = 1.5 \text{ V}$$

At $V_b = 0$:
$$I_b = \frac{1.5}{1000} = 1.5 \text{ mA}$$

We plot the points (1.5 V, 0) and (0, 1.5 mA) on the graph [see Figure 10-4(c)]. The point of intersection establishes the operating point of the circuit.

$$I_b = 0.9 \text{ mA}, \qquad V_b = 0.61 \text{ V}$$

Exercises 10-2 In Exercises 1–6 obtain the solutions of the equations using the graphic technique.

1. $x - y = 1$
   $x + y = 3$

2. $5x + 2y = 10$
   $x - 6y = 14$

3. $2x - 3y = -2$
   $x + 2y = 6$

4. $x + 2y = -6$
   $2x - 3y = 2$

5. $5x - y = 10$
   $3x + 2y = 6$

6. $4x + y = 8$
   $3x - 2y = 6$

7. The resistance of an incandescent light bulb varies with the temperature of the filament, which, in turn, varies with the current flowing through the filament, and thus with the voltage across it. It follows that the resistance of an incandescent bulb is nonlinear. A typical bulb has the following volt-ampere characteristic:

v, V	0	5	10	20	40	60	80	100	120
i, mA	0	100	160	215	300	370	415	470	510

Plot this characteristic on graph paper and determine:

(a) The current that the bulb would conduct if the voltage across the bulb were 7, 35, 50, 95, and 110 V.

(b) The *actual* resistance of the bulb at these operating points.

8. Determine the power which the bulb of Exercise 7 would dissipate at or near 117 V. What is the probable wattage rating of the bulb?

9. The bulb of Exercise 7 is to be operated in series with a 225-Ω resistance across a 115-V source.

(a) Draw the circuit diagram.
(b) Plot the load line for the circuit on the characteristic curve obtained in Exercise 7.
(c) What is the voltage across the bulb? across the resistor?

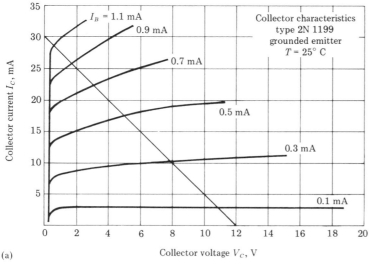

(a)

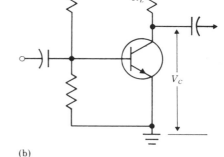

(b)

Figure 10-5

(d) What is the current in the circuit?

(e) What power is dissipated by the bulb? the resistor?

10. The *I-V* characteristics of a 2N1199 transistor were measured and plotted [see Figure 10-5(a)]. The transistor is to be used in the circuit of Figure 10-5(b). The load line for a specific value of R_L is plotted on the device *I-V* characteristic. According to the load line (a) what is V_{CC}? (b) What is R_L?

11. Refer to Exercise 10. What are I_C and V_C when I_B is (a) 0.3 mA? (b) 0.5 mA? (c) 0.4 mA?

12. Refer to Exercise 10. What is the voltage across R_L when I_B is (a) 0.3 mA? (b) 0.7 mA? (c) 0.6 mA?

13. Refer to Figure 10-5. Let $V_{CC} = 10$ V. Construct a load line for $R_L = 0.5$ kΩ. Determine I_C and V_C when I_B is (a) 0.3 mA and (b) 0.5 mA.

14. Refer to Figure 10-5. Let $V_{CC} = 18$ V and $R_L = 0.6$ kΩ. (a) Construct the load line. Find I_C, V_C, and V_L (voltage across R_L) when I_B is (b) 0.1 mA and (c) 0.4 mA.

10-3 SOLVING SIMULTANEOUS EQUATIONS BY SUBSTITUTION AND ELIMINATION

Substitution We may solve sets of two simultaneous equations with two variables (unknowns) by solving one equation for one unknown in terms of the other unknown, and substituting this value into the second equation.

Example 4 Solve, using substitution.

$$2x - y = 3$$
$$x + 2y = 4$$

Since in the second equation the coefficient of x is 1, we solve it easily for x and substitute the value into the first equation. That is, from

$$x + 2y = 4$$

we obtain

$$x = 4 - 2y$$

and substitute into

$$2x - y = 3$$
$$2(4 - 2y) - y = 3$$
$$8 - 4y - y = 3$$
$$-5y = -5$$
$$\underline{y = 1}$$

We substitute $y = 1$ into either equation to obtain x:

$$2x - (1) = 3 \qquad 2x = 4 \qquad \underline{x = 2} \text{ or}$$
$$x + 2(1) = 4 \qquad x = 4 - 2 \qquad \underline{x = 2}$$

Check: $2 \times 2 - 1 = 3 \qquad 4 - 1 = 3$
  $2 + 2 \times 1 = 4 \qquad 2 + 2 = 4$

Elimination Instead of substitution we can use addition or subtraction to eliminate one unknown, so that a single equation in one unknown is left to be solved.

Example 5 Solve, using elimination.

$$2x - y = 3$$
$$x + 2y = 4$$

If we multiply the first equation by 2, then add the equations, we eliminate the y terms.
  Multiply $2x - y = 3$ by 2:

$$4x - 2y = 6$$

Then add $x + 2y = 4$: $\underline{x + 2y = 4}$
  $5x \qquad = 10$
  $\underline{x = 2}$

Then we substitute $x = 2$ into either equation to find y:

$$2 \cdot 2 - y = 3$$
$$-y = 3 - 4$$
$$\underline{y = 1}$$

These values check with the values found for the identical equations in Example 4.

These techniques—substitution and elimination—may be applied very easily and effectively for solving two equations in two unknowns. And although they may correctly be applied to sets of equations of three or even more unknowns, the work rapidly becomes very tedious and time consuming if performed manually for more than two variables. (The method of elimination is often used in programs written for the solution of simultaneous equations by electronic digital computer.) Solution by determinants is generally easier and faster when manual methods (paper, pencil, and calculator) must be employed for the solution of equations involving more than two variables. Solution by determinants is presented in the next section.

Exercises 10-3 In Exercises 1–10 solve the given equations using substitution. Check your solutions.

1. $u + v = 5$
   $u - v = 1$

2. $3x + y = 10$
   $2x + y = 8$

3. $2x + 4y = 10$
   $2x + 3y = -5$

4. $3v - 6i = 9$
   $5v - 7i = 12$

5. $5r - 10R = -15$
   $2r + R = 4$

6. $V + I = 4$
   $V - I = 6$

7. $2v + V = 3$
   $3v - V = 7$

8. $4R_1 + 3R_2 = 2$
   $3R_1 - R_2 = 8$

9. $-I_1 + 2I_2 = 3$
   $2I_1 + I_2 = 4$

10. $2I_1 + 3I_2 = 12$
    $3I_1 + 2I_2 = 8$

11. $5I_1 + 2I_2 = 3$
    $-2I_1 - 3I_2 = 1$

12. $I_1 + 2I_2 = -10$
    $I_1 - 3I_2 = 2$

13. $2I_1 + 3I_2 = 0$
    $I_1 + I_2 = 1$

14. $5x + 7y = 39$
    $x - 7y = -9$

15. $2I_1 - 3I_2 = 49$
    $4I_1 + I_2 = 7$

16. $2I_1 - 1 = I_2$
    $5I_1 + 3 = 2I_2$

17. $I_1 + 4I_2 = 8$
    $2I_1 - 3I_2 = 6$

18. $4I_1 + 2I_2 = 8$
    $9I_1 - 6I_2 = 18$

19. $2I_2 + 64 = 5I_1$
    $4I_2 - 3I_1 = 5$

20. $21 - 7I_2 = 3I_1$
    $64 + 8I_1 = -32I_2$

21–40. In Exercises 1–20 solve using elimination. Check your solutions.

10-4 DETERMINANTS

Let us consider two literal simultaneous equations,

$$ax + by = c \qquad (1)$$
$$dx + ey = f \qquad (2)$$

In these equations the letters a, b, c, d, e, and f may have any value. The solutions of these equations are

$$x = \frac{ce - bf}{ae - bd} \qquad (3)$$

$$y = \frac{af - cd}{ae - bd} \qquad (4)$$

Note that both roots have the same denominator and that the denominator consists only of the coefficients of x and y of the original equations. We define a new expression, *an array*, called the *determinant* of the equations; that is,

$$\begin{vmatrix} a & b \\ d & e \end{vmatrix} = ae - bd$$

Note the relationship of the positions of the coefficients in the determinant to their positions in the original equations,

$$ax + by = c$$
$$dx + ey = f$$

$$\begin{vmatrix} a & b \\ d & e \end{vmatrix}$$

By definition (agreement), when a determinant contains four terms in a two-by-two array, we find its value by multiplying cross terms and subtracting the product that slopes up to the right from the product that slopes down to the right,

$$\begin{vmatrix} a & b \\ d & e \end{vmatrix} \left. \begin{array}{l} -\ bd \text{ is } - \text{ product} \\ \quad (-bd) \\ \\ +\ ae \text{ is } + \text{ product} \\ \quad (+ae) \end{array} \right\} = ae - bd$$

Using this definition of a determinant we now write the solutions [equations (3) and (4)] in determinant form,

$$x = \frac{\begin{vmatrix} c & b \\ f & e \end{vmatrix}}{\begin{vmatrix} a & b \\ d & e \end{vmatrix}} \qquad (5)$$

$$y = \frac{\begin{vmatrix} a & c \\ d & f \end{vmatrix}}{\begin{vmatrix} a & b \\ d & e \end{vmatrix}} \qquad (6)$$

We analyze these forms and the instructions we follow to solve equations by determinants:

1. Write the equations in a vertical pattern so that like variables are in vertical columns.

2.   For the denominators of the roots write the determinant which contains the coefficients of the variables in the same relative positions they occupy in the equations.

3.   Write the numerator of each root as the determinant in which the constant terms of the equations are substituted in the determinant of the denominator for the coefficients of the unknown being solved for.

4.   Evaluate the determinants.

Consider the equations,

$$3u + 2v = 1$$

$$u - 2v = -5$$

We write the roots directly:

$$u = \frac{\begin{vmatrix} 1 & 2 \\ -5 & -2 \end{vmatrix}}{\begin{vmatrix} 3 & 2 \\ 1 & -2 \end{vmatrix}} = \frac{1 \cdot (-2) - 2 \cdot (-5)}{3 \cdot (-2) - 1 \cdot 2} = \frac{-2 + 10}{-6 - 2}$$

$$= \frac{8}{-8} = -1$$

$$v = \frac{\begin{vmatrix} 3 & 1 \\ 1 & -5 \end{vmatrix}}{\begin{vmatrix} 3 & 2 \\ 1 & -2 \end{vmatrix}} = \frac{3 \cdot (-5) - 1 \cdot 1}{3 \cdot (-2) - 1 \cdot 2} = \frac{-15 - 1}{-6 - 2}$$

$$= \frac{-16}{-8} = 2$$

Study this example paying special attention to the positions in the determinants of the numbers obtained from the equations.

   The determinant is always a square. A two-by-two determinant, such as those in the examples, is called a *second-order determinant*. The order designates the number of rows or columns. Thus, a three-by-three determinant is of third order, and so on. Third-order determinants are evaluated in much the same way as second-order determinants (but higher orders require special rules for evaluating and are considered in more advanced texts).

   Consider the third-order determinant

$$\begin{vmatrix} 1 & 2 & -2 \\ 3 & 2 & -1 \\ 5 & 1 & 3 \end{vmatrix}$$

The evaluation of third-order determinants is greatly facilitated by rewriting the first two columns outside the determinant and to the right, as follows

$$\begin{vmatrix} 1 & 2 & -2 \\ 3 & 2 & -1 \\ 5 & 1 & 3 \end{vmatrix} \begin{array}{cc} 1 & 2 \\ 3 & 2 \\ 5 & 1 \end{array}$$

We again form diagonal products; now, each contains three factors. Those sloping up to the right are subtracted from those sloping down to the right. We evaluate the given determinant:

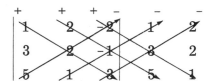

Products sloping down to the right

$$= 1 \cdot 2 \cdot 3 + 2 \cdot (-1) \cdot 5 + (-2) \cdot (3) \cdot (1)$$

Products sloping up to the right

$$-[5 \cdot 2 \cdot (-2) + 1 \cdot (-1) \cdot 1 + 3 \cdot 3 \cdot 2]$$

$$= 6 - 10 - 6 - (-20) - (-1) - 18$$

$$= -7$$

Note that we use the brackets [ ] to show that all the up-to-the-right products are to be subtracted. The sign of each product is changed when we remove the brackets.

Example 6   Use determinants to solve the set of three simultaneous equations.

$$u - 2v \qquad = 2$$

$$3v - w = 1$$

$$2u \qquad + 3w = 1$$

   First, we rewrite the equations, aligning variables:

$$u - 2v \qquad = 2$$

$$3v - w = 1$$

$$2u \qquad + 3w = 1$$

Next we write the determinant of the coefficients. This will be the denominator for all three roots.

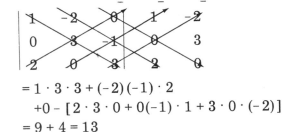

$$= 1 \cdot 3 \cdot 3 + (-2)(-1) \cdot 2$$
$$+ 0 - [2 \cdot 3 \cdot 0 + 0(-1) \cdot 1 + 3 \cdot 0 \cdot (-2)]$$
$$= 9 + 4 = 13$$

Note that zero is the coefficient when a term is missing in an equation.

Now, we write the determinants of the numerators of the roots by substituting the constant terms (from the right of the equal sign) in each equation for the coefficients of the variable being solved for. Thus,

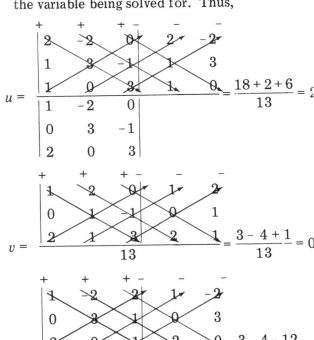

$$u = \frac{18 + 2 + 6}{13} = ?$$

$$v = \frac{3 - 4 + 1}{13} = \frac{0}{13} = 0$$

$$w = \frac{3 - 4 - 12}{13} = \frac{}{13} =$$

Exercises 10-4 Solve and check, using determinants.

1. $3x + y = 10$
   $2x + y = 8$

2. $3i_1 - 6i_2 = 9$
   $5i_1 - 7i_2 = 12$

3. $5v_1 + 2v_2 = 3$
   $-2v_1 - 3v_2 = 1$

4. $2y + 4x = 10$
   $2y + 3x = -5$

5. $5v - 10i = -15$
   $2v + i = 4$

6. $3r - 6R = 9$
   $5r - 7R = 12$

7. $2x + 3y + 6z = 0$
   $x + y + 2z = 1$
   $3x + 2y + 2z = 3$

8. $R_2 - R_3 - 7 = 0$
   $3R_1 - R_2 + R_3 = 2$
   $6R_1 + 2R_2 - R_3 = 4$

9. $v_1 - 2v_2 + 3v_3 = 4$
   $2v_1 + v_2 - 4v_3 = 3$
   $v_1 + 2v_3 = 8$

10. $3(A - B) = 3$
    $3A - 6B + C = 1$
    $2A + B + 4C = 9$

11. $r_1 + r_2 + r_3 = -4$
    $2r_1 + 3r_2 + 4r_3 = 0$
    $-r_1 - r_2 + 2r_3 = 8$

12. $3r - 2s + t = 33$
    $2r + s - t = 7$
    $r + s - 4t = -6$

13. $x + y + z = 20$
    $2(x + z) = 34$
    $3x - 2y + 4z = 57$

14. $3V - 3(v_1 + v_2) = 12$
    $2V + 4v_1 - v_2 = 30$
    $4V - 3v_1 + 4v_2 = -43$

15. $3I_1 + 2I_2 + 2I_3 = 3$
    $2I_1 + 6I_2 + 3I_3 = 0$
    $I_1 + 2I_2 + I_3 = 1$

16. $-4I_1 + 2I_2 + I_3 = 3$
    $2I_1 + I_2 = 8$
    $3I_1 + I_2 - 2I_3 = 4$

17. $I_1 - I_2 + 3I_3 = 2$
    $-I_1 + I_2 - 7 = 0$
    $-I_1 + 2I_2 + 6I_3 = 4$

18. $2I_1 + I_2 - I_3 = 12$
    $4I_1 + 3I_3 = 15$
    $I_1 - 3I_2 - 5I_3 = -3$

10-5 THE BRANCH-CURRENT METHOD

This is one of the several approaches to network analysis which uses Kirchhoff's laws directly. It is structured around the branch currents as the unknown variables. Some terminology definitions will be useful at this point.

branch   A portion of a network through which there is only one path for current to flow, that is, a series path.

node   A *point* in a network where two or more circuit elements or branches are connected together.

loop   A complete circuit path.

The branch-current method consists of the following steps:

1. Write current-law equations for those nodes which are the junction points of three or more branches. However, if there are n such nodes, only $n - 1$ equations are independent and, therefore, needed.
2. Write voltage-law equations for network loops. Again, for L loops, only $L - 1$ equations are independent and needed.
3. Solve the system of simultaneous equations so obtained using your choice of the methods, that is, elimination or determinants. The number of independent equations must equal the number of unknowns, that is, the number of branches.

Two linear equations containing the same variables are independent if their graphs are different or if there is at least one variation in factors. Such equations are dependent if the coefficients of the variables are related by the same factor. For example, the equations

$$2I_1 + 2I_2 - 3I_3 = 6$$

$$6I_1 + 6I_2 - 9I_3 = 18$$

are dependent because corresponding coefficients are related by the factor 3. The equations

$$2I_1 + 2I_2 - 3I_3 = 6$$

$$6I_1 - 6I_2 - 9I_3 = 18$$

are independent since the I_2 terms are related by the factor -3, and all others by 3.

The method of branch currents lends itself to a systematic approach; the outline of one such approach is given below.

1. Draw the circuit diagram and label with all known magnitudes of voltage, current, and resistance. Identify voltage sources with known polarities.
2. Assume directions for all currents, and for the sake of convenience and saving time, be systematic in making assumptions. That is, make no effort to predict the direction of actual currents in the network; rather, assume arbitrarily that all currents flow from left to right and from top to bottom of the diagram, or some other such arbitrary choice. Label the diagram with your assumed directions for all currents. Either electron or conventional current may be used.

3. On the basis of labeled current directions, label all remaining circuit elements for voltage polarity, applying the rule stated in Section 8-2.
4. Write current-law equations for all nodes that join three or more branches and $n - 1$ equations for n such nodes. (Review Section 8-3.)
5. Write voltage-law equations for circuit loops, $L - 1$ equations for L loops. (Review Section 8-3.)
6. Check equations for independence and to insure that there are v independent equations for v variables. Rearrange equations so that corresponding variables are aligned vertically and constants appear on the right side.
7. Solve equations using your choice of methods. (See Sections 10-3 and 10-4.)
8. Check solutions by substitution into *original* equations and preferably, also, into a loop equation not used in obtaining solutions.

Example 7   Use the branch-current method and electron current flow to determine the currents of the network of Figure 10-6(a).

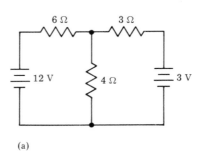

(a)

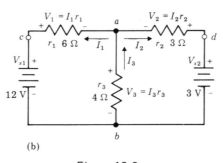

(b)

Figure 10-6

We start by redrawing the circuit, labeling it with assumed current directions and the polarities of IR potentials [see Figure 10-6(b)].

A current-law equation for node a is

$$-I_1 - I_2 + I_3 = 0 \qquad (1)$$

or, multiplying the equation by -1,

$$I_1 + I_2 - I_3 = 0 \qquad (1')$$

Tracing clockwise $bcab$, we find that the voltage-law equation for the left-hand mesh is

$$-V_{s1} + V_1 + V_3 = 0$$

From Ohm's law,

$$V_1 = I_1 r_1, \qquad V_3 = I_3 r_3$$

and therefore the equation for loop $bcab$ becomes

$$-V_{s1} + I_1 r_1 + I_3 r_3 = 0$$

or, substituting known values, we have

$$-12 + 6I_1 + 4I_3 = 0 \qquad (2)$$

Similarly, the equation for the right-hand mesh $badb$ is

$$-V_3 - V_2 + V_{s2} = 0, \quad -I_3 r_3 - I_2 r_2 + V_{s2} = 0$$
$$-4I_3 - 3I_2 + 3 = 0 \qquad (3)$$

We thus have three equations in three unknowns:

$$I_1 + I_2 - I_3 = 0 \qquad (1')$$
$$6I_1 \quad + 4I_3 = 12 \qquad (2)$$
$$3I_2 + 4I_3 = 3 \qquad (3)$$

There are other equations that we could have written. For example, at the node b,

$$I_1 + I_2 = I_3$$

but this equation is actually identical to (1), and it is not independent. Or we could have written an equation for the loop $bcadb$:

$$-V_{s1} + V_1 - V_2 + V_{s2} = 0,$$

but this can be shown to contain both (2) and (3) and therefore is not an independent equation.

Now we solve the three equations in three unknowns using third-order determinants:

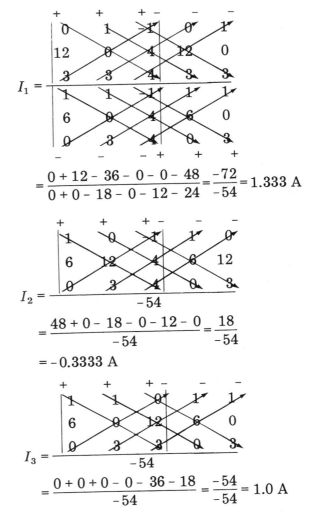

$$I_1 = \frac{0 + 12 - 36 - 0 - 0 - 48}{0 + 0 - 18 - 0 - 12 - 24} = \frac{-72}{-54} = 1.333 \text{ A}$$

$$I_2 = \frac{48 + 0 - 18 - 0 - 12 - 0}{-54} = \frac{18}{-54}$$
$$= -0.3333 \text{ A}$$

$$I_3 = \frac{0 + 0 + 0 - 0 - 36 - 18}{-54} = \frac{-54}{-54} = 1.0 \text{ A}$$

The signs of the current values are indications of their true directions: The directions of I_1 and I_3 are as indicated on the diagram in Figure 10-6(b) because their algebraic signs are + in the solution. The real direction of I_2 is toward node a, rather than away from a, because the sign is $-$.

Check: Substitute the values in the equation for loop $bcadb$ which we decided not to use:

$$-V_{s1} + V_1 - V_2 + V_{s2} = 0$$
$$-V_{s1} + I_1 r_1 - I_2 r_2 + V_{s2} = 0$$
$$-12 + \tfrac{4}{3}(6) - (-\tfrac{1}{3})3 + 3 = 0$$
$$-12 + 8 + 1 + 3 = 0$$
$$0 = 0$$

The branch-current method can be used equally well with conventional current. We demonstrate by analyzing the same network, this time using conventional current.

Example 8 Use the branch-current method and conventional current flow to determine the currents of the network of Figure 10-7(a). To show the application of the rules for marking the circuit, we have redrawn and marked the network in Figure 10-7(b). A current-law equation for node a is:

$$I_1 - I_2 - I_3 = 0$$

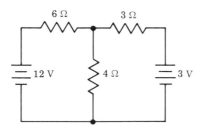

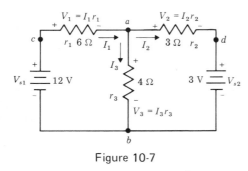

Figure 10-7

Let us write a voltage-law equation for loop $bcab$. We can start at node b and trace clockwise:

$$-12 + 6I_1 + 4I_3 = 0$$

From Ohm's law the voltage across r_1 is $I_1 r_1 = 6I_1$ and the voltage across r_3 is $I_3 r_3 = 4I_3$.

The voltage-law equation for loop $badb$ is

$$-4I_3 + 3I_2 + 3 = 0$$

Rearranging and rewriting equations with unknowns aligned vertically we have

$$I_1 - I_2 - I_3 = 0$$
$$6I_1 \qquad + 4I_3 = 12$$
$$3I_2 - 4I_3 = -3$$

Solving by determinants we have

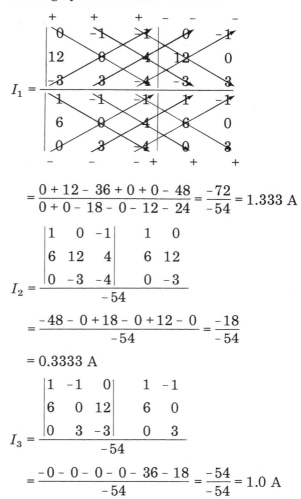

$$= \frac{0 + 12 - 36 + 0 + 0 - 48}{0 + 0 - 18 - 0 - 12 - 24} = \frac{-72}{-54} = 1.333 \text{ A}$$

$$I_2 = \frac{\begin{vmatrix} 1 & 0 & -1 \\ 6 & 12 & 4 \\ 0 & -3 & -4 \end{vmatrix} \begin{matrix} 1 & 0 \\ 6 & 12 \\ 0 & -3 \end{matrix}}{-54}$$

$$= \frac{-48 - 0 + 18 - 0 + 12 - 0}{-54} = \frac{-18}{-54}$$

$$= 0.3333 \text{ A}$$

$$I_3 = \frac{\begin{vmatrix} 1 & -1 & 0 \\ 6 & 0 & 12 \\ 0 & 3 & -3 \end{vmatrix} \begin{matrix} 1 & -1 \\ 6 & 0 \\ 0 & 3 \end{matrix}}{-54}$$

$$= \frac{-0 - 0 - 0 - 0 - 36 - 18}{-54} = \frac{-54}{-54} = 1.0 \text{ A}$$

The signs of the solutions indicate that the assumed and indicated directions of all currents are correct; that is, the directions are the same as those in which real, conventional currents would occur in the actual live circuit represented by the diagram.

Check: Write a voltage-law equation for the loop $bcadb$ and substitute the current values:

$$-12 + 6I_1 + 3I_2 + 3 = 0$$
$$-12 + 6(1.333) + 3(0.3333) + 3 = 0$$
$$-12 + 8 + 1 + 3 = 0$$
$$0 = 0$$

Exercises 10-5   In Exercises 1–8 perform the following: (a) Redraw the network and label with current designations and directions (electron flow) and voltage polarities. (b) Write a set of branch-current method equations for determining the currents of the network. (c) Solve the equations, using the method of your choice.

1. Refer to Figure 10-8. $R_1 = 10\ \Omega$, $R_2 = 15\ \Omega$, and $V_s = 12.5$ V.

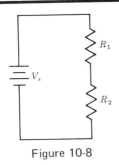

Figure 10-8

2. Refer to Figure 10-8. $R_1 = 25\ \Omega$, $R_2 = 35\ \Omega$, and $V_s = 120$ V.

3. Refer to Figure 10-9. $R_1 = 2.5\ \Omega$, $R_2 = 3\ \Omega, R_3 = 5\ \Omega, R_4 = 6\ \Omega, R_5 = 8\ \Omega$, and $V_s = 16$ V.

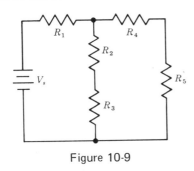

Figure 10-9

4. Refer to Figure 10-9. $R_1 = 4.5\ \Omega$, $R_2 = 7.5\ \Omega, R_3 = 4.7\ \Omega, R_4 = 5\ \Omega$, $R_5 = 7\ \Omega$, and $V_s = 9$ V.

5. Refer to Figure 10-10. $R_1 = 4\ \Omega$, $R_2 = 6\ \Omega, R_3 = 5\ \Omega, V_{s1} = 4.5$ V, and $V_{s2} = 6$ V.

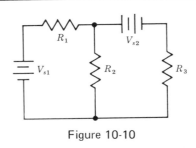

Figure 10-10

6. Refer to Figure 10-10. $R_1 = 8\ \Omega$, $R_2 = 3\ \Omega, R_3 = 4\ \Omega, V_{s1} = 3$ V, and $V_{s2} = 9$ V.

7. Refer to Figure 10-11. $R_1 = 6\ \Omega$, $R_2 = 8\ \Omega, R_3 = 3\ \Omega, V_{s1} = 12$ V, $V_{s2} = 8$ V, and $V_{s3} = 9$ V.

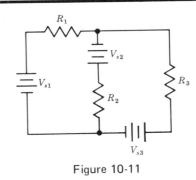

Figure 10-11

8. Refer to Figure 10-11. $R_1 = 2\ \Omega$, $R_2 = 3\ \Omega, R_3 = 6\ \Omega, V_{s1} = 8$ V, $V_{s2} = 9$ V, and $V_{s3} = 5$ V.

9. Refer to Figure 10-12. $R_1 = 4\ \Omega$, $R_2 = 3\ \Omega$, $R_3 = 6\ \Omega$, $R_4 = 7\ \Omega$, $R_5 = 4\ \Omega$, $V_{s1} = 9$ V, and $V_{s2} = 9$ V.
   (a) Redraw the network and label.
   (b) Write a set of equations for determining the five branch currents of the network. Do not solve the equations.

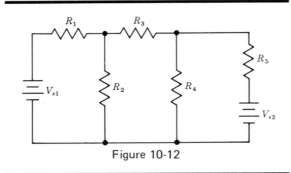

Figure 10-12

10. Refer to Figure 10-12. $R_1 = 5\ \Omega$, $R_2 = 5\ \Omega$, $R_3 = 7\ \Omega$, $R_4 = 3\ \Omega$, $R_5 = 8\ \Omega$, $V_{s1} = 6$ V, and $V_{s2} = 9$ V. Repeat Exercise 9 for this network.

11. Solve the equations obtained in Exercise 9.

12. Solve the equations obtained in Exercise 10.

13–24. Find the required solutions to Exercises 1–12 using conventional current.

10-6 THE LOOP-CURRENT METHOD

As you have discovered, in a system of simultaneous equations the more unknowns, the more equations are required and, consequently, the more tedious and time-consuming the solution. An important characteristic of mathematics is its ability, frequently, to suggest techniques which are time- and laborsaving. The loop method of network analysis as an easier, faster adaption of the branch-current method is one example of this highly desirable feature of mathematics.

If we refer to Figure 10-13(a) and assume branch currents as indicated on the diagram, the current-law equation for node a yields

$$-I_1 + I_2 + I_3 = 0$$

Solving for I_3 yields

$$I_3 = I_1 - I_2 \qquad (1)$$

Equation (1) suggests that I_3, the current in branch ab, need not be considered as a unique third current in the network but simply as the algebraic sum of two superimposed loop currents, I_1 and I_2, flowing in that branch. The circuit is redrawn in Figure 10-13(b) to emphasize this concept. The net result of this view of the relationships involved is to reduce the number of unknown currents and the corresponding equations required for a solution to two. Thus, writing a voltage-law equation for loop 1 ($bcab$), we have

$$-12 + 6I_1 + 4(I_1 - I_2) = 0$$

and for loop 2 ($badb$),

$$-4(I_1 - I_2) + 3I_2 + 3 = 0$$

Rewriting and rearranging with unknowns aligned, we have

$$10I_1 - 4I_2 = 12$$

$$-4I_1 + 7I_2 = -3$$

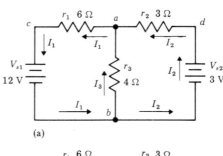

(a)

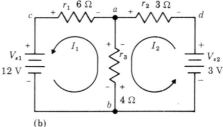

(b)

Figure 10-13

Solving with determinants,

$$I_1 = \frac{\begin{vmatrix} 12 & -4 \\ -3 & 7 \end{vmatrix}}{\begin{vmatrix} 10 & -4 \\ -4 & 7 \end{vmatrix}} = \frac{84 - 12}{70 - 16} = \frac{72}{54} = 1.333 \text{ A}$$

$$I_2 = \frac{\begin{vmatrix} 10 & 12 \\ -4 & -3 \end{vmatrix}}{54} = \frac{-30 + 48}{54} = \frac{18}{54} = 0.3333 \text{ A}$$

$$I_3 = I_1 - I_2 = 1.333 - 0.3333 = 1.00 \text{ A}$$

A systematic approach yields quick results with minimum opportunity for error:

1. Arbitrarily assume all loop currents are clockwise (or counterclockwise, if you prefer). Label circuit diagram with loop-current arrows.

2. Label diagram with voltage polarity markings.

3. Write voltage-law equations for each loop, tracing all loops in some arbitrary direction, the same direction as for loop currents, for example. Write as many equations as there are independent loops. A convenient method for testing for independent loops is to visualize the network as a fishing net. Each "mesh" of the net then is an independent loop. Another helpful analogy is to visualize the network as a window with panes, each pane corresponding to an independent loop.

Exercises 10-6   In Exercises 1–14 perform the given instructions. Use electron flow in your solutions.

1. Work Exercise 3, Exercises 10-5, using the loop method rather than the branch method.

2. Work Exercise 4, Exercises 10-5, using the loop method.

3. Work Exercise 5, Exercises 10-5, using the loop method.

4. Work Exercise 6, Exercises 10-5, using the loop method.

5. Work Exercise 7, Exercises 10-5, using the loop method.

6. Work Exercise 8, Exercises 10-5, using the loop method.

7. Work Exercise 9, Exercises 10-5, using the loop method.

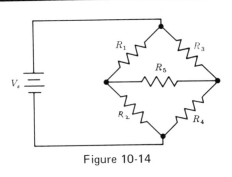

Figure 10-14

8. Work Exercise 10, Exercises 10-5, using the loop method.

9. Solve the equations obtained in Exercise 7.

10. Solve the equations obtained in Exercise 8.

11. Refer to the Wheatstone bridge circuit of Figure 10-14. $V_s = 3$ V, $R_1 = 3 \ \Omega$, $R_2 = 5 \ \Omega$, $R_3 = 5 \ \Omega$, $R_4 = 6 \ \Omega$, and $R_5 = 10 \ \Omega$. Using the loop method, determine the current in R_5.

12. Refer to the Wheatstone bridge circuit of Figure 10-14. $V_s = 1.5$ V, $R_1 = 12 \ \Omega$, $R_2 = 6 \ \Omega$, $R_3 = 8 \ \Omega$, $R_4 = 4 \ \Omega$, and $R_5 = 25 \ \Omega$. Using the loop method, determine the current in R_5.

13. Work Exercise 11, using the branch-current method.

14. Work Exercise 12, using the branch-current method.

15–28. Find the required solutions to Exercises 1–14 using conventional current.

10-7 THE NODE-VOLTAGE METHOD

In the preceding section we saw that mathematics can suggest time- and laborsaving techniques in circuit analyses. Similarly, arranging or looking at circuits in different ways sometimes leads to simplifications in the mathematical approach. For example, the node-voltage method will permit us to analyze the circuit of Figure 10-13 with a single equation in one unknown!

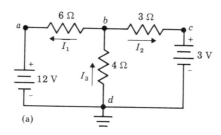

(a)

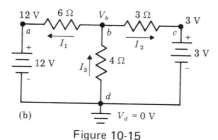

(b)

Figure 10-15

Example 9. Analyze the circuit of Figure 10-13 using the node-voltage method and electron current.

Let us first "ground" the node at the bottom of the circuit [see Figure 10-15(a)] as is actually done in the circuits of most electronic equipment. We will call this node the *reference node* since the voltages of all other nodes in the network will be referred to it. Its voltage is zero, $V_d = 0$. We choose for the reference node one which will provide maximum simplification. Next, we identify the other nodes (review Section 10-5 for description of a node).

Observe nodes a, b, and c in Figure 10-15(a). The voltages of nodes a and c with respect to node d are known: $V_a = 12$ V, $V_c = 3$ V. Some circuit analysts find it helpful to label these other nodes with their known voltages or, if voltage is unknown, with a subscripted symbol, as in Figure 10-15(b). A node whose voltage is unknown is called an *independent* node; a node whose voltage with respect to the reference node is known is called a *dependent* node.

Next, we apply Kirchhoff's current law (KCL) to each independent node; a network with n independent nodes yields n independent equations. For the network of Figure

10-15(b) we write

$$-I_1 - I_2 + I_3 = 0 \text{ or } I_1 + I_2 - I_3 = 0 \quad (1)$$

The current directions were assumed and indicated on the diagram.

A relationship for the current I_1 is expressed using Ohm's law,

$$I_1 = \frac{12 - V_b}{6} \quad (2)$$

The expression $12 - V_b$ represents the potential difference between nodes a and b—the voltage across r_1. The resistance of the branch between a and b is 6 Ω.

Similarly, applying Ohm's law to r_2,

$$I_2 = \frac{3 - V_b}{3} \quad (3)$$

and, finally,

$$I_3 = \frac{V_b - 0}{4} \quad (4)$$

Now, substituting equations (2), (3), and (4) into (1), we have

$$\frac{12 - V_b}{6} + \frac{3 - V_b}{3} - \frac{V_b}{4} = 0$$

A single equation with a single unknown! We solve by first multiplying by 12,

$$24 - 2V_b + 12 - 4V_b - 3V_b = 0$$
$$-9V_b = -36$$
$$V_b = 4 \text{ V}$$

Then

$$I_1 = \frac{12 - 4}{6} = 1.333 \text{ A}$$

$$I_2 = \frac{3 - 4}{3} = -0.3333 \text{ A}$$

$$I_3 = \frac{4}{4} = 1.0 \text{ A}$$

Let us further demonstrate the procedure with a slightly more complicated circuit. We use conventional current in the analysis.

Example 10 Given the circuit of Figure 10-16(a), use the node-voltage method to calculate the current in r_5.

1. Redraw the circuit and label as in Figure 10-16(b). Choose node e as the reference node.

2. Write KCL equations for currents at independent nodes b and c, and solve for unknown voltages.

At node b: $\quad \dfrac{12 - V_b}{2.5} - \dfrac{V_b - V_c}{5} - \dfrac{V_b - 0}{7} = 0$

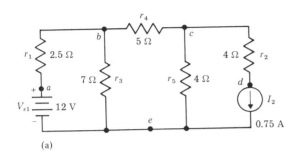

(a)

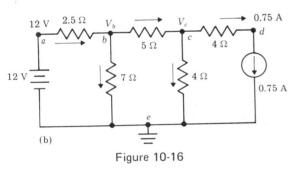

(b)

Figure 10-16

At node c: $\quad \dfrac{V_b - V_c}{5} - \dfrac{V_c - 0}{4} - 0.75 = 0$

Multiplying by 35 and 20, respectively,

$$168 - 14V_b - 7V_b + 7V_c - 5V_b = 0$$
$$4V_b - 4V_c - 5V_c - 15 = 0$$

Collecting terms,

$$26V_b - 7V_c = 168$$
$$4V_b - 9V_c = 15$$

Solving by determinants,

$$V_c = \dfrac{\begin{vmatrix} 26 & 168 \\ 4 & 15 \end{vmatrix}}{\begin{vmatrix} 26 & -7 \\ 4 & -9 \end{vmatrix}} = \dfrac{390 - 672}{-234 + 28} = \dfrac{-282}{-206}$$

$$= 1.369 \text{ V}$$

$$I_{r5} = \dfrac{V_c}{r_5} = \dfrac{1.369}{4} = 0.3423 \text{ A}$$

As with the loop method, we can write a set of steps for a systematic approach.

1. Choose and label as a *reference* node one which is common to all branches or one common to the branch of interest.

2. Label all other nodes: *dependent* nodes with known voltages, *independent* nodes with subscripted symbol.

3. Apply Kirchhoff's current law to each independent node. Express each current, using Ohm's law, as the potential difference between the two nodes divided by the resistance of the branch circuit between them. (Simplify the branch, when necessary, to a single resistance.)

4. Solve the simultaneous equations for the node voltage of interest. Determine currents, as desired.

The node-voltage method is a superior method for many types of circuits. When the method is used repeatedly for more complicated circuits its application is systematized further for rapid and convenient application.

Exercises 10-7 Use the node-voltage method to determine the circuit responses specified in all of the following exercises. *Use electron current in the analyses.*

1. For the circuit of Figure 10-15 assume that the polarity of the 12-V battery is reversed. Find I_1.

2. Determine the current I_2 in the circuit of Figure 10-15 when the polarity of the 3-V battery is reversed.

3. Find the current through V_{s1} in the circuit of Figure 10-16.

4. What is the current in r_4 in the circuit of Figure 10-16?

For Exercises 5–8 refer to Figure 10-12 and use the following values: $R_1 = 4\ \Omega$, $R_2 = 3\ \Omega$, $R_3 = 6\ \Omega$, $R_4 = 7\ \Omega$, $R_5 = 4\ \Omega$, $V_{s1} = 9$ V, and $V_{s2} = 9$ V.

5. Find the current in R_2.
6. Find the current in V_{s2}.
7. Find the current through R_3.
8. How much power does V_{s1} supply?

For Exercises 9 and 10 refer to Figure 10-11 and use the circuit values given in Exercise 7, Exercises 10-5.

9. Determine the power that V_{s1} will have to supply.

10. What power rating would you recommend for R_3?

11–20. Find the required solutions to Exercists 1–10 using conventional current in the node-voltage analyses.

Chapter 11

Applied Trigonometry

Trigonometry is the mathematics of triangles. (The Arecibo Observatory is part of the National Astronomy and Ionosphere Center which is operated by Cornell University under contract with the National Science Foundation.)

11-1 ANGLES AND UNITS

Presentations on alternating-current (ac) circuits include terms such as *phase angle, impedance triangle, sine wave,* and *cycle.* Many of these terms are from the areas of *geometry* and *trigonometry.* (*Trigonometry* means "triangle measurement.") A knowledge of the basic concepts, terminology, and operations of geometry and trigonometry is essential to the understanding of the theory and analysis techniques of ac circuits.

When two lines meet at a point, they form an *angle.* We consider the size of the angle to be the amount of opening between the two lines. An angle is said to be *generated* by *rotating* a line segment about its *endpoint,* or *pivot point,* from an *initial* position to a *terminal* position. In Figure 11-1 the arrow indicates rotation about the pivot point P. The initial position of the line segment is line MP; the terminal position is QP. We call MP the *initial side* and QP the *terminal side.* The size of the angle is the amount of rotation and has nothing to do with the length of the line segments. The pivot point is called the *vertex* of the angle.

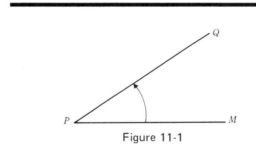

Figure 11-1

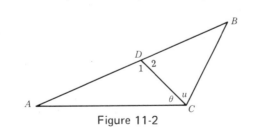

Figure 11-2

Where there is no possibility of confusion, the angle is designated by its vertex point. Thus, the angle in Figure 11-1 is angle P, written $\angle\,P$ or simply P. (The symbol for angle, $\angle$, must not be confused with the symbol for "is less than," $<$.) If there are several angles that could be called $\angle P$, we use a three-letter designation that is more specific; that is, we name one of the sides forming the angle, then the vertex, then the other side bounding the angle. Thus, $\angle P$ in Figure 11-1 could be denoted $\angle MPQ$ or $\angle QPM$. Another way of designating the angle is by a lower-case letter or small number set within the angle. For example, Figure 11-2 includes $\angle 1, \angle 2,$ $\angle u,$ and $\angle \theta$ (lower-case Greek theta). In electronics, θ is the most common symbol for an angle. The Greek letters ϕ (phi), α (alpha), and β (beta) are also used.

Two measurements of angles are commonly used in electronics. These are the *degree* and the *radian.* The symbol for degree is $^\circ$ as in 60°. The abbreviation of radian is *rad.*

The complete rotation of a line segment through a circle produces an angle of 360° or 2π radians ($360^\circ = 2\pi$ rad). Therefore, 1° is $1/360$ of a complete circle, and 1 rad is $1/2\pi$ of a circle. Fractional parts of an angle in degree measure may be specified in decimal notation, for example: $\frac{1}{2}$ degree $= 0.5^\circ$, $2/100$ degree $= 0.02^\circ$. Or fractions of a degree are specified in minutes ($'$) and/or seconds ($''$): $1^\circ = 60$ minutes ($60'$), $1' = 60$ seconds ($60''$), and $1^\circ = 3600''$. The angle $28^\circ\ 17'\ 27''$ is read 28 degrees, 17 minutes, 27 seconds. Most electronic calculators are designed to use decimal degrees. This is the form that will normally be used for calculations in this book.

Although degree measure is used in typical circuit analysis calculations, radian measure is required when some types of mathematical manipulations of equations containing angles as variables are involved. Conversions between unit systems are occasionally required. Some electronic calculators incorporate the capability for changing directly from one form of angular measure to the other. If your calculator has this feature study the instruction manual carefully until you are familiar with the required procedure. Conversions can be made on any calculator by using the following procedures.

To change an angle of $N°$ to radians, multiply by 0.01745:

$$\theta \text{ rad} = N° \times 0.01745 \text{ rad/}°$$
$$(1° = 0.01745 \text{ rad})$$

To change an angle of M rad to degrees, multiply by 57.296:

$$\theta° = M \times 57.296°\text{/rad}$$
$$(1 \text{ rad} = 57.296°)$$

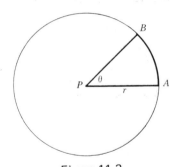

Figure 11-3

Example 1 Perform the indicated conversions.

Degrees to radians:

(a) $45°$

(a) $45° = 45 \times 0.01745 \text{ rad} = 0.7853 \text{ rad}$

(b) $110.75°$

(b) $110.75° = 110.75 \times 0.01745 \text{ rad} = 1.933 \text{ rad}$

(c) $187.9°$

(c) $187.9° = 187.9 \times 0.01745 \text{ rad} = 3.279 \text{ rad}$

(d) $353.25°$

(d) $353.25° = 353.25 \times 0.01745 \text{ rad} = 6.164 \text{ rad}$

Radians to degrees:

(a) 0.2365 rad

(a) $0.2365 \text{ rad} = 0.2365 \times 57.296° = 13.55°$

(b) 1.635 rad

(b) $1.635 \text{ rad} = 1.635 \times 57.296° = 93.68°$

(c) 2.793 rad

(c) $2.793 \text{ rad} = 2.793 \times 57.296° = 160.03°$

(d) 14.95 rad

(d) $14.95 \text{ rad} = 14.95 \times 57.296° = 856.6°$

Convert to decimal degrees:

(a) $17°33'$

(a) $17°33' = 17° + (33/60)° = 17° + 0.55°$
Hence, $17°33' = 17.55°$

(b) $47°53'$

(b) $47°53' = 47° + (53/60)° = 47.88°$

(c) $148°5'32''$

(c) $32'' = (32/60)' = 0.53'$ Hence, $148°5'32'' = 148° + (5.53/60)° = 148.092°$

(d) $350°23'56''$

(d) $350°23'56'' = 350° + [(23 + 56/60)/60]° = 350.4°$

Radian measure relates the measurement of angles to distance along the circumference of a circle. Refer to Figure 11-3, a circle of radius r with center at P. The angle θ is formed by radii PA and PB. If the length of arc AB (written $\overset{\frown}{AB}$), a portion of the circumference of the circle, is one radius: $\overset{\frown}{AB} = r$, then $\theta = 1$ rad.

A radian is defined as that angle which, when placed with its vertex at the center of a circle of radius r, intercepts an arc of length r on the circumference.

Any angle θ formed by the radii of a circle is related to the length l of the arc of the circumference between the radii and the radius r by the formula

$$l = \theta r$$

and, thus,

$$\theta = \frac{l}{r}$$

We say that θ *subtends* the arc of length l, or l subtends θ. Using the formula for l and the fact that a circle subtends 2π rad, we obtain the usual formula for the circumference of a circle:

$$c = 2\pi r$$

Example 2 A radar echo indicates that a ship is 2000 m away. The ship subtends an angle of 0.045 rad at the observation point. What is the length of the ship? (When the angle is small, the arc length is approximately equal to the straight-line distance between the ends of the radii.)

$$l = \theta r = 0.045 \times 2000 = 90 \text{ m}$$

Example 3 A visible artificial earth satellite which is known to have a velocity of 3.2 km/s in its orbit is observed to move through an angle of 15° in 10 s. What is its height above the earth?

First we find l, the length of the arc through which it moves as we watch,

$$l = v \times t = 3.2 \text{ km/s} \times 10 \text{ s} = 32 \text{ km}$$

Then

$$\theta = 15 \times 0.01745 = 0.2618 \text{ rad}$$

Thus,

$$r = \frac{l}{\theta} = \frac{32 \text{ km}}{0.2618} = 122.2 \text{ km} \quad \text{(about 76.4 mi)}$$

Exercises 11-1 In Exercises 1–10 convert the degree measurements to radians.

1. 4°   2. 15°   3. 28°   4. 57°
5. 92°   6. 115°   7. 137.5°
8. 298.36°   9. 193.75°   10. 312.87°

In Exercises 11–20 convert the given angles to decimal form.

11. 15°46′   12. 37°15′   13. 97°35′
14. 137°22′   15. 45°17′   16. 39°5′
17. 15°15′47″   18. 30°18′57″
19. 48°23′38″   20. 29°43′15″

In Exercises 21–40 convert the radian values to degrees.

21. $\pi/8$ rad   22. $\pi/6$ rad   23. $\pi/4$ rad
24. $\pi/3$ rad   25. $\pi/2$ rad   26. π rad
27. $3\pi/2$ rad   28. $5\pi/6$ rad
29. 1.36 rad   30. 4.75 rad   31. 0.37 rad
32. 0.017 rad   33. 0.0056 rad
34. 0.0012 rad   35. 2.61 rad

36. 3.79 rad   37. 5.71 rad   38. 8.92 rad
39. 12.06 rad   40. 427.5 rad

In Exercises 41–53 use appropriate forms of $l = \theta r$ to find the specified quantities.

41. An arc subtends an angle of 18° and the radius of the arc is 5 in. Find the length of the arc.
42. Find the length of an arc that has a 20-cm radius; the angle it subtends is 30°.
43. A 15-cm arc is subtended by an angle of 2 rad. Find the radius.
44. A 4.5-in. arc is subtended by an angle of 3 rad. Find the radius.
45. Find the angle subtended by a 3-in. arc which has a 9-in. radius.
46. A 33-mm arc has a radius of 11 cm. Find the angle subtended by the arc.
47. A ship has a radar echo which subtends an angle of 0.042 rad at a range of 2 km. What is the length of the ship in meters? (See Example 2.)
48. A ship subtends an angle of 0.015 rad at 3.75 km. What is its length in meters?
49. Refer to Example 3 above. A satellite with a velocity of 3.78 km/s moves through an angle of $\pi/10$ rad in 10 s. What is its distance from the earth in meters?
50. A satellite with a velocity of 4.22 km/s moves through an angle of $\pi/4$ rad in 28 s. What is its altitude?
51. An automobile travels 27 mi along a straight highway. What angle at the center of the earth is subtended by the distance traveled? Give the answer in radians and in degrees. Assume that $r_{\text{earth}} = 4000$ mi.
52. A laser beam is confined to a width of 0.02°. What is the diameter of a spot the laser would produce on an object 12,000 mi away?
53. A balloon-type satellite has a 50-ft diameter. Its orbit is a circle around the earth, with an 8400-mi diameter. What angle will the satellite subtend for an observer on earth?

11-2 TERMINOLOGY AND DEFINITIONS

A *right angle* is an angle of 90° or $\pi/2$ rad; it is a quarter of a complete rotation. The sides are said to be perpendicular or *normal* to each other.

Two angles whose sum is 90° are *complements* of each other. Thus, 40° and 50° are complements or *complementary angles*. We also say 40° is the complement of 50°.

If the terminal side of an angle is rotated through 180°, the two sides of the angle form a straight line. The angle is then called a *straight angle*. Its measure is 180° or π rad.

Two angles whose sum is 180° are called *supplements*. Thus, 100° is the supplement of 80°.

An angle less than 90° is called *acute*. An angle between 90° and 180° is called *obtuse*, and a *reflex angle* is one between 180° and 360°.

In Chapter 3 we introduced the (x, y)-coordinate system as a means of graphing equations. The same system simplifies the study of angles and their functions. Figure 11-4 represents the coordinate system, with the quadrants labeled. Note that in the first quadrant, both x and y are positive; in the second, x is negative and y is positive; in the third, both are negative; in the fourth, x is positive and y is negative. An angle is said to be *in standard position* when its initial side is on the positive x-axis and its vertex is at the origin (the intersection of the x- and y-axes). In Figure 11-4, the angle θ is in standard position.

By convention, an angle generated by a counterclockwise rotation is called positive and one generated clockwise is negative. Thus, in Figure 11-5, the angle whose terminal side is OM ($\angle NOM$) is positive, but $\angle NOQ$ is negative. There are two arrows pointing to the terminal side PO; hence $\angle NOP$ could denote either the positive obtuse angle obtained by counterclockwise rotation or the negative reflex angle obtained from clockwise rotation. We can resolve this ambiguity by specifying the "obtuse" angle NOP or the "reflex" angle NOP, or by placing small letters next to the arrows. In the figure, $\angle \alpha$ is the obtuse angle and $\angle \beta$ is the reflex angle. Whenever two or more angles in standard position share a common terminal side, the angles are said to be *coterminal*. Thus, an angle of 60° is coterminal with an angle of minus 300°. Also since a rotation can be more than 360°, an angle of 40° is coterminal with 400° (360° + 40°), 760° (720° + 40°), and so on.

An angle can be constructed or measured with reasonable accuracy with a drawing in-

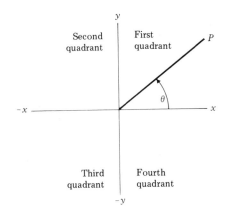

Figure 11-4

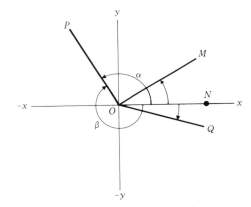

Figure 11-5

strument called a *protractor* [see Figure 11-6(a)]. To measure an angle:

1. Place the protractor over the angle so that its center is exactly over (coincides with) the vertex of the angle.

2. Make one edge of the protractor coincide with one side of the angle.

3. Read the value of the angle where it crosses under the protractor scale. Angle θ in Figure 11-6(a) is 70°.

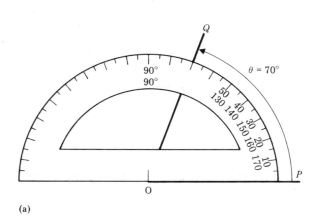

(a)

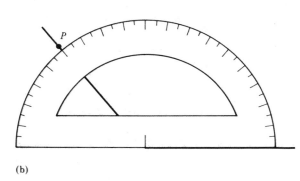

(b)

Figure 11-6

To construct a specified angle:

1.   Draw one side in the desired position and locate the vertex.

2.   Place the protractor so that its center is on the vertex and one edge is along the side of the angle.

3.   Place a dot opposite the desired angle as read from the scale [see Figure 11-6(b)].

4.   Construct the second side of the angle by drawing a line through the vertex and the dot placed in Step 3.

Example 4   Construct the following angles.

(a)   A right angle in standard position.

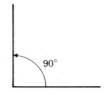

(b)   An obtuse angle in standard position.

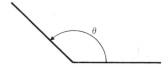

(c)   An angle in standard position which is the complement of 75°.

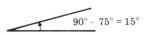

(d)   A negative angle in standard position which is the supplement of 75°.

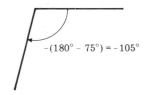

Exercises 11-2   In Exercises 1–4 name the complements of the given angles.

1.   15°, 37°, 45°, 60°
2.   3°, 85°, 15°, 53°
3.   1 rad, 1.08 rad, 0.36 rad, 0.46 rad
4.   0.85 rad, 1.32 rad, 0.37 rad, 1 rad

In Exercises 5–8 name the supplements of the given angles.

5.   15°, 75°, 90°, 127°
6.   3°, 27.6°, 87.3°, 173.8°
7.   1 rad, 1.75 rad, 2.73 rad, 3.05 rad
8.   0.56 rad, 0.037 rad, 1.37 rad, 2.97 rad

In Exercises 9–12 construct the given angles in standard position.

9.   15°, −45°, 148°, −175°
10.   30°, −90°, −180°, −270°
11.   405°, 620°, −515°, −720°
12. $\pi/6$ rad, $\pi/4$ rad, $\pi/2$ rad, π rad

11-3   TRIANGLES

A *triangle* is a plane figure containing three sides and three angles. In Figure 11-7, the figure bounded by the lines OP, OQ, and PQ is a triangle. A triangle is designated by its vertices (in any order) and the symbol $\triangle$. Thus the triangle in Figure 11-7 is $\triangle POQ$ or

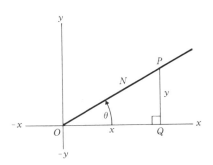

Figure 11-7

$\triangle QOP$ or $\triangle OPQ$, and so on. The word "triangle" or the symbol must be used to differentiate the triangle from the angle with the same name.

If two triangles are identical in all respects, they are said to be *congruent*, and corresponding parts are *equal*. The mathematical sign for congruence is $\cong$. If the angles of one triangle are equal to corresponding angles of a second triangle, the triangles have the same shape but not necessarily the same size; in this case, the triangles are called *similar*. Figure 11-8 shows an example of similar triangles.

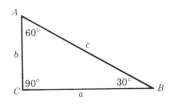

(a)

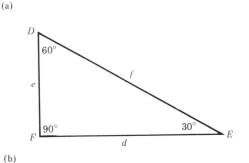

(b)

Figure 11-8

A property of similar triangles is that corresponding sides are in the same proportion; that is, for the triangles in Figure 11-8,

$$\frac{AB}{DE} = \frac{AC}{DF} = \frac{BC}{EF}$$

When one angle of a triangle is $90°$, the triangle is called a *right triangle*. Thus, the triangles in Figure 11-8 are right triangles. Another property of triangles is illustrated in this figure: the sum of the three angles of a triangle is always $180°$. Thus, if we know two of them, we can always find the third.

Right Triangles and the Pythagorean Theorem

In a right triangle, the side opposite the right angle is called the *hypotenuse*. The other two sides are called *legs*. It is common practice to denote the vertices of a triangle by capital letters and to use the corresponding lower-case letter to denote the side opposite each vertex. If the triangle is a right triangle, the letter C is frequently used for the right angle, and A and B are used for the acute angles. Then the hypotenuse would be designated by c and the legs by a and b. [See Figure 11-8(a).] (However, it is not wrong to use other designations for the sides and angles of a triangle.)

The Greek mathematician Pythagoras discovered the principle which is now called the *Pythagorean theorem*. This principle states that *the square of the hypotenuse equals the sum of the squares of the other two sides*. Thus, in Figure 11-8,

$$c^2 = a^2 + b^2 \quad \text{and} \quad c = \sqrt{a^2 + b^2}$$

Also

$$b = \sqrt{c^2 - a^2} \quad \text{or} \quad a = \sqrt{c^2 - b^2}$$

If $a = 3$ and $b = 4$, then

$$c^2 = 9 + 16 = 25$$

and therefore

$$c = \sqrt{25} = 5$$

A triangle with sides $a = 3$, $b = 4$, $c = 5$ (or $a = 4$, $b = 3$, $c = 5$) is called a 3-4-5 triangle. If the sides are multiples of these numbers, the triangle is still called a 3-4-5 triangle. Examples are: 6, 8, 10 or 9, 12, 15. You can use this information to find the unknown third side of a 3-4-5 triangle without resorting to the evaluation of a radical.

The three sides of a right triangle are not always integers. Thus, if $a = 4$ and $b = 5$, then $c^2 = 16 + 25 = 41$, and $c = \sqrt{41}$. However, there are some groups of integers which do satisfy the condition $c^2 = a^2 + b^2$. Examples are: $c = 13$, $a = 5$, $b = 12$, and $c = 25$, $a = 7$, $b = 24$.

Two important right triangles are the 45° triangle and the 30°-60°-90° triangle. The 45° triangle has two 45° angles in addition to the right angle; its two legs are equal and therefore it is also an *isosceles triangle*. The hypotenuse is equal to $\sqrt{2}$ times the leg since, if $a = b = 1$, $c^2 = 1 + 1 = 2$, and therefore

$$c = \sqrt{2} = 1.414$$

We see from the above equation that if the leg of a 45° triangle is given, we can obtain the hypotenuse by simply multiplying the given length by $\sqrt{2}$. For example, if one leg of a 45° triangle is 5, the hypotenuse is $5\sqrt{2}$ or 7.07.

In a 30-60-90 triangle, the two acute angles are 30° and 60°. The side opposite the 30° angle is the shorter leg and is one-half the length of the hypotenuse; the other leg is equal to $\sqrt{3}$ times the shorter leg. Thus, if the shorter leg is 10, then the hypotenuse is 20, and the long leg is $10\sqrt{3}$.

$$20^2 = (10\sqrt{3})^2 + (10)^2, \qquad 400 = 300 + 100$$

The relationships for the 45° and the 30-60-90 triangles are illustrated in Figure 11-9.

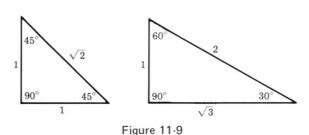

Figure 11-9

Example 5   The legs of a right triangle are 3.6 and 4.7. Find the hypotenuse.

The appropriate formula is

$$c = \sqrt{a^2 + b^2}$$

We substitute

$$c = \sqrt{(3.6)^2 + (4.7)^2}$$

If you have a calculator with the $x^2, \sqrt{x}$, and $M+$ functions the result can be evaluated with the following sequence of operations:

Enter 3.6, x^2, $M+$, enter 4.7, x^2, $M+$, $M{\to}X$ (or MR), $\sqrt{x}$. Readout: 5.920. Thus,

$$\underline{c = 5.92}$$

Study the instruction manual for your calculator to determine the recommended procedure for this kind of problem.

Example 6   The hypotenuse of a right triangle is 28 and one leg is 13. Find the other leg.

$$a = \sqrt{c^2 - b^2} = \sqrt{28^2 - 13^2}$$

Calculator operations: Enter 28, x^2, $M+$, enter 13, x^2, $M-$, $M{\to}X$ (or MR), $\sqrt{x}$. Readout: 24.799. Thus,

$$\underline{a = 24.80}$$

The Impedance Triangle   In ac circuits the effects of inductance and capacitance are called *inductive reactance X_L* and *capacitive reactance X_C*, respectively. The total effect of resistance and reactance is called *impedance*. The three entities—resistance, reactance, and impedance—are related mathematically in a right triangle called the *impedance triangle*. The legs are R and X, and Z is the hypotenuse. The angle made by R and Z is designated θ. In an impedance triangle constructed with X_L, θ is positive [see Figure 11-10(a)]; θ is negative in triangles constructed with X_C, (X_C is negative) [see Figure 1-10(b)]. All three quantities (X_C, X_L, and Z) have the same unit as resistance—the *ohm*.

Example 7   An impedance consists of $R = 15.8 \ \Omega$ and $X_L = 25.3 \ \Omega$.

(a) Construct the impedance triangle.

(a)

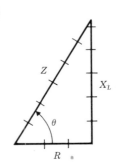

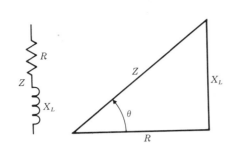

(a)

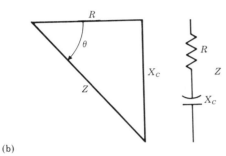

(b)

Figure 11-10

(b) Measure θ.   (b) $\theta = 58°$
(c) Calculate Z.   (c) $Z = \sqrt{R^2 + X^2} =$
$\sqrt{15.8^2 + 25.3^2}$
$Z = 29.83\ \Omega$

Example 8 An impedance consists of $R = 55\ \Omega$ and $X_C = -35\ \Omega$.

(a) Construct the im-   (a)
pedance triangle.

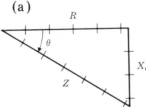

(b) Measure θ.   (b) $\theta = -32°$
(c) Calculate Z.   (c) $Z = \sqrt{R^2 + X^2} =$
$\sqrt{55^2 + (-35)^2}$
$Z = 65.19\ \Omega$

Example 9 An impedance Z is equal to 5 kΩ and includes $R = 3.9$ kΩ and an inductive reactance X_L.

(a) Calculate X_L.
(a) $X_L = \sqrt{Z^2 - R^2} =$
$\sqrt{(5 \times 10^3)^2 - (3.9 \times 10^3)^2}$
$X_L = 3.129$ kΩ

(b) Construct the impedance triangle.
(b)

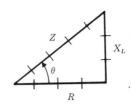

(c) Measure θ.
(c) $\theta = 39°$

Exercises 11-3 In Exercises 1–10 c is the hypotenuse of a right triangle and a and b are the legs. (a) Find the missing side. (b) Sketch the triangle to scale. Study your calculator instruction manual and utilize the recommended procedures in your calculations.

1. $a = 6$, $b = 11$ 2. $a = 0.65$, $b = 0.28$
3. $a = 255$, $b = 137$ 4. $a = 3550$, $b = 7480$
5. $a = 12$, $c = 45$ 6. $b = 78$, $c = 165$
7. $a = 0.485$, $c = 1.038$ 8. $a = 0.0129$, $c = 0.0279$
9. $a = 37,300$, $b = 7,550$ 10. $b = 475,000$, $c = 625,000$

In Exercises 11–20 R, X_L (or X_C), and Z represent the parts of an impedance triangle. (a) Calculate the missing element. (b) Construct the impedance triangle. (c) Measure θ, the angle between Z and R.

11. $R = 4\ \Omega$, $X_L = 5\ \Omega$
12. $R = 15\ \Omega$, $X_C = 9\ \Omega$
13. $R = 75\ \Omega$, $X_C = 225\ \Omega$
14. $R = 87\ \Omega$, $X_L = 12\ \Omega$
15. $R = 15\ \Omega$, $Z = 28\ \Omega$
16. $X_L = 55\ \Omega$, $Z = 120\ \Omega$
17. $R = 2.4$ kΩ, $Z = 4.4$ kΩ
18. $X_C = 15$ kΩ, $Z = 40$ kΩ
19. $R = 5$ kΩ, $X_L = 90$ kΩ
20. $R = 1.5$ MΩ, $X_C = 50$ kΩ

21. A television serviceman must get up on a roof to service an antenna. He places his 30-ft ladder so that the upper end just reaches the edge of the roof. The foot of the ladder rests on the ground 9 ft from the wall. What is the height of the roof?

22. A guy wire that is 55 m long is attached 1 m from the top of a broadcasting antenna. The guy is anchored 39 m from the base of the antenna. What is the height of the antenna?

23. Town B is 12 mi from town A as the crow flies. Town C is 7.5 mi from A on an alternative route from A to B. The alternative route is straight from A to C, then makes a 90° turn at C, and proceeds on a straight line from C to B. How far is town C from town B?

24. You plan to guy your TV antenna tower at a point 3 m above its base. You will guy in three directions and will secure the guys to the roof, at points 5 m from the tower base. How much guy wire will you need?

25. A boy flying a kite on a 100-m string, unwound to full length, loses the kite when its tail becomes entangled in high-voltage wires. The power line is 17 m up. What is the horizontal distance to the power line from the boy?

11-4 TRIGONOMETRIC FUNCTIONS— DEFINITIONS

In Figure 11-11 an angle θ is shown in standard position. From a point P on its terminal side, a perpendicular is dropped to the x-axis. The little square at the foot of the perpendicular indicates that the angle at Q is a right angle. If the point P has coordinates (x, y), then the length of the perpendicular PQ is y and the length OQ is x. [For example, if P is the point (4, 3), then $PQ = 3$ and $OQ = 4$.] The distance from the origin to the point P is

called the *radius vector* and is denoted by r. No matter which point along the terminal side is chosen to be P, the ratio between any two of the three values x, y, and r is constant for a fixed angle θ. *These ratios have special names and have been tabulated for all angles.* Thus, if we know any two of the values, or the ratio of two of the values, we can find the angle θ from the tables. These ratios are called *trigonometric functions* and are defined as follows:

$$\text{sine } \theta = \frac{y}{r}$$

$$\text{cosine } \theta = \frac{x}{r}$$

$$\text{tangent } \theta = \frac{y}{x}$$

$$\text{cotangent } \theta = \frac{x}{y} = \frac{1}{\text{tangent } \theta}$$

$$\text{secant } \theta = \frac{r}{x} = \frac{1}{\text{cosine } \theta}$$

$$\text{cosecant } \theta = \frac{r}{y} = \frac{1}{\text{sine } \theta}$$

For simplicity, we use three-letter abbreviations without periods for the functions: sin θ, cos θ, tan θ, cot θ, sec θ, csc θ. In electronics only the sine, cosine, and tangent functions are widely used. These are the only functions provided on calculators. Therefore, we will work only with sin θ, cos θ, and tan θ in this book.

Example 10 A radius vector passes through the point P (6, 8).

(a) Construct the angle θ.
(a)

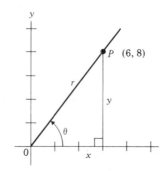

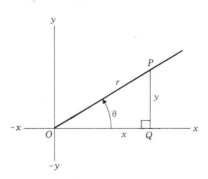

Figure 11-11

(b) Determine r using the Pythagorean theorem.

(b) $r = \sqrt{6^2 + 8^2} = \underline{10}$

(c) Calculate $\sin \theta$, $\cos \theta$, and $\tan \theta$.

(c) $\sin \theta = \dfrac{y}{r} = \dfrac{8}{10} = \underline{0.8}$

$\cos \theta = \dfrac{x}{r} = \dfrac{6}{10} = \underline{0.6}$

$\tan \theta = \dfrac{y}{x} = \dfrac{8}{6} = \underline{1.333}$

Example 11 If $\sin \theta = 0.74$ find $\cos \theta$ and $\tan \theta$.

Since $\sin \theta = y/r = 0.74 = 74/100$ we may assume that $y = 74$ and $r = 100$ and calculate x using the Pythagorean Theorem,

$$x = \sqrt{r^2 - y^2} = \sqrt{100^2 - 74^2}$$
$$x = \underline{67.26}$$

The point $(67.26, 74)$ is on the terminal side

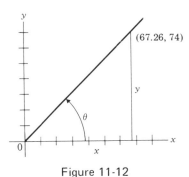

Figure 11-12

(see Figure 11-12). Using this information we calculate the other functions.

$$\cos \theta = \frac{x}{r} = \frac{67.26}{100} = \underline{0.6726}$$

$$\tan \theta = \frac{y}{x} = \frac{74}{67.26} = \underline{1.10}$$

Note that since division by zero is undefined the tangent function is undefined when $x = 0$.

Exercises 11-4 In Exercises 1–10 a radius vector passes through the given points. Deter-

mine $\sin \theta$, $\cos \theta$, and $\tan \theta$ for the angle formed by the radius vector and the positive x-axis.

1. $(6, 8)$ 2. $(4, 5)$ 3. $(12, 7)$
4. $(2, 7)$ 5. $(\sqrt{3}, 5)$ 6. $(7, 3\sqrt{2})$
7. $(\sqrt{2}, \sqrt{3})$ 8. $(0.5, 0.09)$ 9. $(\tfrac{1}{2}, \tfrac{1}{4})$
10. $(458, 58)$

In Exercises 11–20 use the Pythagorean theorem, the technique of Example 11, and the given function to find the indicated functions.

11. $\sin \theta = 0.74$, find $\cos \theta$ and $\tan \theta$.
12. $\sin \theta = 0.33$, find $\cos \theta$ and $\tan \theta$.
13. $\cos \theta = 0.91$, find $\sin \theta$ and $\tan \theta$.
14. $\cos \theta = 0.15$, find $\sin \theta$ and $\tan \theta$.
15. $\tan \theta = 0.15$, find $\sin \theta$ and $\cos \theta$.
16. $\tan \theta = 0.5$, find $\sin \theta$ and $\cos \theta$.
17. $\tan \theta = 1.0$, find $\sin \theta$ and $\cos \theta$.
18. $\tan \theta = 2.5$, find $\sin \theta$ and $\cos \theta$.
19. $\sin \theta = 1.0$, find $\cos \theta$ and $\tan \theta$.
20. $\cos \theta = 1.0$, find $\sin \theta$ and $\tan \theta$.

11-5 VALUES OF TRIGONOMETRIC FUNCTIONS

The values of the trigonometric functions of angles from $0°$ to $90°$ have been calculated and tabulated. These tables are available in mathematics textbooks and specialized books of mathematics tables. Many electronic calculators also incorporate the capability of providing the sine, cosine, and tangent values of any angle between $0°$ and $90°$. In some instances calculators provide function values for any angle. Study the instruction manual for your calculator to learn its capabilities and limitations and how to obtain the trigonometric functions of an angle.

Table II in the Appendix lists the values of functions for angles between $0°$ and $90°$. To find the function of an angle, read down the left column (or up the right column for angles greater than $45°$) until you come to the number corresponding to the desired angle. Read across on this line until you come to the column of the desired function. Obtain the value of the function. (Note that for angles greater than $45°$ the function column headings are along the bottom of the table.) For example, $\sin 51.4° = 0.78152$, $\cos 51.4° = 0.62388$, and $\tan 51.4° = 1.2527$.

Tables and calculators also enable us to find an angle when we know the value of a function. On a calculator we generally "call up" the angle by entering the value of the function, say 0.5 corresponding to $\sin \theta$, and then touching a key labeled "$\sin^{-1} x$," "arcsin x," or the combination "arc" and "sin x." With a table, we look to find the listed value of the known function and then read the value of the angle which corresponds to the function value. Continuing with the example 0.5 = $\sin \theta$, we study the values in the "sin" columns until we find "0.5000." We note that the angle for $\sin \theta = 0.5000$ is 30.0°.

Example 12   Find the indicated functions of the given angles. Find the function first on your calculator and then check by looking in the table in the Appendix.

(a) $\sin 25°$ (a) $\sin 25° =$ 0.42262 (calc.), 0.42262 (table)

(b) $\cos 47°$ (b) $\cos 47° =$ 0.682 (calc.), 0.68200 (table)

(c) $\tan 54.5°$ (c) $\tan 54.5° =$ 1.4019 (calc.), 1.4019 (table)

Example 13   Find the angles corresponding to the given functions. Use your calculator first, then check by looking in the table in the Appendix.

(a) $\sin \theta =$ 0.5000   (a) $\sin \theta = 0.5000$ $\theta = 30°$ (calc.), 30.0° (table)

(b) $\cos \theta =$ 0.9646   (b) $\cos \theta = 0.9646$ $\theta = 15.29°$ (calc.), 15.3° (table)

(c) $\tan \theta =$ 0.8040   (c) $\tan \theta = 0.8040$ $\theta = 38.8°$ (calc), 38.8° (table)

Exercises 11-5   In Exercises 1–10 find the sine, cosine, and tangent functions of the given angles.

1. 15°, 30°, 45°   2. 5°, 25°, 85°
3. 1°, 7°, 12°   4. 28°, 55°, 87°
5. 3.5°, 23.6°, 37.1°
6. 0.1°, 31.8°, 73.7°

7. 0.70°, 1.2°, 2.3°
8. 73.2°, 86.1°, 89.9°
9. 12.35°, 47.65°, 78.69°
10. 0.67°, 1.57°, 3.91°

In Exercises 11–20 find the angles corresponding to the given functions.

11. $\sin \theta = 0.5$ 12. $\sin \theta = 0.1513$
13. $\cos \theta = 0.9646$ 14. $\cos \theta = 0.8949$
15. $\tan \theta = 0.8040$ 16. $\sin \theta = 0.9646$
17. $\cos \theta = 0.3371$ 18. $\tan \theta = 33.69$
19. $\cos \theta = 0.0889$ 20. $\sin \theta = 0.0715$

11-6 TRIGONOMETRIC SOLUTIONS FOR RIGHT TRIANGLES

"Solving a triangle" means to find its unknown elements—sides or angles. In Section 11-3 we learned how to find an unknown side of a right triangle by an algebraic method called the Pythagorean theorem. In this section we will develop the technique for solving a right triangle by using the trigonometric functions—$\sin \theta$, $\cos \theta$, and $\tan \theta$.

Counting its three angles we say that a triangle has six parts—three angles and three sides. If we know three parts of a right triangle, including one side, we can solve for the other three parts with the aid of the functions. (The right angle is one of three required parts.)

Let us consider again the right triangle we used to define the trigonometric functions—a triangle formed by a radius vector making an acute angle in standard position and a perpendicular from that vector to the x-axis [see Figure 11-13(a)].

We construct an identical triangle in Figure 11-13(b) but label it so that C is the right angle, A corresponds to θ, c is equivalent to r, b to x, and a to y. Starting with the original definitions we develop the following additional definitions:

$$\sin A = \frac{y}{r} = \frac{a}{c} \qquad \cos A = \frac{x}{r} = \frac{b}{c}$$

$$\tan A = \frac{y}{x} = \frac{a}{b}$$

Let us now rotate the triangle so that angle B is in the standard position instead of angle A

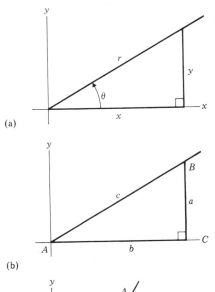

(a)

(b)

(c)

Figure 11-13

[see Figure 11-13(c)]. Now we have

$$\sin B = \frac{y}{r} = \frac{b}{c} \qquad \cos B = \frac{x}{r} = \frac{a}{c}$$

$$\tan B = \frac{y}{x} = \frac{b}{a}$$

A careful study of these relations will lead to the following generalized definitions of the trigonometric functions of either of the acute angles of *any right triangle in any position.*

$$\sin A \text{ (or } B) = \frac{\text{opposite side}}{\text{hypotenuse}}$$

$$\cos A \text{ (or } B) = \frac{\text{adjacent side}}{\text{hypotenuse}}$$

$$\tan A \text{ (or } B) = \frac{\text{opposite side}}{\text{adjacent side}}$$

Let us note another important fact which appears in the above definitions. *Sin* A = a/c, *cos* B = a/c, *and therefore sin* A = *cos* B. Since in every right triangle one angle is 90° and the sum of the angles is 180°, the sum of the acute angles is always 90°. Therefore, $B = 90° - A$, and $\sin A = \cos (90° - A)$. We summarize:

In a right triangle with acute angles *A* **and** *B*:

$A + B = 90°$

$\sin A = \cos B, \qquad \cos A = \sin B$

$\sin A = \cos (90° - A), \qquad \cos A = \sin (90° - A)$

The relations for the solution of right triangles can now be developed by simple algebraic manipulations of the definitions. Since $\sin A = a/c$,

$$a = c \sin A \qquad (1)$$

$$c = \frac{a}{\sin A} \qquad (2)$$

and from $\cos A = b/c$,

$$b = c \cos A \qquad (3)$$

$$c = \frac{b}{\cos A} \qquad (4)$$

then from $\tan A = a/b$,

$$a = b \tan A \qquad (5)$$

$$b = \frac{a}{\tan A} \qquad (6)$$

Similarly, $\sin B = b/c$,

$$b = c \sin B \qquad (7)$$

$$c = \frac{b}{\sin B} \qquad (8)$$

$\cos B = a/c$,

$$a = c \cos B \qquad (9)$$

$$c = \frac{a}{\cos B} \qquad (10)$$

$\tan B = b/a$,

$$b = a \tan B \qquad (11)$$

$$a = \frac{b}{\tan B} \qquad (12)$$

Example 14   Refer to the triangle of Figure 11-13. Let $A = 30°$, and $a = 5$. Find B, b, and c.

Using equation (2) we solve for c,

$$c = \frac{a}{\sin A} = \frac{5}{\sin 30°} = \frac{5}{0.5} = 10$$

and from (6),

$$b = \frac{a}{\tan A} = \frac{5}{\tan 30°} = \frac{5}{0.5775} = 8.66$$

We find B using the relation $B = 90° - A$,

$$B = 90° - 30° = 60°$$

Exercises 11-6   In Exercises 1–14 the given values are for the triangle of Figure 11-13. Find the unknown elements (sides and angles) of the triangle using trigonometric functions.

1. $A = 30°, a = 5$ 2. $A = 70°, b = 10$
3. $B = 15°, b = 6$ 4. $B = 45°, a = 8$
5. $A = 28.5°, c = 7$ 6. $A = 36.7°, c = 15$
7. $B = 41.7°, c = 35$ 8. $B = 78.3°, c = 50$
9. $a = 3, b = 4$ 10. $a = 8, b = 15$
11. $a = 10, c = 15$ 12. $a = 15, c = 21$
13. $b = 40, c = 70$ 14. $b = 50, c = 125$

In Exercises 15–18 use trigonometric functions to find your solutions.

15. Exercise 22, Exercises 11-3
16. Exercise 23, Exercises 11-3
17. Exercise 24, Exercises 11-3
18. Exercise 25, Exercises 11-3

11-7 INVERSE FUNCTIONS

In performing the evaluations of the preceding section on your calculator you have used another important concept in trigonometry—the *inverse function*. For example, after evaluating the ratio a/c you have a number which represents $\sin A$ ($a/c = \sin A$). You know a/c but not A. In such situations it would be convenient to say "A is the angle whose sine is a/c" in shorthand. We use the inverse function to do just that:

$$A = \arcsin \frac{a}{c}$$

is read "A is the angle whose sine is a/c."

That is, "arc" indicates "the angle whose _____ ." Remember: $\arcsin \frac{1}{2}$ means "the angle whose sine is $\frac{1}{2}$." What does $\tan (\arcsin \frac{1}{2})$ mean? It means: "find the tangent of the angle whose sine is $\frac{1}{2}$." On a calculator we enter 0.5 ($=\frac{1}{2}$), touch "$\arcsin x$" (or $\sin^{-1} x$), and obtain the readout "30." Then we touch "$\tan x$" and read "0.57735." Or,

$$\tan (\arcsin \tfrac{1}{2}) = 0.5774$$

The inverse function may also be written $\sin^{-1} \frac{1}{2}$ instead of $\arcsin \frac{1}{2}$. The form was popular at one time and is again with many calculator manufacturers. However, it is sometimes erroneously confused with a negative exponent. As an exponent it would signify the following:

$$(\sin \tfrac{1}{2})^{-1} = \frac{1}{\sin \tfrac{1}{2}}$$

This is *not* the meaning of the inverse function.

Example 15   Evaluate the following inverse functions (find a positive, acute angle).

(a) $\arcsin 0.5$ (a) $\arcsin 0.5 = 30°$
(b) $\arccos 0.866$ (b) $\arccos 0.866 = 30°$
(c) $\tan^{-1} 1$ (c) $\tan^{-1} 1 = 45°$

Example 16   Evaluate the following functions.

(a) $\tan (\arcsin 0.5)$ (a) $\tan (\arcsin 0.5) = 0.5774$
(b) $\cos (\tan^{-1} 1)$ (b) $\cos (\tan^{-1} 1) = 0.7071$
(c) $\sin (\cos^{-1} 0.75)$ (c) $\sin (\cos^{-1} 0.75) = 0.6614$

Example 17   In Figure 11-13, $a = 4$ and $b = 3$. Find c and A. Use the inverse function to simplify the calculation.

$$c = \frac{a}{\sin A}, \qquad A = \arctan \frac{a}{b}$$

Thus,

$$c = \frac{a}{\sin (\arctan a/b)} = \frac{4}{\sin (\arctan \tfrac{4}{3})}$$

Calculator steps for Example 17:

Step	Action	Readout	Symbolic result
1.	enter 4	4	a is in register
*2. (optional)	touch $M+$ (or $X \to M$)	4	a is in memory
3.	touch $\div$	4	$a \div$
4.	enter 3	3	$a \div b$
5.	touch $=$	1.3333333	$a \div b = \tan A$
6.	touch $arctan\ x$ (or $\tan^{-1} x$ or $F, \tan^{-1} x$)	53.13009	$A = \arctan \dfrac{a}{b}$
7.	Write down "$A = 53.13°$" for the first part of required solution.		
8.	touch $sin\ x$	0.8	$\sin A = \sin \left(\arctan \dfrac{a}{b}\right)$
9.	touch $1/x$	1.25	$1/\sin A$
10.	touch $\times$	1.25	$(1/\sin A) \times$
*11.	touch MR (or $M \to X$) or enter 4	4	place a in register
12.	touch $=$	5.0	$c = \dfrac{a}{\sin (\arctan a/b)}$
13.	Write down "$c = 5.0$" for the second part of the solution.		

*Steps 2 and 11 can be performed on calculators with memory capability. Storing in memory saves the keystrokes of reentering a which may be a significant saving if a is a long number.

Example 18 In Figure 11-13 $b = 12$ and $c = 18$. Find a and A.

$$a = c \sin A = c \sin \left(\arccos \frac{b}{c}\right)$$

$$a = 18 \sin \left(\arccos \frac{12}{18}\right) = 13.42$$

$$A = \arccos \frac{12}{18} = 48.19°$$

Exercises 11-7 In Exercises 1-12 evaluate the given inverse functions (find a positive acute angle).

1. arcsin 0.5
2. arcsin 0.1
3. $\sin^{-1}$ 0.375
4. $\sin^{-1}$ 0.937
5. arccos 0.866
6. arccos 0.156
7. $\cos^{-1}$ 0.5
8. $\cos^{-1}$ 0.912
9. arctan 0.136
10. arctan 2.5
11. $\tan^{-1}$ 3.65
12. $\tan^{-1}$ 0.785

In Exercises 13-24 evaluate the given functions.

13. sin (arccos 0.75)
14. sin ($\tan^{-1}$ 1.0)
15. sin (arctan 0.952)
16. sin ($\cos^{-1}$ 0.0915)
17. cos (arctan 1.0)
18. cos (arcsin 0.25)
19. cos ($\sin^{-1}$ 0.875)
20. cos ($\tan^{-1}$ 3.75)
21. tan (arcsin 0.357)
22. tan (arccos 0.753)
23. tan ($\sin^{-1}$ 0.0819)
24. tan ($\cos^{-1}$ 0.0653)

In Exercises 25–30 the values given refer to the triangle of Figure 11-13. Find the unknown elements of the triangle. Take advantage of inverse functions to reduce the number of steps required in your solutions. (See Examples 17 and 18.)

25. $a = 3, b = 4$
26. $a = 8, b = 15$
27. $a = 10, c = 15$
28. $a = 15, c = 21$
29. $b = 40, c = 70$
30. $b = 50, c = 125$

11-8 APPLICATION OF TRIGONOMETRIC FUNCTIONS TO THE IMPEDANCE TRIANGLE

In electronics a widely used application of the solution of right triangles using trigonometric functions is to the impedance triangle (review Section 11-3). We list the relevant definitions.

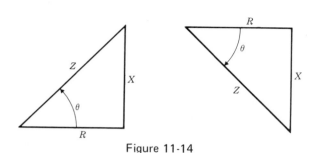

Figure 11-14

$$\sin \theta = \frac{X}{Z} \qquad (13)$$

$$\cos \theta = \frac{R}{Z} \qquad (14)$$

$$\tan \theta = \frac{X}{R} \qquad (15)$$

$$R = Z \cos \theta \qquad (16)$$

$$X = Z \sin \theta \qquad (17)$$

$$Z = \frac{R}{\cos \theta} = \frac{X}{\sin \theta} \qquad (18)$$

It should be noted that these formulas enable us to find the magnitudes of R, X, and Z. Consideration of the algebraic sign of X will be included in a subsequent section.

Example 19 An impedance includes the elements $R = 4\ \Omega$ and $X = 3\ \Omega$. Find Z and θ using trigonometric functions.

From (15),

$$\tan \theta = \frac{X}{R} = \frac{3}{4} = 0.75, \qquad \theta = 36.87°$$

and from (18),

$$Z = \frac{R}{\cos \theta} = \frac{4}{\cos 36.87°} = \frac{4}{0.8}, \qquad Z = 5.0\ \Omega$$

It is important that calculations such as this be performed efficiently and rapidly since they must be performed frequently in electronics. Your calculator may have one or more special features which will enable you to save keystrokes. Study its instruction manual carefully for any appropriate suggestions. We can apply the techniques of Examples 17 and 18, Section 11-7, to good advantage in saving time and effort in obtaining solutions involving the impedance triangle.

Example 20 Let $X = 12\ \Omega$, and $Z = 18\ \Omega$. Find R and θ. Take advantage of the concept of inverse functions and special features of your calculator to achieve the solutions with a minimum of keystrokes.

From equation (16),

$$R = Z \cos \theta = Z \cos \left(\arcsin \frac{X}{Z} \right)$$

$$R = 18 \cos \left(\arcsin \frac{12}{18} \right)$$

Calculator steps for Example 20:

Step	Action	Readout	Symbolic result
1.	enter X (12)	12	X is placed in register
2.	touch $\div$	12	$X \div$
3.	enter Z (18)	18	$X \div Z$
4. (optional)	touch $M+$ (or equivalent)	18	Z is stored in memory
5.	touch =	0.6666666	$X \div Z = \sin \theta$
6.	arcsin x (or $\sin^{-1} x$)	41.81031	$\theta = \arcsin \frac{X}{Z}$
7. Write: "$\theta = 41.81°$" for first part of solution.			
8.	touch $\cos x$	0.745356	$\cos \left(\arcsin \frac{X}{Z} \right)$
9.	touch $\times$	0.745356	$\cos \left(\arcsin \frac{X}{Z} \right) \cdot$
10.	enter Z or touch MR	18	$\cos \left(\arcsin \frac{X}{Z} \right) \cdot Z$
11.	touch =	13.416408	$R = Z \cos \left(\arcsin \frac{X}{Z} \right)$
12. Write: "$R = 13.42\ \Omega$" for the second part of the required solution.			

Exercises 11-8 In Exercises 1-18 the values given are parts of impedance triangles. Find the unknown values R, X, Z, or θ.

1. $X = 12\ \Omega$, $Z = 18\ \Omega$
2. $X = 3\ \Omega$, $Z = 5\ \Omega$
3. $X = 15\ \Omega$, $Z = 20\ \Omega$
4. $X = 5\ k\Omega$, $Z = 9\ k\Omega$
5. $R = 4\ \Omega$, $Z = 5\ \Omega$
6. $R = 10\ \Omega$, $Z = 15\ \Omega$
7. $R = 4.8\ k\Omega$, $Z = 8.5\ k\Omega$
8. $R = 15\ k\Omega$, $Z = 22\ k\Omega$
9. $R = 4\ \Omega$, $X = 3\ \Omega$
10. $R = 35\ \Omega$, $X = 15\ \Omega$
11. $R = 5.6\ k\Omega$, $X = 10\ k\Omega$
12. $R = 240\ k\Omega$, $X = 90\ k\Omega$
13. $R = 12\ \Omega$, $\theta = 30°$
14. $R = 1.8\ k\Omega$, $\theta = 75°$
15. $X = 25\ \Omega$, $\theta = 15°$
16. $X = 75\ k\Omega$, $\theta = 87°$
17. $Z = 50\ \Omega$, $\theta = 10°$
18. $Z = 1.5\ M\Omega$, $\theta = 78°$

11-9 SIGNS OF TRIGONOMETRIC FUNCTIONS

In the preceding sections we have dealt only with positive acute angles ($0° \leqslant \theta \leqslant 90°$). What are the functions of other angles? Let us refer again to the x–y-coordinates and the concept of angles in the standard position (see Figure 11-15). The four parts of the co-ordinate diagram are called *quadrants* and are generally identified as *first quadrant*, *second quadrant*, and so forth, or *I*, *II*, *III*, and *IV* as shown in Figure 11-15. Not unexpectedly, then, an acute positive angle, we say, is "a first quadrant angle" because its terminal side lies in that quadrant. *Other angles in standard position are similarly identified by the positions of their terminal sides.* To understand how the signs of the trigonometric functions are determined let us restate their definitions and relate these, in turn, to the coordinate diagram.

$$\sin\theta = \frac{y}{r}, \qquad \cos\theta = \frac{x}{r}, \qquad \tan\theta = \frac{y}{x}$$

The radius vector r is always considered to be positive. The signs of the functions, then, are determined by the signs of y and x. And the position of the terminal side determines whether these are positive or negative. We summarize information concerning the signs of y, x, and the functions according to the position of the terminal side.

Table 11-1

Quantity	Quadrant in which terminal side is located			
	I	II	III	IV
x	+	−	−	+
y	+	+	−	−
$\sin\theta = \dfrac{y}{r}$	+	+	−	−
$\cos\theta = \dfrac{x}{r}$	+	−	−	+
$\tan\theta = \dfrac{y}{x}$	+	−	+	−

It is essential that the information in Table 11-1 be memorized so as to be available for immediate recall when needed. The chart in Figure 11-16 is provided for further assistance in this endeavor.

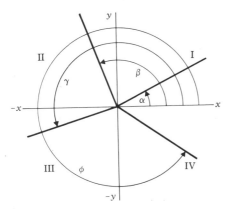

Figure 11-15

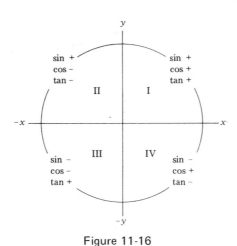

Figure 11-16

Example 21   Find the algebraic signs of the following functions.

(a) sin 30°, cos – 30°, sin 145°, tan – 145°, sin – 196°, tan 196°, cos 287°

(b) cos 98°, tan 120°, sin – 30°, sin 199°, cos 240°, tan – 50°, tan – 195°, cos – 130°, cos – 210°

    (a) All functions listed are positive.

    (b) All functions listed are negative.

Exercises 11-9   In Exercises 1–8 state the algebraic signs of the given functions.

1. sin 30°,
   tan – 30°,
   cos 125°

2. tan 175°,
   sin 210°,
   cos 330°

3. sin – 65°,
   cos – 198°,
   tan – 155°

4. cos 165°,
   tan 250°,
   sin – 330°

5. cos 45°,
   tan 148°,
   sin 320°

6. cos – 60°,
   tan 240°,
   sin – 220°

7. tan 210°,
   sin 355°,
   cos – 310°

8. tan – 15°,
   cos 150°,
   sin 185°

In Exercises 9–16 state the algebraic signs of the three functions—sine, cosine, tangent—of the angles whose terminal sides pass through the given points.

9. (3, 4)   10. (–6, 1)   11. (–4, –8)

12. (–5, 4)   13. (6, –3)   14. (5, –7)

15. (–6, –7)   16. (12, –15)

11-10 FUNCTIONS OF ANY ANGLE

Recall that whenever two (or more) angles in standard position share a common terminal side, the angles are said to be *coterminal* (review Section 11-2). Therefore, any angle in standard position will be coterminal with a positive angle between 0° and 360°. Its trigonometric functions will be the same as the coterminal angle. And if we know how to find the functions of angles between 0° and 360° we can find the functions of any angle.

Related Angles   The trigonometric function of an angle between 0° and 360° is determined by the position of its terminal side. However, most tables and many calculators provide only the functions of positive angles between 0° and 90°. The functions of other angles are found by assigning the appropriate algebraic sign to the functions of the acute angle formed by the terminal side of the angle and the x-axis (see Figure 11-17). Every angle has such an associated acute angle called its *related angle*. In Figure 11-17 the related angles are α_1 for θ_1 (they are identical), α_2 for θ_2, α_3 for θ_3, and α_4 for θ_4. Related angles are found as follows:

Second quadrant: $\alpha_2 = 180° - \theta_2$

Third quadrant: $\alpha_3 = \theta_3 - 180°$

Fourth quadrant: $\alpha_4 = 360° - \theta_4$

Referring to Section 11-9, we list the signs of the functions of the various related angles.

Related Angle	sine	Function cosine	tangent
α_2	+	–	–
α_3	–	–	+
α_4	–	+	–

Example 22   Find the three trigonometric functions (sin, cos, and tan) for the following angles.

(a) 125°   (a) $\alpha_{125°} = 180 - 125° = 55°$
   sin 125° = sin 55° = 0.8192
   cos 125° = – cos 55° =
   – 0.5736
   tan 125° = – tan 55° =
   – 1.428

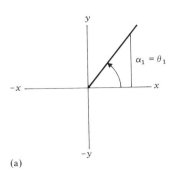

(a)

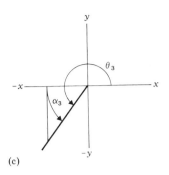

(c)

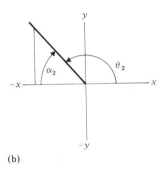

(b)

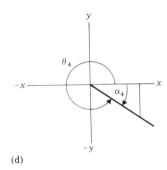

(d)

Figure 11-17

(b) 240°

(b) $\alpha_{240°} = 240° - 180° = 60°$
$\sin 240° = -\sin 60° = -0.866$
$\cos 240° = -\cos 60° = -0.5$
$\tan 240° = \tan 60° = 1.732$

(c) 315°

(c) $\alpha_{315°} = 360 - 315° = 45°$
$\sin 315° = -\sin 45° = -0.7071$
$\cos 315° = \cos 45° = 0.7071$
$\tan 315° = -\tan 45° = -1.0$

(d) -65°

(d) $\alpha_{-65°} = 65°$
$\sin -65° = -\sin 65° = -0.9063$
$\cos -65° = \cos 65° = 0.4226$
$\tan -65° = -\tan 65° = -2.145$

When we wish to find the function of an angle greater than 360° we (1) find the angle less than 360° with which it is coterminal, and (2) proceed as in Example 22 above. To find the coterminal angle θ' less than 360° we subtract an appropriate multiple of 360°:

$$\theta' = \theta - n360°$$

For example, to find θ' for 850°,

$$\theta' = 850° - 2 \cdot 360° = 130°$$

for 1400°,

$$\theta' = 1400° - 3 \cdot 360° = 320°$$

Example 23 Find the functions of the following angles.

(a) 590°

(a) $\theta' = 590° - 360° = 230°$
$\alpha_3 = 230° - 180° = 50°$
$\sin 590° = -\sin 50° = -0.766$
$\cos 590° = -\cos 50° = -0.6428$
$\tan 590° = \tan 50° = 1.192$

(b) 1550°

(b) $1550° - 4 \cdot 360° = 110°$
$\alpha_2 = 180° - 110° = 70°$
$\sin 1550° = \sin 70° = 0.9397$
$\cos 1550° = -\cos 70° = -0.342$
$\tan 1550° = -\tan 70° = -2.747$

It is now possible to consider the signs of the reactance values and θ in the impedance triangle. We may think of the impedance triangle as one with θ in standard position. Then R and Z are always positive, X_L and the associated θ are positive, and X_C and the associated θ are negative.

Example 24   An impedance is $Z = 50\ \Omega$ with $\theta = -40°$. Find R and X. Is the X-quantity inductive (X_L) or capacitive (X_C)?

$$R = Z \cos \theta = 50 \cos -40° = 38.3\ \Omega$$
$$X = Z \sin \theta = 50 \sin -40° = -32.14\ \Omega$$

Since the X-value is negative it is an X_C, that is, it is the reactance of a capacitor.

Functions of Angles in Radians   In electricity/electronics it is sometimes convenient to find the functions of angles expressed in radian measure (review Section 11-1). It is customary not to indicate the unit of the angle when writing the expressions of functions of angles in radians. Thus,

"tan 1.5," means "tan 1.5 rad."

The notation should not be confused with inverse functions:

"$\tan^{-1} 1.5$" means "the angle whose tangent is 1.5."

Many calculators of the type recommended for use with this book include the feature which permits one to find the functions of angles in radians directly. (Study the instruction manual for your calculator to determine if it has this feature.) Otherwise, it is necessary to convert the angle to degrees (see Section 11-1).

Example 25   Find the required trigonometric functions of the angles whose values are given in radians.

(a) $\sin 0.5$

(b) $\cos \dfrac{\pi}{4}$

($\pi = 3.1415926$)

(c) $\tan \dfrac{\pi}{3}$

(a) $\sin 0.5 = 0.4794$

(b) $\cos \dfrac{\pi}{4} = 0.7071$

(c) $\tan \dfrac{\pi}{3} = 1.732$

Exercises 11-10   In Exercises 1–12 find the three trigonometric functions—sin, cos, and tan—of the given angles.

1. 125°   2. 240°   3. 315°   4. -65°
5. -150°   6. 176°   7. 191°
8. -198°   9. 330°   10. 285°
11. 205°   12. -345°

In Exercises 13–24 find the three trigonometric functions—sin, cos, and tan—of the given angles.

13. 590°   14. 1550°   15. 374°
16. 465°   17. 554°   18. 672°
19. -383°   20. -473°   21. -547°
22. -653°   23. -873°   24. -1237°

In Exercises 25–36 the given values are for impedance triangles. (a) Find the missing values—R, X, Z, or θ—and (b) state whether reactances are inductive or capacitive.

25. $Z = 50\ \Omega, \theta = -40°$ 26. $Z = 20\ \Omega, \theta = 15°$

27. $Z = 15\ \text{k}\Omega, \theta = 35°$ 28. $Z = 70\ \text{k}\Omega, \theta = -65°$

29. $R = 8\ \Omega, X_C = -15\ \Omega$ 30. $R = 18\ \Omega, X_C = -4\ \Omega$

31. $R = 2.7\ \text{k}\Omega, X_L = 900\ \Omega$ 32. $R = 56\ \text{k}\Omega, X_L = 90\ \text{k}\Omega$

33. $R = 5\ \Omega, \theta = 38°$ 34. $R = 12\ \Omega, \theta = -72°$

35. $X_L = 15\ \Omega, \theta = 12°$ 36. $X_C = -40\ \Omega, \theta = -22°$

In Exercises 37–48 find the three trigonometric functions—sin, cos, and tan—of the angles whose values are given in radians.

37. 1   38. $\pi/6$ 39. $-\pi/8$ 40. 1.8
41. 2.75   42. $3\pi/2$ 43. 5.49
44. 6.05   45. 2π 46. -1.83
47. -4.92   48. -5.65

Chapter 12
Sine Waves, Vectors, and Phasors

Heinrich Rudolph Hertz (1857–1894), a German physicist, was the first known person to produce in the laboratory electromagnetic energy in the form which we would now call short radio waves. The hertz, *the unit of frequency in electricity/electronics, is named in his honor. (One hertz equals one cycle per second.) (The Bettman Archive)*

12-1 Graph of the Sine Function

Graphs of the trigonometric functions demonstrate very clearly the variations of the functions. The graph of the sine function is of special interest in electricity/electronics because so many ac voltages and currents can be represented by an equation of the type, $v = V_m \sin \theta$ (read "v is equal to V_m times the sine of the variable angle θ"), and a graph of the sine function reveals to us pictorially the way in which these quantities vary. The oscilloscope is an electronic instrument which can plot the graphs of these quantities instantaneously.

Example 1   Construct the graph of $y = \sin x$.

The sine function may be plotted on the rectangular coordinate system (review Section 3-2). As with other equations we first prepare a table of values. The first table at the bottom of the page is for the equation $y = \sin x$ with x expressed in radians.

Next we plot the values, obtaining the graph of Figure 12-1.

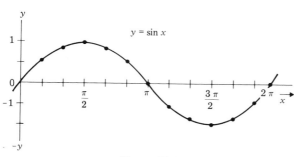

Figure 12-1

Next, we plot the values, obtaining the graph of Figure 12-1.

Study carefully the shape or form of the curve of Figure 12-1. We may change the basic equation $y = \sin x$ in several different ways but as long as it contains the sine function the graph will display the same shape.

Example 2   Construct the graph of $y = 2 \sin x$.

Our new table of values will be identical to that for $y = \sin x$ except that each ordinate will be multiplied by 2. This is the table at the bottom of the page. The graph is shown in Figure 12-2.

Note that in Figure 12-2 the greatest value of the curve is 2 instead of 1. The coefficient a in the generalized equation $y = a \sin x$ is called the *amplitude* of y and is always equal to the *maximum* or *peak value* of the curve and of y. The corresponding equations for ac voltage and current are $v = V_m \sin \theta$ and $i = I_m \sin \theta$. Here V_m and I_m are the peak values of the voltage and current v and i. Such electrical quantities are called *sinusoidal* or *sine-wave* phenomena because of the patterns of their variations.

The value of y, v, or i, and so forth, for any specific value of the independent variable—x, or θ, and so on—is called the *instantaneous value*. For example, the instantaneous value of $v = 10 \sin \theta$ V when $\theta = \pi/6$ rad is $v = 10 \sin \pi/6 = 5$ V. We say the graph of $v = 10 \sin \theta$ is a plot of its instantaneous values.

Example 3   An ac signal voltage can be represented by the equation $v = 100 \sin \theta$ mV.

x	0	$\dfrac{\pi}{6}$	$\dfrac{\pi}{3}$	$\dfrac{\pi}{2}$	$\dfrac{2\pi}{3}$	$\dfrac{5\pi}{6}$	π	$\dfrac{7\pi}{6}$	$\dfrac{4\pi}{3}$	$\dfrac{3\pi}{2}$	$\dfrac{5\pi}{3}$	$\dfrac{11\pi}{6}$	2π
y	0	0.5	0.87	1	0.87	0.5	0	-0.5	-0.87	-1	-0.87	-0.5	0

x, rad	0	$\dfrac{\pi}{6}$	$\dfrac{\pi}{3}$	$\dfrac{\pi}{2}$	$\dfrac{2\pi}{3}$	$\dfrac{5\pi}{6}$	π	$\dfrac{7\pi}{6}$	$\dfrac{4\pi}{3}$	$\dfrac{3\pi}{2}$	$\dfrac{5\pi}{3}$	$\dfrac{11\pi}{6}$	2π
y	0	1	1.73	2	1.73	1	0	-1	-1.73	-2	-1.73	-1	0

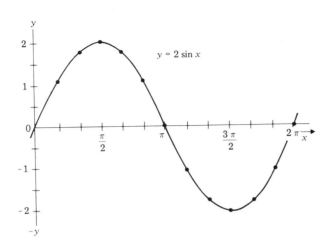

Figure 12-2

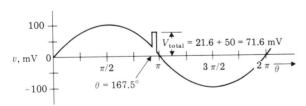

Figure 12-3

Once we are familiar with the sinusoidal pattern we may sketch the graphs of these functions quickly using only the points where the function is zero or maximum.

Example 4   Sketch the curve for the voltage $v = 3 \sin \theta$ V.

We construct the table of values:

θ, rad	0	$\dfrac{\pi}{2}$	π	$\dfrac{3\pi}{2}$	2π	$\dfrac{5\pi}{2}$	3π	$\dfrac{7\pi}{2}$	4π
v, V	0	3	0	-3	0	3	0	-3	0

and sketch the curve.

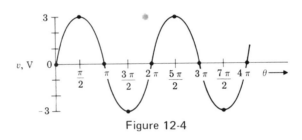

Figure 12-4

(a) What is the value of v at the instant that $\theta = 167.5°$?

(b) A pulse of dc voltage $V = 50$ mV will be added to v at the moment when θ is $167.5°$. What is the total voltage at that instant?

(c) Sketch the waveform and pulse.

(a) For $\theta = 167.5°$, $v = 100 \sin 167.5° = \underline{21.64 \text{ mV}}$

(b) $V_{\text{total}} = v + V = 21.64 + 50 = \underline{71.64 \text{ mV}}$.

(c)

It should be noted that if we extended the table of values and the graph in Example 4 to larger values of θ the pattern would simply repeat itself over and over again. The set of values of the dependent variable—y, v, i, and so forth—included in one complete "unit" of the pattern is called a *cycle* (for example, between $\theta = 0$ and $\theta = 2\pi$, or between $\theta = \pi/2$ and $\theta = 5\pi/2$, and so forth). The "length" of the cycle, along the axis of the independent variable, is called its *period*. The period of $v = 3 \sin \theta$ (see Figure 12-4) is 2π rad.

Graph of $y = a \sin bx$ The effect of multiplying the independent variable by a number b is to change the number of cycles of the basic pattern for a given range of values of x; in short, to change the period. We demonstrate with an example.

Example 5   Plot the equations $y = \sin x$ and $y = \sin 2x$ on the same set of coordinates.

We start by constructing the tables of values.

x	0	$\dfrac{\pi}{8}$	$\dfrac{\pi}{4}$	$\dfrac{3\pi}{8}$	$\dfrac{\pi}{2}$	$\dfrac{5\pi}{8}$	$\dfrac{3\pi}{4}$	$\dfrac{7\pi}{8}$	π	$\dfrac{9\pi}{8}$	$\dfrac{5\pi}{4}$	$\dfrac{11\pi}{8}$	$\dfrac{3\pi}{2}$
y	0		0.7		1		0.7		0		-0.7		-1
$2x$	0	$\dfrac{\pi}{4}$	$\dfrac{\pi}{2}$	$\dfrac{3\pi}{4}$	π	$\dfrac{5\pi}{4}$	$\dfrac{3\pi}{2}$	$\dfrac{7\pi}{4}$	2π	$\dfrac{9\pi}{4}$	$\dfrac{5\pi}{2}$	$\dfrac{11\pi}{2}$	3π
y	0	0.7	1	0.7	0	-0.7	-1	-0.7	0	0.7	1	0.7	0

Then we plot the values. See Figure 12-5.

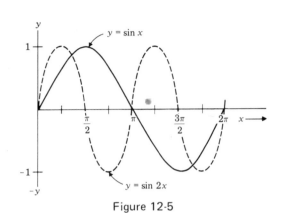

Figure 12-5

Exercises 12-1 In Exercises 1–4 construct the graphs of the given functions by completing the following table of values and plotting the values on a set of rectangular coordinates. (See Example 1.)

x	$-\pi$	$-\dfrac{3\pi}{4}$	$-\dfrac{\pi}{2}$	$-\dfrac{\pi}{4}$	0	$\dfrac{\pi}{4}$	$\dfrac{\pi}{2}$	$\dfrac{3\pi}{4}$	π	$\dfrac{5\pi}{4}$	$\dfrac{3\pi}{2}$	$\dfrac{7\pi}{4}$	2π	$\dfrac{9\pi}{4}$	$\dfrac{5\pi}{2}$	$\dfrac{11\pi}{4}$	3π
y																	

1. $y = \sin x$ 2. $y = -\sin x$
3. $y = \cos x$ 4. $y = -\cos x$

In Exercises 5–14 make sketches of the given functions. Use the technique for quick sketching illustrated in Example 4.

   5. $y = \sin x$ 6. $y = 2 \sin x$
   7. $v = 10 \sin \theta$ V   8. $i = 4 \sin \theta$ mA
   9. $y = -1.5 \sin x$ 10. $v = -4.5 \sin \theta$ V
11. $y = 3 \cos x$ 12. $i = 5 \cos \theta$ mA
13. $v = 15 \sin \theta$ μV   14. $i = -50 \sin \theta$ μA

15–24. State the amplitude (maximum or peak value) of the quantities in Exercises 5–14.

In Exercises 25–30 find the instantaneous values of v for the indicated values of θ. (See Example 3.)

25. $v = 100 \sin \theta$ V for $\theta = 5°$, $167.5°$, and $270°$

26. $v = 10 \sin \theta$ mV for $\theta = 27°$, $180°$, and $330°$

27. $v = -4.5 \sin \theta$ V for $\theta = 60°$, $135°$, and $300°$

28. $v = 15 \sin \theta$ μV for $\theta = -\dfrac{\pi}{4}, \dfrac{\pi}{6}$, and $\dfrac{3\pi}{2}$

29. $v = 145 \sin \theta$ mV for $\theta = -60°$, $120°$, and $225°$

30. $v = -25 \sin \theta$ V for $\theta = 72.5°$, $171.7°$, and $247.8°$

12-2 THE ROTATING RADIUS VECTOR

In Section 11-1 we stated that an angle is "generated" by "rotating" a line segment from an initial position around a pivot point. Let us expand this idea by using a line segment that is to be the radius of a circle. We call the line segment now a *radius vector*. It is designated with an arrow symbol $\overrightarrow{AB}$ or with a letter in boldface type **R**. Its length is R units. (Remember: We use R here to represent "radius" not "resistance.") The end opposite the arrow is called the *initial point* which is placed at the center of the circle [see Figure 12-6(a)]. We will rotate R

counterclockwise at a constant angular velocity of ω (Greek "omega") radians per second (rad/s). After t s, R will have generated the angle θ,

$$\theta = \omega t.$$

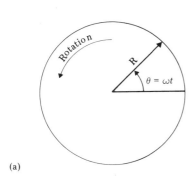

(a)

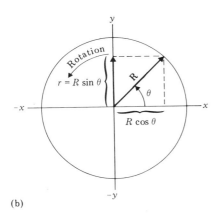

(b)

Figure 12-6

If we construct a set of rectangular coordinates so that the origin is located at the center of the circle [see Figure 12-6(b)] we see that R has a vertical or y-component equal to $R \sin \theta$ or $R \sin \omega t$. If we designate this y-component r, then

$$r = R \sin \omega t \qquad (1)$$

Note the similarity to the equation $y = a \sin x$. Indeed, we can use the radius-vector idea as the basis of a geometric method for constructing a sine-wave graph.

Example 6   Generate the graph of $r = R \sin \omega t$ using the geometric method based on the concept of the rotating vector.

Let us first restate the basic facts we will be using:
(a) R rotates counterclockwise at the constant angular velocity ω.
(b) R has a constant length, R.

1.   Start by constructing R in the desired initial position and of the desired length (see Figure 12-7).
2.   Next, draw a circle with the radius equal to R and the center at the initial point of R.
3.   Divide the circle into a convenient number of equal arcs, say 12, and label p_0, p_1, p_2, and so on.
4.   Construct a set of x-, y-coordinates (positive x-axis only) to the right of the circle (see Figure 12-7). Divide the x-axis with 12 equally spaced markers, and label t_0, t_1, t_2, and so on.
5.   Project the points p_0, p_1, p_2, and so on, from the circle parallel to the x-axis. Project the t-points vertically. The points of intersection of corresponding projections represent points on the sine curve. Sketch in the sine curve.

The results for three different starting positions of R are shown in Figures 12-7(a), (b), and (c).

Let us now relate the concepts illustrated in the sine wave generated by a rotating vector to electrical theory.

1.   The equation for the y-component of the radius vector R is analogous to the equation for the instantaneous value of a sinusoidal alternating voltage (or current):

$$r = R \sin \omega t$$

$$v = V_m \sin \omega t$$

$$i = I_m \sin \omega t$$

2.   The amplitude (maximum or peak value) of a sinusoidal quantity corresponds to the length of the radius vector. The instantaneous value corresponds to the y-component of the radius vector. This component is described as the projection of R on the vertical axis.
3.   One revolution of the radius vector "generates" one cycle (see Section 12-1) of the sine wave. One revolution, and one cycle, correspond to $\theta = 2\pi$ rad. If we let

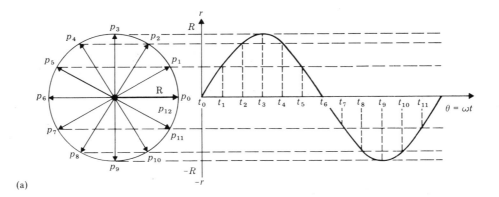

(a)

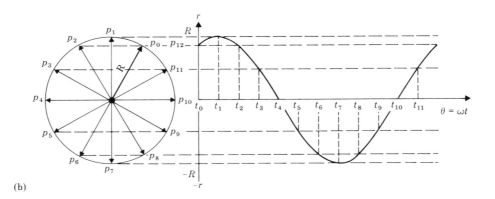

(b)

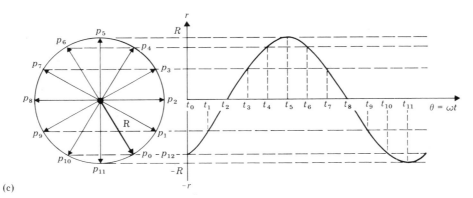

(c)

Figure 12-7

n = number of revolutions per second, then

$$\omega = 2\pi n \text{ rad/s}$$

In electricity we let f = number of cycles per second. Then, since 1 revolution = 1 cycle, $f = n$, and

$$\omega = 2\pi f \text{ rad/s}$$

Note that

$$f = \frac{\omega}{2\pi}$$

The equations become

$$r = R \sin 2\pi n t$$

$$v = V_m \sin 2\pi f t$$

$$i = I_m \sin 2\pi f t$$

4. The period of the sine wave is the time T for one cycle, that is, it is the time required to generate $\theta = 2\pi$ rad. Thus, for one cycle,

$$\theta = 2\pi f T = 2\pi$$

$$fT = 1$$

and

$$T = \frac{1}{f}$$

or

$$f = \frac{1}{T}$$

The period T *is equal to the reciprocal of the frequency* f *(number of cycles per second) and the frequency is equal to the reciprocal of the period. The unit for* T *is seconds (s) and for* f *is cycles per second (cps) or hertz (Hz):*

$$1 \text{ Hz} = 1 \text{ cps}$$

5. Shifting the initial position of the radius vector shifts the position of the sine curve. The shift in position is called the *phase angle* of the function. The phase angle is designated variously as θ or ϕ (Greek phi). Its effect appears in the equation as a modification of the angle,

$$r = R \sin (2\pi f t \pm \theta)$$

A + θ shifts the sine curve to the left and is

called a *leading phase angle*; a $-\theta$ shifts it to the right and is called a *lagging phase angle*. The equation for Figure 12-7(b) is

$$r = R \sin (2\pi f t + 60°)$$

and for Figure 12-7(c) it is

$$r = R \sin (2\pi f t - 60°)$$

6. *In summary, a sinusoidal quantity, such as ac voltage or current, can be represented as a rotating vector.*

We cannot emphasize too strongly the importance of these statements. We will make further use of them in subsequent sections.

Exercises 12-2 In Exercises 1–10 use the geometric method of the rotating vector to sketch the graphs of the given equations.

1. $y = \sin x$
2. $v = \sin x$
3. $y = 2 \sin x$
4. $i = 3 \sin x$
5. $v = 2 \sin (\omega t + 45°)$
6. $i = 3 \sin (\omega t + 60°)$
7. $v = 3 \sin (\omega t - 30°)$
8. $i = 1.5 \sin (\omega t - 45°)$
9. $i = 4 \sin (\omega t + 90°)$
10. $v = 4 \sin (\omega t - 90°)$

12-3 COMBINING TRIGONOMETRIC CURVES

In electronics we often need to find the sum of two or more sinusoidal voltages or currents. We find the resulting graph of the sum of two sine waves by first sketching the graphs of each and then adding the y-values. The method is called *addition of ordinates.* We illustrate the process.

Example 7 Sketch the graph of the sum voltage

$$v_T = v_A + v_B = 10 \sin \omega t + 5 \sin \omega t$$

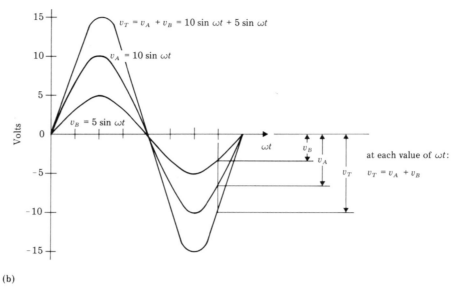

Figure 12-8

Although the addition-of-ordinates method is reliable, it is too tedious and time consuming for many practical applications in electronics. The sum of two sine waves of the same frequency is another sine wave also of the same frequency (see Figure 12-8). This fact enables us to use the mathematics of vectors for combining sinusoidal quantities of the same frequency. (We saw in Section 12-2 that a sinusoidal quantity can be represented by a rotating vector.) Vector addition is a much faster method than addition of ordinates. It is presented in the next section.

Exercises 12-3 In Exercises 1–10 use the method of addition of ordinates to sketch the given combined curves.

1. $y = 2 + \sin x$
2. $y = -3 + \sin x$
3. $y = \sin x + 2 \sin x$
4. $y = 2 \sin x + 3 \sin x$
5. $v = 2 \sin \omega t + \sin \omega t$ V
6. $v = 2 \sin \omega t + 3 \sin \omega t$ V
7. $i = \sin \omega t + 1.5 \sin \omega t$ mA
8. $i = 3 \sin \omega t - 2 \sin \omega t$ mA
9. $v = -5 \sin \omega t + 3 \cos \omega t$ V
10. $i = 6 \cos \omega t - 4 \sin \omega t$ mA

12-4 VECTORS AND VECTOR ADDITION

In our study of the phenomena of the physical world we find that in some instances quantities can be completely expressed in simple statements consisting of numbers and units. For example, we may talk about 27 marbles, 953 people, a speed of 55 mph, a 15-gal tank, a length of 4 ft, and so on. In the precise language of science, such quantities are called *scalar* quantities. The term "scalar" means that they possess only *magnitude*. However, there are many situations for which magnitude does not provide a complete description. Suppose we know that two people are pushing a box across the floor, each with a force of 35 lb. Is this information sufficient if we want to know where the box will come to rest? Our information is specific, just as the preceding examples were, and yet it lacks something—it does not tell us the directions in which the two people are pushing! Thus, we see that in this instance, *magnitude only* is not sufficient to describe completely the "quantity" involved; we must also include information about direction.

A quantity which has both *magnitude* and *direction* is called a *vector quantity*, as opposed to the scalar quantity defined above which has only magnitude. When the direction of a quantity can be only one or the other of two directions, for example temperature, a scalar quantity, it is convenient to use plus and minus signs to represent the two directions. But when the direction can be *any direction*, we must have a more definite method of specifying direction. One such method is graphic; *we draw a line whose length represents the magnitude of the vector quantity and whose direction, as indicated by an arrowhead, represents the direction of the vector quantity*. The line is called a *vector*. Having direction and a definite length, a vector has a beginning called its *initial point* and an ending called its *terminal point* (the end with the arrowhead). For example, Figure 12-9(a) represents diagrammatically the box of our example and the "pushes" of the two people, and Figure 12-9(b) contains the two *vectors* which represent the forces the two people are exerting on the box. The push A is represented by the vector $\overrightarrow{OA}$, the push of B by $\overrightarrow{OB}$.

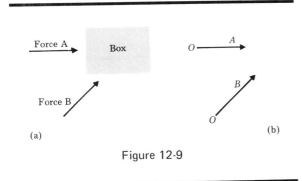

Figure 12-9

Let us digress briefly from our example to clarify the matter of notation. Since both scalar quantities and vector quantities are commonly represented by letter symbols, we shall need a means to distinguish between the two. In print, a scalar quantity is usually represented by an *italic* letter, whereas a vector quantity is represented by a roman **boldface** letter. The absolute value or magnitude of the vector quantity is a scalar quantity and is represented by a capital italic form of the boldface letter. Mathematically, we say that $F = |\mathbf{F}|$.

In handwritten work, the notations are often different because it is not feasible to

try to show italic and boldface letters. Vector quantities are sometimes represented in the form $\dot{\mathbf{F}}$, or $\overline{\mathbf{F}}$, or $\hat{\mathbf{F}}$, or still other forms. The most common convention for handwritten work is to use a capital letter for a vector quantity and the same letter with the vertical-bars notation to show when only magnitude is intended.

To return to our example, if only one force is exerted on the box, for example, force A in Figure 12-10(a), we would expect the box to move in the direction of the force, and similarly, for force B in Figure 12-10(b). But if both forces are acting at the same time, in what direction will the box move? The answer is that the box will move in the direction of what is called the *resultant force*.

We can obtain *resultant force* (or simply *resultant*) of two or more forces by a graphic procedure called *vector addition*. We find the resultant of the forces A and B of Figures 12-9(a) and (b) by first drawing two vectors representing the forces such that they have a common initial point and the $\angle\,\theta$ between them, as shown in Figure 12-10(c). We add the vectors $\overrightarrow{OA}$ and $\overrightarrow{OB}$ by completing the parallelogram of which the two vectors are two adjacent sides. The resultant vector, $\overrightarrow{OC}$, representing the resultant, is the *diagonal* of the parallelogram; its initial point is common to that of the original vectors. See Figure 12-10(d). It is important to note that this procedure determines not only the *length* but also the *direction* of the resultant vector because the lengths and directions of the original vectors determine the shape of the parallelogram. This method can be used as a practical analytical tool for solving any vector addition problem: we draw the known vectors to scale and at the proper angle with respect to each other, we complete the parallelogram, and we measure the resultant vector length and direction.

A common shortcut method of vector addition is based on the parallelogram method just discussed. In Figure 12-10(d), in completing the vector parallelogram, we had to draw line AC parallel to and of the same length as $\overrightarrow{OB}$, the vector for the force B. If instead of drawing vector $\overrightarrow{OB}$, we had drawn the vector for force B at the terminal point of A, we could have produced the diagram of Figure 12-10(e). As a memory aid, we note that the *tail* of vector B is located at the *head* of vector A (represented by the arrowhead).

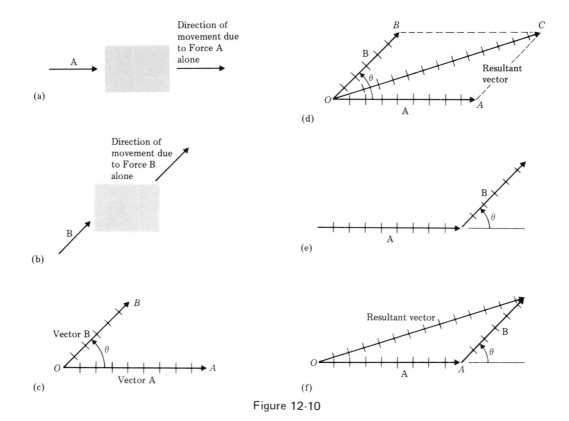

Figure 12-10

Thus, we add the vectors and determine the resultant vector by simply completing the triangle. See Figure 12-10(f). Such a triangle is called a *vector triangle*. Careful construction with straightedge, protractor, and scale will yield results of satisfactory accuracy for many engineering problems.

In many instances, the directions of two vectors of interest are at right angles to each other. The vector triangle of two vectors whose directions are perpendicular to each other is obviously a right triangle and as such can be "solved" not only by the graphic method above but also by either the Pythagorean theorem or the trigonometric method described in Chapter 11. An example will illustrate the procedure.

Example 8 A body has two forces acting on it: force **A** = 45 kg, acting due east, and force **B** = 30 kg, acting due north (see Figure

12-11). Find the magnitude and direction of the resulting force (a) graphically, and (b) with the use of trigonometric functions.

(a)

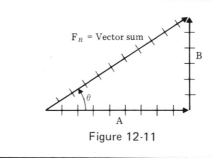

Figure 12-11

(b) $F_R = \dfrac{B}{\sin \theta} = \dfrac{B}{\sin \left(\arctan \dfrac{B}{A}\right)}$

$\qquad = \dfrac{30}{\sin \left(\arctan \dfrac{30}{45}\right)}$

$\qquad = \underline{54.08 \text{ kg}}$

Vector Subtraction When a vector is to be subtracted we reverse its direction and then add.

Example 9 Vector **A** is 15 due west, vector **B** is 25 due north. Subtract **B** from **A** (Figure 12-12).

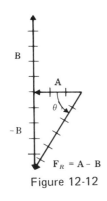

Figure 12-12

$F_R = \dfrac{A}{\cos \left(\arctan \dfrac{B}{A}\right)} = \dfrac{15}{\cos \left(\arctan \dfrac{25}{15}\right)}$

$\qquad = \underline{29.15}$

$\theta = \arctan \dfrac{25}{15} = \underline{59.04° \quad \text{(S of W)}}$

x- and y-Components Vectors not perpendicular to each other can be added by resolving them into their x- and y-components and then combining these components.

Example 10 Refer to Figure 12-13. **A** = 5, $\theta_A = 15°$; **B** = 6, $\theta_B = 65°$. Find **S**, the vector sum of **A** and **B**.

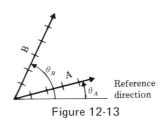

Figure 12-13

We construct the vectors on a set of x-, y-coordinates (see Figure 12-14). We study the diagram and note the following:

x-components:

$A_x = A \cos \theta_A = 5 \cos 15° = 4.830$
$B_x = B \cos \theta_B = 6 \cos 65° = 2.536$
$S_x = A_x + B_x = 7.366$

y-components:

$A_y = A \sin \theta_A = 5 \sin 15° = 1.294$
$B_y = B \sin \theta_B = 6 \sin 65° = 5.438$
$S_y = A_y + B_y = 6.732$

Therefore,

$S = \dfrac{S_y}{\sin \left(\arctan \dfrac{S_y}{S_x}\right)} = \dfrac{6.732}{\sin \left(\arctan \dfrac{6.732}{7.366}\right)}$

$S = \underline{9.979}$

$\theta = \arctan \dfrac{6.732}{7.366} = \underline{42.43°}$

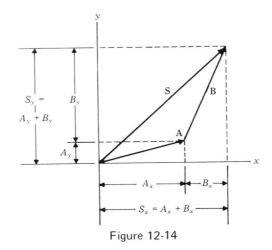

Figure 12-14

Exercises 12-4   In Exercises 1–10 the given values are for vectors **A** and **B**, respectively. **A** is acting due east and **B** is acting due north. Find **S**, the vector sum of **A** and **B**, using the graphic method only (scale **S** and measure θ with a protractor).

1. 3, 4   2. 8, 13   3. 12, 7   4. 15, 11
5. 6, –4   6. 10, –17   7. 2.5, –1.5
8. 3.5, –6.5   9. 8.75, 6.25
10. 4.25, –5.75

11–20.   Refer to the instructions for Exercises 1–10. Find **S** (magnitude and direction) for Exercises 1–10 using the trigonometric method.

In Exercises 21–30 the given values are for A, θ_A, B, and θ_B, respectively (see Example 10). Find S and θ_S using the technique of Example 10.

21. 5, 15°; 6, 65°   22. 3, 12°, 5, 55°
23. 8, 68°; 3, 27°   24. 8, 45°; 12, 65°
25. 15, 45°;   26. 45, –15°;
        15, –45°   55, 45°
27. 78, –34.5°; 86, –15.7°
28. 0.568, 78.8°; 0.469, 12.7°
29. 0.04637, 37.68°; 0.01637, –73.68°
30. 2.756 × 10³, –53.76°; 6.937 × 10³, 2.56°

12-5 PHASORS

We now return to the task of finding the sum of sinusoidal quantities such as ac voltage and current. It was shown in Section 12-2 that

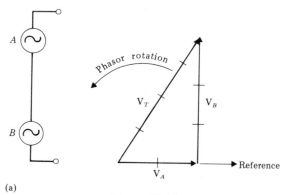

(a)

Figure 12-15

for such quantities which have the same frequency, the qualities of amplitude, instantaneous values, and relative phase angle could be represented by a geometric device called a rotating radius vector, or simply a rotating vector. Section 12-4 demonstrated how vectors (and radius vectors are included) can be added. Can we correctly represent ac voltages and currents with vectors and use vector addition to find their sums?

Speaking very precisely, the vectorlike arrows that we will use to represent the magnitude, direction, and relative phase angle of sinusoidal quantities are not truly vectors. To keep things straight they are called *phasors* (from *ph*ase vect*ors*). Vector arithmetic, however, can be applied to phasors.

Example 11   The output of generator **A** is $v_A = 10 \sin 10t$ V and is to be added to the output of generator **B**, $v_B = 15 \sin (10t + 90°)$ V. Find $v_T = v_A + v_B$.

We will use $\mathbf{V}_A$ as the reference phasor since its phase angle is 0°. We draw the phasor (vector) $\mathbf{V}_A$ in the positive x direction, the reference direction for phasors—the direction for phasors with 0° phase angle. We construct the phasor $\mathbf{V}_B$ at the head of $\mathbf{V}_A$ and at right angles to it (the phase angle of v_B is +90°). See Figure 12-15. The diagram is called a *phasor diagram*. Referring to the diagram we write

$$V_T = \frac{V_B}{\sin \left(\arctan \dfrac{V_B}{V_A}\right)}$$

$$= \frac{15}{\sin \left(\arctan \dfrac{15}{10}\right)}$$

$$= \underline{18.03 \text{ V}}$$

$$\theta_T = \arctan \frac{15}{10} = \underline{56.31°}$$

$$v_T = 18.03 \sin (10t + \theta_T) \text{ V}$$
$$= \underline{18.03 \sin (10t + 56.31°) \text{ V}}$$

The method of finding the sum of two or more sinusoidal voltages or currents having the same frequency by the phasor method depends on several important basic ideas:

1.   We represent each quantity as a rotating phasor. The speed of rotation is the same for all since all have the same frequency.

2. To construct the phasor diagram, we "stop" the rotation (as with a flash photograph) when $t = 0$. The relative positions of the "still" phasors are determined by their phase angles since they rotate at the same speed.

3. Conventional direction of rotation is counterclockwise. Reference position is the $+x$-axis. Positive phase angles are measured CCW from the reference position.

4. The length of a phasor is proportional to the amplitude of the voltage or current it represents.

5. Phasors are combined using vector arithmetic (see Section 12-4).

When the phase angle difference of two phasors is other than 90°, the phasor sum is obtained by the use of the sum of x- and y-components.

Example 12 Find the phasor sum of the phasors $v_1 = 10 \sin \left(377t + \dfrac{\pi}{12} \right)$ V and $v_2 = 8 \sin \left(377t - \dfrac{\pi}{3} \right)$ V. (Note that the phase angles are given in radians.)

We first construct the phasor diagram as an aid in visualizing the relationships (see Figure 12-16). Then we apply the method of the sum of x- and y-components:

$$V_{T_x} = V_{1_x} + V_{2_x}$$

$$= 10 \cos \frac{\pi}{12} + 8 \cos \left(-\frac{\pi}{3} \right)$$

$$= 9.659 + 4.0$$

$$= 13.66 \text{ V}$$

$$V_{T_y} = V_{1_y} + V_{2_y}$$

$$= 10 \sin \frac{\pi}{12} + 8 \sin \left(-\frac{\pi}{3} \right)$$

$$= 2.588 - 6.928$$

$$= -4.34 \text{ V}$$

$$V_T = \frac{V_{T_y}}{\sin \left(\arctan \dfrac{V_{T_y}}{V_{T_x}} \right)} = \frac{-4.34}{\sin \left(\arctan \dfrac{-4.34}{13.66} \right)}$$

$$= \underline{14.33 \text{ V}}$$

$$\theta_T = \arctan \frac{V_{T_y}}{V_{T_x}} = \arctan \frac{-4.34}{13.66}$$

$$= -0.3077 \text{ rad} = \underline{-17.63°}$$

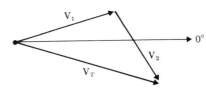

Figure 12-16

Exercises 12-5 In Exercises 1–14 the given phasor quantities are to be added. Find V_T or I_T and θ_T using the phasor sum method.

1. $v_1 = 2 \sin \omega t$ V, $v_2 = 3 \sin (\omega t + 90°)$ V
2. $v_1 = 1.5 \sin (\omega t - 30°)$ V, $v_2 = 2 \sin \omega t$ V
3. $i_1 = 4 \sin \left(\omega t + \dfrac{\pi}{2} \right)$ A, $i_2 = 1.75 \sin \omega t$ A

4. $i_1 = 5 \sin \left(\omega t - \dfrac{\pi}{4} \right)$ A, $i_2 = 4.5 \sin \omega t$ A

5. $i_1 = 2.5 \sin \left(\omega t + \dfrac{\pi}{3} \right)$ A,

   $i_2 = 3 \sin \left(\omega t - \dfrac{\pi}{6} \right)$ A

6. $v_1 = 2.5 \sin \left(\omega t - \dfrac{\pi}{4} \right)$ V,

   $v_2 = 6 \sin \left(\omega t + \dfrac{\pi}{4} \right)$ V

7. $v_1 = 75 \sin \omega t$ mV,
   $v_2 = 97 \sin (\omega t + 48°)$ mV

8. $v_1 = 58 \sin (\omega t - 62.5°)$ μV,
   $v_2 = 36 \sin \omega t$ μV

9. $i_1 = 12.5 \sin (\omega t + 30°)$ mA,
   $i_2 = 28.7 \sin \omega t$ mA

10. $i_1 = 6.75 \sin \omega t$ μA,
    $i_2 = 2.35 \sin (\omega t + 90°)$ μA

11. $v_1 = 150 \sin \left(\omega t + \dfrac{\pi}{2} \right)$ kV,

    $v_2 = 225 \sin \left(\omega t + \dfrac{\pi}{4} \right)$ kV

12. $i_1 = 278 \sin \left(\omega t - \dfrac{\pi}{3} \right)$ mA,

    $i_2 = 315 \sin \left(\omega t + \dfrac{\pi}{12} \right)$ mA

13. $v_1 = 0.685 \sin \left(\omega t - \dfrac{\pi}{3} \right)$ V,

    $v_2 = 0.194 \sin \left(\omega t - \dfrac{\pi}{6} \right)$ V

14. $i_1 = 0.087 \sin \left(\omega t + \dfrac{\pi}{4}\right)$ A,

$i_2 = 0.118 \sin \left(\omega t + \dfrac{\pi}{2}\right)$ A

12-6 TIME AS THE VARIABLE

Trigonometric functions are the functions of (or vary with) angles. In the previous sections of this chapter we have plotted a function against an angle, usually in radians. However, in the everyday world of electronics we generally think of a voltage such as $v = 10 \sin 377t$ as a function of time (it varies with time). Is it possible or correct to speak of the sine function of time? It is correct in this instance because time (represented by t) is the part that varies in the "argument" $377t$ in the function "sin $377t$."

Recall from Section 12-2 that the generalized statement for a sinusoidal voltage is $v = V_m \sin(2\pi ft \pm \theta)$. Further, $2\pi f$ is equal to ω, an angular velocity in radians per second. Thus, $2\pi ft$ has the units of an angle: rad/s × s = radians. Hence, "$377t$" actually represents an angle which varies with time, and in expressions of the form $v = 10 \sin 377t$ it is appropriate to consider v as a function of time. Graphs of such functions are commonly plotted with t as the independent variable (see Figure 12-17).

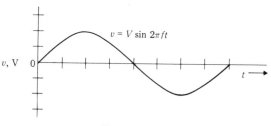

Figure 12-17

Frequency and Period Calculations The oscilloscope displays time-varying voltages or currents as a function of time—pulse waveforms as well as sinusoids (see Figure 12-18). When any phenomenon—pulses or cycles—repeats itself, frequency (f) is defined as the

repetition rate per second (PRR = pulse repetition rate is widely used for pulses). Period (T) is the time required for one occurrence. Mathematically,

$$f = \frac{1}{T}, \qquad T = \frac{1}{f}$$

We use calibrated oscilloscope displays to determine period and frequency.

Example 13 The scale factor for horizontal scale (t) of the oscilloscope graticule in Figure 12-18(a) is 1 cm = 5 ms. Determine T and f for the signal displayed.

One cycle extends over 5 cm; thus,

$$T = 5 \text{ cm} \times 5 \text{ ms/cm} = 25 \text{ ms}$$

and

$$f = \frac{1}{T} = \frac{1}{25 \times 10^{-3}} = 40 \text{ Hz}$$

(Review Section 12-2 concerning unit for frequency: 1 cycle per second = 1 hertz.)

Example 14 The horizontal scale factor for the pulse signal display in Figure 12-18(b) is 1 cm = 0.2 μs. Determine T and f or PRR for the signal.

The distance from the beginning of one pulse to the beginning of the next pulse is 2.5 cm. Thus,

$$T = 2.5 \text{ cm} \times 0.2 \ \mu\text{s/cm} = 0.5 \ \mu\text{s}$$
$$\text{PRR} = f = \frac{1}{T} = \frac{1}{0.5 \times 10^{-6}} = 2 \times 10^6 \text{ Hz}$$
$$= \underline{2 \text{ MHz}}$$

Wavelength Calculations When a high-frequency signal is being transmitted on a transmission line it travels at a finite speed—the speed of light, $c = 3.0 \times 10^8$ m/s. A sine-wave pattern travels along the line at this speed. One cycle of this spatial pattern is now called a *wavelength* λ (Greek lambda). The actual length of a wave is related to its frequency f and velocity c,

$$\lambda = \frac{c}{f}, \qquad f = \frac{c}{\lambda}$$

Example 15 An antenna for a broadcasting station, whose operating frequency is 1500

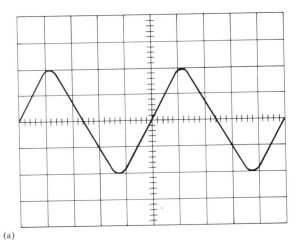

(a)

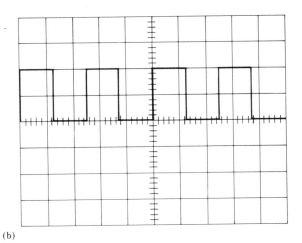

(b)

Figure 12-18

kHz, is $\frac{1}{4}\lambda$ in length. What is its length in meters?

$$\lambda = \frac{c}{f} = \frac{3 \times 10^8}{1.5 \times 10^6} = 2 \times 10^2 \text{ m}$$

$$\tfrac{1}{4}\lambda = 50 \text{ m}$$

Example 16 Amateur radio operators (hams) are permitted to transmit in certain "bands" of frequencies. These are commonly identified by their average wavelength. What nominal frequency would be associated with the "15-meter band?"

$$f = \frac{c}{\lambda} = \frac{3 \times 10^8}{15} = 20 \text{ MHz}$$

Exercises 12-6 In Exercises 1–6 the given values are signal frequencies. Determine T for each.

1. 60, 120, and 1200 Hz
2. 2, 5, and 15 kHz
3. 30, 50, and 200 kHz
4. 0.54, 0.87, and 1.5 MHz
5. 10, 25, and 50 MHz
6. 250, 650, and 1,250 MHz

In Exercises 7–12 the given values are the periods of various signals. Determine f for each.

7. 0.01, 0.01667, and 0.005 s
8. 1, 0.7, and 0.08 ms
9. 250, 75, and 16 ms
10. 7.5, 4.8, and 1.5 μs
11. 35, 18, and 3.5 ns
12. 250, 375, and 96 ps

In Exercises 13–18 the given values are for the horizontal (time base) axis of the oscilloscope display of Figure 12-18(a). Determine T and f for the displays.

13. 50 ms/cm
14. 0.2 ms/cm
15. 0.1 ms/cm
16. 20 μs/cm
17. 5 μs/cm
18. 0.2 μs/cm

In Exercises 19–24 the given values are scale factors for the horizontal (time base) axis of the oscilloscope display of Figure 12-18(b). Determine T and the PRR for the pulse displays.

19. 2 s/cm
20. 10 ms/cm
21. 0.5 ms/cm
22. 50 μs/cm
23. 10 μs/cm
24. 0.5 μs/cm

In Exercises 25–28 the given values are the operating frequencies of radio stations. Determine the wavelengths of the signals (in meters).

25. 980 kHz
26. 3.565 MHz
27. 90.7 MHz
28. 776 MHz

29. A half-wave ($\frac{1}{2}\lambda$) antenna is 19.8 m in length. What is the frequency for which it is designed to be used?

30. An AM broadcasting station antenna is exactly $\frac{1}{4}\lambda$ in height at 50.34 m. What is the station's operating frequency?

12-7 PEAK, INSTANTANEOUS, EFFECTIVE, AND AVERAGE VALUES

As we have seen, a so-called sinusoidal voltage varies with time according to the equation

$$v = V_m \sin (2\pi ft \pm \theta)$$

The corresponding equation for a sinusoidal current is

$$i = I_m \sin (2\pi ft \pm \theta)$$

Peak Value   The factor V_m or I_m is the *amplitude* of the quantity. It is also called the *maximum* or *peak value* as it is the value at the maximum point or peak of the sine curve. The *peak-to-peak value*, V_{p-p}, is easily read from an oscilloscopic display of v [see Figure 12-18(a)], and

$$V_{p-p} = 2 V_m, \qquad V_m = V_{p-p}/2$$

Example 17   The scale factor for the vertical axis of the graticule of Figure 12-18(a) is 2 mV/cm. (a) What is V_{p-p} for the signal displayed? (b) What is V_m? (c) What is the equation for v if the horizontal scale factor is 0.2 ms/cm?

(a) V_{p-p} = (4 cm) (2 mV/cm) = 8 mV
(b) $V_m = V_{p-p}/2$ = 8/2 = 4 mV
(c) T = (5 cm) (0.2 ms/cm) = 1 ms

$$f = \frac{1}{T} = \frac{1}{1 \times 10^{-3}} = 1 \text{ kHz}$$

$$v = V_m \sin 2\pi ft = 4 \sin 6280t \text{ mV}$$

Instantaneous Value   The value of a sinusoidal quantity at a particular point in the cycle is important in predicting the performance of certain types of electronics circuits. The basic equation defines the instantaneous value and is evaluated for the desired angle.

Example 18   The triggering circuit of an oscilloscope is found to be starting the trace

(triggering) at 67° for the signal v = 55 sin ωt mV. What is the "triggering level" in millivolts?

$$v_t = 55 \sin 67° = \underline{50.63 \text{ mV}}$$

(Triggering occurs when ωt = 67°.)

Example 19   A signal v = 10 sin ωt V is applied to the input of a controlled rectifier circuit which starts to conduct when its input voltage is 0.5 V. At what angle in the signal cycle will the circuit be "turned on?"

We define θ_c as the angle at which v equals or exceeds the threshold voltage, 0.5 V. Thus,

$$0.5 = 10 \sin \theta_c$$

$$\theta_c = \arcsin \frac{0.5}{10} = \underline{2.87°}$$

Effective Value   The effective value of a sine wave or any other complex waveform is determined by its heating effect.

The effective value of any voltage or current is equal to the amplitude of a pure dc voltage which produces the same average heating effect as the unknown voltage or current.

Since heating effect is proportional to the square of voltage or current ($P = V^2/R = I^2 R$) the effective value is obtained by a mathematical process which involves squaring instantaneous values of the quantity. The effective value is the root-mean-square (rms) value:

1. Instantaneous values are squared.
2. The average or "mean" of the squared values is found.
3. The root of the mean is taken (hence, "root mean square").

For a pure sine wave quantity the effective value is equal to 0.7071 times the peak value:

$$V_{eff} = 0.7071 \, V_m, \qquad I_{eff} = 0.7071 \, I_m$$
$$V_m = 1.414 \, V_{eff}, \qquad I_m = 1.414 \, I_{eff}$$

By convention, when the value of a sinusoidal quantity is stated it is understood that the value is the effective value unless otherwise indicated. The symbols V and I when used to represent sine wave voltage and current imply the effective value, peak values require the subscripts m or p: V_m or V_p and I_m or I_p.

Average Value The average value of any set of values is equal to the sum of all the values divided by the number of values. Symbolically, the average value $\overline{X}$ (read X bar) of a set of n values of x is

$$\overline{X} = \frac{\Sigma x}{n}$$

This kind of mathematical analysis has been performed on the sine wave, and for one-half cycle of $v = V_m \sin \omega t$,

$$V_{ave} = \frac{2}{\pi} V_m = 0.6366 \ V_m$$

Why is $V_{ave} = 0$ over a full cycle? For a half-wave rectified voltage $V_{ave} = 0.3183 \ V_m$ and for full-wave rectification $V_{ave} = 0.6366 \ V_m$.

Example 20 An electronic light dimmer reduces the voltage available to an electric bulb (and therefore reduces the light output) by rectifying (converting to dc voltage) only a portion of each alternation (half-cycle) of an ac sine wave. Estimate V_{eff} and V_{ave} for the voltage across the bulb if the circuit is adjusted to rectify line voltage $V = 120°$ V during the portion of each cycle from $75°$ to $180°$ and from $255°$ to $360°$ [see Figure 12-19(a)]. The rectified voltage is shown in Figure 12-19(b).

We will estimate V_{eff} and V_{ave} by sampling values at $15°$ intervals between $0°$ and $180°$. Since both alternations are being used the overall average and effective values will be the same as those for a half-cycle.

First we determine V_m,

$$V_m = 1.414 \ V = 1.414 \times 120$$
$$= 169.7 \text{ V}$$

Next we construct a table of the instantaneous values at $15°$ intervals over the half-cycle.

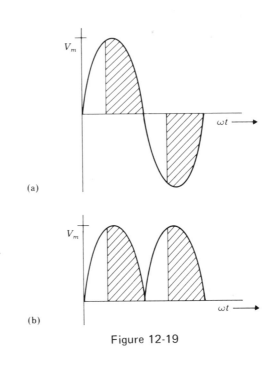

(a)

(b)

Figure 12-19

θ	$\sin \theta$	v	v^2	
$15°$	0.2588	0	0	
$30°$	0.50	0	0	Rectification
$45°$	0.7071	0	0	begins at $75°$
$60°$	0.8660	0	0	
$75°$	0.9659	163.9	2.69×10^4	
$90°$	1.0	169.7	2.88×10^4	
$105°$	0.9659	163.9	2.69×10^4	
$120°$	0.8660	147.0	2.16×10^4	
$135°$	0.7071	120.0	1.44×10^4	
$150°$	0.50	84.9	0.72×10^4	
$165°$	0.2588	43.9	0.19×10^4	
$180°$	0	0	0	
Totals		893.3	12.77×10^4	

$$V = \sqrt{\frac{12.77 \times 10^4}{12}} = \underline{103 \text{ V}}$$

$$V_{ave} = \frac{893.3}{12} = \underline{74.4 \text{ V}}$$

Exercises 12-7   In Exercises 1–10 the given values are the vertical and horizontal scale factors, respectively, for the oscilloscope graticule of Figure 12-18(a). For the signal displayed, determine (a) V_{p-p}, (b) V_m, (c) V (effective value), and (d) the equation for the instantaneous value v (see Example 17).

1. 2 mV/cm and 0.2 ms/cm
2. 50 V/cm and 50 ms/cm
3. 0.1 V/cm and 2 μs/cm
4. 0.2 V/cm and 0.1 μs/cm
5. 0.5 V/cm and 0.2 μs/cm
6. 1 V/cm and 50 μs/cm
7. 2 V/cm and 2 ms/cm
8. 5 V/cm and 5 ms/cm
9. 10 V/cm and 10 ms/cm
10. 20 V/cm and 20 ms/cm

In Exercises 11–14 the equation for the signal voltage being applied to an oscilloscope is given as well as the angle of that signal at which the oscilloscope is triggering. Determine the triggering level. (See Example 18.)

11. $v = 55 \sin \omega t$ mV, 67°
12. $v = 3.5 \sin \omega t$ V, 2.5°
13. $v = 45 \sin \omega t$ mV, 40°
14. $v = 225 \sin \omega t$ mV, 27°

In Exercises 15–18 the equation for the input voltage to a controlled rectifier circuit is given as well as the value of v_c, the threshold voltage for turn on of the circuit. Determine θ_c, the electrical angle for each cycle when the circuit turns on. (See Example 19.)

15. $v = 10 \sin \omega t$ V, 0.5 V
16. $v = 170 \sin \omega t$ V, 40 V
17. $v = 800 \sin \omega t$ mV, 150 mV
18. $v = 675 \sin \omega t$ mV, 95 mV

For Exercises 19–22 refer to Example 20 and the following. The portion of the alternations of the sine wave during which conduction of the rectifier occurs, as well as the effective value of the applied voltage, are given. Determine (a) V, and (b) V_{ave} for the voltage across the light bulb.

19. 75° to 180°, 255° to 360°, 120 V
20. 105° to 180°, 285° to 360°, 120 V
21. 120° to 180°, 300° to 360°, 120 V
22. 0° to 180° only, 120 V

Chapter 13

Complex Numbers and the j-Operator, with Applications

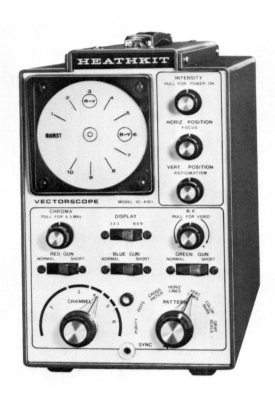

A "vectorscope" is used in conjunction with a color bar generator to check the phase angle and amplitude of the color phasors in a color video test signal. (Courtesy of Heath Company)

The analyses of alternating-current (ac) circuits can require a considerably greater amount of time and effort than those of dc circuits. We must consider "impedance triangles" rather than simple resistances, and voltages and currents that are complicated by phase angles. Mathematics has come to our rescue again with a development which is used to simplify, organize, and expedite the calculations of ac-circuit analysis. The development is called *complex numbers*. It is utilized in electric circuit analysis through the use of a mathematical device called the *j-operator*.

13-1 IMAGINARY NUMBERS

When a number is multiplied by itself the result is always a positive number. Thus,

$$(+2)(+2) = +4 \quad \text{and} \quad (-2)(-2) = +4$$

Consequently, we recognize that

$$\sqrt{+4} = +2 \quad \text{or} \quad -2$$

that is,

$$\sqrt{4} = \pm 2$$

What, then, do we do with

$$\sqrt{-4} = ?$$

First, we can factor the $\sqrt{4}$ and write

$$\sqrt{-4} = \sqrt{4}\sqrt{-1} = 2\sqrt{-1}$$

Mathematicians call $\sqrt{-1}$ an *imaginary number* and represent it with the letter i,

$$i = \sqrt{-1}$$

Since i is already used in electronics to represent current, we use j instead of i,

$$j = \sqrt{-1}$$

Thus,

$$\sqrt{-4} = 2\sqrt{-1} = 2j \quad \text{or} \quad j2$$

The letter j, widely called the *j-operator* in electricity/electronics, is a factor and can be manipulated like any other factor. For example,

$$j^2 = \sqrt{-1} \cdot \sqrt{-1} = -1$$
$$j^3 = j^2 \cdot j = -1 \cdot j = -j$$
$$j^4 = j^2 \cdot j^2 = (-1)(-1) = 1$$

Example 1 Use the *j*-operator to express the result of taking the indicated roots.

(a) $\sqrt{-9}$ (a) $\sqrt{-9} = 3\sqrt{-1} = j3$

(b) $\sqrt{-25}$ (b) $\sqrt{-25} = 5\sqrt{-1} = j5$

(c) $\sqrt{-36}$ (c) $\sqrt{-36} = 6\sqrt{-1} = j6$

Example 2 Perform the indicated operations.

(a) $(2j)^3$ (a) $(2j)^3 = 8 \cdot j \cdot j^2 = \underline{-8j}$
                $(j^2 = -1)$

(b) $(j)^6$ (b) $(j)^6 = j^2 \cdot j^2 \cdot j^2 =$
                $(-1)(-1)(-1) = \underline{-1}$

(c) $(j)^7$ (c) $(j)^7 = j^6 \cdot j = -1 \cdot j = \underline{-j}$

Exercises 13-1 In Exercises 1–10 use the *j*-operator to express the result of taking the indicated roots.

1. $\sqrt{-9}$ 2. $\sqrt{-4}$ 3. $\sqrt{-16}$ 4. $\sqrt{-25}$

5. $\sqrt{-2}$ 6. $\sqrt{-3}$ 7. $\sqrt{-7}$ 8. $\sqrt{-81}$

9. $\sqrt{-a^4}$ 10. $\sqrt{-a^4 b^2}$

In Exercises 11–20 perform the indicated operations.

11. (a) $(j)^2$ (b) $(j)^4$ 12. (a) $(j)^3$ (b) $(j)^5$

13. (a) $(j)^5$ (b) $(j)^7$ 14. (a) $(j)^4$ (b) $(j)^6$

15. (a) $(j)^4$ (b) $(j)^8$ 16. (a) $(j)^5$ (b) $(j)^9$

17. $j^3 \cdot j^5$ 18. j^5/j^3

19. j^8/j^4 20. j^{14}/j^{11}

13-2 COMPLEX NUMBERS

A complex number is the sum of a real and an imaginary number. Either part or both parts can be negative or positive. Thus, the following are complex numbers:

$$3 + j4, \quad -2 + j, \quad -1 - j2, \quad 4 - j3$$

Note that in the second complex number, j is by itself, which means it is multiplied by unity.

Complex numbers may also be literal. For example,

$$2a + jb, \quad 3 - j2y$$

A complex number is never said to be positive or negative, but its real and imaginary parts are so labeled. Thus, in $2 + j3$, the real part, 2, and the imaginary part, $3j$, are both positive. In $3a - j2b$, the real part is $+3a$, and the imaginary part is $-j2b$.

Two complex numbers are equal if their real parts are equal and their imaginary parts are equal. Thus, if

$$a + jb = 2 - j3$$

then $a = 2$ and $b = -3$. In effect, when two complex numbers are equal, we have two equations, one equating the real parts and the other the imaginary parts. Thus, solving the equation $3 - jy = x + j2$ for x and y, we find that the real parts give $x = 3$, and the imaginary parts give $y = -2$. It is even possible for simultaneous equations to appear as a single complex equation.

Example 3   Solve for x and y: $x + y + jx = 4 + jy + j2$.

To solve for x and y, first let the real numbers be equal; we get $x + y = 4$. Then we let the imaginaries be equal, to get

$$jx = jy + j2$$

We divide by j:

$$x = y + 2$$

From our real-number equation, we know that

$$x + y = 4, \quad \text{so} \quad y = 4 - x$$

We substitute this value of y into our imaginary-number equation:

$$x = 4 - x + 2, \quad 2x = 6, \quad \underline{x = 3}$$

Hence,

$$\underline{y = 1}$$

Exercises 13-2   Solve for x and y.

1. $x + jy = 2 - j3$ 2. $4 + j5 = x + jy$
3. $x + 2 = j4 + jy$ 4. $x + j6 = jy - 5$
5. $x + y + jx = 4 + jy + j2$
6. $x - j4 + jy = 2 - y + jx$

13-3 COMPLEX PLANE

In Chapter 3, we discussed the (x, y)-coordinate system, in which any point represents a pair of values, one for x and one for y. If we plot real values along the x-axis and imaginary values along the y-axis, *any complex number* can be represented on the coordinate system. Since a real number can be

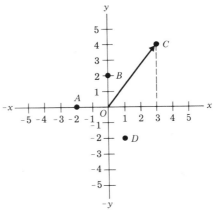

Figure 13-1

considered a complex number in which the imaginary part is zero, and similarly, since an imaginary number is a complex number in which the real part is zero, this coordinate system includes *all points:* real, imaginary, and complex.

Figure 13-1 shows the coordinate system used to represent complex numbers. It is exactly the same as the rectangular coordinate system introduced in Chapter 3. However, when we use it to represent complex numbers, we call it the *complex plane*. In Figure 13-1, point A represents the real value -2 or the complex number $-2 + j0$. All real values lie on the x-axis, negative to the left of the origin and positive to the right. The point B represents the imaginary number $j2$. All imaginary numbers lie on the y-axis. Any point that does not lie on one of the axes represents a complex number. Thus, point C represents $3 + j4$, and point D represents $1 - j2$.

As we shall see, the coordinate system of the complex plane has very useful practical applications.

13-4 VECTORS AND PHASORS

Point C in Figure 13-1 is connected to the origin by an arrow with its head at C. From Chapter 12, we recall that such a line segment is a *vector* and that a vector has both magnitude and direction. The magnitude is the length OC, and the direction is indicated by

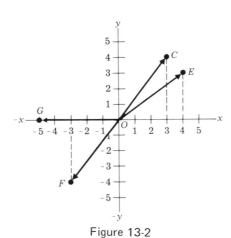

Figure 13-2

the line and its arrowhead. We see, then, *an important property of complex numbers: they can be used to represent magnitude and direction.*

Figure 13-2 shows point C at $3 + j4$, as well as point E at $4 + j3$. These vectors are equal in length; each is the hypotenuse of a right triangle in which the absolute values of the x- and y-dimensions are the lengths of the other two sides. Thus,

$$\overrightarrow{OC}^2 = 3^2 + 4^2 = 9 + 16 = 25, \qquad OC = 5$$

The magnitude of vector $\overrightarrow{OC}$ is 5, and

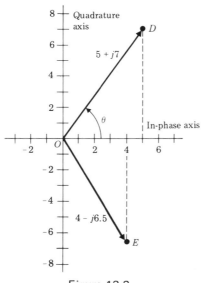

Figure 13-3

$$\overrightarrow{OE}^2 = 4^2 + 3^2 = 16 + 9 = 25, \qquad OE = 5$$

The point F is at $-3 - j4$. Thus,

$$\overrightarrow{OF}^2 = -3^2 + (-4)^2 = 9 + 16 = 25, \qquad OF = 5$$

The vector $\overrightarrow{OF}$ is also 5. Point G is at the point $-5 + j0$. The magnitude of vector $\overrightarrow{OG}$ is $|-5| = 5$. Therefore, all four vectors shown are of the same magnitude and differ only in direction. However, complex notation enables us to indicate this difference in direction.

Alternating currents and voltages are vector-like quantities in which relative phase angles are important. The magnitude and phase angle of such quantities can be represented by directed line segments. Impedance and admittance are not vector quantities, but are conveniently handled by vectorlike diagrams and operations. In either case, the magnitude and direction sense of a quantity can be represented very simply and effectively by complex numbers. As we learned in Chapter 12, the symbols for vector and phasor quantities appear in boldface type while symbols for the magnitudes of these quantities appear in italics.

Figure 13-3 shows an ac-voltage phasor with phase angle θ drawn on the complex plane so that its initial point is at the origin. It is apparent that the magnitude and direction of the line segment can represent the magnitude and relative phase angle of the phasor.

We can think of the phasor as consisting of two components: a real component and an imaginary component. In practice, the real component is called the *in-phase component*, since it is always measured along the horizontal axis which is conventionally chosen as the in-phase (zero phase angle) reference. The imaginary component is always ±90° out of phase and is commonly referred to as the *quadrature component*. ("Quadrature" is derived from the Latin *quadratus*, meaning "made to fit into a square." We use it to mean "at 90°.") Thus, a "complex" voltage or current can be specified as

$$\mathbf{V} = V_{\text{in phase}} \pm j V_{\text{quadrature}}$$

or

$$\mathbf{I} = I_{\text{in phase}} \pm j I_{\text{quadrature}}$$

The magnitude of a complex quantity is obtained from

$$V = \sqrt{V^2_{\text{in phase}} + V^2_{\text{quad}}}$$

or

$$I = \sqrt{I^2_{\text{in phase}} + I^2_{\text{quad}}}$$

In Figure 13-3 if the phasor $\overrightarrow{OD}$ represents a voltage, we can write

$$V_D = 5 + j7 \text{ V}$$
$$V_D = \sqrt{5^2 + 7^2} = 8.6 \text{ V}$$

Keep in mind that the j-operator is merely a label and does not figure in the solution of the equation. It identifies the axis of the number it accompanies. We find the voltage by the Pythagorean theorem, since the voltage vector is the hypotenuse of the triangle formed by the +5-V side on the real-number axis and the +7-V side on the imaginary-number axis. Similarly, phasor $\overrightarrow{OE}$ is specified as

$$V_E = 4 - j6.5$$
$$V_E = \sqrt{4^2 + 6.5^2} = 7.6 \text{ V}$$

The phasors $\overrightarrow{OD}$ and $\overrightarrow{OE}$ in Figure 13-3 could just as well have represented complex currents, in which case they could have been specified as

$$I_D = 5 + j7 \text{ A}$$

and

$$I_E = 4 - j6.5 \text{ A}$$

In Figure 13-4(a), we shall let the horizontal axis represent resistance and the vertical axis, reactance; thus, any point on the complex plane represents an impedance composed of resistance and reactance. For example, point A which has coordinates (8, 6) represents an impedance of $R = 8 \ \Omega$ and $X = 6 \ \Omega$. The length of the line segment $\overrightarrow{OA}$, which is the hypotenuse of a triangle whose sides are $|R|$ and $|X|$, represents the magnitude of the impedance:

$$Z_A = \sqrt{R^2 + X^2} = \sqrt{8^2 + 6^2} = 10 \ \Omega$$

In complex form, Z_A is written as

$$Z_A = R_A + jX_A, \quad \text{or} \quad Z_A = 8 + j6 \ \Omega$$

The circuit which this impedance represents is shown in Figure 13-4(b).

Similarly, point B represents the impedance

$$Z_B = R_B - jX_B, \quad Z_B = 6 - j8 \ \Omega,$$
$$Z_B = \sqrt{6^2 + 8^2} = 10 \ \Omega$$

The circuit is shown in Figure 13-4(c).

Since it is obviously possible for jX to be negative as well as positive, what is the significance of the sign of the j-term? We should recall that the impedance triangle looks like

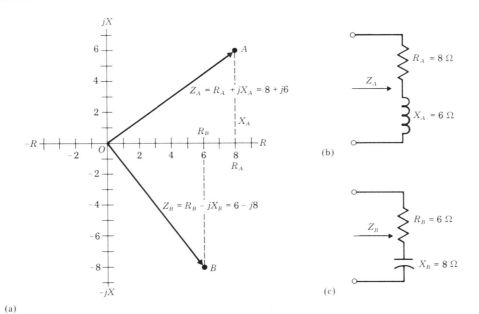

Figure 13-4

the triangle $OR_A A$ when the reactance is inductive in a series circuit, and that it looks like the triangle $OR_B B$ when the reactance is capacitive; it is logical that $+jX$ conventionally represents inductive reactance, and $-jX$ represents capacitive reactance. Thus,

$$\mathbf{Z} = R + jX_L \quad \text{or} \quad \mathbf{Z} = R - jX_C.$$

Exercises 13-4   In Exercises 1–10, draw a diagram on the complex plane representing the complex quantities. Determine the magnitude of the quantity.

1. $3 + j5$ V   2. $7 + j11$ A   3. $1 - j6$ A
4. $15 + j3 \ \Omega$. Draw the circuit diagram which this quantity represents.
5. $110 + j30$ V   6. $15 + j20$ mA
7. $2.5 + j3.8 \ \mathrm{k\Omega}$. Draw the circuit.
8. $85 - j5$ kV
9. $450 - j1150 \ \Omega$. Draw the circuit.
10. $750 - j150 \ \mu$A

13-5 ADDITION AND SUBTRACTION

In adding or subtracting complex numbers, we combine reals with reals, and imaginaries with imaginaries. We use the same rules of signs that we used for pure real numbers. Thus,

$$(3 + j2) + (4 - j3) = (3 + 4) + j(2 - 3)$$

$$= 7 - j$$

$$(-u + j2v) - (-2u + jv) = (-u + 2u) + j(2v - v)$$

$$= u + jv$$

If a circuit has voltage drops of $15 + j12$, $32 - j25$, $16 - j50$, and $10 + j2$ V, the sum of the voltage drops is obtained by adding:

$$\begin{array}{r} 15 + j12 \\ 32 - j25 \\ 16 - j50 \\ 10 + j2 \\ \hline 73 - j61 \text{ V} \end{array}$$

If the impedances in a series circuit are $4800 + j1000$, $1800 - j7200$, and $3300 + j10,000 \ \Omega$, the total impedance is the sum

$$\begin{array}{r} 4800 + j1000 \\ 1800 - j7200 \\ 3300 + j10000 \\ \hline 9900 + j3800 \ \Omega \end{array}$$

If we are to add a real number to a complex number, we assume that the real number is complex, with the imaginary part equal to zero. That is,

$$(3) + (2 + j5) = (3 + j0) + (2 + j5) = 5 + j5$$

Likewise,

$$j6 + (3 - j2) = (0 + j6) + (3 - j2) = 3 + j4$$

Graphically, the addition of complex quantities creates a new vector whose real component is the sum of the real (horizontal) components of the vectors added and whose imaginary component is the sum of the imaginary (vertical) components. Figure 13-5 shows the graphic representation of the vector sum $(5 + j2) + (1 + j4) = 6 + j6$. Similarly, Figure 13-6 is a graph of the vector sum $(5 + j2) + (2 - j6) = 7 - j4$.

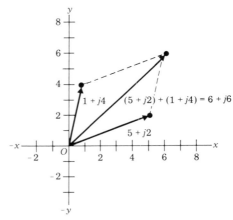

Figure 13-5

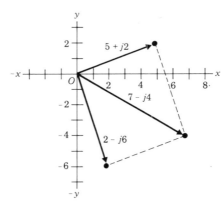

Figure 13-6

Example 4 An ac circuit has the following voltage drops: $3 + j4$ V, $-j7$ V, $6 + j2$ V, and $8 - j2$ V. Find V_{total}.

$$\begin{array}{r} 3 + j4 \\ - j7 \\ 6 + j2 \\ 8 - j2 \\ \hline 17 - j3 \text{ V} \end{array}$$

Example 5 A series circuit contains the following impedances: $200\ \Omega$, $300 + j500\ \Omega$, $300 - j200\ \Omega$, $j700\ \Omega$, and $100 + j100\ \Omega$. Find Z_T. ($Z_T = Z_1 + Z_2 + Z_3 + Z_4 + Z_5$.)

$$\begin{array}{r} 200 \\ 300 + j500 \\ 300 - j200 \\ + j700 \\ 100 + j100 \\ \hline 900 + j1100 \ \Omega \end{array}$$

Exercises 13-5 In Exercises 1–4 add the numbers.

1. $3 + j5$ and $2 + j2$ 2. $4 + j6$ and $2 - j5$
3. $12 - j5$ and $3 - j7$ 4. $6 + j3$ and $7 - j9$

In Exercises 5–8 subtract the numbers.

5. $2 + j7$ from $5 + j8$ 6. $3 - j5$ from $8 - j9$
7. $5 - j8$ from $2 + j3$ 8. $8 + j2$ from $5 - j7$

In Exercises 9–12 simplify by collecting like terms.

9. $3 + 4 - j5 + 6 + j7$
10. $j6 - 3 + j5 - j9 + 5$
11. $-5 + j7 - j9 + 3 - j6 + 11$
12. $R_1 + jX_1 + R_2 - jX_2$

In Exercises 13–16 the given values represent voltage drops in a series circuit. Find V_T if $V_T = V_1 + V_2 + V_3 + \cdots$.

13. $12 + j3$ V, 15 V, $3 - j23$ V, $+ j2$ V
14. $250 - j150$ mV, $75 + j15$ mV, $-j35$ mV, 150 mV
15. $5\ \mu V$, $4 - j7\ \mu V$, $-j11\ \mu V$, $2 - j3\ \mu V$
16. $12 - j6$ kV, $4 + j8$ kV, $5 - j9$ kV, $6 + j2$ kV, 15 kV, $-j12$ kV

In Exercises 17 and 18 the given values represent the branch currents in a parallel circuit. Determine I_T.

17. $3 - j4$ A, $5 + j3$ A, $2 + j8$ A

18. $16 + j24$ mA, $2 - j3$ mA, 20 mA, $2 - j5$ mA, $-j2$ mA, $3 + j5$ mA

In Exercises 19–22 the given values are for the impedances of a series circuit. Find Z_T.

19. $3 + j5\ \Omega$, $6 - j2\ \Omega$
20. $5 + j2$ kΩ, $3 - j8$ kΩ, $2 - j2$ kΩ
21. $15 - j2$ kΩ, $8 + j15$ kΩ, $j8$ kΩ, $9 - j3$ kΩ, 10 kΩ
22. $1.5 - j0.7$ MΩ, $2.5 + j1.5$ MΩ, 1.8 MΩ, $0.9 + j2.5$ MΩ

In ac circuits the analogy for the conductance of dc circuits is *admittance*, represented by Y.

$$Y = G + jB \text{ S (siemens)}$$

where

G = conductance in siemens

and

B = susceptance in siemens

In parallel circuits admittances add like impedances in series:

$$Y_T = Y_1 + Y_2 + Y_3 + \cdots$$

In Exercises 23 and 24 the given values are for the admittances of a parallel circuit. Find Y_T.

23. $3 + j5$ mS, $2 - j7$ mS, $3 - j2$ mS
24. $8 - j7\ \mu S$, $7 + j3\ \mu S$, $-j5\ \mu S$, $4 + j2\ \mu S$, $3\ \mu S$, $3 - j2\ \mu S$

13-6 POLAR FORM OF COMPLEX NUMBERS

In Figure 13-7 the point P represents the complex number $x + jy$. This notation is called the *rectangular form* of the vector $\overrightarrow{OP}$, since we plot the point P in the conventional manner on a rectangular coordinate system.

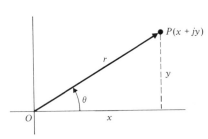

Figure 13-7

We can describe the vector $\overrightarrow{OP}$ by another means, however. When we know the length of $\overrightarrow{OP}$ and the angle between $\overrightarrow{OP}$ and the x-axis, we can represent vector $\overrightarrow{OP}$ by $r\underline{/\theta}$; this expression is the *polar form* of the vector $\overrightarrow{OP}$. The letter r is the *absolute value* of the number (the length of the vector), and the angle θ is the *argument* of the complex number. It is important to note that $r\underline{/\theta}$ is read "*r* at an angle θ"; it does *not* mean "r times θ." In this sense, the polar form is not a true mathematical expression; it is shorthand notation.

Since $r\underline{/\theta}$ specifies the same thing as $x + jy$, there must be some relationship between the quantities. From the Pythagorean theorem,

$$r^2 = x^2 + y^2$$

From trigonometric considerations,

$$\tan \theta = \frac{y}{x}, \qquad x = r \cos \theta, \qquad y = r \sin \theta$$

Therefore,

$$x + jy = r(\cos \theta + j \sin \theta)$$

This combination of forms is called the *trigonometric form* of the vector.

Example 6 Represent $4 + j3$ in polar form. Thus,

$$r = \sqrt{16 + 9} = 5$$
$$\tan \theta = \tfrac{3}{4}, \qquad \theta = 36.87°$$
$$4 + j3 = 5\underline{/36.87°}$$

Example 7 Represent $3 - j2$ in trigonometric form.

$$r = \sqrt{9 + 4} = \sqrt{13} = 3.606, \qquad \tan \theta = -\tfrac{2}{3}$$

Since $3 - 2j$ is in the fourth quadrant and the reference angle is $33.69°$, θ must be $360° - 33.69° = 326.31°$. The trigonometric form is

$$3.606\,(\cos 326.31° + j \sin 326.31°)$$

Applying the transformation relationships of this section to electrical quantities yields the following expressions:

$$
\begin{aligned}
\mathbf{V} &= V\underline{/\theta} = V_x + jV_y = V \cos \theta + jV \sin \theta \\
\mathbf{I} &= I\underline{/\theta} = I_x + jI_y = I \cos \theta + jI \sin \theta \\
\mathbf{Z} &= Z\underline{/\theta} = R + jX = Z \cos \theta + jZ \sin \theta \\
\mathbf{Y} &= Y\underline{/\theta} = G + jB = Y \cos \theta + jY \sin \theta
\end{aligned}
$$

Calculator Conversions Many types of calculators incorporate key fucntions which enable the user to make the conversions between the polar and rectangular forms of complex numbers with a minimum of keystrokes. If you have a calculator with key functions labeled with $\rightarrow P$, $\rightarrow R$, or something similar, your calculator includes this conversion feature. Study the instruction manual carefully, and practice the illustrated procedure until you make the conversions—polar to rectangular and rectangular to polar—quickly and easily.

If your calculator does not have the special direct conversion feature, but does incorporate the trigonometric functions, you can learn to make conversions with only slightly more effort. The procedure involves the application of the definitions of the forms.

Example 8 Convert $3 + j4$ to its polar form.

We use the definitions:

$$x + jy = r\underline{/\theta}$$

$$\theta = \arctan \frac{y}{x}$$

$$r = \frac{y}{\sin\left(\arctan \dfrac{y}{x}\right)}$$

Step	Action	Readout
1	enter y: 4	$(y=)$ 4
2	touch $M+$ (or $X{\rightarrow}M$) (store y)	$(y=)$ 4
3	touch $\div$	$(y=)$ 4
4	enter x:3	$(x=)$ 3
5	touch $=$	$\left(\dfrac{y}{x} =\right)$ 1.3333333

6	touch *arctan*	$\left(\arctan \dfrac{y}{x} =\right)$ 53.13009
7	write down value of θ :	$\underline{\hspace{3cm}} \underline{/53.13^\circ}$
8	touch *sin*	$\left[\sin \left(\arctan \dfrac{y}{x}\right) = \right] 0.8$
9	touch $1/x$	$\left[1/\sin \left(\arctan \dfrac{y}{x}\right) = \right] 1.25$
10	touch $\times$	$\left[\times \; 1/\sin \left(\arctan \dfrac{y}{x}\right) = \right] 1.25$
11	touch *MR* (or enter y)	$(y=)\; 4$
12	touch =	$\left[r = y/\sin \left(\arctan \dfrac{y}{x}\right) = \right] 5$
13	complete writing polar form: $\underline{5/53.13^\circ}$	

Study and practice the above procedure until you both understand it and can perform it with ease. Practice the following conversions using the same steps:

$$4 + j3 = 5 \underline{/36.87^\circ}$$
$$5 + j8 = 9.434 \underline{/57.99^\circ}$$
$$15 + j7 = 16.55 \underline{/25.02^\circ}$$
$$4 - j3 = 5 \underline{/-36.87^\circ}$$
$$8 - j9 = 12.04 \underline{/-48.37^\circ}$$
$$6.25 - j2.75 = 6.828 \underline{/-23.75^\circ}$$

Example 9 Convert $5 \underline{/53.13^\circ}$ to rectangular form.

We use the definitions: $x = r \cos\theta$, $y = r \sin\theta$

$$x + jy = r \underline{/\theta}$$

Step	Action	Readout
1	enter θ: 53.13°	53.13
2	touch $M+$ (store θ)	53.13
3	touch *cos* ($\cos\theta$ =)	0.60
4	touch $\times$ ($\cdot \cos\theta$)	0.60
5	enter r: 5	5.0
6	touch = ($r \cdot \cos\theta$ =)	3.0
7	write: $3.0 + j$ _____	
8	touch *MR* (recall θ)	53.13
9	touch *sin* ($\sin\theta$ =)	0.799999
10	touch $\times$ ($\cdot \sin\theta$)	0.799999
11	enter r: 5	5.
12	touch = ($r \cdot \sin\theta$ =)	3.999995
13	complete writing the rectangular form: $3.0 + j4.0$	

Practice the following conversions using the preceding steps.

$$6 \underline{/30^\circ} = 5.196 + j3.0$$
$$8 \underline{/75^\circ} = 2.071 + j7.727$$
$$7 \underline{/15^\circ} = 6.761 + j1.812$$
$$9 \underline{/-35^\circ} = 7.372 - j5.162$$
$$10 \underline{/-78^\circ} = 2.079 - j9.781$$
$$12 \underline{/-12^\circ} = 11.74 - j2.495$$

Example 10 An ac generator produces the voltage $120 \underline{/30^\circ}$ V. Express the voltage in trigonometric and rectangular form.

$$120 \underline{/30^\circ} = 120 \cos 30^\circ + j120 \sin 30^\circ$$
$$= 103.9 + j60 \text{ V}$$

Example 11 An impedance consists of $R = 35\ \Omega$ and $X_L = 75\ \Omega$. Express the impedance in rectangular and polar form.

$$\mathbf{Z} = 35 + j75 = 82.76 \underline{/64.98^\circ}\ \Omega$$

Example 12 An impedance is given as $2.5 \underline{/-70^\circ}$ kΩ. Express in rectangular form and identify the R- and X-components.

$$\mathbf{Z} = 2.5 \underline{/-70^\circ}\ \text{k}\Omega = R + jX$$
$$= \underline{0.8551 - j2.349}\ \text{k}\Omega$$

Therefore,

$$R = \underline{0.8551\ \text{k}\Omega}, \qquad X_C = \underline{2.349\ \text{k}\Omega}$$

Exercises 13-6 In Exercises 1–8 convert to polar form.

1. $3 + j4$ 2. $8 + j5$ 3. $2 + j3$
4. $15 + j12$ 5. $6 - j7$ 6. $9 - j3$
7. $8 - j6$ 8. $3 - j9$

In Exercises 9–14 convert the given impedances to polar form.

9. $6 + j7\ \Omega$ 10. $15 - j6\ \Omega$
11. $3.6 - j5.2\ k\Omega$ 12. $9.1 + j5.5\ k\Omega$
13. $1.5 + j0.82\ M\Omega$ 14. $0.75 - j2.4\ M\Omega$

In Exercises 15–20 convert the given voltages
to polar form.

15. $6 - j2$ V   16. $4 + j8$ V   17. 10 mV
18. $j15$ mV   19. $2.75 - j1.55\ \mu V$
20. $12.5 + j37.5$ mV

In Exercises 21–26 convert the given currents
to polar form.

21. $1 + j2$ A   22. $3 - j$ A
23. $15 + j42$ mA   24. $45 - j9$ mA
25. $175 + j15\ \mu A$ 26. $235 - j685\ \mu A$

In Exercises 27–48 convert to rectangular
form.

27. $5\underline{/53.13°}$ 28. $5\underline{/-36.87°}$
29. $9\underline{/-15°}$ 30. $12\underline{/75°}$ 31. $15\underline{/25.7°}$ V
32. $120\underline{/-42.85°}$ 33. $35\underline{/17°}$ mV
34. $175\underline{/-23°}$ mV   35. $12\underline{/-56°}$ A
36. $0.25\underline{/5.5°}$ A   37. $25\underline{/12.5°}$ mA
38. $450\underline{/-2.75°}\ \mu A$ 39. $15\underline{/15°}\ \Omega$
40. $50\underline{/-65°}\ \Omega$ 41. $5.6\underline{/-57°}$ kΩ
42. $45\underline{/88°}$ kΩ 43. $1.5\underline{/-1.5°}$ MΩ
44. $2.25\underline{/47.5°}$ MΩ 45. $3\underline{/15°}$ mS
46. $5.6\underline{/-35°}$ mS   47. $45\underline{/-27.5°}\ \mu S$
48. $125\underline{/18.8°}\ \mu S$

13-7 MULTIPLICATION OF COMPLEX NUMBERS

Finding the product of complex numbers is
direct and easily accomplished when the
numbers are in polar form.

To find the product of two or more complex numbers:
1. Convert the numbers to polar form: $r\underline{/\theta}$
2. Find the magnitude r_p of the product by finding the product of the *r*-values of the factors:
$$r_p = r_1 \cdot r_2 \cdot r_3 \cdots$$
3. Find the angle θ_p for the product by finding the algebraic sum of the angles of the factors:
$$\theta_p = \theta_1 + \theta_2 + \theta_3 + \cdots$$
$$r_p\ \underline{/\theta_p} = r_1 \cdot r_2 \cdot r_3 \cdots\ \underline{/\theta_1 + \theta_2 + \theta_3 + \cdots}$$

Example 13 Find the indicated products.

(a) $5\underline{/45°} \cdot 4\underline{/-23°}$
(a) $5\underline{/45°} \cdot 4\underline{/-23°} = 5 \cdot 4\underline{/45° - 23°} = 20\underline{/22°}$
(b) $(3 + j5)(2 - j4)$
(b) We convert to polar form: $3 + j5 = 5.831\underline{/59.04°}$ $2 - j4 = 4.472\underline{/-63.43°}$
Then multiply: $5.831\underline{/59.04°}$
$4.472\underline{/-63.43°} = 26.08\underline{/-4.39°}$

Example 14 An ac current $I = 15\underline{/15°}$ mA is flowing in the impedance $Z = 4 + j9$ kΩ. Find V, the voltage drop across the impedance.

First, we convert the impedance to polar form:

$Z = (4 + j9) \times 10^3 = 9.849\underline{/66.04°} \times 10^3\ \Omega$

Then substitute into the Ohm's law formula:

$V = IZ = (15\underline{/15°} \times 10^{-3})(9.849\underline{/66.04°} \times 10^3)$
$= 147.7\underline{/81.04°}$ V

Finding the product of complex numbers in rectangular form is possible although very cumbersome. However, such products are sometimes desired for algebraic derivations. We apply the principles used to find the products of binomials (if necessary, review Section 3-4),

$$(a + jb)(c + jd) = a(c + jd) + jb(c + jd)$$
$$= ac + jad + jbc - bd$$

(*Note:* $jb \cdot jd = j^2 bd$, but $j^2 = -1$. Thus, jb $jd = -bd$.)

Collecting the real and the imaginary terms,

$$(a + jb)(c + jd) = (ac - bd) + j(ad + bc)$$

Example 15 Find the indicated products using the rectangular form.

(a) $(R + jX)(R - jX)$

(a)
$$
\begin{array}{r}
R + jX \\
R - jX \\
\hline
R^2 + jRX \\
\quad - jRX - j^2X^2 \\
\hline
R^2 \qquad\quad + X^2
\end{array}
$$

(b) $(4 + j3)(3 - j2)$

(b)
$$
\begin{array}{r}
4 + j3 \\
3 - j2 \\
\hline
12 + j9 \\
\quad - j8 - j^2 6 \\
\hline
12 + j + 6 = 18 + j
\end{array}
$$

Exercises 13-7 In Exercises 1–10 find the indicated products using the polar form.

1. $5\underline{/45°} \cdot 4\underline{/-23°}$
2. $6\underline{/15°} \cdot 3\underline{/12°}$
3. $2.5\underline{/-35°} \cdot 4.6\underline{/-15°}$
4. $3.76\underline{/-8°} \cdot 4\underline{/0°}$
5. $(3.8\underline{/21°})(2 + j3)$
6. $(4 + j7)(6.85\underline{/-27.5°})$
7. $(15 - j12)(8.95\underline{/48.5°})$
8. $(27.5 + j13.6)(25.7\underline{/-36.7°})$
9. $(3 + j5)(4 - j2)$
10. $(5.69 - j2.73)(4.68 + j12.95)$

In Exercises 11–16 the given values are for **I** and **Z**, respectively. Find **V** using the formula **V = IZ**.

11. $5\underline{/15°}$ A, $2.5\underline{/0°}$ Ω
12. $4.5\underline{/-55°}$ A, $5.7\underline{/42°}$ Ω
13. $2 + j7$ mA, $0.75\underline{/-25°}$ kΩ
14. $0.065\underline{/0°}$ mA, $4.8 - j3.6$ kΩ
15. $3.5\underline{/36°}$ μA, $1.5 - j0.5$ MΩ
16. $275\underline{/-16.5°}$ μA, $0.91 + j1.2$ MΩ

In Exercises 17–20 the given values are for **V** and **Y**, respectively. Find **I** using the formula **I = VY**.

17. $5\underline{/15°}$ V, $0.2\underline{/-12°}$ S
18. $120\underline{/0°}$ V, $0.65\underline{/18°}$ S
19. $2.5\underline{/-58°}$ V, $35\underline{/0°}$ mS
20. $15\underline{/30°}$ V, $275\underline{/-15°}$ μS

In Exercises 21–32 find the indicated products using multiplication in rectangular form.

21. $(a + jb)(c + jd)$
22. $(x + jy)(r - js)$
23. $(R + jX)(R - jX)$
24. $(R + jX)(R + jX)$
25. $(R_1 + jX_1)(R_2 + jX_2)$
26. $(I + ji)(I - ji)$
27. $(3 + j4)(2 - j5)$
28. $(4 + j7)(5 - j7)$
29. $(4 + j7)(4 + j7)$
30. $5(4 - j5)$
31. $j3(5 + j2)$
32. $j6(2 - j5)$

13-8 DIVISION OF COMPLEX NUMBERS

To find the quotient of one complex number divided by a second complex number:
1. Convert the numbers to polar form: $r\underline{/\theta}$.
2. Find r_q of the quotient by dividing the r-value of the first number by the r-value of the second number:

$$r_q = \frac{r_1}{r_2}$$

3. Find the angle of the quotient θ_q by subtracting algebraically the angle θ_2 of the second number (the divisor), from the angle θ_1 of the first number (the dividend):

$$\theta_q = \theta_1 - \theta_2, \qquad r_q \underline{/\theta_q} = \frac{r_1}{r_2} \underline{/\theta_1 - \theta_2}$$

Example 16 Perform the indicated division.

(a) $5\underline{/15°}/3\underline{/-20°}$

(a) $5\underline{/15°}/3\underline{/-20°} = \tfrac{5}{3}\underline{/15° - (-20°)}$

$\qquad\qquad\qquad = 1.667\underline{/35°}$

(b) $(6 + j7)/2.75\underline{/65°}$

(b) We first convert $6 + j7$ to polar form,

$$6 + j7 = 9.22\underline{/49.4°}$$

Then divide

$$9.22\underline{/49.4°}/2.75\underline{/65°}$$
$$= (9.22/2.75)\underline{/49.4° - 65°}$$
$$= 3.353\underline{/-15.6°}$$

Example 17 The voltage V across an impedance Z is $V = 25\underline{/48°}$ V when the current in Z is $4.65\underline{/14.8°}$ mA. Find **Z**.

Use the Ohm's law formula **Z = V/I**.

$$\mathbf{Z = V/I} = 25\underline{/48°}/4.65\underline{/14.8°} \times 10^{-3}$$
$$= 5.376\underline{/33.2°} \text{ k}\Omega$$

Division of complex numbers may also be accomplished when they are in rectangular form. The procedure is much more cumbersome than when the polar form is used. It requires the use of a form called the *complex conjugate*. Division in rectangular form is utilized in certain algebraic derivations.

Complex Conjugates The word *conjugate* is from a Latin root which means "joined together." In mathematics *conjugate* refers to expressions which are joined together because they are almost, but not quite, identical.

The most common mathematical conjugates are binomial expressions, expressions containing two terms. If two binomials have identical terms but differ only in the signs separating the terms, they are *conjugates*.

Since complex numbers are binomials, it is possible for complex numbers to be conjugates or, more specifically, *complex conjugates*. Thus $a + jb$ and $a - jb$ are complex conjugates; each is the conjugate of the other. Similarly, $4 - j5$ and $4 + j5$ are complex conjugates.

Conjugates and their properties have several useful functions in the manipulations of complex numbers. We shall see that conjugate electrical quantities produce important and unique results in electrical circuits. Hence, we need to study and understand complex conjugates.

The sum of complex conjugates is a real number. Thus, when we add

$$\begin{array}{r} x + jy \\ x - jy \\ \hline 2x + j0 \end{array}$$

the sum is $2x + j0$. The imaginary parts cancel.

The product of conjugates is a real number. For example, multiply

$$(a + jb)(a - jb)$$

Recall that $(u + v)(u - v) = u^2 - v^2$. Then

$$(a + jb)(a - jb) = a^2 - (jb)^2$$
$$= a^2 - (-1)b^2 = a^2 + b^2$$

Similarly,

$$(4 + 3j)(4 - 3j) = 16 + 9 = 25$$

Division by a complex number in rectangular form is performed by multiplying the numerator and denominator by the conjugate of the denominator, then simplifying the expression which is the result.

Example 18 Divide $5 + j4$ by $1 - j3$.

$$\frac{5 + j4}{1 - j3} = \frac{(5 + j4)(1 + j3)}{(1 - j3)(1 + j3)} = \frac{5 + j4 + j15 - 12}{1 + 9}$$
$$= \frac{-7 + j19}{10} = -\frac{7}{10} + j\frac{19}{10}$$

Example 19 Divide: (a) $5/j2$; (b) $(3 + j2)/j4$

To evaluate division by an imaginary number we multiply the numerator and denominator by j,

(a) $\dfrac{5}{j2} = \dfrac{5 \cdot j}{j2 \cdot j} = \dfrac{5j}{-2} = -j2.5$

(b) $\dfrac{3 + j2}{j4} = \dfrac{(3 + j2)(j)}{j4 \cdot j} = \dfrac{3j - 2}{-4} = -j\dfrac{3}{4} + \dfrac{2}{4}$
$$= \dfrac{1}{2} - j\dfrac{3}{4}$$

Example 20 Divide: $(R + jX)/(R - jX)$

$$\frac{R + jX}{R - jX} = \frac{(R + jX)(R + jX)}{(R - jX)(R + jX)} = \frac{R^2 - X^2 + j2RX}{R^2 + X^2}$$

Exercises 13-8 In Exercises 1–8 perform the indicated division using the polar form.

1. $5\underline{/15°}/3\underline{/-20°}$

2. $3.5\underline{/20°}/2.5\underline{/-12°}$

3. $4.8\underline{/-45°}/2.6\underline{/-65°}$

4. $6.75\underline{/-15°}/8.35\underline{/-75°}$

5. $2.98\underline{/30.5°}/(6 + j5)$

6. $36.85\underline{/17.5°}/(27 - j33)$

7. $(48 - j3)/63.65\underline{/26.83°}$

8. $(0.65 - j0.57)/0.86\underline{/5.97°}$

In Exercises 9–12 the given values are for **V** and **I**, respectively. Find **Z** using the Ohm's law formula **Z** = **V**/**I**.

9. $8\underline{/30°}$ V, $2.5\underline{/20°}$ A

10. $12.5\underline{/-15°}$ V, $3.5\underline{/-50°}$ A

11. $12 + j125$ mV, $38\underline{/56°}$ μA

12. $368\underline{/-15°}$ mV, $65 - j158$ mA

In Exercises 13–16 the given values are for **V** and **Z**, respectively. Find **I** using the formula **I** = **V**/**Z**.

13. $8\underline{/30°}$ V, $3\underline{/5°}$ Ω

14. $12.5\underline{/-15°}$ V, $4.7\underline{/-48°}$ Ω

15. $125\underline{/85°}$ mV, $470 + j0$ Ω

16. $-j368$ mV, $27\underline{/-28°}$ kΩ

In Exercises 17–30 perform the indicated divisions using the rectangular form.

17. $\dfrac{a + jb}{c + jd}$ 18. $\dfrac{a + jb}{c - jd}$ 19. $\dfrac{R + jX}{R - jX}$

20. $\dfrac{R - jX}{R + jX}$ 21. $\dfrac{R + jX}{jX}$ 22. $\dfrac{R - jX}{jX_1}$

23. $\dfrac{5 - j2}{j}$ 24. $\dfrac{16 - j20}{j4}$ 25. $\dfrac{4 + j8}{1 + j3}$

26. $\dfrac{5 + j4}{3 - j4}$ 27. $\dfrac{3 - j}{2 - j3}$ 28. $\dfrac{6 - j4}{5 + j}$

29. $\dfrac{9 + j8}{3 - j7}$ 30. $\dfrac{12 - j2}{4 - j8}$

Chapter 14

Methods of Analysis for ac Circuits

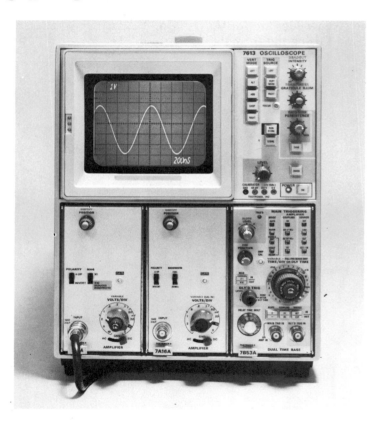

An oscilloscope plots ac phenomena instantaneously, enabling the user to obtain information on amplitude, frequency, and phase angle. (Courtesy of Tektronix, Inc.)

The basic format of the techniques used in dc-circuit analysis will not have to be modified for ac circuits. It is necessary simply to broaden the underlying principles and laws, such as Ohm's and Kirchhoff's laws, to encompass complex electrical quantities, and then apply the techniques suggested by these principles to the complex circuits. Hence, in this chapter we shall, for the most part, demonstrate and practice the use of complex quantities in the formulas and techniques with which we are already familiar.

14-1 BASIC PRINCIPLES

Let us proceed with ac-circuit analysis by first carefully restating the basic principles which apply to all electrical circuits. The forms of Ohm's law for ac circuits are:

INSTANTANEOUS VALUES

$$(1) \ i = \frac{v}{Z}, \quad (2) \ v = iZ, \quad (3) \ Z = \frac{v}{i}$$

PHASOR VALUES

$$(1) \ I = \frac{V}{Z}, \quad (2) \ V = IZ, \quad (3) \ Z = \frac{V}{I}$$

It should be noted that the quantities I, V, and Z are *complex* quantities.

Kirchhoff's laws must also be restated.

VOLTAGE LAW

INSTANTANEOUS VALUES. At any instant, the algebraic sum of the instantaneous values of the applied emf's and the potential drops (iz) around a closed circuit is zero.

PHASOR VALUES. The phasor sum of the applied phasor emf's and the phasor potential drops around a closed circuit is zero.

CURRENT LAW

INSTANTANEOUS VALUES. At any instant, the sum of the instantaneous values of all currents flowing toward a point is equal to the sum of the instantaneous values of all currents flowing away from the point.

PHASOR VALUES. The phasor sum of all phasor currents flowing toward a point is equal to the sum of all phasor currents flowing away from the point.

In this chapter, we shall deal primarily with phasor values.

Example 1 A series ac circuit includes the following voltage drops: $V_1 = 1.9 + j7.7$ V, $V_2 = 3\underline{/-55°}$ V, and $V_3 = 4\underline{/-15°}$ V. (a) Construct the phasor diagram. (b) Find V_T using $V_T = V_1 + V_2 + V_3$. Express result in polar and rectangular form.

(a) To assist in constructing the phasor diagram we convert V_1 to polar form:

$$1.9 + j7.7 = 7.931\underline{/76.14°} \text{ V}$$

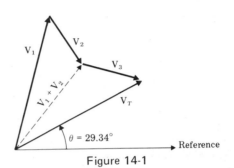

Figure 14-1

(b) We convert V_2 and V_3 to rectangular form so that we may add:

V_1: $7.931\underline{/76.14°}$ V $= 1.9 \quad + j7.7$ V
V_2: $\quad 3\underline{/-55°}$ V $= 1.721 - j2.457$
V_3: $\quad 4\underline{/-15°}$ V $= 3.864 - j1.035$
$V_T = V_1 + V_2 + V_3 = \underline{7.485 + j4.208}$
$\qquad\qquad = \underline{8.587 \ \underline{/29.34°} \text{ V}}$

Example 2 Branches 1, 2, and 3 of an ac circuit meet at node a. $I_1 = 4.5\underline{/15°}$ mA, $I_2 = 3.6\underline{/-23°}$ mA, and $I_3 = I_1 + I_2$. (a) Find I_3. Express in polar and rectangular form. (b) Construct the phasor diagram.

(a) We convert the currents to rectangular form so that we may add them,

$I_1 = 4.5\underline{/15°}$ mA $= 4.347 + j1.165$ mA
$I_2 = 3.6\underline{/-23°}$ mA $= 3.314 - j1.407$ mA
$I_3 = I_1 + I_2 \qquad = \underline{7.661 - j0.242 \text{ mA}}$
$\qquad\qquad = \underline{7.665 \ \underline{/-1.81°} \text{ mA}}$

(b)

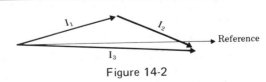

Figure 14-2

Exercises 14-1 In Exercises 1–10 the given values represent voltage drops in a series ac circuit. (a) Find V_T, the phasor sum of the given voltages. Express the result in both polar and rectangular form. (b) Construct the phasor diagram.

1. $3\underline{/15°}$ V and $4\underline{/45°}$ V
2. $4\underline{/55°}$ V and $2\underline{/10°}$ V
3. $3.5\underline{/62°}$ V and $2.5\underline{/-25°}$ V
4. $2.8\underline{/30°}$ V and $3.4\underline{/-65°}$ V
5. $4.2\underline{/-35°}$ V and $4.8\underline{/68°}$ V
6. $1.8\underline{/-5°}$ V and $2.6\underline{/-40°}$ V
7. $3\underline{/75°}$ V, $2-j3$ V, and $2\underline{/-20°}$ V
8. $4\underline{/-72°}$ V, $2\underline{/36°}$ V, and $1+j2$ V
9. $2.4\underline{/-5°}$ V, $1.6\underline{/-12°}$ V, and $0.2+j4$ V
10. $4.8\underline{/5°}$ V, $6\underline{/15°}$ V, and $2-j6$ V

In Exercises 11–20 the given values represent currents flowing in the branches of an ac circuit toward a common node. (a) Find I_T, the phasor sum of the currents. Express the result in both polar and rectangular form. (b) Draw the phasor diagram of the currents.

11. $24\underline{/25°}$ mA and $36\underline{/-10°}$ mA
12. $16\underline{/85°}$ mA and $8\underline{/-28°}$ mA
13. $340\underline{/-5°}$ µA and $620\underline{/-26°}$ µA
14. $220\underline{/-12°}$ µA and $180\underline{/-27°}$ µA
15. $0.86\underline{/-35°}$ mA and $0.68\underline{/42°}$ mA
16. $0.048\underline{/-56°}$ mA and $0.034\underline{/37°}$ mA
17. $2.4\underline{/65°}$ mA, $3.8\underline{/-12°}$ mA, and $1.8\underline{/-24°}$ mA
18. $4.6\underline{/0°}$ mA, $2.4+j3.2$ mA, and $3.8\underline{/-54°}$ mA
19. $36\underline{/90°}$ mA, $28\underline{/-38°}$ mA, and $30-j20$ mA
20. $280\underline{/-90°}$ µA, $360\underline{/0°}$ µA, and $120+j290$ µA

14-2 IMPEDANCE

The opposition to current flow is called *resistance* (R) in dc circuits. It is called *impedance* (Z) in ac circuits. Impedance is a complex or phasor quantity. It can be specified completely only with a magnitude and a "direction"—a phase angle. The directional quality arises because of *reactance*. *Reactance is the opposition of inductance or capacitance to alternating current.* Impedance is specified as follows:

$$Z\underline{/\theta} \ \Omega \quad \text{or} \quad R+j(X_L-X_C)$$

Inductive Reactance The opposition of inductance L to current flow in ac circuits is called *inductive reactance* X_L. The opposition is directly proportional to the quantity of inductance L and the frequency f of the current,

$$X_L = 2\pi f L \ \Omega$$

f = Frequency of current in hertz, Hz

L = Inductance in henries, H

The voltage drop across an inductive reactance leads the current by 90° (I lags V by 90°). Thus, in complex notation,

$$X_L = X_L\underline{/90°} = jX_L \ \Omega$$

Capacitive Reactance The opposition of capacitance to current flow is called *capacitive reactance* X_C. (Although a capacitor is in fact an open circuit and cannot pass current, it charges and discharges as the ac current alternates and thus appears to pass current.) The opposition to flow is inversely proportional to the capacitance C and the frequency f of the current,

$$X_C = \frac{1}{2\pi f C} \ \Omega$$

f = Frequency of current, Hz

C = Capacitance in farads, F

The voltage across the capacitor lags the current by 90° (I leads V by 90°). Thus, in complex notation,

$$X_C = X_C\underline{/-90°} = -jX_C \ \Omega$$

Example 3 A 5-kΩ resistor is connected in series with a pure inductance of $L = 2$ H. The circuit will be used as an ac circuit in which $f = 240$ Hz (see Figure 14-3). (a) Determine X_L and Z. Specify Z in polar and rectangular form. (b) Construct the impedance triangle.

(a) $X_L = 2\pi f L = (2\pi)(240)(2) = \underline{3016}$ Ω

$Z = R + jX_L$

$= 5 + j3.016$ kΩ $= 5.839\underline{/31.1°}$ kΩ

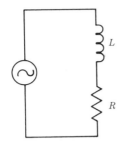

Figure 14-3

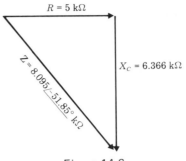

Figure 14-6

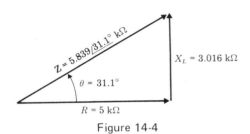

Figure 14-4

Example 4 A 5-kΩ resistor is connected in series with a capacitor, $C = 0.01 \ \mu$F. The circuit will be used with a generator for which $f = 2.5$ kHz. (a) Determine X_C and **Z**. Express **Z** in both polar and rectangular form. (b) Sketch the impedance triangle.

(a) $X_C = \dfrac{1}{2\pi fC} = \dfrac{0.15915}{fC}$

$ = \dfrac{0.15915}{(2.5)(10^3)(0.01)(10^{-6})}$

$ = 6.366 \text{ k}\Omega$

$\mathbf{Z} = R - jX_C = 5 - j6.366 \text{ k}\Omega$

$\phantom{\mathbf{Z}} = \underline{8.095/\!\!-51.85°\text{ k}\Omega}$

(b) See Figure 14-6.

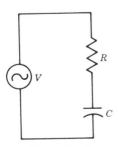

Figure 14-5

Example 5 Derive the following equations: (a) L in terms of f and X_L; (b) f in terms of L and X_L; (c) C in terms of f and X_C; and (d) f in terms of C and X_C.

(a) $X_L = 2\pi fL, \quad \dfrac{X_L}{2\pi f} = \dfrac{2\pi fL}{2\pi f}, \quad L = \dfrac{X_L}{2\pi f}$

(b) $X_L = 2\pi fL, \quad \dfrac{X_L}{2\pi L} = \dfrac{2\pi fL}{2\pi L}, \quad \underline{f = \dfrac{X_L}{2\pi L}}$

(c) $X_C = \dfrac{1}{2\pi fC}, \quad \dfrac{X_C C}{X_C} = \dfrac{C}{2\pi fCX_C},$

$\qquad C = \dfrac{1}{2\pi fX_C}$

(d) $X_C = \dfrac{1}{2\pi fC}, \quad \dfrac{X_C f}{X_C} = \dfrac{f}{2\pi fCX_C},$

$\qquad \underline{f = \dfrac{1}{2\pi CX_C}}$

Exercises 14-2 In Exercises 1–10 the given values represent the resistance, inductance, and frequency, respectively, for an ac circuit. (a) Determine **Z**. Give result in both rectangular and polar form. (b) Sketch the impedance triangle.

1. 5 kΩ, 2 H, 240 Hz
2. 5 kΩ, 1 H, 1 kHz
3. 200 Ω, 3 mH, 10 kHz
4. 400 Ω, 6 mH, 15 kHz
5. 8 kΩ, 250 μH, 10 MHz
6. 10 kΩ, 25 μH, 80 MHz
7. 4 kΩ, 35 μH, 100 MHz
8. 1.2 kΩ, 50 μH, 10 MHz
9. 150 Ω, 1.5 H, 60 Hz
10. 120 Ω, 10 H, 120 Hz

In Exercises 11–20 the given values represent the resistance, capacitance, and frequency, respectively, for an ac circuit. (a) Determine Z. Give result in both rectangular and polar form. (b) Sketch the impedance triangle.

11. 50 Ω, 10 μF, 60 Hz
12. 100 Ω, 8 μF, 120 Hz
13. 2 kΩ, 0.1 μF, 1 kHz
14. 2 kΩ, 0.005 μF, 10 kHz
15. 1.5 kΩ, 200 pF, 1 MHz
16. 3.3 kΩ, 50 pF, 850 kHz
17. 48 kΩ, 0.005 μF, 0.8 kHz
18. 150 Ω, 0.002 μF, 200 kHz
19. 82 Ω, 0.001 μF, 1.2 MHz
20. 560 Ω, 0.005 μF, 300 kHz
21. Derive an equation for f: (a) in terms of X_L and L, and (b) in terms of X_C and C.
22. Derive an equation: (a) for L in terms of f and X_L, and (b) for C in terms of f and X_C.

In Exercises 23–26 the given values are for Z and f. Determine R, X_L, and L.

23. $5\underline{/30°}$ kΩ, 1 kHz
24. $8\underline{/75°}$ kΩ, 15 kHz
25. $1.5\underline{/15°}$ kΩ, 600 kHz
26. $1.8\underline{/50°}$ kΩ, 5 MHz

In Exercises 27–30 the given values are for Z and f. Determine R, X_C, and C.

27. $6\underline{/-45°}$ kΩ, 1 kHz
28. $3\underline{/-78°}$ kΩ, 12 kHz
29. $1.2\underline{/-20°}$ kΩ, 750 kHz
30. $0.82\underline{/-60°}$ kΩ, 10 MHz

14-3 SERIES CIRCUITS

When we apply Ohm's law and Kirchhoff's voltage law to a series circuit containing several impedances we can state the following relationships:

$$Z_T = \frac{V_T}{I}$$

$$V_T = V_1 + V_2 + V_3 + \cdots$$

$$V_1 = IZ_1, \quad V_2 = IZ_2, \quad V_3 = IZ_3, \text{ and so on}$$

Thus,

$$V_T = IZ_T = IZ_1 + IZ_2 + IZ_3 + \cdots$$

Then

$$Z_T = Z_1 + Z_2 + Z_3 + \cdots$$

Since each Z is a complex quantity of the form $Z = R + jX$, we can state that

$$Z_T = R_T + jX_T$$

and

$$R_T = R_1 + R_2 + R_3 + \cdots,$$

$$X_T = X_1 + X_2 + X_3 + \cdots$$

or, in general,

$$R_T = \Sigma R_n, \qquad X_T = \Sigma X_n$$

That is, the total impedance of a series circuit is a complex quantity composed of R_T, the sum of all resistances in the circuit, and X_T, the *algebraic sum* of all reactances. In the other complex form, we can express the total impedances as

$$Z_T = Z_T\underline{/\arctan(X_T/R_T)}$$

Also, letting $\theta = \arctan(X_T/R_T)$, we see that

$$Z_T = Z_T \cos\theta + jZ_T \sin\theta$$

Thus

$$R_T = Z_T \cos\theta \quad \text{and} \quad X_T = Z_T \sin\theta$$

Example 6 In the circuit of Figure 14-7 $R = 65$ Ω, $X_L = 95$ Ω, and $V_T = 120$ V$\underline{/0°}$ V (effective). Determine (a) the impedance Z, (b) the current in the circuit, (c) the voltages across R and X_L, and (d) P, the power dissipated in the circuit. (e) Construct the phasor diagram for V_R, V_L, V_T, and I.

(a) $Z = R + jX_L = 65 + j95 = \underline{115.1\underline{/55.62°}}$ Ω

(b) $I = \dfrac{V_T}{Z} = \dfrac{120\underline{/0°}}{115.1\underline{/55.62°}} = \underline{1.043\underline{/-55.62°}}$ A

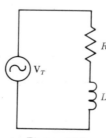

Figure 14-7

(c) $\mathbf{V}_R = \mathbf{IR} = (1.043\underline{/-55.62°})(65)$
$= 67.8\underline{/-55.62°}$ V
$\mathbf{V}_L = \mathbf{IX}_L = (1.043\underline{/-55.62°})(95\underline{/90°})$
$= 99.09\underline{/34.38°}$ V

(d) The power in an ac circuit is found using the formula $P = V_T I \cos\theta$, where V_T is the total voltage across the circuit (the *effective value*), I is the total *effective* current, and θ is the phase angle between V_T and I, $\theta = \theta_I - \theta_{V_T}$.

$P = V_T I \cos\theta$
$= (120)(1.043)(\cos - 55.62°)$
$= \underline{70.68 \text{ W}}$

(e)

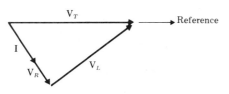

Figure 14-8

Example 7 In the circuit of Figure 14-9 $R = 680\ \Omega$, $X_C = 300\ \Omega$, and $\mathbf{V}_T = 12\underline{/0°}$ V. Determine (a) $\mathbf{Z}$, (b) $\mathbf{I}$, (c) $\mathbf{V}_R$ and $\mathbf{V}_C$, and (d) P. (e) Construct the phasor diagram.

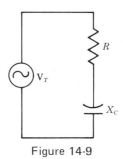

Figure 14-9

(a) $\mathbf{Z} = R - jX_C = 680 - j300\ \Omega$
$= 743\underline{/-23.81°}$

(b) $\mathbf{I} = \dfrac{\mathbf{V}_T}{\mathbf{Z}} = \dfrac{12\underline{/0°}}{743\underline{/-23.81°}} = 16.15\underline{/23.81°}$ mA

(c) $\mathbf{V}_R = \mathbf{IR} = (16.15\underline{/23.81°})(10^{-3})(680)$
$= \underline{10.98\underline{/23.81°} \text{ V}}$
$\mathbf{V}_C = (16.15\underline{/23.81°})(10^{-3})(300\underline{/-90°})$
$= \underline{4.845\underline{/-66.19°} \text{ V}}$

(d) $P = V_T I \cos\theta$
$= (12)(16.15)(10^{-3})(\cos 23.81°)$
$= 0.1773 \text{ W} = \underline{177.3 \text{ mW}}$

(e)

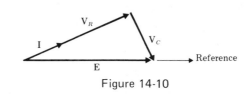

Figure 14-10

Example 8 In the circuit of Figure 14-11(a), $\mathbf{Z}_1 = 5 + j7\ \Omega$, $\mathbf{Z}_2 = 12 - j6\ \Omega$, $\mathbf{Z}_3 = 3 + j14\ \Omega$, and $\mathbf{V}_T = 120 + j0$ V. Find $\mathbf{Z}_T$; $\mathbf{I}$; θ, the angle between $\mathbf{V}_T$ and $\mathbf{I}$; $\mathbf{V}_1$, $\mathbf{V}_2$, and $\mathbf{V}_3$; and P.

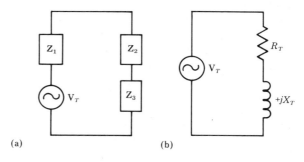

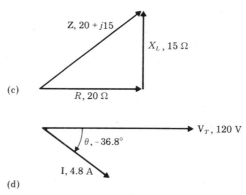

Figure 14-11

$$Z_T = Z_1 + Z_2 + Z_3 = R_T + jX_T$$
$$= R_1 + R_2 + R_3 + j(X_1 + X_2 + X_3)$$
$$= 5 + 12 + 3 + j[7 + (-6) + 14]$$
$$= 20 + j15 \ \Omega$$

The equivalent-circuit diagram for Z_T is shown in Figure 14-11(b); the impedance diagram of Z_T is shown in Figure 14-11(c). We have

$$I = \frac{V_T}{Z_T} = \frac{120 + j0}{20 + j15}$$

Since we can most easily divide complex quantities when they are in polar or exponential form, we will transform V_T and Z_T to polar form:

$$V_T = 120\underline{/0°}, \quad Z_T = 25\underline{/36.87°}$$

Thus,

$$I = \frac{120\underline{/0°}}{25\underline{/36.87°}} = 4.8\underline{/-36.87°} \ A$$

The phasor diagram of V_T and I is shown in Figure 14-11(d). It is obvious from the diagram that I lags V_T by 36.87°. Thus, if θ is the angle of I with respect to V_T, then $\theta = -36.87°$. Or, if θ_{V_T} is the relative phase angle of V_T, and θ_I is the relative phase angle of I, then

$$\theta = \theta_I - \theta_{V_T} = -36.87° - 0 = -36.87°$$

Next, we determine V_1, V_2, and V_3:

$$V_1 = IZ_1$$

For ease in multiplication, we transform Z_1 to polar form:

$$Z_1 = 5 + j7 = 8.602\underline{/54.46°} \ \Omega$$

Then

$$V_1 = (4.8\underline{/-36.87°})(8.602\underline{/54.46°})$$
$$= 41.29\underline{/17.59°} \ V$$

Similarly,

$$V_2 = IZ_2 = (4.8\underline{/-36.87°})(13.42\underline{/-26.57°})$$
$$= 64.42\underline{/-63.44°} \ V$$

and

$$V_3 = IZ_3 = (4.8\underline{/-36.87°})(14.32\underline{/77.91°})$$
$$= 68.74\underline{/41.04°} \ V$$

We can check our work by substituting the voltages into the Kirchhoff voltage law equation for the circuit,

$$V_T = V_1 + V_2 + V_3$$

For addition, the complex voltages should be in rectangular form.

$$V_1 = 41.29\underline{/17.59°} = 39.36 + j12.48 \ V$$
$$V_2 = 64.42\underline{/-63.44°} = 28.8 - j57.62 \ V$$
$$V_3 = 68.74\underline{/41.04°} = 51.85 + j45.13 \ V,$$
$$V_T \qquad\qquad = 120.01 - j \ 0.01$$
$$= 120 \ V$$

The average power in an ac circuit is equal to the product of the *in-phase* components of total voltage and current; that is,

$$P = V_T I \cos \theta$$

where θ is the angle between V_T and I found above, and

$$P = 120 \cdot 4.8 \cdot \cos(-36.87°) = 576 \cdot 0.8$$
$$= 461 \ W$$

Exercises 14-3 In Exercises 1–4 the given values are for R, X_L, and V_T, respectively, of a series ac circuit. Determine (a) Z, (b) I, (c) V_R and V_L, and (d) P. (e) Sketch the phasor diagram, including on it V_T, V_R, V_L, and I. (See Example 6.)

1. 75 Ω, 60 Ω, 115$\underline{/0°}$ V
2. 3300 Ω, 6800 Ω, 12$\underline{/0°}$ V
3. 27 kΩ, 5 kΩ, 50$\underline{/0°}$ mV
4. 65 Ω, 9.5 Ω, 120$\underline{/0°}$ V

In Exercises 5–8 the given values are for R, X_C, and V_T, respectively, of a series ac circuit. Determine (a) Z, (b) I, (c) V_R and V_C, and (d) P. (e) Sketch the phasor diagram, including on it V_T, V_R, V_C, and I. (See Example 7.)

5. 50 Ω, 475 Ω, 375$\underline{/0°}$ V
6. 2.2 MΩ, 7.5 MΩ, 750$\underline{/0°}$ mV
7. 150 kΩ, 90 kΩ, 275$\underline{/0°}$ μV
8. 680 Ω, 300 Ω, 12$\underline{/0°}$ V

In Exercises 9–13 the given values are for a series ac circuit. Determine Z and I.

	R	X_L	X_C	$V_T(\theta = 0°)$
9.	240 Ω	130 Ω	450 Ω	120 V
10.	15 Ω	370 Ω	330 Ω	12 mV
11.	0.68 MΩ	2.5 MΩ	1.2 MΩ	450 V
12.	9.1 kΩ	10 kΩ	4 kΩ	45 V
13.	100 kΩ	500 kΩ	1.0 MΩ	1600 V

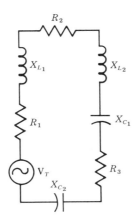

Figure 14-12

14. In the circuit of Figure 14-12 V_T = 56$\underline{/0°}$ V, R_1 = 55 Ω, R_2 = 68 Ω, R_3 = 82 Ω, X_{L_1} = 75 Ω, X_{L_2} = 100 Ω, X_{C_1} = 140 Ω, and X_{C_2} = 120 Ω.
  (a) Reduce the circuit to an equivalent circuit of one R and one X.
  (b) Determine Z_T.
  (c) Determine I.

In Exercises 15–18 the given values are for a series ac circuit. Determine Z_T and I.

	V_T	Z_1	Z_2	Z_3
15.	120$\underline{/0°}$ V	1.5 Ω	1.5 + j2.5 Ω	10 + j6 Ω
16.	25$\underline{/0°}$ mV	2 kΩ	$-j$1.8 kΩ	2.7 kΩ
17.	3.8$\underline{/0°}$ μV	1.5 MΩ	j600 kΩ	500 kΩ + j700 kΩ
18.	3.5$\underline{/-50°}$ V	1200 − j400 Ω	200 + j50 Ω	875 − j250 Ω

19. A circuit contains the impedances Z_1 = 250$\underline{/0°}$ kΩ, Z_2 = 175$\underline{/45°}$ kΩ, and Z_3 = 325$\underline{/-75°}$ kΩ in series. The voltage across Z_2 is V_2 = 131$\underline{/60°}$ V. Determine V_1 and V_3.

20. The voltage across a circuit containing two impedances is measured and found to be 12.5 V. The current measures 175 mA and leads the voltage by 23°. The impedance Z_1 is known to be 31$\underline{/61°}$ Ω. Determine Z_2. Draw the circuit diagram showing the makeup of the two impedances.

14-4 COMPLEX VOLTAGE DIVIDERS: THE RATIO METHOD IN SERIES AC CIRCUITS

If we take the ratio of one impedance of an ac series circuit to the total impedance, we get the expression Z_1/Z_T. Multiplying the numerator and denominator by I gives IZ_1/IZ_T. Since multiplying in this fashion does not change the value of a fraction, we can state that

$$\frac{Z_1}{Z_T} = \frac{IZ_1}{IZ_T}$$

But $IZ_1 = V_1$ and $IZ_T = V_T$; hence,

$$\frac{V_1}{V_T} = \frac{Z_1}{Z_T} \quad \text{or} \quad V_1 = \frac{Z_1}{Z_T} V_T$$

Similarly,

$$V_2 = \frac{Z_2}{Z_T} V_T \quad \text{and} \quad V_3 = \frac{Z_3}{Z_T} V_T$$

These relationships are stated in words as:

The phasor voltage drop across any imped-ance in a series ac circuit is equal to the ratio of that impedance to the total impedance of the circuit times the total voltage.

Applying this technique to the example of the preceding section [see Figure 14-11(a), we obtain

$$V_1 = \frac{Z_1}{Z_T} V_T = \frac{8.602\underline{/54.46°}}{25\underline{/36.87°}} \, 120\underline{/0°}$$
$$= 41.29\underline{/17.59°} \text{ V}$$

$$V_2 = \frac{13.42\underline{/-26.57°}}{25\underline{/36.87°}} \, 120\underline{/0°}$$
$$= 64.42\underline{/-63.44°} \text{ V}$$

$$V_3 = \frac{14.32\underline{/77.91°}}{25\underline{/36.87°}} \, 120\underline{/0°} = 68.74\underline{/41.04°} \text{ V}$$

Thus, just as with dc circuits, we determine the voltage across an element in an ac circuit by means of the ratio method, a handy device which avoids the intermediate step of determining the current in the circuit. The ratio method lets us estimate the relative magnitudes of voltage drops in a series circuit. Use the ratio technique wherever possible; its usefulness increases with increased skill in applying it.

Exercises 14-4   In Exercises 1–14 you are referred to the exercises of Section 14-3. Find the required voltages using the ratio method.

1–4. Find V_R and V_L for the circuits of Exercises 1–4, Section 14-3.

5–8. Find V_R and V_C for the circuits of Exercises 5–8, Section 14-3.

9–13. Find V_R, V_L, and V_C for the circuits of Exercises 9–13, Section 14-3.

14. Find V_{R_1}, V_{L_1}, and V_{C_1}, for the circuit of Exercise 14, Section 14-3.

For Exercises 15–18 find the values of V_L. Use the ratio method. Refer to Figure 14-13.

	V_g	Z_i	Z_s	Z_L
15.	$10\underline{/0^\circ}$ V	$1.5\ \Omega$	$1.5 + j2.5\ \Omega$	$10 + j6\ \Omega$
16.	$275\underline{/0^\circ}$ mV	$2\ k\Omega$	$-j1.8\ k\Omega$	$2.7\ k\Omega$
17.	$25\underline{/25^\circ}\ \mu$V	$1.5\ M\Omega$	$j600\ k\Omega$	$500\ k\Omega + j700\ k\Omega$
18.	$0.27\underline{/-18^\circ}$ V	$25\ k\Omega$	$18\ k\Omega - j30\ k\Omega$	$8\ k\Omega - j15\ k\Omega$

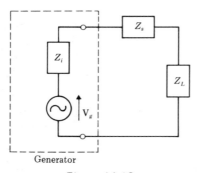

Figure 14-13

14-5 PARALLEL CIRCUITS

By definition, a parallel circuit is one in which the voltage is the same across all elements. Thus, the circuit of Figure 14-14 is a parallel circuit containing the parallel impedances Z_1, Z_2, and Z_3. Also the total current in a parallel circuit is equal to the phasor sum of the branch currents. Thus,

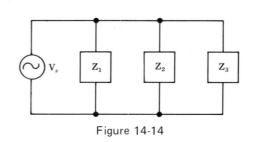

Figure 14-14

$$I_T = I_1 + I_2 + I_3$$

But

$$I_T = \frac{V_s}{Z_T},$$

$$I_1 = \frac{V_s}{Z_1},$$

$$I_2 = \frac{V_s}{Z_2}, \quad \text{and so on}$$

Hence,

$$I_T = \frac{V_s}{Z_T} = \frac{V_s}{Z_1} + \frac{V_s}{Z_2} + \frac{V_s}{Z_3}$$

or

$$\frac{1}{Z_T} = \frac{1}{Z_1} + \frac{1}{Z_2} + \frac{1}{Z_3} \qquad (1)$$

The reciprocal of the total impedance of a parallel circuit is equal to the *complete sum* of the reciprocals of the individual impedances. When we perform the indicated operations, we obtain an expression of the form

$$Z_T = Z_{eq} = R_{eq} + jX_{eq}$$

The term Z_{eq} in this expression is sometimes referred to as the *series equivalent of a parallel circuit*. By definition:

If Z_{eq} is substituted for the parallel circuit, no matter how complicated, it draws the same current from a given voltage source and at the same phase angle.

If there are only two impedances, the expression becomes

$$\frac{1}{Z_{eq}} = \frac{1}{Z_1} + \frac{1}{Z_2}$$

Converting the terms on the right-hand side of the equals sign to a common denominator, we obtain

$$\frac{1}{Z_{eq}} = \frac{Z_2 + Z_1}{Z_1 Z_2}$$

Inverting, we get

$$Z_{eq} = \frac{Z_1 Z_2}{Z_1 + Z_2} \qquad (2)$$

We can easily see the similarities between these expressions for parallel impedances in an ac circuit and the expressions for parallel resistances in a dc circuit. However, we must remember that impedance may be a complex quantity and hence the mathematical operations in (1) and (2) above must be performed in accordance with the rules for complex quantities.

Example 9 In the circuit of Figure 14-14 $Z_1 = 25 + j45$, $Z_2 = 15 + j0$, $Z_3 = 0 - j80$, and $V_s = 75 + j0 \ \mu V$. Determine Z_{eq}, I_1, I_2, I_3, I_T. Furthermore, determine θ, the angle between V_s and I_T, and the total power in the circuit. First we calculate Z_{eq}:

$$\frac{1}{Z_{eq}} = \frac{1}{Z_1} + \frac{1}{Z_2} + \frac{1}{Z_3} = \frac{1}{25 + j45} + \frac{1}{15} + \frac{1}{-j80}$$

Converting to polar form, we get

$$\frac{1}{Z_{eq}} = \frac{1}{51.48\underline{/60.95°}} + \frac{1}{15\underline{/0°}} + \frac{1}{80\underline{/-90°}}$$

$$= 0.0194\underline{/-60.95°} + 0.0667\underline{/0°}$$

$$+ 0.0125\underline{/90°}$$

$$= 0.0094 - j0.017 + 0.0667 + j0 + 0$$

$$+ j0.0125$$

$$= 0.0761 - j0.0045 = 0.0762\underline{/-3.38°}$$

$$Z_{eq} = \frac{1}{0.0762\underline{/-3.38°}} = \underline{13.12\underline{/3.38°} \ \Omega}$$

Next we determine I_1, I_2, I_3, and I_T:

Determine I_1:

$$I_1 = \frac{V_s}{Z_1} = \frac{75 \times 10^{-6}\underline{/0°}}{51.48\underline{/60.95°}} = \underline{1.457\underline{/-60.95°} \ \mu A}$$

$$I_2 = \frac{75 \times 10^{-6}\underline{/0°}}{15\underline{/0°}} = \underline{5\underline{/0°} \ \mu A}$$

$$I_3 = \frac{75 \times 10^{-6}\underline{/0°}}{80\underline{/-90°}} = \underline{0.9375\underline{/90°} \ \mu A}$$

$$I_T = \frac{V_s}{Z_{eq}} = \frac{75 \times 10^{-6}\underline{/0°}}{13.12\underline{/3.38°}} = \underline{5.716\underline{/-3.38°} \ \mu A}$$

The phasor diagram for the currents is shown in Figure 14-15. We compute θ:

$$\theta = \theta_{I_T} - \theta_{V_S} = \underline{-3.38°}$$

The total current lags the applied voltage by 3.38°.

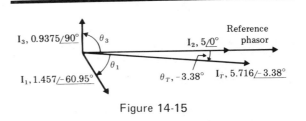

Figure 14-15

Power is found using

$$P = V_s I_T \cos \theta$$

$$P = (75 \times 10^{-6})(5.716 \times 10^{-6})(\cos - 3.38°)$$

$$= \underline{428 \ pW}$$

When one or more branches of a parallel circuit contain series impedances, we must reduce these impedances to a total equivalent impedance for each branch before combining the branches as parallel circuits. A numerical example will demonstrate the procedure.

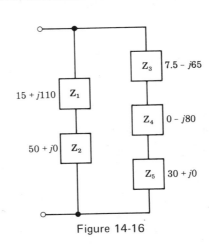

Figure 14-16

Example 10   In Figure 14-16 $Z_1 = 15 + j110$, $Z_2 = 50 + j0$, $Z_3 = 7.5 - j65$, $Z_4 = 0 - j80$, and $Z_5 = 30 + j0\ \Omega$. Determine the Z_{eq} of the two-branch parallel circuit.

$$Z_{12} = Z_1 + Z_2 = 15 + j110 + 50 + j0$$
$$= 65 + j110 = 127.8\underline{/59.42°}\ \Omega$$

$$Z_{345} = Z_3 + Z_4 + Z_5$$
$$= 7.5 - j65 + 0 - j80 + 30 + j0$$
$$= 37.5 - j145 = 149.8\underline{/-75.5°}\ \Omega$$

$$Z_{eq} = \frac{Z_{12}Z_{345}}{Z_{12} + Z_{345}}$$
$$= \frac{(127.8\underline{/59.42°})(149.8\underline{/-75.5°})}{65 + j110 + 37.5 - j145}$$
$$= \frac{1.914 \times 10^4\underline{/-16.08°}}{102.5 - j35}$$
$$= \frac{1.914 \times 10^4\underline{/-16.08°}}{108.3\underline{/-18.85°}}$$
$$= 176.7\underline{/2.77°} = 176.5 + j8.539\ \Omega$$

Exercises 14-5   In Exercises 1–6 the given values are for impedances in parallel. Determine Z_{eq}. Express in polar and rectangular form.

1. $5\underline{/25°}\ \Omega$, $6\underline{/-15°}\ \Omega$
2. $4\underline{/48°}\ \Omega$, $3\underline{/5°}\ \Omega$
3. $3.5\underline{/27°}\ k\Omega$, $1.2\underline{/-45°}\ k\Omega$
4. $4 + j6\ k\Omega$, $5 - j3\ k\Omega$
5. $4\underline{/48°}\ k\Omega$, $3\underline{/5°}\ k\Omega$, $9\underline{/-15}\ k\Omega$
6. $4 + j6\ k\Omega$, $5 - j3\ k\Omega$, $2 - j8\ k\Omega$

In Exercises 7–12 the given values are for voltages to be applied to the specified parallel circuits. Determine the individual branch currents and I_T. Express results in both rectangular and polar form.

7. Apply $120\underline{/0°}$ V to the circuit of Exercise 1.
8. Apply $50\underline{/0°}$ V to the circuit of Exercise 2.
9. $30\underline{/0°}$ V, Exercise 3.
10. $12\underline{/0°}$ mV, Exercise 4.
11. $475\underline{/0°}$ mV, Exercise 5.
12. $750\underline{/0°}$ μV, Exercise 6.

In Exercises 13 and 14 the given values are for the circuit of Figure 14–16. Determine Z_{eq} for the circuit. Express in polar and rectangular form.

13. $Z_1 = 5 + j4\ \Omega$, $Z_2 = 5 + j0\ \Omega$, $Z_3 = 3 + j2\ \Omega$, $Z_4 = 0 - j8\ \Omega$, and $Z_5 = 5 + j0\ \Omega$
14. $Z_1 = 5\underline{/30°}\ k\Omega$, $Z_2 = 7\underline{/-65°}\ k\Omega$, $Z_3 = 3\underline{/12°}\ k\Omega$, $Z_4 = 4\underline{/0°}\ k\Omega$, and $Z_5 = 2 + j5\ k\Omega$

14-6 COMPLEX CURRENT DIVIDERS: THE RATIO METHOD IN PARALLEL AC CIRCUITS

The ratio method for parallel circuits, developed in Section 7-5, is applicable to ac circuits. We must substitute complex impedances Z for R (see Figure 14-17):

$$I_1 = \frac{Z_2}{Z_1 + Z_2}\ I_T$$

$$I_T = \frac{Z_1 + Z_2}{Z_2}\ I_1$$

$$I_T = \frac{Z_1 + Z_2}{Z_1}\ I_2$$

$$I_2 = \frac{Z_1}{Z_1 + Z_2}\ I_T$$

Example 11   In the circuit of Figure 14-17 $Z_1 = 5 + j4\ \Omega$, $Z_2 = 4 - j7\ \Omega$, and $I_1 = 15\underline{/-30°}$ A. Determine I_2 and I_T.

We first convert Z_1 and Z_2 to polar form,

$$Z_1 = 5 + j4 = 6.403\underline{/38.66°}\ \Omega,$$
$$Z_2 = 4 - j7 = 8.062\underline{/-60.26°}\ \Omega$$

Next, we substitute into the formula,

$$I_T = \frac{Z_1 + Z_2}{Z_2}\ I_1 = \frac{5 + j4 + 4 - j7}{8.062\underline{/-60.26°}} \cdot 15\underline{/-30°}$$
$$= \frac{9.487\underline{/-18.43°}}{8.062\underline{/-60.26°}} \cdot 15\underline{/-30°}$$
$$= 17.65\underline{/11.83°}\ A = 17.28 + j3.618\ A$$

$$I_2 = \frac{Z_1}{Z_2}\ I_1 = \frac{6.403\underline{/38.66°}}{8.062\underline{/-60.26°}} \cdot 15\underline{/-30°}$$
$$= 11.91\underline{/68.92°} = 4.284 + j11.11\ A$$

We check our result using $I_T = I_1 + I_2$:

$$I_1 = 15\underline{/-30°} = 12.99 - j7.5$$
$$I_T = 12.99 - j7.5 + 4.284 + j11.11$$
$$= 17.27 + j3.61\ A$$

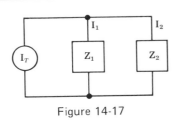

Figure 14-17

Exercises 14-6 In Exercises 1–10, apply the ratio method to obtain the solution.

1. A parallel circuit consists of $Z_1 = 5\underline{/25°}\ \Omega$ and $Z_2 = 6\underline{/-15°}\ \Omega$. If $I_1 = 2\underline{/-25°}$ A, find I_2 and I_T.

2. For the circuit of Exercise 1 $I_T = 4\underline{/-10°}$ A. Determine I_1 and I_2.

3. A parallel circuit contains two impedances: $Z_1 = 4\underline{/48°}\Omega$ and $Z_2 = 3\underline{/5°}\Omega$. If $I_2 = 2\underline{/-5°}$ mA, find I_1 and I_T.

4. For the circuit of Exercise 3 $I_T = 5\underline{/-20°}$ mA. Find I_1 and I_2.

5. The two impedances, $Z_1 = 3.5\underline{/27°}\ k\Omega$ and $Z_2 = 1.2\underline{/-45°}\ k\Omega$, are in parallel, and $I_T = 6.5\underline{/30°}$ mA. Determine I_1 and I_2.

6. For the circuit of Exercise 5 $I_2 = 3\underline{/45°}$ mA. Determine I_1 and I_T.

7. A parallel circuit consists of the impedances, $Z_1 = 50\underline{/15°}\ \Omega$ and $Z_2 = 35 - j60\ \Omega$. The circuit is supplied by $I_T = 2\underline{/0°}$ A. Find I_1 and I_2.

8. For the circuit of Exercise 7 $I_2 = 3.7\underline{/40°}$ A. Find I_1 and I_T.

9. A circuit has the impedances, $Z_1 = 3.3\underline{/-30°}\ k\Omega$ and $Z_2 = 8.8\underline{/25°}\ k\Omega$ connected in parallel. The current I_1 is found to be $3.5 + j4.5$ mA. Find I_2 and I_T.

10. For the circuit of Exercise 9 $I_T = 5.5 + j1.5$ mA. Find I_1 and I_2.

14-7 ADMITTANCE IN COMPLEX NOTATION

In working with parallel circuits it is sometimes convenient to use the *admittance* concept. *By definition, the admittance of a given circuit is the reciprocal of the impedance.* Thus,

$$Y = \frac{1}{Z}$$

The impedance is of the form

$$Z = R + jX$$

and the admittance is

$$Y = G + jB$$

where G is the conductance of the element and B is the susceptance of the element.

It is important to point out a mistake frequently made in conversions from impedance to admittance: *G is not simply the reciprocal of R, and B is not simply the reciprocal of X!* The true relationships are derived as follows:

$$Y = \frac{1}{Z} = \frac{1}{R + jX} = \frac{1}{R + jX} \frac{R - jX}{R - jX}$$

Hence,

$$Y = \frac{R}{R^2 + X^2} - j\frac{X}{R^2 + X^2} = G + jB$$

Thus,

$$G = \frac{R}{R^2 + X^2} = \frac{R}{Z^2} \quad \text{and} \quad B = -\frac{X}{R^2 + X^2} = -\frac{X}{Z^2}$$

where G and B are in siemens units if R and X or Z are in ohms.

Using the admittance form to compute the effect of elements on a circuit is of advantage when there are many branches in parallel, as in the circuit of Figure 14-18. Since for a single branch

$$I = V_s Y$$

the total current in the circuit is

$$I_T = V_s Y_T = V_s Y_1 + V_s Y_2 + V_s Y_3 + V_s Y_4 + \cdots$$

Dividing by V_s, we get

$$Y_T = Y_1 + Y_2 + Y_3 + Y_4 + \cdots$$

We see, then, that:

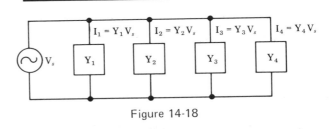

Figure 14-18

The total admittance of a multibranch parallel network is the complex sum of the individual branch admittances.

Also, when we take ratios, we get

$$\frac{I_1}{I_T} = \frac{V_s Y_1}{V_s Y_T} \quad \text{or} \quad I_1 = \frac{Y_1}{Y_T} I_T$$

$$I_2 = \frac{Y_2}{Y_T} I_T \qquad Y_3 = \frac{I_3}{I_T} Y_T$$

and so forth. The ratio technique for admittances and currents in parallel circuits is similar to that for impedances and voltages in series circuits. Let us illustrate the admittance concept by means of numerical examples.

Example 12 In the circuit of Figure 14-18 $Y_1 = 250 - j115$ mS, $Y_2 = 91$ mS, $Y_3 = j270$ mS, $I_4 = 275\underline{/25°}$ mA, and $V_s = 750\underline{/0°}$ mV. Determine Y_4, I_1, I_2, I_3, I_T, Y_T, θ, and the total average power dissipated in the circuit.

Determine Y_4:

$$Y_4 = \frac{I_4}{V_s} = \frac{275 \times 10^{-3}\underline{/25°}}{750 \times 10^{-3}\underline{/0°}}$$
$$= \underline{366.7\underline{/25°} = 332 + j155 \text{ mS}}$$

Determine I_1:

$$I_1 = Y_1 V_s = (250 - j115)(750 + j0) \times 10^{-6}$$
$$= \underline{187.5 - j86.2 \text{ mA}}$$

Determine I_2:

$$I_2 = (91 + j0)(750 + j0) \times 10^{-6}$$
$$= \underline{68.2 + j0 \text{ mA}}$$

Determine I_3:

$$I_3 = (0 + j270)(750 + j0) \times 10^{-6}$$
$$= \underline{j202.5 \text{ mA}}$$

Determine I_T:

$$I_T = I_1 + I_2 + I_3 + I_4$$
$$= 187.5 - j86.2 + 68.2 + j202.5 + 249 + j116$$
$$= \underline{504.7 + j232.3 = 555.6\underline{/24.72°} \text{ mA}}$$

Determine Y_T:

$$Y_T = Y_1 + Y_2 + Y_3 + Y_4$$
$$= 250 - j115 + 91 + j270 + 332 + j155$$
$$= \underline{673 + j310 \text{ mS}}$$

Our calculations for I_T can be checked as follows:

$$I_T = Y_T V_s$$
$$= (673 + j310)(750 + j0) \times 10^{-6}$$
$$= \underline{504.8 + j232.5 \text{ mA}} \quad (Check)$$

Determine θ: The angle θ is the angle between the voltage phasor and the phasor for the total current, with the voltage phasor as the reference:

$$\theta = \theta_{I_T} - \theta_{V_s} = 24.72° - 0° = \underline{24.72°}$$

The current leads the voltage by $24.72°$.

Determine P_T:

$$P_T = V_s I_T \cos\theta$$

But $I_T \cos\theta$ is simply the in-phase component of I_T, 505 mA; hence,

$$P = (750)(505) \times 10^{-6} = \underline{379 \text{ mW}}$$

Example 13 In the circuit of Figure 14-19 $I_1 = 22.5\underline{/-65°}$ mA, $Y_1 = 375\underline{/-65°}$ μS, and $Y_2 = 625\underline{/85°}$ μS. Determine I_2.

$$I_2 = \frac{Y_2}{Y_1} I_1$$
$$= \frac{625\underline{/85°}}{375\underline{/-65°}} \cdot 22.5\underline{/-65°}$$
$$= \underline{37.5\underline{/85°} \text{ mA}}$$

Exercises 14-7 In Exercises 1–5 convert to admittance.

1. $5 + j7$ Ω   2. $27 - j75$ kΩ
3. $4.8 - j1.2$ kΩ   4. $0.05 + j1.5$ MΩ
5. $275\underline{/-18°}$ Ω

In Exercises 6–10 convert to impedance.

6. $6 + j11$ S   7. $1.5 - j5.6$ mS
8. $125\underline{/75°}$ μS   9. $250 + j0$ mS

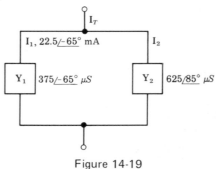

Figure 14-19

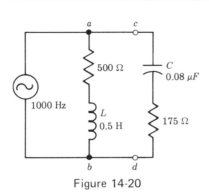

Figure 14-20

10. $0 + j7.5 \ \mu$S

11. In a circuit similar to that of Figure 14-18 $Y_1 = 12.5 - j6.8$, $Y_3 = 5.6 + j25$, $Y_4 = 18/\underline{-48°}$, and $Y_T = 35$. All values are in millisiemens. Determine Y_2, and express it in both rectangular and polar forms.

12. Assume that the circuit of Exercise 11 is energized with a source $V_s = 30$ V. Determine each of the branch currents and the total current.

13. In a circuit similar to that of Figure 14-19 $I_1 = 45/\underline{-65°}$ mA, $I_2 = 37.5/\underline{42°}$ mA, and $Y_1 = 3.5/\underline{-65°}$ mS. Determine I_T, Y_2, and Y_T, using the ratio method.

14. For the circuit of Figure 14-20 determine Y_{ab}, Y_{cd}, and Y_T.

15. For the circuit of Exercise 14, $I_T = 15/\underline{63.67°}$ mA. Determine I_{ab} and I_{cd}.

16. Find Y_1, Y_2, and Y_T for the circuit of Exercise 1, Section 14-6.

17. A parallel circuit contains three impedances: $Z_1 = 0.75/\underline{65°}$ MΩ, $Z_2 = 1.1/\underline{15°}$ MΩ, and $Z_3 = 0.18/\underline{-40°}$ MΩ. Find Y_1, Y_2, and Y_3.

18. If the circuit of Exercise 17 is energized with the voltage $V_s = 27/\underline{0°}$ mV, find I_1, I_2, I_3, and I_T.

14-8 A WORD ABOUT NOTATION

We have now been thoroughly exposed to the concept of complex quantities, quantities that have both magnitude and direction. To distinguish between a complex quantity and its absolute magnitude, we have used symbols in boldface type to denote complex or phasor quantity, and lightface *italics* to denote absolute magnitude. From this point on, however, we shall abandon this procedure, and use italics for *complex quantities;* and, whenever it is necessary to indicate that magnitude only is meant, we shall employ the standard notation for absolute value, that is, $|V|$, $|Z|$, and so on.

14-9 SERIES-PARALLEL CIRCUITS

In the strictest sense, it is not possible to make practical circuits which are purely resistive, inductive, or capacitive. Actual combination circuits may resemble the circuit of Figure 14-21(a), in which Z_1, Z_2, and Z_3 may each contain at least two, and possibly all three, of the basic passive circuit elements: R, L, and C.

Let us suppose that for a circuit such as that of Figure 14-21, we need to determine quantities such as I_T, I_2, I_3, V_1, and so on. How do we do this mathematically? To determine I_T, we could reduce the circuit to a single equivalent impedance just as we reduced the corresponding dc circuit to a single equivalent resistance. That is, we have to find the series equivalent of the (Z_2, Z_3) parallel circuit and then combine this equivalent with Z_1. The overall procedure should be familiar. The only "new" thing about the analysis of this circuit is that, of course, impedances are complex quantities and must be manipulated according to the rules of complex numbers. Let us try the procedure with a numerical example.

Example 14 The circuit of Figure 14-21(a) consists of impedances of the following values:

$$Z_1: \quad R_1 = 125 \ \Omega, \quad X_{L_1} = 800 \ \Omega,$$
$$X_{C_1} = 600 \ \Omega$$
$$Z_2: \quad R_2 = 15 \ \Omega, \quad X_{L_2} = 135 \ \Omega$$
$$Z_3: \quad R_3 = 7.5 \ \Omega, \quad X_{C_3} = 110 \ \Omega;$$
$$V_s = 120 \text{ V}$$

See Figure 14-21(b). Determine I_T, I_2, and I_3; θ, the power-factor angle of the complete circuit; V_1, the voltage across Z_1; and P, the total average power dissipated in the circuit.

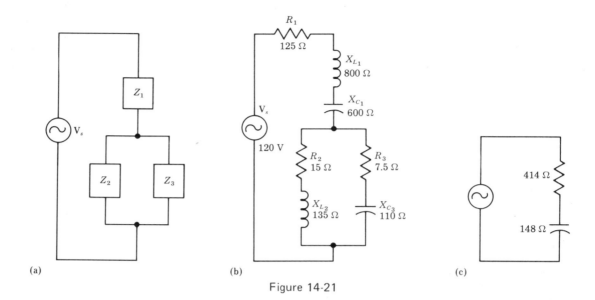

Figure 14-21

Determine I_T:

$$I_T = \frac{V_s}{Z_T}$$

$$Z_T = Z_1 + \frac{Z_2 Z_3}{Z_2 + Z_3}$$

$$= 125 + j(800 - 600)$$

$$+ \frac{(15 + j135)(7.5 - j110)}{15 + j135 + 7.5 - j110}$$

$$= 125 + j200$$

$$+ \frac{(135.8\underline{/83.66°})(110.3\underline{/-86.1°})}{33.63\underline{/48.01°}}$$

$$= 125 + j200 + 283.6 - j343.4$$

$$= 408.6 - j143.4$$

$$= 433\underline{/-19.34°} \ \Omega$$

The circuit reduced to one resistance and one reactive element is shown in Figure 14-21(c). Then

$$I_T = \frac{120\underline{/0°}}{433\underline{/-19.34°}} = 0.277\underline{/19.34°}$$

$$= 0.261 + j0.092 \text{ A}$$

Determine I_2:

$$I_2 = \frac{Z_3}{Z_2 + Z_3} I_T = \frac{110.3\underline{/-86.1°}}{33.63\underline{/48.01°}} \ 0.277\underline{/19.34°}$$

$$= 0.909\underline{/-114.77°} = -0.381 - j0.825 \text{ A}$$

Determine I_3:

$$I_3 = \frac{Z_2}{Z_2 + Z_3} I_T = \frac{135.8\underline{/83.66°}}{33.63\underline{/48.01°}} \ 0.277\underline{/19.34°}$$

$$= 1.12\underline{/55.4°} = 0.642 + j0.917 \text{ A}$$

When we combine I_2 and I_3 to check for I_T, we get

$$I_T = I_2 + I_3 = -0.381 - j0.825 + 0.642 + j0.917$$

$$= 0.261 + j0.092 \text{ A}$$

Determine θ:

$$\theta = \theta_{I_T} - \theta_{V_s} = 19.34° - 0° = \underline{19.34°}$$

Determine V_1:

$$V_1 = I_T Z_1 = 0.277\underline{/19.34°} \cdot 235.8\underline{/57.99°}$$

$$= \underline{65.32\underline{/77.33°} \text{ V}}$$

Determine P:

$$P = |V_s| \, |I_T| \cos \theta = 120 \cdot 0.261 = \underline{31 \text{ W}}$$

14-9 EXERCISES

1. In a circuit similar to that of Figure 14-21 $Z_1 = 15 + j75 \ \Omega$, $Z_2 = 50 + j0 \ \Omega$, and $Z_3 = 0 - j90 \ \Omega$. Draw the circuit diagram showing individual R's, L's, and C's. Determine the equivalent Z for the complete circuit. Draw the diagram of the equivalent circuit.

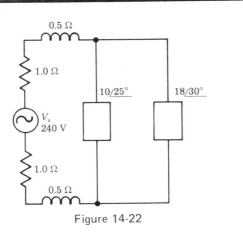

Figure 14-22

2. A signal voltage of 750 μV is applied to the circuit of Exercise 1. Determine I_T, V_1, and V_3. Draw a phasor diagram for these values, using the applied voltage as a reference.

3. For the circuit of Figure 14-22 determine the voltage across the parallel loads.

4. Determine the power factor of the circuit which the generator sees in the circuit of Figure 14-22. (Power factor = $\cos \theta$, $\theta = \theta_{I_T} - \theta_{Vs}$).

5. For the circuit of Figure 14-22 determine what happens to the load voltage and to the θ of the circuit when a 300-μF capacitor is added in parallel with the loads ($f = $ 60 Hz).

6. Figure 14-23 represents an ac generator with internal impedance Z_i feeding two parallel loads consisting of complex imped-

ances Z_1 and Z_2. Given: V_s = 100 V, Z_i = 0.4 + j0.3 Ω, Z_1 = 10 + j6 Ω, and Z_2 = 15 – j25 Ω. Determine I_T, $V_1 = V_2$, the total power supplied by the generator, and the power factor of the load seen by the generator. Do not fail to use phasor diagrams in your analysis.

7. In the circuit of Figure 14-23 V_s = 250 mV, Z_i = 50 + j10 Ω, Z_1 = 150 + j90 Ω, and Z_2 = 275 – j150 Ω. Determine the voltage across the resistance portion of Z_2.

8. Repeat Exercise 7, assuming that the frequency is doubled.

14-10 THE LADDER METHOD

We have seen that a complex circuit can be analyzed by the technique of reducing it to an equivalent circuit of a single impedance. This reduction permits us to find the total current, for instance, if we know the applied voltage. Knowing the total current, we can calculate voltage drops, branch currents, and so on. For some circuits there is no better method of analysis than this. However, when this reduction method is applied to a circuit of the type shown in Figure 14-24, commonly called a *ladder circuit*, it becomes unnecessarily tedious. For circuits of this type, it is preferable to use the *ladder method*.

Let us apply the ladder method to the circuit of Figure 14-24, using numerical values. For simplicity of calculations, the impedances are considered to be pure resistances. The key to the method is our making the assumption of the value of a current in a strategic branch of the circuit. We start by assuming

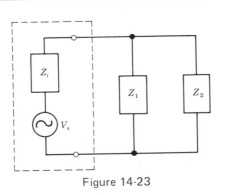

Figure 14-23

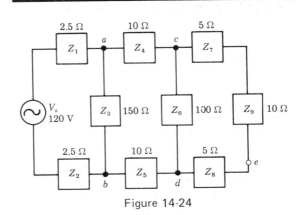

Figure 14-24

that a current I flows through the branch ced. Thus,

$$I_{ced} = I, \qquad V_{cd} = 20I$$

$$I_{cd} = \frac{20I}{100} = 0.2I, \qquad I_{ac} = I_{cd} + I_{ced} = 0.2I + I$$
$$= 1.2I = I_{db}$$

$$V_{ab} = V_{cd} + V_{ac} + V_{db}$$
$$= 20I + 10(1.2I) + 10(1.2I) = 44I$$

$$I_{ab} = \frac{V_{ab}}{150} = \frac{44I}{150} = 0.293I$$

$$I_T = I_{ab} + I_{ac} = 0.293I + 1.2I = 1.493I$$

$$V_s = I_T Z_1 + V_{ab} + I_T Z_2$$
$$= 2.5(1.493I) + 44I + 2.5(1.493I)$$
$$= 51.5I$$

$$I = \frac{V_s}{51.5} = \frac{120}{51.5} = 2.33 \text{ A}$$

We can now find any of the intermediate values simply by substituting the value for I in the appropriate expression. For example, let us find V_{cd}:

$$V_{cd} = 20I = (20)(2.33) = 46.6 \text{ V}$$

The ladder method can be usefully applied in the analysis of the feedback network in a *phase-shift* oscillator. The network is shown in Figure 14-25. If V_o represents the output of the oscillator, then V_a must be 180° out of phase with V_o, or $V_a = -kV_o$.

Our strategy is to start the analysis with the voltage V_a; then

$$I_a = \frac{V_a}{R}$$

$$V_b = V_a + I_a Z_C = V_a - jX_C I_a$$
$$= V_a - j\frac{X_C V_a}{R} = V_a\left(1 - j\frac{X_C}{R}\right)$$

$$I_b = I_a + \frac{V_b}{R} = V_a\left(\frac{1}{R} + \frac{1}{R} - j\frac{X_C}{R^2}\right)$$
$$= V_a\left(\frac{2}{R} - j\frac{X_C}{R^2}\right)$$

$$V_c = V_b + I_b Z_C = V_a\left(1 - j\frac{X_C}{R} - j\frac{2X_C}{R} - \frac{X_C^2}{R^2}\right)$$
$$= V_a\left(1 - \frac{X_C^2}{R^2} - j\frac{3X_C}{R}\right)$$

$$I_c = I_b + \frac{V_c}{R}$$
$$= V_a\left(\frac{2}{R} - j\frac{X_C}{R^2} + \frac{1}{R} - \frac{X_C^2}{R^3} - j\frac{3X_C}{R^2}\right)$$
$$= V_a\left(\frac{3}{R} - \frac{X_C^2}{R^3} - j\frac{4X_C}{R^2}\right)$$

$$V_o = V_c + I_c Z_C =$$
$$V_a\left(1 - \frac{X_C^2}{R^2} - j\frac{3X_C}{R} - j\frac{3X_C}{R} + j\frac{X_C^3}{R^3} - \frac{4X_C^2}{R^2}\right)$$
$$= V_a\left(1 - \frac{5X_C^2}{R^2} - j\frac{6X_C}{R} + j\frac{X_C^3}{R^3}\right)$$

To achieve our objective, we must make certain that the imaginary terms equal zero; thus, we see that

$$\frac{X_C^3}{R^3} = \frac{6X_C}{R}, \qquad \frac{X_C^2}{R^2} = 6, \qquad \frac{X_C}{R} = \sqrt{6}$$

The frequency at which this occurs can be determined by

$$\frac{1}{2\pi f C R} = \sqrt{6}, \qquad f = \frac{1}{2\pi\sqrt{6}\,RC}$$

The relationship between V_a and V_o is

$$V_o = V_a(1 - 5 \cdot 6) = -29\,V_a$$

or

$$V_a = -\tfrac{1}{29}V_o, \qquad k = \tfrac{1}{29}$$

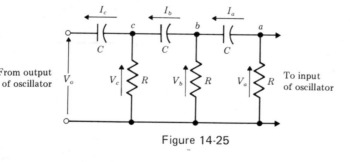

Figure 14-25

EXERCISES 14-10

1. Determine I_L and V_L in Figure 14-26; the impedances are all pure resistances and have the following values: $Z_1 = 1000\ \Omega$, $Z_2 = 680\ \Omega$, $Z_3 = 125\ \Omega$, $Z_4 = 1500\ \Omega$, $Z_5 = Z_6 = 250\ \Omega$, $Z_L = 500\ \Omega$, and $V_s = 15\underline{/0^\circ}$ mV.

2. Repeat Exercise 1 for the condition $Z_3 = -j500\ \Omega$.

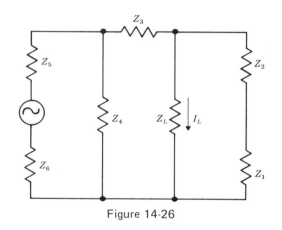

Figure 14-26

3. Using the ladder method, determine the impedance of the circuit of Figure 14-27. Determine the current in the coil circuit.

4. A fourth R and C branch is added to the feedback circuit of a phase-shift oscillator. The equation $V_a = -kV_o$ describes the input and output voltages of the circuit. Determine the value of k.

5. What is the expression for the frequency of operation of the condition described in Exercise 4?

14-11 OTHER ANALYSIS TECHNIQUES

All of the techniques for the analysis of dc circuits and networks described in Chapters 8, 9, and 10 are applicable to ac circuits. In ac applications, of course, phasor quantities are used along with the algebra of complex numbers (Chapter 13). The superposition theorem applied to an ac network will be demonstrated first.

Example 15 Find the currents $I_1, I_2,$ and I_3 for the network of Figure 14-28 using the superposition theorem.

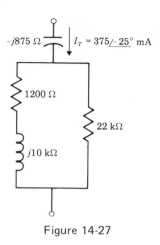

Figure 14-27

The general procedure that will be followed for the solution is:

1. Find first-source currents and second-source currents.
2. Then superpose the currents in the original network and combine vectorally.

The first-source network is shown in Figure 14-28(b).

$$Z_{T1} = Z_1 + Z_2 \parallel Z_3 = 5 + j4 + \frac{(4 - j3)(3 + j6)}{4 - j3 + 3 + j6}$$

$$= 5 + j4 + \frac{(5\underline{/-36.87^\circ})(6.708\underline{/63.43^\circ})}{7.616\underline{/23.20^\circ}}$$

$$= 5 + j4 + 4.396 + j0.258$$

$$= 9.396 + j4.258 = 10.32\underline{/24.38^\circ}\ \Omega$$

$$I_{T1} = I_{11} = \frac{V_{s1}}{Z_{T_1}} = \frac{10\underline{/0^\circ}}{10.32\underline{/24.38^\circ}}$$

$$= 0.969\underline{/-24.38^\circ}\ \text{A} = 0.883 - j0.400\ \text{A}$$

$$I_{21} = \frac{Z_3}{Z_2 + Z_3} \cdot I_{11}$$

$$= \frac{6.708\underline{/63.43^\circ}}{7.616\underline{/23.20^\circ}} \cdot 0.969\underline{/-24.38^\circ}$$

$$= 0.853\underline{/15.85^\circ} = 0.821 + j0.233\ \text{A}$$

$$I_{31} = \frac{Z_2}{Z_2 + Z_3} \cdot I_{11}$$

$$= \frac{5\underline{/-36.87^\circ}}{7.616\underline{/23.20^\circ}} \cdot 0.969\underline{/-24.38^\circ}$$

$$= 0.636\underline{/-84.45^\circ} = 0.062 - j0.633\ \text{A}$$

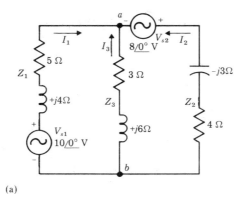

(a)

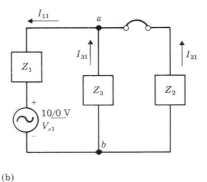

(b)

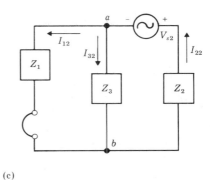

(c)

Figure 14-28

Now the second-source network is analyzed [see Figure 14-28(c)]:

$$Z_{T2} = Z_2 + Z_1 \parallel Z_3 = 4 - j3 + \frac{(5+j4)(3+j6)}{5+j4+3+j6}$$

$$= 4 - j3 + \frac{(6.403\underline{/38.66°})(6.708\underline{/63.43°})}{12.81\underline{/51.34°}}$$

$$= 4 - j3 + 2.121 + j2.597$$

$$= 6.121 - j0.403 \ \Omega = 6.134\underline{/-3.77°} \ \Omega$$

$$I_{T2} = I_{22} = \frac{V_s}{Z_{T2}} = \frac{8\underline{/0°}}{6.134\underline{/-3.77°}}$$

$$= 1.304\underline{/3.77°} \ A = 1.301 + j0.086 \ A$$

$$I_{12} = \frac{Z_3}{Z_1 + Z_3} \cdot I_{22}$$

$$= \frac{6.708\underline{/63.43°}}{12.81\underline{/51.34°}} \cdot 1.304\underline{/3.77°}$$

$$= 0.683\underline{/15.86°} = 0.657 + j0.187 \ A$$

$$I_{32} = \frac{Z_1}{Z_1 + Z_3} \cdot I_{22}$$

$$= \frac{6.403\underline{/38.66°}}{12.81\underline{/51.34°}} \cdot 1.304\underline{/3.77°}$$

$$= 0.652\underline{/-8.91} = 0.644 - j0.101 \ A$$

The current I_1 is found by superposing I_{11} and I_{12},

$$I_1 = I_{11} + I_{12} = -(0.883 - j0.400)$$
$$- (0.657 + j0.187)$$

$$I_1 = -(1.540 - j0.213) = -(1.555\underline{/-7.87°}) \ A$$

(Negative signs are used with the values of I_{11} and I_{12} in the equation for I_1 because the direction of I_1 in Figure 14-28(a) is the opposite of the directions of I_{11} and I_{12}. The direction of I_1 would be away from node a in the real circuit.)

$$I_2 = I_{21} + I_{22} = 0.821 + j0.233 + 1.301$$
$$+ j0.086$$

$$= 2.122 + j0.319 = 2.146\underline{/8.55°} \ A$$

$$I_3 = I_{31} + I_{32} = 0.062 - j0.633$$
$$- (0.644 - j0.101)$$

$$= -(0.582 + j0.532) = -(0.789\underline{/42.43°}) \ A$$

Check: $I_1 + I_2 + I_3 = 0$

$$-(1.540 - j0.213) + (2.122 + j0.319)$$
$$- (0.582 + j0.532)$$

$$= -1.540 + 2.122 - 0.582 + j0.213 + j0.319$$
$$- j0.532$$

$$= 0 + j0$$

Example 16 Find the currents I_1, I_2, and I_3 for the network of Figure 14-28(a) using the loop-current method (review Section 10-6).

First, the circuit is redrawn and marked with loop-current arrows (see Figure 14-29). Although the current is alternating, + and – voltage markings corresponding to instantaneous voltage polarities aid in writing loop equations.

Loop equations are written:

$$(Z_1 + Z_3)I_{L1} - Z_3 I_{L2} = V_{s1}$$
$$-Z_3 I_{L1} + (Z_2 + Z_3)I_{L2} = V_{s2}$$

We substitute known values into the equations and set up determinants for their solution:

$$(5 + j4 + 3 + j6)I_{L1} - (3 + j6)I_{L2} = 10\underline{/0°}$$
$$-(3 + j6)I_{L1} - (4 - j3 + 3 + j6)I_{L2} = 8\underline{/0°}$$

$$I_{L1} = \frac{\begin{vmatrix} 10\underline{/0°} & -6.708\underline{/63.43°} \\ 8\underline{/0°} & 7.616\underline{/23.20°} \end{vmatrix}}{\begin{vmatrix} 12.81\underline{/51.34°} & -6.708\underline{/63.43°} \\ -6.708\underline{/63.43°} & 7.616\underline{/34.20°} \end{vmatrix}}$$

$$= \frac{76.16\underline{/23.20°} + 53.66\underline{/63.43°}}{97.45\underline{/74.54°} - 45\underline{/126.86°}}$$

$$= \frac{70 + j30 + 24 + j48}{26 + j94 + 27 - j36} = \frac{94 + j78}{53 + j58}$$

$$= \frac{122.1\underline{/39.69°}}{78.59\underline{/47.59°}}$$

$$= 1.554\underline{/-7.9°} \text{ A} = 1.539 - j0.214 \text{ A}$$

$$I_{L2} = \frac{\begin{vmatrix} 12.81\underline{/51.34°} & 10\underline{/0°} \\ -6.708\underline{/63.43°} & 8\underline{/0°} \end{vmatrix}}{78.59\underline{/47.59°}}$$

$$= \frac{102.48\underline{/51.39°} + 67.08\underline{/63.43°}}{78.59\underline{/47.59°}}$$

$$= \frac{63.96 + j80.09 + 30 + j60}{78.59\underline{/47.59°}}$$

$$= \frac{93.96 + j140.08}{78.59\underline{/47.59°}}$$

$$= \frac{1.687\underline{/56.15°}}{78.59\underline{/47.59°}} = 2.147\underline{/8.56°}$$

$$= 2.123 + j0.32 \text{ A}$$

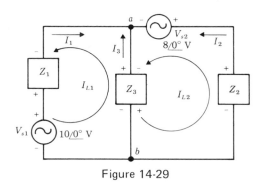

Figure 14-29

Finally, branch currents are obtained using the loop currents:

$$I_1 = -I_{L1} = -(1.554\underline{/-7.9°}) \text{ A}$$
$$I_2 = I_{L2} = 2.147\underline{/8.56°} \text{ A}$$
$$I_3 = I_{L1} - I_{L2} = 1.539 - j0.214$$
$$\quad - (2.123 + j0.32)$$
$$= -(0.584 + j0.534) = -(0.791\underline{/42.44°}) \text{ A}$$

The values for I_1, I_2, and I_3 check with those found in Example 15 within the expected degree of accuracy.

Example 17 Determine the Thévenin equivalent circuit for the network of Figure 14-28(a) considering Z_3 as the load impedance. Find I_3.

A first step is to find Z_{th} as the impedance looking into the network at nodes a and b with Z_3 disconnected and V_{s1} and V_{s2} shorted (see Figure 14-30):

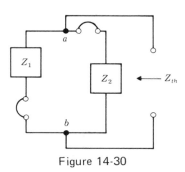

Figure 14-30

$$Z_{th} = Z_1 \parallel Z_2 = \frac{(6.403\underline{/38.66°})(5\underline{/-36.87°})}{5+j4+4-j3}$$

$$= 3.535\underline{/-4.55°} = 3.524 - j0.280 \ \Omega$$

Next, the open-circuited V_{ab} is calculated:

$$V_{ab} = -V_1 + V_{s1} = -\frac{Z_1}{Z_1+Z_2}(V_{s1}+V_{s2}) + V_{s1}$$

$$= -\frac{6.403\underline{/38.66°}}{9.055\underline{/6.34°}} \cdot 18\underline{/0°} + 10\underline{/0°}$$

$$= -(10.756 + j6.805) + 10 + j0$$

$$= -0.756 - j6.805$$

$$= -(6.848\underline{/83.66°}) \ V$$

The Thévenin equivalent circuit is shown in Figure 14-31. Now I_3 can be calculated:

$$I_3 = -\frac{V_{th}}{Z_{th}+Z_3} = -\frac{6.848\underline{/83.66°}}{3.524 - j0.280 + 3 + j6}$$

$$= -\frac{6.848\underline{/83.66°}}{8.676\underline{/41.24°}}$$

$$= -(0.789\underline{/42.42°}) \ A$$

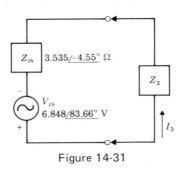

Figure 14-31

Exercises 14-11 In Exercises 1–8 refer to Figure 14-32 and the following data: $V_{s1} = 8\underline{/0°}$ V, $V_{s2} = 6\underline{/25°}$ V, $Z_1 = 2+j0 \ \Omega$, $Z_2 = 4+j0 \ \Omega$, and $Z_3 = 3+j0 \ \Omega$.

1. Find I_1, I_2, and I_3 (Figure 14-32) using the superposition theorem.

2. Analyze the circuit of Figure 14-32 (find I_1, I_2, and I_3) using the loop-current method. (Hint: Use a phasor diagram as an aid in combining superposed currents.)

3. Repeat Exercise 1 using the node-voltage method.

4. Use the branch-current method to analyze the circuit of Figure 14-32.

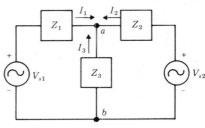

Figure 14-32

5. Determine the Thévenin equivalent circuit for Figure 14-32 considering Z_3 as the load impedance. Find I_3.

6. Find the Norton equivalent circuit for the network of Figure 14-32 considering Z_2 as the load impedance. Find I_2.

7. Find the Millman equivalent of the network of Figure 14-32 for nodes a-b. Find I_{load} if $Z_{load} = 3+j4 \ \Omega$ is connected between nodes a and b.

8. Assume that V_{s2} in the network of Figure 14-32 is changed from the given value to $V_{s2} = 8\underline{/-35°}$. Find I_3 using your favorite method of analysis.

In Exercises 9–16 refer to Figure 14-33 and the following data: $V_{s1} = 12\underline{/0°}$ V, $V_{s2} = 10\underline{/0°}$ V, $Z_1 = 6+j2$ kΩ, $Z_2 = 4+j3$ kΩ, and $Z_L = 5-j2$ kΩ.

9. Using the superposition theorem find the branch currents of the circuit of Figure 14-33.

10. Apply the loop-current method to the network of Figure 14-33 to find the branch currents.

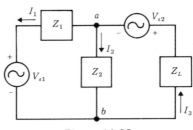

Figure 14-33

11. Repeat Exercise 10 using the node-voltage method.

12. Find the branch currents of the network of Figure 14-33 using the branch-current method.

13. Find the Thévenin equivalent circuit for Figure 14-33 using Z_L as the load impedance.

14. See Exercise 13. Find I_L for Z_L equal to (a) $5 - j2$ kΩ, (b) $5 + j3$ kΩ, and (c) Z_{th}.

15. Find the Norton equivalent circuit for Figure 14-33 considering Z_L as the load impedance. Find I_L.

16. Find the Millman equivalent circuit of Figure 14-33 for nodes a-b.

Chapter 15
Quadratic Equations

A parabola (the curve of a quadratic equation) is the ideal shape for focusing the electromagnetic energy received by a radiotelescope. (Alan T. Moffet)

15-1 DEFINITION

When an unknown appears raised to the second power (that is, it is squared) in an equation the equation is called a *quadratic equation*. These are quadratic equations:

$$x^2 + 4x + 8 = 4$$

$$3x^2 - 27 = 0$$

$$I^2R - VI + P = 0$$

In previous chapters methods have been developed for solving linear equations—equations in which the unknown appears to the first power. Recall that linear equations have only one root or solution. By contrast, quadratic equations may have two solutions.

Quadratic equations arise from some of the relations of electronic circuits. A basic knowledge of this form of equation and its solution is useful in practical electronics.

15-2 PURE QUADRATIC EQUATIONS

When a quadratic equation contains only two terms—the term containing the unknown and a constant term—the equation is called a *pure quadratic*. The solution of a pure quadratic is simple and direct.

To find the roots of a pure quadratic equation:
1. Transpose the quadratic term to the left side of the equal sign and the constant term to the right.
2. Divide by the coefficient of the quadratic term.
3. Take the square root of both sides of the equation.

Example 1 Find the roots of the following pure quadratics.

(a) $3x^2 - 27 = 0$

(a) $3x^2 = 27, x^2 = \dfrac{27}{3} = 9, \sqrt{x^2} = \sqrt{9}, x = \pm 3$

(b) $x^2 = 5$

(b) $\sqrt{x^2} = \sqrt{5}, x = \pm\sqrt{5}$

(c) $ax^2 - b = c$ (for x)

(c) $ax^2 = b + c,$

$$x^2 = \frac{b+c}{a} \qquad \sqrt{x^2} = x = \pm \sqrt{\frac{b+c}{a}}$$

(d) $I^2R = P$ (for I)

(d) $I^2 = \dfrac{P}{R},$

$$\sqrt{I^2} = I = \pm \sqrt{\frac{P}{R}}$$

Exercises 15-2 In Exercises 1–10 solve for x.

1. $x^2 = 4$ 2. $4x^2 = 64$
3. $3x^2 - 243 = 0$ 4. $8x^2 - 18 = 0$
5. $16x^2 - 25 = 0$ 6. $14x^2 - 50 = 0$
7. $x^2 - 4c = 0$ 8. $ax^2 - 4b = 4c$
9. $3x^2 - 27b = 81c$ 10. $a^4x^2 - 72b = 36c$
11. Solve the power formula $P = I^2R$ for I.
12. Solve the formula $P = V^2/R$ for V.
13. Solve the power formula $P = V^2G$ for V.
14. Solve $P = I^2/G$ for I.

15-3 SOLUTION BY THE QUADRATIC FORMULA

The general form of a quadratic equation is

$$ax^2 + bx + c = 0$$

The solution of any quadratic equation of the form $ax^2 + bx + c = 0$ can be obtained using the formula

$$x = \frac{-b \pm \sqrt{b^2 - 4ac}}{2a}$$

or the equivalent

$$x = -\frac{b}{2a} \pm \sqrt{\frac{b^2}{4a^2} - \frac{c}{a}}$$

This is called the general formula for quadratic equations.

The general formula is very important and must be memorized. It can be used to solve any quadratic equation. It lends itself to evaluation with a hand-held electronic calculator.

Example 2 Solve the following quadratic equations using the quadratic formula.

(a) $x^2 + 2x - 3 = 0$

(b) $3m^2 - 6m + 1 = 0$

(c) $y^2 - 2py + 3q = 0$ (for y)

(d) $z^2 + 2z + 2 = 0$

(a) Recalling the general form, $ax^2 + bx + c = 0$, we note that here $a = 1$, $b = 2$, and $c = -3$. Thus,

$$x = \frac{-2 \pm \sqrt{4 + 12}}{2} = \frac{-2 \pm \sqrt{16}}{2} = \frac{-2 \pm 4}{2}$$

$$x = \frac{-2 + 4}{2} = \underline{1} \quad \text{and} \quad x = \frac{-2 - 4}{2} = \underline{-3}$$

(b) For $3m^2 - 6m + 1$, $a = 3$, $b = -6$, and $c = 1$. Then

$$m = \frac{+6 \pm \sqrt{36 - 12}}{6} = \frac{6 \pm \sqrt{24}}{6}$$

$$= \frac{6 \pm 2\sqrt{6}}{6} = \frac{3 \pm \sqrt{6}}{3}$$

$$m = \frac{3 + \sqrt{6}}{3} \quad \text{and} \quad m = \frac{3 - \sqrt{6}}{3}$$

(c) For $y^2 - 2py + 3q = 0$, $a = 1$, $b = -2p$, $c = 3q$. From the formula,

$$y = \frac{2p \pm \sqrt{4p^2 - 12q}}{2} = \frac{2p \pm 2\sqrt{p^2 - 3q}}{2}$$

$$= p \pm \sqrt{p^2 - 3q}$$

Thus,

$$y = p + \sqrt{p^2 - 3q} \quad \text{and} \quad y = p - \sqrt{p^2 - 3q}$$

(d) For $z^2 + 2z + 2 = 0$, $a = 1$, $b = 2$, $c = 2$, and

$$z = \frac{-2 \pm \sqrt{4 - 8}}{2} = \frac{-2 \pm \sqrt{-4}}{2} = \frac{-2 \pm 2\sqrt{-1}}{2}$$

$$= -1 \pm \sqrt{-1}$$

Thus,

$$z = -1 + \sqrt{-1} \quad \text{and} \quad z = -1 - \sqrt{-1}$$

These are the correct roots even though $\sqrt{-1}$ is not a real number (see Chapter 13).

The Discriminant In the formula, the quantity under the radical sign is $b^2 - 4ac$. This is called the *discriminant* of the quadratic. If this quantity is zero, the solution reduces to $x = -b/2a$. Both roots are equal. This means that the quadratic is a perfect square. If the discriminant is positive (that is, $b^2 > 4ac$), then there are two real roots. If the discriminant is negative, then the quantity under the radical is negative. There is no real number which when multiplied by itself gives a negative value. Thus, a negative discriminant indicates that the roots are not real numbers. These "unreal," or nonreal, numbers are imaginary numbers (review Section 13-1).

Extraneous and Meaningless Roots Let us solve the equation

$$\frac{x^2 + 3x - 4}{x - 1} = 1$$

We obtain

$$x^2 + 3x - 4 = x - 1$$

$$x^2 + 2x - 3 = 0$$

$$(x + 3)(x - 1) = 0$$

$$x = -3, 1$$

If we check these roots in the original equation, substituting $x = -3$ yields

$$\frac{9 - 9 - 4}{-3 - 1} = 1, \quad 1 = 1$$

and substituting $x = 1$ gives

$$\frac{1 + 3 - 4}{1 - 1} = 1, \quad \frac{0}{0} = 1$$

Division by zero is not permitted in our number system since it is undefined. Therefore, only $x = -3$ is a valid root to this equation. When an equation is multiplied by an expression containing the unknown it *may not* remain equivalent to the original equation. That is, some or all of the roots of the new equation may not be roots of the original equation. Such roots are called *extraneous roots*.

In other instances, where equations describe physical relationships, some of the resultant roots may be physically meaningless. Let us consider an example.

Example 3 The length of a rectangle is 5 in. greater than the width. The area is 24 sq in. Find the dimensions of the rectangle.

Let w denote the width. Then the length is $w + 5$, and the area is the product of the two:

$$w(w + 5) = 24 \quad \text{or} \quad w^2 + 5w - 24 = 0$$

Substituting $a = 1$, $b = 5$, and $c = -24$ into the quadratic formula,

$$w = \frac{-5 \pm \sqrt{25 + 96}}{2} = \frac{-5 \pm 11}{2}$$

The roots are thus $w = -8$ and $w = 3$. Obviously, even though the root $w = -8$ will satisfy the equation it is not a solution to the physical problem since the rectangle must have positive dimensions. The root $w = 3$ gives the solution: the width is 3 in. and the length is 5 more, or 8 in. The root $w = -8$ is a meaningless root for this problem.

Checking the roots of quadratic equations is very important since a root satisfying the equation may not be a possible solution to the problem. The roots must be checked against the conditions of the original problem.

Application: Parallel Resonance Resonance is a special phenomenon of circuits containing both L and C which occurs only at one specific frequency. At resonance a parallel circuit will appear to be a pure resistance.

Example 4 Find the value of C for which the circuit of Figure 15-1 will be resonant at a frequency of 1 kHz.

We start with the relation for the total admittance,

$$Y = \frac{1}{2 - jX_C} + \frac{1}{4 + j3}$$

$$= \frac{1}{2 - jX_C} \cdot \frac{2 + jX_C}{2 + jX_C} + \frac{1}{4 + j3} \cdot \frac{4 - j3}{4 - j3}$$

$$\text{(Review Section 14-8.)}$$

$$= \frac{2 + jX_C}{4 + X_C^2} + \frac{4 - j3}{25} = \left(\frac{2}{4 + X_C^2} + \frac{4}{25} \right)$$

$$+ j \left(\frac{X_C}{4 + X_C^2} - \frac{3}{25} \right)$$

At resonance the circuit will have no reactive component; the j part of the admittance will equal zero. Thus,

$$\frac{X_C}{4 + X_C^2} - \frac{3}{25} = 0, \quad 3(4 + X_C^2) = 25X_C$$

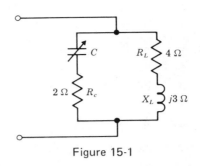

Figure 15-1

or

$$3X_C^2 - 25X_C + 12 = 0, \quad X_C = \frac{25 \pm \sqrt{625 - 144}}{6}$$

$$X_C = \frac{25 \pm \sqrt{481}}{6} = \frac{25 \pm 21.93}{6}$$

$$= 7.82 \ \Omega \quad \text{or} \quad 0.51 \ \Omega$$

Therefore, substituting into the equation,

$$X_C = \frac{1}{2\pi fC} \quad \text{with } f = 1 \text{ kHz}$$

$$C = \frac{0.15915}{fX_C} = \frac{0.15915}{1000 \times 7.82} = 20.3 \ \mu\text{F}$$

or

$$C = \frac{0.15915}{1000 \times 0.51} = 312 \ \mu\text{F}$$

Example 5 Find the value of L for which the circuit of Figure 15-2 will be resonant at $f = 100$ kHz.

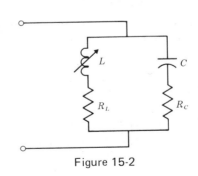

Figure 15-2

$$Y = \frac{1}{5 + jX_L} + \frac{1}{6 - j7}$$

$$= \frac{1}{5 + jX_L} \cdot \frac{5 - jX_L}{5 - jX_L} + \frac{1}{6 - j7} \cdot \frac{6 + j7}{6 + j7}$$

$$= \frac{5 - jX_L}{25 + X_L^2} + \frac{6 + j7}{36 + 49} = \left(\frac{5}{25 + X_L^2} + \frac{6}{85} \right)$$

$$+ j \left(\frac{7}{85} - \frac{X_L}{25 + X_L^2} \right)$$

Setting the j part equal to zero,

$$\frac{7}{85} - \frac{X_L}{25 + X_L^2} = 0, \qquad \frac{7}{85} = \frac{X_L}{25 + X_L^2}$$

$$X_L^2 - \frac{85}{7} X_L + 25 = 0$$

$$X_L = \frac{12.14 \pm \sqrt{147.4 - 100}}{2}$$

$$= \frac{12.14 \pm 6.89}{2}$$

$$X_L = 9.515 \ \Omega \quad \text{or} \quad 2.625 \ \Omega$$

Using the formula $X_L = 2\pi fL$,

$$L = \frac{X_L}{2\pi f} = \frac{9.515}{2\pi \times 10^5} = 15.14 \ \mu\text{H}$$

or

$$L = \frac{2.625}{2\pi \times 10^5} = 4.178 \ \mu\text{H}$$

Exercises 15-3 In Exercises 1–8 solve the given equations using the quadratic formula. Check for extraneous roots.

1. $x^2 + 6x + 5 = 0$ 2. $x^2 + 8x = 20$
3. $15x^2 - 19x + 6 = 0$ 4. $4x^2 + 19x = 30$
5. $x^2 - 5x - 9 = 0$ 6. $x^2 + x - 7 = 0$
7. $2x^2 - 3x + 4 = 0$ 8. $3x^2 - 2x + 6 = 0$
9. The length of a rectangle is 6 cm greater than its width. The area is 40 cm². Find the dimensions.
10. The width of a rectangle is 3 cm less than its length. The area is 70 cm². Find the dimensions.
11. The base of a triangle is 3 cm more than its height. Find the base when the area is 5 cm². ($A = \frac{1}{2}bh$.)
12. The height of a triangle is 4 cm more than the base. Find the height if the area is 16 cm².
13. A parallel circuit like that of Figure 15-1 is to be resonant at 500 kHz. Find the required value of C if $R_C = 3 \ \Omega$, $R_L = 2 \ \Omega$, and $X_L = 30 \ \Omega$.

14. See Exercise 13. Find C if $f = 90$ MHz, $R_C = 2 \ \Omega$, $R_L = 1 \ \Omega$, and $X_L = 5 \ \Omega$.
15. A parallel circuit like that of Figure 15-2 is to be resonant at $f = 1$ MHz. Find the required value of L if $R_C = 20 \ \Omega$, $R_L = 6 \ \Omega$, and $X_C = 10 \ \Omega$.
16. See Exercise 15. Find L if $R_C = 52 \ \Omega$, $R_L = 8 \ \Omega$, $X_C = 4 \ \Omega$, and $f = 10$ MHz.
17. See Exercises 13 and 14. Derive a general formula for X_C at resonance in terms of R_C, R_L, and X_L.
18. See Exercises 15 and 16. Derive a general formula for X_L at resonance in terms of R_C, R_L, and X_C.

15-4 THE GRAPH OF A QUADRATIC FUNCTION

Section 3-2 dealt with the graphs of equations containing two variables. Examples there included the equations $y = 2x$ and $I = 0.5V$. These are literal equations which have been solved for one variable (the one on the left) in terms of another variable. The variable on the left is called the *dependent variable* and the one on the right, the *independent variable*. Engineers and mathematicians sometimes prefer to call such statments *functions*. That is, $y = 2x$ is said to be *a function of x*, $I = 0.5V$ is *a function of V*, $y = r \sin \theta$ is a trigonometric function of θ, and so forth. There is a shorthand notation which they use which means "a function of." It is $f(x)$. Thus, $y = f(x)$ is read "y is a function of x," $f(x) = 2x$ is read "the function of x is $2x$," $f(V) = 0.5V$ is read "the function of V is $0.5V$," and $f(\theta) = r \sin \theta$ is read "the function of θ is $r \sin \theta$."

When the function (the equation) is a linear equation it is called, appropriately, *a linear function*. When the equation is a quadratic equation its function is *a quadratic function*. In the example

$$y = ax^2 + bx + c$$

the dependent variable y is equal to an expression which fits the form of a quadratic equation. We could write

$$y = f(x)$$

$$f(x) = ax^2 + bx + c$$

That is, $y = ax^2 + bx + c$ is a quadratic function.

The graph of a function can be plotted by obtaining a set of values from the function (equation):

1. Choose arbitrary values for the independent variable.

2. Then substitute each of these values into the equation (function) and calculate the corresponding value of the dependent variable.

Each type of function has a typically shaped graph. Section 3-2 demonstrated that the graph of a linear equation (or function) is a straight line. The graph of a quadratic function is a curved line called a *parabola*.

The *roots* (sometimes called the *zeros*) of a quadratic function are the points on the graph where the curve crosses the x-axis or the axis of the independent variable; that is, the points at which y, the dependent variable, equals zero. These points are also called the *intercepts*.

Example 6 (a) Obtain a set of values for the quadratic function $y = -x^2 + 6x$. (b) Plot the graph of the function. (c) Name the zeros or roots.

(a) We first construct the table of values.

x	-2	-1	0	1	2	3	4	5	6	7	8
y	-16	-7	0	5	8	9	8	5	0	-7	-16

Next, plotting the values produces the curve of Figure 15-3.

(b)   Figure 15-3

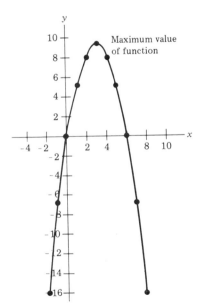

Figure 15-3

(c) The roots are $x = 0$ and $x = 6$.

Exercises 15-4 In Exercises 1–6 plot the quadratic functions on graph paper.

1. $y = x^2 + 2x + 1$
2. $y = -x^2 + 2x + 1$
3. $y = -x^2 - 3x + 2$
4. $y = x^2 + 2x - 3$
5. $y = -2x^2 + 2x - 3$
6. $y = 3x^2 - 4x + 2$

15-5 EXTREME VALUES

The curve of Figure 15-3 is similar to the cross-sectional view of an upside-down bowl. It rises to its greatest or *maximum value* and then decreases again. The fact that a function of this type has a maximum value is of considerable importance in several practical applications. In some instances, the "bowl" is right side up and the function has a least or *minimum value*. The extreme maximum or minimum value of a quadratic function can be predicted using an algebraic technique. The following facts are utilized.

1. The general formula for the roots of the general quadratic $ax^2 + bx + c = 0$ is

$$x = \frac{-b \pm \sqrt{b^2 - 4ac}}{2a}$$

2. The general formula also yields the roots or zeros of the general quadratic function $y = ax^2 + bx + c$, that is, it gives the values of x when $y = 0$. These are the values of x where the graph of the function crosses the x-axis.

3. If a is positive, the graph of the function opens upward, $\cup$, and the extreme value of y is its least value, a minimum. If a is negative, the graph opens downward, $\cap$, and the extreme value of y is its greatest value, a maximum.

4. The graph (and function) reaches its turning point or extreme value midway between the roots. This point is always when $x = -b/2a$.

5. The value of the function at its extreme is

$$y_{ex} = c - \frac{b^2}{4a}$$

Example 7 For the given quadratic functions (a) determine the extreme values of the functions, (b) state whether the extreme value is a maximum or a minimum value, and (c) determine the corresponding x-coordinate.

(i) $y = -x^2 + 2x + 1$

(a) $y_{ex} = c - \dfrac{b^2}{4a} = 1 - \dfrac{2^2}{4(-1)} = 2$

(b) y_{ex} is a maximum value because a is negative.

(c) $x = -b/2a = -2/-2 = 1$

(ii) $y = x^2 + 10x - 6$

(a) $y_{ex} = c - \dfrac{b^2}{4a} = -6 - \dfrac{100}{4} = -31$

(b) y_{ex} is a minimum because a is positive.

(c) $x = -b/2a = -10/2 = -5$

Exercises 15-5   For the given quadratic functions in Exercises 1–6 (a) determine the extreme values of the functions, (b) state whether the extreme values are maxima or minima, and (c) determine the corresponding x-coordinates.

1. $y = -x^2 + 2x + 1$
2. $y = x^2 + 2x + 1$
3. $y = x^2 + 2x - 3$
4. $y = -x^2 - 3x + 2$
5. $y = 3x^2 - 4x + 2$
6. $y = -2x^2 + 2x - 3$

The power in the load R_L of a series circuit (see Figure 15-4) is given by the quadratic

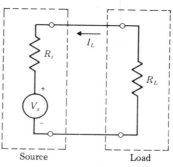

Source   Load

Figure 15-4

function

$$P_L = -I_L^2 R_i + V_s I_L$$

In Exercises 7–10 determine (a) maximum P_L, and the corresponding values of (b) I_L, and (c) R_L. (Use the relation $I_L = V_s/(R_i + R_L)$ to determine R_L.) The given values are for V_s and R_i, respectively.

7. 10 V, 5 Ω   8. 12 V, 2.4 kΩ

9. 2 V, 600 Ω   10. 6 V, 500 kΩ

11. Using the concept of the maximum value of a quadratic function prove the statement "Maximum power is transferred to a load, R_L, when $R_L = R_i$." (See Figure 15-4.)

Chapter 16

Logarithms and Applications

The decibel *(dB), an application of logarithms, is an everyday word in broadcast station control rooms. (Christelon/Jeroboam, Inc.)*

16-1 DEFINITION OF A LOGARITHM

Chapter 2 presented exponents and the rules for their use in considerable detail. Included there was a notation of the form x^n called *exponential notation*. Furthermore, in applications of scientific and powers-of-ten notation, exponential notation has been used extensively in a particular set in which $x = 10$ and n is either a positive or negative integer. There is another algebraic form which uses exponents, a form in which exponents are called *logarithms*.

To define a logarithm (log) let us first examine the exponential form,

$$N = b^x$$

in which the exponent x is a variable and can be any real number, and b can be any positive number other than 1. The equation says that "any number N can be obtained by raising the base b to an appropriate power as expressed by the exponent x." The equivalent logarithmic form is

$$x = \log_b N$$

that is, "x is the logarithm, to the base b, of the number N."

> **The logarithm of a number N is the exponent expressing the power to which a base quantity must be raised to equal the given number.**

Example 1 State the following relationships (equations) in logarithmic form.

(a) $100 = 10^2$ | (a) $\log_{10} 100 = 2$
(b) $8 = 2^3$ | (b) $\log_2 8 = 3$
(c) $2 = 10^{0.3010}$ | (c) $\log_{10} 2 = 0.3010$
(d) $5 = 10^{0.6990}$ | (d) $\log_{10} 5 = 0.6990$
(e) $2000 = 2 \times 10^3$
$\qquad = 10^{0.3010} \times 10^3 = 10^{3.3010}$
(e) $\log_{10} 2000 = 3.3010$
(f) $y = e^t$
(f) $\log_e y = t$
(g) $\dfrac{I}{I_0} = e^{t/RC}$
(g) $\log_e \dfrac{I}{I_0} = t/RC$

Example 2 State the following logarithmic equations in exponential form.

(a) $\log_{10} 1000 = 3$
(a) $10^3 = 1000$
(b) $\log_2 16 = 4$
(b) $2^4 = 16$
(c) $\log_{10} 20,000 = 4.3010$
(c) $10^{4.3010} = 20,000$
(d) $\log_e 6.75 = 1.9095$
(d) $e^{1.9095} = 6.75$
$\qquad$ ($e = 2.7183$ is the base of a set of logarithms called *natural logarithms*.)
(e) $\log_e y = t$
(e) $e^t = y$
(f) $\log_e \dfrac{I}{I_0} = \dfrac{t}{RC}$
(f) $e^{t/RC} = \dfrac{I}{I_0}$

The applications of logarithms in electronics require frequent conversions between the exponential and logarithmic forms. It is important that you practice such conversions so that you may develop speed and confidence in performing them.

Exercises 16-1 In Exercises 1–10 convert the given equations to logarithmic form.

1. $2^2 = 4$, $3^3 = 27$
2. $4^{0.5} = 2$, $9^{0.5} = 3$
3. $10^{0.5} = 3.1623$, $16^{0.25} = 2$
4. $25^{0.5} = 5$, $10^{0.3010} = 2$
5. $10^{0.4771} = 3$, $10^{2.4771} = 300$
6. $10^{0.6021} = 4$, $10^{3.6021} = 4000$
7. $V = p^x$, $I = q^t$
8. $y = e^{t/t_0}$, $x = e^{t/RC}$ [See Example 2(d).]
9. $e^{1.609} = 5$, $e^{3.912} = 50$
10. $e^{2.0794} = 8$, $e^{4.3820} = 80$
11. $2^{-2} = \frac{1}{4}$, $4^{-2} = \frac{1}{16}$
12. $5^{-2} = \frac{1}{25}$, $25^{-0.5} = \frac{1}{5}$
13. $10^{-2} = 0.01$, $10^{-3} = 0.001$
14. $10^{-0.301} = 0.5$, $10^{-0.602} = 0.25$

In Exercises 15–28 convert the given logarithmic equations to exponential notation.

15. $\log_4 2 = 0.5$, $\log_9 3 = 0.5$

16. $\log_2 4 = 2$, $\log_3 27 = 3$
17. $\log_{25} 5 = 0.5$, $\log_{10} 4000 = 3.6021$
18. $\log_{10} 300 = 2.4771$, $\log_{10} 4 = 0.6021$
19. $\log_{10} 2 = 0.3010$, $\log_{10} 3 = 0.4771$
20. $\log_{10} 10 = 1$, $\log_{10} 1 = 0$
21. $\log_{10} 6 = 0.7782$, $\log_{10} 6000 = 3.7782$
22. $\log_e 5 = 1.609$, $\log_e 50 = 3.912$
23. $\log_{10} 0.5 = -0.3010$, $\log_{10} 0.25 = -0.6021$
24. $\log_2 0.5 = -1.0$, $\log_2 0.25 = -2.0$
25. $\log_e 0.5 = -0.6931$, $\log_e 0.25 = -1.3863$
26. $\log_e 0.1 = -2.3026$, $\log_e 0.01 = -4.6052$
27. $\log_p V = x$, $\log_q I = t$
28. $\log_e y = t/t_0$, $\log_e V = t/RC$

16-2 LAWS FOR EXPONENTS APPLIED TO LOGARITHMS

A logarithm is an exponent and the laws for exponents (review Sections 2-3, 2-6, and 2-7) apply.

Products Restating the law for products involving exponents we have

$$b^u \cdot b^v = b^{u+v}$$

Letting $N = b^u$ and $M = b^v$, then

$$N \cdot M = b^u \cdot b^v = b^{u+v}$$

and converting to logarithmic form, we have

$$\log_b N = u, \qquad \log_b M = v,$$

$$\log_b (N \cdot M) = u + v$$

Thus,

$$\log_b (N \cdot M) = \log_b N + \log_b M$$

The logarithm of the product of two numbers is equal to the sum of the logarithms of the numbers.

$$\log_b (N \cdot M) = \log_b N + \log_b M$$

Quotients Proceeding in a similar fashion from the law for quotients involving exponents,

$$\frac{b^u}{b^v} = b^{u-v}$$

we write

$$\frac{N}{M} = \frac{b^u}{b^v} = b^{u-v}$$

and, then, converting to logarithmic form,

$$\log_b \frac{N}{M} = u - v$$

But

$$u - v = \log_b N - \log_b M$$

Thus,

$$\log_b \frac{N}{M} = \log_b N - \log_b M$$

The logarithm of the quotient of two numbers is equal to the logarithm of the numerator minus the logarithm of the denominator.

$$\log_b \frac{N}{M} = \log_b N - \log_b M$$

Power of a Power Let us review the law for raising a power to a power:

$$(b^u)^v = b^{u \cdot v}$$

Thus, if

$$N = b^u$$

then the vth power of N is

$$N^v = (b^u)^v = b^{u \cdot v}$$

But

$$\log_b N = u$$

and

$$\log_b N^v = uv$$

Therefore,

$$\log_b N^v = v \log_b N$$

> The logarithm of the vth power of a number is equal to v times the logarithm of the number.
>
> $$\log_b N^v = v \log_b N$$

Since it has not been stated to the contrary, v may be a fraction as well as an integer and this property of a logarithm may be used to find the root, as well as the power, of a number.

There are two additional properties of logarithms that we should be familiar with and use when they are helpful. First, since

$$b = b^1, \quad \text{then} \quad \log_b b = 1$$

The logarithm to the base b of b is 1.

Then, since any number to the zero power is 1,

$$b^0 = 1$$

Then

$$\log_b 1 = 0$$

The logarithm of 1 to any base is 0.

Example 3   Express the given logarithms as a sum, difference, or product. Simplify the result, if possible.

(a) $\log_b IR$

(a) $\log_b IR = \log_b I + \log_b R$

(b) $\log_{10} V/R$

(b) $\log_{10} V/R = \log_{10} V - \log_{10} R$

(c) $\log_{10} V^2$

(c) $\log_{10} V^2 = 2 \log_{10} V$

(d) $\log_5 (5 \cdot 6)$

(d) $\log_5 (5 \cdot 6) = \log_5 5 + \log_5 6 = 1 + \log_5 6$

(e) $\log_{10} 3000$

(e) $\log_{10} 3000 = \log_{10} (3 \times 10^3) = \log_{10} 3 + \log_{10} 10^3 = \log_{10} 3 + 3$

Example 4   Express each of the following as single logarithms.

(a) $\log_b x + \log_b y$

(a) $\log_b x + \log_b y = \log_b xy$

(b) $\log_{10} P_2 - \log_{10} P_1$

(b) $\log_{10} P_2 - \log_{10} P_1 = \log_{10} \dfrac{P_2}{P_1}$

(c) $\log_2 3 + \log_2 6$

(c) $\log_2 3 + \log_2 6 = \log_2 18$

(d) $\frac{1}{2} \log_{10} P - \frac{1}{2} \log_{10} R$

(d) $\frac{1}{2} \log_{10} P - \frac{1}{2} \log_{10} R = \frac{1}{2} \log_{10} \dfrac{P}{R} = \log_{10} \sqrt{\dfrac{P}{R}}$

Example 5   Solve for I in terms of P and R, if

$$\log_b I = \tfrac{1}{2} \log_b P - \tfrac{1}{2} \log_b R$$

We combine the two terms in the right member,

$$\log_b I = \log_b \sqrt{\dfrac{P}{R}}$$

If the logarithms of two numbers are equal, the numbers must be equal; thus,

$$I = \sqrt{\dfrac{P}{R}}$$

Exercises 16-2   In Exercises 1–20 express the logarithms as a sum, difference, or product. Simplify when possible.

1. $\log_b pq$ 2. $\log_{10} VG$ 3. $\log_2 VI$
4. $\log_{10} 10I$ 5. $\log_{10} x/y$
6. $\log_{10} P_o/P_i$ 7. $\log_2 V/5$
8. $\log_3 9/2$ 9. $\log_5 V^2$ 10. $\log_{10} I^2$
11. $\log_b x^5$ 12. $\log_{10} 176^5$
13. $\log_b x^2 y$ 14. $\log_{10} V^2/R$
15. $\log_2 I^2 R$ 16. $\log_{10} \sqrt{P/R}$
17. $\log_2 \sqrt{PR}$ 18. $\log_3 \sqrt[3]{V^2}$
19. $\log_2 \dfrac{V_1^2 R_2}{V_2^2 R_1}$ 20. $\log_{2.7} \dfrac{I_1^2 R_1}{I_2^2 R_2}$

In Exercises 21–32 express the given quantities as single logarithms.

21. $\log_b I + \log_b R$
22. $\log_{10} x + \log_{10} y$
23. $\log_2 R + \log_2 r$
24. $\log_{10} 460 + \log_{10} 375$
25. $\log_2 b - \log_2 a$
26. $\log_{10} V - \log_{10} R$

27. $\log_{2.7} I - \log_{2.7} i$

28. $\log_5 75 - \log_5 26$

29. $\frac{1}{2} \log_{10} P + \frac{1}{2} \log_{10} R$

30. $2 \log_{10} P_1 - 2 \log_{10} P_2$

31. $\frac{4}{3} \log_{2.7} x + \frac{2}{3} \log_{2.7} y$

32. $2 \log_{10} I_1 + \log_{10} R_1 - 2 \log I_2 - \log R_2$

33. If $\log_b V = \log_b I + \log_b R$, solve for V in terms of I and R.

34. Given: $\log_{10} I = \log_{10} V - \log_{10} R$. Solve for I.

35. Given: $\log_a P = 2 \log_a V - \log_a R$. Solve for P.

36. If $\log_b V = \frac{1}{2} \log_b P + \frac{1}{2} \log_b R$, solve for V.

37. Given: $\log_b v = \log_b V - t_x$. Solve for v. (*Hint*: $-t_x = \log_b b^{-t_x}$)

38. If $\log_e i = \log_e I - t/RC$, solve for i.

16-3 COMMON LOGARITHMS

Logarithms to the base 10 are the most frequently used logarithms for computation and are known as *common logarithms*. The base subscript is normally not written with the abbreviation "log" in the expressions for common logarithms (because this form of logarithm is so common). Thus "$\log x$" means "$\log_{10} x$."

Since our number system is a "base 10" system, common logarithms have unique characteristics. Recall that any number, however large or small, can be expressed in scientific notation as a number between 1 and 10 times a power of ten. For example,

$$486{,}000 = 4.86 \times 10^5$$

$$0.0000486 = 4.86 \times 10^{-5}$$

We take the logarithm of these numbers,

$$\log 486{,}000 = \log 4.86 + \log 10^5$$

$$= 5 + \log 4.86$$

$$\log 0.0000486 = \log 4.86 + \log 10^{-5}$$

$$= -5 + \log 4.86$$

Thus, we see that a table of logarithms to the base 10, usable for any number, need include only the logarithms of numbers between 1 and 10 to the desired degree of precision (four places, or six places, and so forth). Table I in the Appendix is such a table of four-place common logarithms.

In general, if we express a number in scientific notation in the form

$$N = M \times 10^c$$

where M is a number between 1 and 10, then

$$\log N = c + \log M$$

In this expression, c, which is equal to the exponent of ten for the number in the scientific notation, is called the *characteristic* of the logarithm. Log M is the logarithm of a number between 1 and 10 and is the part found in the table. It is called the *mantissa* of the logarithm. In the following examples we demonstrate how to use the table to find $\log M$.

Example 6   Find log 486,000 using Table I.

We start by writing the number in scientific notation:

$$486{,}000 = 4.86 \times 10^5$$

Next, in Table I we look down the column headed N until we come to 48, the first two significant digits of the number. Then we look across the top of the table to find the column headed by the third significant digit, 6. Looking down this column to the row on which 48 was located we find "6866." Thus,

$$\log 4.86 = 0.6866$$

and

$$\log 4.86 \times 10^5 = 5 + 0.6866 = \underline{5.6866}$$

Example 7   Find log 0.0000486.

First, we write

$$0.0000486 = 4.86 \times 10^{-5}$$

Then

$$\log 0.0000486 = \log 4.86 \times 10^{-5}$$

$$= -5 + \log 4.86$$

but we have already "looked up" log 4.86 in Example 6. Therefore, we write

$$\log 0.0000486 = -5 + 0.6866 = \underline{-4.3134}$$

Example 8   Find log 7120.

$$\log 7120 = \log 7.12 \times 10^3$$

$$\log 7.12 \times 10^3 = 3 + 0.8525 = \underline{3.8525}$$

If a number has four or more significant digits, the task of finding the logarithm in a four-place table is more difficult and involves the process called *interpolation*. However, a far superior method of finding logarithms, in terms of both accuracy and time saving, is by means of an electronic calculator with the logarithm function. The function is represented by a key typically labeled "log *x*."

Logarithms by Electronic Calculator   In an electronic calculator with the logarithm function, you have at your command, at the touch of the [log *x*] key, a table look-up service which will "look up" the logarithm of the number you have entered on the unit. The "service" provides the characteristic of the logarithm as well as the mantissa. Mantissas may contain only four digits on the least expensive calculators, or up to nine or more digits on more expensive units.

On some inexpensive calculators only natural logarithms (natural logarithms are presented in Section 16-5) are available directly. However, these can easily be converted to common logarithms using the formula

$$\log X = \ln X \div 2.30259$$

(The notation ln *x* means "the natural logarithm of *x*, or $\log_e x$." *e* = 2.718281.) Study the instruction manual for your calculator and learn the procedure for finding the common logarithm of a number. Check yourself with the following examples.

$$\log 4.6895 = 0.6711$$
$$\log 2.6895 \times 10^5 = 5.4297$$
$$\log 586395 = 5.7682$$
$$\log 3.8968 \times 10^{-3} = -2.4093$$
$$\log 0.009315 = -2.0308$$
$$\log 416.59 \times 10^{-4} = -1.3803$$

Exercises 16-3   In Exercises 1–12 use your calculator to find the logarithms (base 10) of the given numbers.

1. 4.56, 4560, 456,000
2. 2.23, 23.2, 232
3. 0.468, 0.00468, 46.8
4. 0.0579, 0.000579, 579
5. 6.18, 618, 0.00618
6. 654,000, 0.654, 0.0000654
7. 0.00826, 8.26×10^3, 82,600
8. 91.7, 0.0917, 91.7×10^{-5}
9. 9.53, 95.3, 0.0953
10. 1.08, 0.00108, 108
11. 0.0000191, 1.09×10^{-8}, 1091
12. 0.235, 0.000235, 23,500

In Exercises 13–24 the logarithms of the given numbers are stated along with the numbers. Write the base-10 exponential equivalent of the numbers. (Example: log 0.00468 = −2.3298; $0.00468 = 10^{-2.3298}$.)

13. $\log 45{,}860 = 4.6614$
14. $\log 45.86 = 1.6614$
15. $\log 0.000468 = -3.3298$
16. $\log 0.465 = -0.3325$
17. $\log 2{,}358{,}000 = 6.3725$
18. $\log 235.8 = 2.3725$
19. $\log 6059 = 3.7824$
20. $\log 2.79 \times 10^{-5} = -4.5544$
21. $\log 605{,}900 = 5.7824$
22. $\log 72.35 \times 10^{-6} = -4.1406$
23. $\log 975.8 = 2.9894$
24. $\log 9.758 \times 10^8 = 8.9894$

In Exercises 25–34 the given numbers are stated in common notation and in the equivalent base-10 exponential form. Write the logarithms of the numbers using the information given. (Example: 4586 = $10^{3.6614}$; log 4586 = 3.6614.)

25. $458{,}600 = 10^{5.6614}$
26. $23{,}580 = 10^{4.3725}$
27. $0.000468 = 10^{-3.3298}$
28. $0.00465 = 10^{-2.3325}$
29. $2.79 \times 10^{-3} = 10^{-2.5544}$
30. $605{,}900 = 10^{5.7824}$
31. $72.35 \times 10^{-8} = 10^{-6.1406}$
32. $1.059 \times 10^2 = 10^{2.0249}$
33. $9.758 \times 10^1 = 10^{1.9894}$
34. $37.19 \times 10^3 = 10^{4.5704}$

16-4 ANTILOGARITHMS

The preceding sections have developed the concept of logarithms and how to use the look-up service of a calculator to obtain the logarithm of any number. Before we can use logarithms in their usual applications we must be able to find the number N which corresponds to a given logarithm L. The process is called finding the *antilogarithm* (antilog) of L. If

$$L = \log_b N$$

then

$$N = \text{antilog}_b L = b^L$$

Let us remember that, for common logarithms,

$$N = 10^L$$

Hence, if our calculator has the function 10^x, we find the antilogarithm simply by entering L into the calculator register and then touching the 10^x key.

Example 9 Find the antilogs of the following logarithms.

(a) 3.7824   (b) 8.9894
(c) -2.1316   (d) -3.5302

(a) $10^{3.7824} = \underline{6059}$
(b) $10^{8.9894} = \underline{9.759 \times 10^8}$
(c) $10^{-2.1316} = \underline{0.007386}$
(d) $10^{-3.5302} = \underline{2.9499 \times 10^{-4}}$

If your calculator incorporates the functions labeled "ln x" and "e^x" but not "log x" and "10^x" you can find the common logarithm of any number by using the ln x function and the formula (see Section 16-3),

$$\log x = \ln x \div 2.30259$$

Similarly, you can find the antilog $L = N$ of a logarithm on these calculators using the following procedure.

Let

$$L = \ln x = \log x \cdot 2.30259$$

Then

$$N = e^L$$

That is, multiply the $\log_{10} N$ by 2.30259 to obtain the natural logarithm of N, ln N. Then, find the antilogarithm of ln N by touching the e^x key.

Example 10 Find the antilogs of the following base-10 logarithms by using (a) the formula ln $x = \log x \cdot 2.30259$, and then (b) the e^x key of your calculator.

(a) 3.7824   (b) 8.9894
(c) -2.1316   (d) -3.5302

(a) $e^{3.7824 \cdot 2.30259} = \underline{6059}$
(b) $e^{8.9894 \cdot 2.30259} = \underline{9.759 \times 10^8}$
(c) $e^{-2.1316 \cdot 2.30259} = \underline{0.007386}$
(d) $e^{-3.5302 \cdot 2.30259} = \underline{2.9498 \times 10^{-4}}$

Exercises 16-4 In Exercises 1–10 use the 10^x function of your calculator to convert the given expressions to common notation. (Example: $10^{3.7824} = 6059$; $10^{3.7824} = e^{3.7824 \cdot 2.30259}$.)

1. $10^{3.7543}$, $10^{-4.2457}$
2. $10^{3.0899}$, $10^{-3.9101}$
3. $10^{4.0251}$, $10^{-2.9749}$
4. $10^{6.6614}$, $10^{-4.3386}$
5. $10^{2.7447}$, $10^{-2.2553}$
6. $10^{0.9488}$, $10^{-2.0512}$
7. $10^{0.3654}$, $10^{-0.3654}$
8. $10^{7.3973}$, $10^{1.5638}$
9. $10^{0.05638}$, $10^{-1.3572}$
10. $10^{11.6832}$, $10^{-3.6895}$

In Exercises 11–20 the given numbers are common logarithms. Use your calculator to find the antilogarithms.

11. 3.7824, -2.1316   12. 8.9894, -3.5302
13. 4.6614, -2.3386   14. 6.0251, -4.9749
15. 4.9281, -3.0719   16. 2.7543, -5.2457
17. 0.0899, -0.9101   18. 2.7251, -2.2749
19. 3.9488, -5.0512   20. 6.7447, -5.2553

16-5 LOGARITHMS TO BASES OTHER THAN 10

For a number of applications of logarithms in mathematics the *irrational number e* = 2.718281 . . . is used as the base (an irrational number is one which is not equivalent to one integer divided by another). Logarithms to the base e may be written $\log_e N$ but are more commonly written *ln N*, and are called *natural logarithms*.

The number e plays a part in many natural phenomena. (The Greek letter epsilon, ϵ, is also frequently used instead of the lower-case Roman *e* to represent the base of natural logarithms.) For example, the charging of a capacitor in a dc circuit can be represented mathematically with the use of *e*. The decay of the current in an inductance can be similarly represented.

Change of Base   It is sometimes necessary to change logarithms from one base to another. For example, some calculators provide for ln *x* but not log *x*, while log *x* is required for applications such as decibels in electronics. Suppose, then, that we want to find $\log_a N$ but only have the means to find logarithms to base *b*. If

$$\log_a N = u, \quad \text{then } a^u = N$$

Taking logarithms to the base *b* of $a^u = N$, we have

$$u \log_b a = \log_b N$$

and, solving for *u* ($u = \log_a N$, as desired),

$$u = \frac{\log_b N}{\log_b a}$$

Thus, if *a* = 10 and *b* = *e*,

$$u = \log N = \frac{\ln N}{\ln 10} = \frac{\ln N}{2.30259}$$

and if *a* = *e* and *b* = 10,

$$u = \ln N = \frac{\log N}{\log e} = \frac{\log N}{0.43429}$$

We summarize:

$$\log_a N = \frac{\log_b N}{\log_b a}$$

$$\log N = \frac{\ln N}{2.30259}$$

$$\ln N = \frac{\log N}{0.43429}$$

Example 11   Use your calculator to find the natural logarithms.

(a) ln 468   (a) ln 468 = <u>6.14847</u>

(b) ln 0.00576   (b) ln 0.00576 = -<u>5.15682</u>

Example 12   Find the antilogarithms of the natural logarithms.

(a) 2.15378 = ln *N*

(a) If $x = \ln N, N = e^x$
antiln 2.15378 = $e^{2.15378}$ = <u>8.617</u>

(b) -3.61589 = ln *N*

(b) antiln -3.61589 = $e^{-3.61589}$ = <u>0.02689</u>

Example 13   Use the formula log *x* = ln *x* ÷ 2.30259 to find the common logarithms.

(a) log 650, ln 650 = 6.47697

(a) log 650 = <u>2.8129</u>

(b) log 0.0479, ln 0.0479 = -3.03864

(b) log 0.0479 = -<u>1.3197</u>

Example 14   Use the formula $\log_a x = \dfrac{\log_b x}{\log_b a}$ to find the logarithms.

(a) $\log_2 16$

(a) $\log_2 16 = \dfrac{\log 16}{\log 2} = \dfrac{1.2041}{0.3010} = \underline{4.0}$

(b) $\log_{16} 4695$

(b) $\log_{16} 4695 = \dfrac{\ln 4695}{\ln 16} = \dfrac{8.4543}{2.7726} = \underline{3.0492}$

Exercises 16-5   In Exercises 1–10 use your calculator to find the natural logarithms of the given numbers.

1. 413, 0.00413
2. 0.561, 5610
3. 219,000, 0.0219
4. 159, 0.159
5. 0.001076, 107.6
6. 701.7, 0.07017
7. 8715, 87.15
8. 5.35, 53,500
9. 632, 0.0632
10. 37,710, 377.1

In Exercises 11–20 use your calculator to find the antilogarithm of the given natural logarithms. (If ln $N = x$, $N = e^x$.)

11. 5.93251, 10.53768
12. -2.76145, 6.44889
13. 10.88744, 1.67709
14. 9.07280, 4.46763
15. -2.65683, 6.55351
16. -6.83450, 4.67842
17. -1.83885, 5.06890
18. 12.29683, -3.82127
19. 8.63231, -0.57803
20. 6.02345, -5.48948

In Exercises 21–26 use the formula $\log x =$ $\ln x \div 2.30259$ to find the common logarithms.

21. $\log 35$, $\ln 35 = 3.5553$
22. $\log 750$, $\ln 750 = 6.62007$
23. $\log 0.56$, $\ln 0.56 = -0.57982$
24. $\log 0.00108$, $\ln 0.00108 = -6.83079$
25. $\log 2300$, $\ln 2300 = 7.74066$
26. $\log 6.7 \times 10^4$, $\ln 6.7 \times 10^4 = 11.11245$

In Exercises 27–32 use the formula $\ln x =$ $\log x \div 0.43429$ to find the natural logarithms.

27. $\ln 55$, $\log 55 = 1.74036$
28. $\ln 5500$, $\log 5500 = 3.74036$
29. $\ln 0.0215$, $\log 0.0215 = -1.66756$
30. $\ln 6.79 \times 10^5$, $\log 6.79 \times 10^5 = 5.83187$
31. $\ln 8.03 \times 10^{-4}$, $\log 8.03 \times 10^{-4} =$ -3.09528
32. $\ln 7.07 \times 10^{-2}$, $\log 0.0707 = -1.15058$

In Exercises 33–36 use the formula $\log_a N = \dfrac{\log_b N}{\log_b a}$ to find the logarithms.

33. $\log_2 8$
34. $\log_8 4096$
35. $\log_{16} 256$
36. $\log_{25} 1575$

16-6 APPLICATIONS OF LOGARITHMS IN ELECTRONICS

Logarithms were developed primarily as an aid to manual computation. This application has been largely superseded because of the universal availability of the pocket calculator (however, calculators may use logarithms internally in performing their calculations). Nevertheless, logarithms are still applied extensively in electronics because (a) the natural relationship between quantities is logarithmic, or (b) the use of the logarithmic relationship is the most meaningful way to define a unit. An example of (a) is in the formula for the characteristic impedance of an air-insulated parallel-conductor transmission line (see Figure 16-1),

$$Z_0 = 276 \log \frac{b}{a}$$

where $Z_0 =$ Characteristic impedance
$b =$ Center-to-center distance between conductors
$a =$ Radius of conductor (in same units as b)

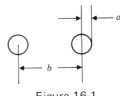

Figure 16-1

An example of application (b) is the decibel unit, the subject of Section 16-7.

Example 15   Calculate the characteristic impedance of a parallel-conductor transmission line in which No. 18 gauge wire is spaced 5 cm from center to center (diam. No. 18 = 0.102 cm. See Table VI in Appendix.)

$$Z_0 = 276 \log \frac{5}{0.102/2} = 276 \log 98.04$$
$$= \underline{549.6 \ \Omega}$$

Example 16   The characteristic impedance of an air-insulated coaxial transmission line (see Figure 16-2) can be calculated using the formula

$$Z_0 = 138 \log \frac{b}{a}$$

where $b =$ Inside diameter of outer conductor
$a =$ Outside diameter of inner conductor

Find Z_0 for a line in which $b = 1.5$ cm, $a = 0.5$ cm.

$$Z_0 = 138 \log \frac{1.5}{0.5} = 138 \log 3 = \underline{65.8 \ \Omega}$$

Example 17   A coaxial cable having a $Z_0 = 75 \ \Omega$ is desired. The center conductor is to be No. 14 wire. What must be the inside diameter of the outer conductor?

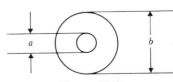

Figure 16-2

We start with the formula for Z_0,

$$Z_0 = 138 \log \frac{b}{a}$$

Then

$$\log \frac{b}{a} = \frac{Z_0}{138}$$

And, taking the antilog of both members of the equations,

$$\frac{b}{a} = \text{antilog} \frac{Z_0}{138}$$

$$b = a \text{ antilog} \frac{Z_0}{138}$$

From Table VI, the diameter of No. 14 wire is 0.163 cm. Thus,

$$b = 0.163 \text{ antilog} \frac{75}{138} = 0.163 \times 10^{0.5435}$$

$$= 0.163 \times 3.495 = \underline{0.5697 \text{ cm}}$$

Exercises 16-6 In Exercises 1–6 find the characteristic impedance of parallel-conductor air-insulated transmission lines with the given values of center-to-center separation, and radius, respectively.

1. 1 in., 0.016 in.
2. 3 cm, 0.103 cm
3. 0.75 in., 0.020 in.
4. 7.5 cm, 0.081 cm
5. 4 in., No. 14 gauge
6. 10 cm, No. 16 gauge

In Exercises 7–12 find the characteristic impedance of air-insulated coaxial transmission lines with the given values for the dimensions b and a, respectively (see Example 16).

7. 0.82 in., 0.074 in.
8. 0.68 in., 0.022 in.
9. 1.17 cm, 0.081 cm
10. 0.927 cm, 0.163 cm
11. 1.5 cm, 0.326 cm
12. 2.5 cm, 0.93 cm
13. An air-insulated parallel-conductor transmission line is to be constructed with No. 14 gauge wire. A Z_0 of 300 Ω is desired. What center-to-center spacing should be used? Express your result in centimeters.

14. Repeat Exercise 13 for a $Z_0 = 600$ Ω using No. 12 gauge wire.
15. What is the inside diameter of the outer conductor of an air-insulated coaxial line if $Z_0 = 52$ Ω and the center conductor is No. 18 wire?
16. Repeat Exercise 15 for $Z_0 = 75$ Ω and a center conductor of No. 12 wire.

16-7 DECIBELS

The most common application of logarithms in electronics is in the use of the *decibel* unit, defined as follows:

$$\textit{Number of decibels} = 10 \log_{10} \frac{P_1}{P_2}$$

where P_1 and P_2 are two signal power levels to be compared.

In the early days of telephone a unit was needed to indicate the relative signal power levels at various points in a telephone system. The unit desired was to correspond closely with the way the human ear perceives the relative power levels. The basic unit defined for this purpose was the *bel* named in honor of the inventor of the telephone.

$$\text{Number of bels} = \log \frac{P_1}{P_2}$$

The decibel is a smaller, more convenient submultiple unit.

The decibel unit was defined with the logarithm function because the human sense of hearing, like other senses, responds logarithmically, rather than directly, to the relative levels of stimuli. For example, the level of sound energy from a jet airplane may be 10,000 times that of a hand clap. However, our perception of a comparison of the "loudness" of the sounds is described more meaningfully by a comparison expressed as the logarithm of the ratio of the levels,

$$N_{\text{dB}} = 10 \log 10,000 = 40$$

than by the direct ratio—10,000.

Decibels are commonly used to express the gain of some electronic device such as an amplifier. If we define the power gain of an amplifier as

$$\text{Power gain} = A_P = \frac{P_{\text{out}}}{P_{\text{in}}}$$

then the gain can be expressed in decibels as follows:

$$\text{Power gain in decibels} = 10 \log \frac{P_o}{P_i}$$

$$= 10 \log A_p \quad (1)$$

We know that electrical power can be expressed in terms of the magnitude of the voltage or current and resistance. (All V and I terms below represent magnitude only.)

$$P = \frac{V^2}{R} \quad (2)$$

$$P = I^2 R \quad (3)$$

If we substitute equation (2) in the formula for decibels, equation (1), we have

$$\text{dB gain} = 10 \log \frac{P_o}{P_i} = 10 \log \frac{V_o^2}{R_o} \div \frac{V_i^2}{R_i}$$

or

$$\text{dB gain} = 10 \log \frac{V_o^2}{V_i^2} \frac{R_i}{R_o}$$

$$= 10 \log \frac{V_o^2}{V_i^2} + 10 \log \frac{R_i}{R_o} \quad (4)$$

If the output resistance R_o equals the input resistance R_i, that is, if $R_o = R_i$ in equation (4), then the ratio R_i/R_o is 1. Since the log of 1 is zero, the second term of (4) drops out. Then we have

$$\text{dB gain} = 10 \log \frac{V_o^2}{V_i^2} = 10 \log \left(\frac{V_o}{V_i}\right)^2 \quad (5)$$

But since

$$\log X^2 = 2 \log X$$

equation (5) becomes

$$\text{dB} = 20 \log \frac{V_o}{V_i} \quad (6)$$

Remember that equation (6) applies *only* when $R_o = R_i$; otherwise equation (4) or equation (1) must be used.

If we had substituted equation (3) in equation (1) and proceeded in the same manner, we would have

$$G_{dB} = 10 \log \frac{I_o^2 R_o}{I_i^2 R_i} \quad (7)$$

$$= 10 \log \frac{I_o^2}{I_i^2} + 10 \log \frac{R_o}{R_i} \quad (8)$$

$$= 20 \log \frac{I_o}{I_i} + 10 \log \frac{R_o}{R_i} \quad (9)$$

Again, if $R_o = R_i$, then

$$10 \log \frac{R_o}{R_i} = 0$$

and

$$G_{dB} = 20 \log \frac{I_o}{I_i} \quad (10)$$

We summarize the results.

$$\text{Gain in dB} = 10 \log_{10} \frac{P_o}{P_i} \quad (1)$$

$$= 20 \log_{10} \frac{|V_o|}{|V_i|} + 10 \log \frac{R_i}{R_o} \quad (11)$$

$$= 20 \log_{10} \frac{|I_o|}{|I_i|} + 10 \log \frac{R_o}{R_i} \quad (9)$$

$$= 20 \log_{10} \frac{|V_o|}{|V_i|} \left.\begin{array}{c} \\ \\ \end{array}\right\} \text{ when } R_o = R_i$$

$$= 20 \log_{10} \frac{|I_o|}{|I_o|} \quad (10)$$

Table 16-1 shows the relationship between some voltage/current ratios, power ratios, and decibels.

Gain Less Than 1 Not all electronic devices, circuits, or systems produce gain—an increase in the power level of the signal.

When $P_o = P_i$, power gain is 1: $G = P_o/P_i = 1$, and the gain in decibels is 0:

$$G_{dB} = 10 \log 1 = 0$$

When $P_o < P_i$, then $0 <$ power gain < 1, and $G_{dB} = 10 \log (P_o/P_i)$ is less than 0—it is negative.

The gain in decibels is negative when the power ratio is less than 1 because the logarithm of a number less than 1 is negative.

Table 16-1
Decibels, and Current, Voltage, and Power Ratios

dB	Current and Voltage Ratio		Power Ratio		dB	Current and Voltage Ratio		Power Ratio	
	Gain	Loss	Gain	Loss		Gain	Loss	Gain	Loss
0.1	1.01	0.989	1.02	0.977	8.0	2.51	0.398	6.31	0.158
0.2	1.02	0.977	1.05	0.955	8.5	2.66	0.376	7.08	0.141
0.3	1.03	0.966	1.07	0.933	9.0	2.82	0.355	7.94	0.126
0.4	1.05	0.955	1.10	0.912	9.5	2.98	0.335	8.91	0.112
0.5	1.06	0.944	1.12	0.891	10.0	3.16	0.316	10.00	0.100
0.6	1.07	0.933	1.15	0.871	11.0	3.55	0.282	12.6	0.079
0.7	1.08	0.923	1.17	0.851	12.0	3.98	0.251	15.8	0.063
0.8	1.10	0.912	1.20	0.832	13.0	4.47	0.224	19.9	0.050
0.9	1.11	0.902	1.23	0.813	14.0	5.01	0.199	25.1	0.040
1.0	1.12	0.891	1.26	0.794	15.0	5.62	0.178	31.6	0.032
1.1	1.13	0.881	1.29	0.776	16.0	6.31	0.158	39.8	0.025
1.2	1.15	0.871	1.32	0.759	17.0	7.08	0.141	50.1	0.020
1.3	1.16	0.861	1.35	0.741	18.0	7.94	0.126	63.1	0.016
1.4	1.17	0.851	1.38	0.724	19.0	8.91	0.112	79.4	0.013
1.5	1.19	0.841	1.41	0.708	20.0	10.00	0.100	100.0	0.010
1.6	1.20	0.832	1.44	0.692	25.0	17.7	0.056	3.16×10^2	3.16×10^{-3}
1.7	1.22	0.822	1.48	0.676	30.0	31.6	0.032	10^3	10^{-3}
1.8	1.23	0.813	1.51	0.661	35.0	56.0	0.018	3.16×10^3	3.16×10^{-4}
1.9	1.24	0.803	1.55	0.646	40.0	100.0	0.010	10^4	10^{-4}
2.0	1.26	0.794	1.58	0.631	45.0	177.0	0.006	3.16×10^4	3.16×10^{-5}
2.2	1.29	0.776	1.66	0.603	50.0	316	0.003	10^5	10^{-5}
2.4	1.32	0.759	1.74	0.575	55.0	560	0.002	3.16×10^5	3.16×10^{-6}
2.6	1.35	0.741	1.82	0.550	60.0	1,000	0.001	10^6	10^{-6}
2.8	1.38	0.724	1.90	0.525	65.0	1,770	0.0006	3.16×10^6	3.16×10^{-7}
3.0	1.41	0.708	1.99	0.501	70.0	3,160	0.0003	10^7	10^{-7}
3.2	1.44	0.692	2.09	0.479	75.0	5,600	0.0002	3.16×10^7	3.16×10^{-8}
3.4	1.48	0.676	2.19	0.457	80.0	10,000	0.0001	10^8	10^{-8}
3.6	1.51	0.661	2.29	0.436	85.0	17,700	0.00006	3.16×10^8	3.16×10^{-9}
3.8	1.55	0.646	2.40	0.417	90.0	31,600	0.00003	10^9	10^{-9}
4.0	1.58	0.631	2.51	0.398	95.0	56,000	0.00002	3.16×10^9	3.16×10^{-10}
4.2	1.62	0.617	2.63	0.380	100.0	100,000	0.00001	10^{10}	10^{-10}
4.4	1.66	0.603	2.75	0.363	105.0	177,000	0.000006	3.16×10^{10}	3.16×10^{-11}
4.6	1.70	0.589	2.88	0.347	110.0	316,000	0.000003	10^{11}	10^{-11}
4.8	1.74	0.575	3.02	0.331	115.0	560,000	0.000002	3.16×10^{11}	3.16×10^{-12}
5.0	1.78	0.562	3.16	0.316	120.0	1,000,000	0.000001	10^{12}	10^{-12}
5.5	1.88	0.531	3.55	0.282	130.0	3.16×10^6	3.16×10^{-7}	10^{13}	10^{-13}
6.0	1.99	0.501	3.98	0.251	140.0	10^7	10^{-7}	10^{14}	10^{-14}
6.5	2.11	0.473	4.47	0.224	150.0	3.16×10^7	3.16×10^{-8}	10^{15}	10^{-15}
7.0	2.24	0.447	5.01	0.199	160.0	10^8	10^{-8}	10^{16}	10^{-16}
7.5	2.37	0.422	5.62	0.178	170.0	3.16×10^8	3.16×10^{-9}	10^{17}	10^{-17}

However, tables of logarithms contain only the mantissas of numbers greater than 1 (they contain no negative mantissas). Note that if

$$L = \log \frac{P_o}{P_i} = \log P_o - \log P_i$$

then when $P_o < P_i$, L will not be found in a table. However, if we invert the ratio P_o/P_i,

we have

$$\log \frac{P_i}{P_o} = \log P_i - \log P_o = -(\log P_o - \log P_i)$$

and, therefore,

$$\log \frac{P_i}{P_o} = -\log \frac{P_o}{P_i} \quad \text{or} \quad \log \frac{P_o}{P_i} = -\log \frac{P_i}{P_o}$$

Thus, if $P_o < P_i$ we can find $\log P_o/P_i$ using tables simply by finding $\log P_i/P_o$ and multiplying the result by -1.

A gain of less than 1 is referred to as a loss. Losses always have negative decibel values. Prior to the advent of the pocket calculator it has been common practice to use the following formula for losses:

$$\text{Loss in dB} = -10 \log \frac{P_i}{P_o}$$

The formula is still valid. However, many calculators provide negative logarithms (logarithms of numbers between 0 and 1) making it possible to use the standard formula, even for losses:

$$G_{dB} = 10 \log \frac{P_o}{P_i}$$

If G_{dB} is positive, there is a power gain; if G_{dB} is negative, there is a power loss.

Example 18 Calculate G_{dB} for the given input-output relationships.

(a) $P_o = 20$ W, $P_i = 5$ mW

(a) $G_{dB} = 10 \log \dfrac{P_o}{P_i} = 10 \log \dfrac{20}{5 \times 10^{-3}}$

$= \underline{36.02 \text{ dB (gain)}}$

(b) $V_o = 10$ V, $R_o = 8\ \Omega$;
$V_i = 20$ mV, $R_i = 2$ kΩ

(b) $G_{dB} = 20 \log \dfrac{V_o}{V_i} + 10 \log \dfrac{R_i}{R_o}$

$= 20 \log \dfrac{10}{20 \times 10^{-3}} + 10 \log \dfrac{2 \times 10^3}{8}$

$= 53.98 + 23.98 = \underline{77.96 \text{ dB (gain)}}$

(c) $I_o = 1.5$ mA, $R_o = 4$ kΩ;
$I_i = 5$ mA, $R_i = 7.5$ kΩ

(c) $G_{dB} = 20 \log \dfrac{I_o}{I_i} + 10 \log \dfrac{R_o}{R_i}$

$= 20 \log \dfrac{1.5}{5} + 10 \log \dfrac{4}{7.5}$

$= -10.46 + (-2.73)$

$= \underline{-13.19 \text{ dB (loss)}}$

Example 19 An attenuator has a gain of -6 dB. What is the ratio of P_o to P_i?

$$G_{dB} = 10 \log \frac{P_o}{P_i}, \qquad \log \frac{P_o}{P_i} = -\frac{6}{10} = -0.6$$

Taking the antilog of both sides of the equation,

$$\frac{P_o}{P_i} = \text{antilog}(-0.6) = 10^{-0.6} = \underline{0.25}$$

Example 20 An amplifier has a gain of 15 dB. If the input is 0.50 W, what is P_o?

$$G_{dB} = 10 \log \frac{P_o}{P_i}, \qquad \log \frac{P_o}{P_i} = \frac{G_{dB}}{10} = \frac{15}{10} = 1.5$$

$$\frac{P_o}{P_i} = \text{antilog } 1.5 = 10^{1.5} = 31.62$$

$$P_o = 31.62 P_i = 31.62(0.50) = \underline{15.81 \text{ W}}$$

Example 21 An amplifier has a gain of 35 dB. What is V_i when $V_o = 475$ mV? Assume $R_o = R_i$.

$$G_{dB} = 20 \log \frac{V_o}{V_i}, \qquad \log \frac{V_o}{V_i} = \frac{G_{dB}}{20}$$

$$\frac{V_o}{V_i} = \text{antilog } \frac{G_{dB}}{20}, \qquad V_i = \frac{V_o}{\text{antilog}(G_{dB}/20)}$$

$$V_i = \frac{475}{\text{antilog } 35/20} \text{ mV} = \frac{475}{56.23} = \underline{8.45 \text{ mV}}$$

Exercises 16-7 In Exercises 1–4 convert the given sets of power ratios to decibels.

1. 20, 120, 450, 1500
2. 15, 75, 500, 1200
3. 0.05, 0.5, 5, 50
4. 0.01, 0.1, 10, 1000

In Exercises 5–8 the given values are for V_o/V_i, or I_o/I_i, for electronic circuits. If $R_o = R_i$ in all cases, find G_{dB} for each.

5. 20, 120, 450, 1500
6. 15, 75, 500, 1200
7. 0.05, 0.5, 5, 50
8. 0.01, 0.1, 10, 1000

In Exercises 9–12 the given values are for G_{dB} for electronic circuits. Determine the related ratios, P_o/P_i, and $V_o/V_i = I_o/I_i$, for each.

9. (a) 15 (b) 40   10. (a) 55 (b) 67
11. (a) -15 (b) -5.5 12. (a) -12 (b) -35

In Exercises 13–18 the given values are for the output power P_o and input power P_i, respectively, of electronic circuits. Calculate G_{dB}.

13. 10 W, 0.2 W

14. 450 mW, 550 μW

15. 15 mW, 80 mW

16. 0.5 W, 3.75 W

17. 0.65 mW, 1.85 mW

18. 200 W, 0.65 μW

In Exercises 19–22 the given values are for output voltage V_o and resistance R_o, and for input voltage V_i and resistance R_i, respectively, of electronic circuits. Calculate G_{dB}.

19. 22 V, 16 Ω; 15 mV, 1.8 kΩ

20. 15 V, 8 Ω; 8.5 mV, 1.2 kΩ

21. 0.275 mV, 600 Ω; 2.35 mV, 300 Ω

22. 850 μV, 300 Ω; 2.75 mV, 300 Ω

In Exercises 23–26 the given values are for output current I_o and output resistance R_o, and input current I_i and input resistance R_i, respectively, of electronic circuits. Calculate G_{dB}.

23. 1.375 A, 16 Ω; 8.333 μA, 1.8 kΩ

24. 1.875 A, 8 Ω; 7.083 μA, 1.2 kΩ

25. 2.833 μA, 300 Ω; 9.167 μA, 300 Ω

26. 0.4583 μA, 600 Ω; 7.833 μA, 300 Ω

In Exercises 27–30 the given values are for G_{dB} and V_i, or I_i, respectively, for electronic circuits. Find the related V_o or I_o values. Assume $R_o = R_i$.

27. 30 dB, 2.55 mV   28. 45 dB, 6.75 mA

29. -18 dB, 15 mV

30. -28 dB, 18.65 mA

In Exercises 31–34 the given values are for G_{dB} and V_o, or I_o, respectively, for electronic circuits. Find the related V_i or I_i values. Assume $R_o = R_i$.

31. 42 dB, 12 V   32. 57 dB, 1.75 A

33. -15 dB, 15.7 mA   34. -23 dB, 1.65 mV

35. An amplifier has a maximum output of 30 W. A selector switch allows the user to reduce the output in 5 dB steps. What is P_o for the following settings of the switch? (a) 0 dB, (b) -5 dB, (c) -10 dB, and (d) -30 dB.

36. An audio generator has an output level control switch which changes the output voltage in 10 dB steps. If 0 dB = 1 V rms,

find V_o for (a) +10 dB, (b) -10 dB, (c) -30 dB, and (d) -50 dB.

37. The output of a hi-fi set increases from 0.5 W to 10 W when the volume control is turned from its minimum setting to the maximum setting. Describe the effect of the volume control in decibels (that is,

$$N_{dB} = 10 \log \frac{P_{max}}{P_{min}} = ?).$$

38. The output of a radio broadcasting station is changed from an effective 25 kW to 80 kW when the station switches from its daytime omnidirectional antenna to its nighttime directional antenna. What is the gain of the directional over the omnidirectional antenna?

39. If "stero separation" means "10 log P_A / P_B, when there is signal on channel B but none on channel A," what is the stereo separation of a receiver when $P_A = 2.5$ mW, with its input shorted, and $P_B = 10$ W?

40. The noise output power of an amplifier with input shorted is 45 μW. The signal output power for a standard input signal and the same gain setting is 18 W. What is the signal-to-noise ratio in decibels for the conditions described?

41. The outputs of certain types of electronic circuits are said to be attenuated at the rate of 6 dB/octave. ("Octave" describes a frequency ratio of 2 to 1.) Explain the meaning of the phrase. What power ratio does the attenuation of 6 dB represent? What voltage ratio? Give examples of these ratios.

16-8 DECIBEL REFERENCE LEVELS AND DECIBEL METERS

The decibel is a relative unit since it simply compares two quantities by ratio. It is often transformed into a definite unit by the use of a *reference level*. A popular reference, one used in the radio and telephone industries, is the level of 1.0 mW produced in a 600-Ω circuit. The result of a gain or loss calculation made using 1 mW as the reference power level is labeled *dBm* (*m* from *milliwatt*).

Example 22   The output of a tape deck produces 80 mW in the 600-Ω input circuit of a studio control panel. What is P_o of tape deck in dBm?

$$N_{dBm} = 10 \log \frac{P_o}{1 \text{ mW}} = 10 \log \frac{80 \text{ mW}}{1 \text{ mW}}$$

$$= 10 \log 80 = \underline{19 \text{ dBm}}$$

Example 23 The output of a microphone is -60 dBm. What is P_o in microwatts?

$$N_{dBm} = 10 \log \frac{P_o}{1 \text{ mW}}$$

$$\log \frac{P_o}{1 \text{ mW}} = \frac{N_{dBm}}{10} = \frac{-60}{10} = -6$$

$$\frac{P_o}{1 \text{ mW}} = \text{antilog} -6 = 10^{-6},$$

$$P_o = 1 \times 10^{-3} \times 10^{-6} = \underline{0.001 \ \mu\text{W}}$$

Most multimeters have a scale calibrated in decibels (see Figure 16-3). Since the meter is actually measuring ac voltage when the decibel scale is used and decibels are related to power ratios, the dB scale must be specified for use with a specific dB reference level and a specific resistance in order to have any meaning. A common specification for such meters is: 0 dB is 1 mW in 600 Ω. Therefore, since $P = V^2/R$, we have for 1 mW,

$$V = \sqrt{PR} = \sqrt{(1 \times 10^{-3})(600)}$$

$$= \sqrt{0.6} = 0.775 \text{ V}$$

A reading of 0 dB corresponds to 0.775 V on the 2.5-V scale.

If the measurement provides a reading of 0 dB but the meter is on the 50-V range, what is the actual dB value?

We find the answer as follows: The 50-V range represents a $50/2.5 = 20$ multiplier for voltages. That is, since 0 dB corresponds to 0.775 V on the 2.5-V range, it represents $20 \times 0.775 = 15.5$ V on the 50-V range. Or since dB $= 20 \log (V_2/V_1)$, we have

$$dB_{15.5 V} = 20 \log 20 = (20)(1.3010)$$

$$= 26.02 \text{ dB}$$

Since our answer, 26 dB, was obtained simply from the voltage ratio of the 50-V range to the 2.5-V range, it follows that on the 50-V range the dB value for any point on the dB scale will be 26 dB more than is indicated. Multimeters are usually provided with a table (on the meter face) which lists a conversion dB factor for each voltage range which must be added to the meter dB indication to give the correct dB measurement.

When such meters are used to measure dB's (voltage) across resistances other than the one specified by the meter manufacturer, the dB indications will not be correct. That is, 0.775 V across 3600 Ω will not produce the same power as 0.775 V across 600 Ω. However, from the basic formulas for calculating dB, we can derive a correction factor which can be applied to the meter-indicated dB value to obtain a true value.

ACV Range	ADD dB
2.5	+0
10	+12
50	+26
250	+40
1000	+52

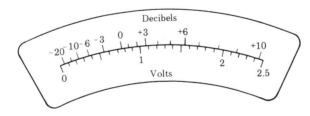

Figure 16-3

What is the dB value given for the 2.5-V point on the meter?

$$P_{2.5} = \frac{V^2}{R} = \frac{6.25}{600} = 10.42 \text{ mW}$$

$$dB = 10 \log \frac{10.42}{1} = 10 \cdot 1.018 = 10.18 \text{ dB}$$

Equation (4) from Section 16-7 yields the relation

$$dB = 20 \log \frac{V_2}{V_1} + 10 \log \frac{R_1}{R_2}$$

The reading taken from the dB scale of the meter gives the first term of this equation.

The second term then represents a correction factor which can be calculated. The resistance R_1 is the resistance specified as a reference by the meter manufacturer—600 Ω in our example above. That is,

$$\text{true dB} = \text{dB}_{\text{meter}} + \text{CF}$$

and

$$\text{CF} = 10 \log \frac{R_{\text{reference for meter}}}{R_{\text{circuit}}}$$

If $R_{\text{ref}} > R_{\text{circuit}}$, then $\text{CF} > 0$; if $R_{\text{ref}} < R_{\text{circuit}}$, then $\text{CF} < 0$.

Example 24   A multimeter with dB scale (reference: 1 mW, 600 Ω) reads +5 dB across a 2400-Ω resistance, with 10 ACV range selected. Find (a) true dB; (b) power in the 2400-Ω resistance.

   (a) The interpretation of the meter reading is

$$\text{dB}_{\text{meter}} = \text{dB}_{\text{reading}} + \text{Scale correction}$$
$$= +5 + 12 = 17 \text{ dB}$$

[We obtain the +12 dB correction for the 10-V range from the table on the meter scale (see Figure 16-3).]

Next, we calculate CF:

$$\text{CF} = 10 \log \frac{R_{\text{ref}}}{R_{\text{crct}}} = 10 \log \frac{600}{2400} = 10(-0.6)$$
$$= -6 \text{ dB}$$

Therefore,

$$\text{true dB} = \text{dB}_{\text{meter}} + \text{CF} = 17 + (-6) = \underline{11 \text{ dB}}$$

(b) We start with the basic formula for decibels to find P_{2400},

$$N_{\text{dB}} = 10 \log \frac{P_{2400}}{P_{\text{reference}}} = 11$$

Thus,

$$\log \frac{P_{2400}}{1 \text{ mW}} = \frac{11}{10} = 1.1$$

and taking the antilogarithm of both members of the equation,

$$\frac{P_{2400}}{1 \text{ mW}} = \text{antilog } 1.1 = 10^{1.1} = 12.59$$

then

$$P_{2400} = 12.59 \times 1 \text{ mW} = \underline{12.59 \text{ mW}}$$

Exercises 16-8   In Exercises 1–4 convert the given sets of dBm values to the equivalent values of power in watts (or a convenient prefixed unit of watts).

1. 10 dBm, 75 dBm, −20 dBm

2. −15 dBm, 45 dBm, 5 dBm

3. −1.5 dBm, 37 dBm, 12.5 dBm

4. 0.8 dBm, −1.7 dBm, 105 dBm

In Exercises 5–8 convert the given values of power to the equivalent dBm values.

5. 5 mW, 10 W, 0.75 mW

6. 20 W, 450 μW, 15 mW

7. 5 μW, 150 nW, 15 kW

8. 12 μW, 750 nW, 200 W

In Exercises 9–14 the given values are, in the order given, for (1) the dB scale readings, (2) the ac voltage ranges used, and (3) the values of the load resistances across which the readings of a multimeter were taken. The meter has the common reference values: 0 dB = 1 mW, R_{ref} = 600 Ω (see Figure 16-3). Determine (a) true dB values, and (b) the power in the load resistances.

9. 0 dB, 2.5 V, 300 Ω

10. +7 dB, 10 V, 1500 Ω

11. +4 dB, 50 V, 500 Ω

12. +8 dB, 250 V, 20 kΩ

13. −2 dB, 1000 V, 50 kΩ

14. −4 dB, 1000 V, 100 kΩ

15. In your work you find yourself consistently measuring the dB level of signals across 1500-Ω impedances on the 50-V range of a multimeter which has the typical reference values: 1 mW, 600 Ω. Work out a correction factor you could apply to the dB readings which would cover both the range change and the difference in resistance values.

16. Refer to Exercise 15. You frequently use the 10-V range across impedances of 750 Ω. What is the total correction factor for all readings taken under these conditions?

17. A certain FM receiver is advertised to have a sensitivity of 6 μV across 300 Ω (it will provide a useful output if the voltage across its antenna terminals is only 6 μV). Express the sensitivity in dBm ($P = V^2/R$).

18. A certain shortwave receiver has a sensitivity of $2 \mu V$ across 75 Ω. Express sensitivity in dBm (see Exercise 17).

19. The power of the signal received from a lunar probe is -58 dBm. What is the signal power expressed in watts?

20. A test technician testing production models of a certain hi-fi amplifier records the following readings taken from the dB scale of a multimeter:

Amplifier	A	B	C	D	E
dB reading	28	27.5	26.5	26	29

The dB scale is designed to indicate dBm when used to measure voltage across a 600-Ω load. The test setup measures voltages across an 8-Ω dummy speaker load. The amplifiers are rated at 40 W. Which ones, if any, should the technician reject for having a low output? If the technician were recording output voltages, what would he list in the table of measured values?

16-9 OVERALL GAIN CALCULATIONS WITH DECIBELS

One of the advantages of calculations with dB units derives from the basic laws of exponents, and hence of logarithms. That is, we know that if

$$M = b^x \quad \text{and} \quad N = b^y, \quad \text{then} \quad (M)(N) = b^{x+y}$$

but since

$$x = \log_b M \quad \text{and} \quad y = \log_b N$$

then

$$\log_b (M)(N) = x + y$$

Let us assume that we have several electronic devices in series, as shown in Figure 16-4, so that the output of A is the input of B, and so on. Then if the power-ratio gain of each is G_A, G_B, G_C, and so on, the total gain of the system is

$$G_T = G_A \cdot G_B \cdot G_C \cdot \ldots \cdot G_n$$

That is, to find the total gain of a system we must find the product of the gains of each of the elements. If the gains are expressed in decibels, however, we need only to add:

$$dB_T = dB_A + dB_B + dB_C + \cdots + dB_n$$

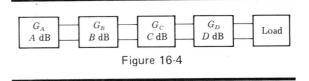

Figure 16-4

Example 25 Assume that the elements of the system of Figure 16-4 have the following specifications, and determine the power output of the system:

$$A, \text{microphone:} \quad dBm = -3.5$$
$$B, \text{preamplifier:} \quad dB = +12.5$$
$$C, \text{300-ft coaxial cable:} \quad dB = -6.5$$
$$D, \text{amplifier:} \quad dB = +37.5$$

The total absolute gain of the system is

$$-3.5 + 12.5 - 6.5 + 37.5 = 40 \text{ dBm}$$

(We can specify dBm, since the output of the microphone is given in dBm, and the other dB values relate outputs to inputs.) We now determine the power output P_o:

$$dB = 10 \log \frac{P_o}{P_i}$$
$$40 = 10 \log \frac{P_o}{1 \text{ mW}}$$
$$\log \frac{P_o}{1 \text{ mW}} = 4$$
$$P_o = (10^4)(1 \text{ mW}) = \underline{10 \text{ W}}$$

Exercises 16-9

1. The output of a record player is -45 dBm. The other components of a high-fidelity music reproduction system have the following gain characteristics: preamplifier, 35 dB; equalization network, -10 dB; volume control, -5 dB; amplifier, 55 dB. Determine the output of the system in dBm and watts.

2. In the system of Exercise 1 how much more gain would be required of the amplifier in order to achieve an output of 200 W? Give your answer in dB.

3. The output of a tape recorder is -15 dBm. The output is fed through an equalizer circuit with an insertion loss of -20 dB and a volume control, $G = -8$ dB, to an amplifier. The output of the amplifier with a normal signal from the tape recorder is 200 W. What is the output in

dBm? What is the gain of the amplifier in dB?

4. In the system of Exercise 3, if the output of the tape recorder is increased to -5 dBm, what is the output of the system in watts if full amplifier gain is used? What should be the amount of loss of a volume control to reduce output to 200 W?

5. An AM radio receiver produces 4 W of audio output. The elements in the signal path of the receiver have the following gains: antenna, +6 dB; if amplifier, +20 dB; detector, -4 dB; volume control, -10 dB; audio-frequency amplifier, 57 dB. Estimate the level of radio-frequency energy that exists at the antenna of the receiver.

6. In a certain area the field strength (output of a standard test antenna) of television channel 4 is -3 dBmV (0 dBmV = 1000 μV across 75 Ω of impedance). A viewer is planning his antenna system to include an antenna with a gain of 7 dB, 30 m of 75-Ω coax cable for lead-in with a loss of 0.1 dB/m, and a two-set signal splitter with a loss of 3 dB. Determine, in dBmV and μV, the amplitude of the channel 4 signal at the input terminals of his TV receivers.

7. Refer to Exercise 6. If the viewer must have a signal of 6 dBmV at the input terminals of his TV receiver in order to have a satisfactory picture, what gain should a proposed antenna preamplifier provide to achieve the required signal level?

8. A certain radio broadcasting studio is required to maintain an audio-signal level of 0 dBm at the input of the telephone line which carries the program signal to the remote transmitter. The typical studio hookup feeds the signal from a microphone through a cable to a preamp, then to a fader, a mixer, a master level control, and a program amplifier, and then to the line. If the maximum output of a microphone is -77 dBm, what must be the minimum overall gain of the studio chain of components in order to achieve the required line input signal? If the preamp and program amplifier provide a total fixed gain of 90 dB, what total reduction in gain must the fader, mixer, and master control be able to provide? Assume 0-dB loss in mike cable.

9. Refer to Exercise 8. The mixer console requires an input (maximum) on the mike line of -22 dBm. What minimum gain should the microphone preamp have?

16-10 THE BODE DIAGRAM

The *Bode diagram*, or *chart*, a graphical device introduced by the American engineer H. W. Bode, displays graphically both the amplitude and phase relationships of the transfer characteristic of a circuit or system. These quantities are plotted against the logarithm of the frequency, since working with logs enables us to display the quantities over a frequency range of several decades. (A decade is a range of frequencies in which the highest frequency is ten times the lowest frequency, for example, one decade comprises the frequencies between 1000 Hz and 10,000 Hz.) The amplitude relationship is normally shown in decibels.

An example will demonstrate the technique of plotting the Bode diagram.

Example 26 Consider the circuit of Figure 16-5. Plot the amplitude and phase angle of the relationship V_o/V_i in the form of a Bode plot.

By the ratio method, we know that

$$V_o = \frac{-jX_C}{R - jX_C} V_i \quad \text{or} \quad \frac{V_o}{V_i} = \frac{-jX_C}{R - jX_C}$$

Dividing the numerator and denominator by $-jX_C$, we get

$$\frac{V_o}{V_i} = \frac{1}{1 - \frac{R}{jX_C}} = \frac{1}{1 + j\frac{R}{1/\omega C}}$$

or

$$\frac{V_o}{V_i} = \frac{1}{1 + j\omega RC} \qquad (1)$$

The product RC represents the time constant of the circuit, which we will represent here by the symbol T. Thus, we can express equation (1) as

$$\frac{V_o}{V_i} = \frac{1}{1 + j\omega T} \qquad (2)$$

From equation (2) we can develop expressions that enable us to calculate readily the magnitude and phase angle of the transfer

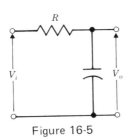

Figure 16-5

function $G = V_o/V_i$:

$$G = \frac{V_o}{V_i} = \frac{1}{\sqrt{1^2 + (\omega T)^2} \; \underline{/\arctan \omega T}}$$

$$= \frac{1}{\sqrt{1 + (\omega T)^2}} \; \underline{/-\arctan \omega T}$$

Thus,

$$|G| = \left| \frac{V_o}{V_i} \right| = \frac{1}{\sqrt{1 + (\omega T)^2}} \qquad (3)$$

or

$$G_{dB} = 20 \log \frac{1}{\sqrt{1 + (\omega T)^2}} \qquad (3')$$

and

$$\phi = -\arctan \omega T \qquad (4)$$

where ϕ is the *phase angle* of the transfer function.

Remember that equations (3) and (4) are only for a circuit of the type shown in Figure 16-5. Equations for other types of circuits can be developed in a similar fashion.

Values for plotting the Bode diagram for this circuit can be obtained by assuming

values for ωT, substituting them in equations (3') and (4), and performing the indicated operations. Organizing the procedure with the aid of a tabular format, we obtain the results shown in Table 16-2. The plot of the values obtained from the calculations summarized in Table 16-2 is shown in Figure 16-6.

Several observations are useful in the interpretation of Bode diagrams and lead to a shortcut method of construction.

1. The curve of the amplitude component of a transfer function of the general form $1/[1 + j(\omega T)]$ approaches two intersecting straight lines, called *asymptotes*, and can be approximated by these lines.

(a) One of the lines is horizontal, that is, its slope is 0. (*Slope* means "amount of upward or downward slant—the deviation from the horizontal."

$$\text{Slope} = \frac{\text{Amount of rise or fall}}{\text{Horizontal distance}}$$

(b) The slope of the second line is ±20 dB/decade or ±6 dB/octave. (An octave is a range of frequencies having a high-to-low frequency ratio of 2 to 1.)

(c) The two lines intersect at a frequency at which $\omega T = 1$. The frequency at intersection is called the *corner* or *break frequency*, denoted f_c. Thus,

$$T = \frac{1}{\omega_c} = \frac{1}{2\pi f_c} \quad \text{and} \quad f_c = \frac{1}{2\pi T}$$

(d) The *actual* amplitude of the transfer function deviates 3 dB from the *approximate amplitude* (as represented by the straight lines) at the corner frequency, and it deviates 1 dB from the approximate amplitude at 1 octave above and below the corner frequency.

2. The phase-angle component of the transfer function approaches $0°$ as the amplitude approaches 0 dB, and it approaches $\pm 90°$ as the amplitude approaches -40 dB. Other intermediate relationships are:

V_o/V_i, dB	-20	-10	-7	-3*	-1	-0.043	0
ϕ, deg	±84	±71	±64	±45	±27	±5.7	0

*At $\omega T = 1$.

Now let us apply the shortcut method.

Table 16-2

ωT	$\sqrt{1 + (\omega T)^2}$	$\left\|\dfrac{V_o}{V_i}\right\|$	Decibels	ϕ
0.01	1	1	0	0
0.03	1	1	0	0
0.1	1.005	0.995	-0.043	-5.7°
0.3	1.04	0.96	-0.38	-16°
0.5	1.12	0.983	-1.0	-27°
1.0	1.41	0.707	-3	-45°
2.0	2.24	0.446	-7	-64°
3.0	3.16	0.316	-10	-71°
10.0	10.05	0.099	-20	-84°
30.0	30	0.033	-30	-88°
100.0	100	0.01	-40	-89°

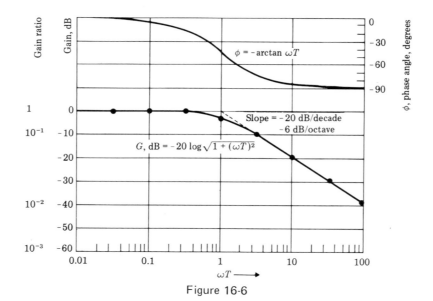

Figure 16-6

Example 27   Calculate and plot, as a function of frequency, the transfer function of the circuit of Figure 16-5. Examine the function for two decades above and below the corner frequency; $C = 0.001\ \mu\text{F}$ and $R = 20\ \text{k}\Omega$.

$$T = RC = 20 \times 10^3 \times 1 \times 10^{-9} = 2 \times 10^{-5}$$

The corner frequency f_c is obtained from the relationship

$$T = \frac{1}{\omega_c} = \frac{1}{2\pi f_c}$$

$$f_c = \frac{1}{6.28 \times 2 \times 10^{-5}} \approx 8\ \text{kHz}$$

Plot the Bode diagram using the following procedure (refer to Figure 16-7).

1.   Using four-cycle semilog graph paper, draw a horizontal line, at the 0-dB level,

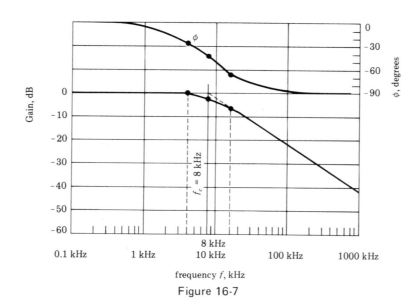

Figure 16-7

beginning at $f = f_c = 8$ kHz and extending to the left.

2. Starting at the point (8 kHz, 0 dB), draw a line with a -20-dB/decade (-6-dB/octave) slope, extending it to 800 kHz.

3. Locate a point (-3 dB, 8 kHz). (Actual G_{dB} is down 3 dB from approximate value at corner frequency.) Similarly, locate points 1 dB less than the approximate value of the gain at one octave above and below f_c; that is, locate the points (-1 dB, 4 kHz) and (-7 dB, 16 kHz).

4. Draw a smooth curve through these points and tangent to the asymptotes at $f \ll f_c$ and $f \gg f_c$. The curve is the amplitude plot of the desired Bode diagram.

5. To construct the phase-angle curve for the transfer function, plot on the same paper the following angular values to an appropriate scale and against the frequency axis used for the amplitude plot. (See observation 2 that precedes this example.)

f	ϕ, degrees
at $G = 0$ dB	$\phi = 0$
-0.043	-5.7
-1	-27
-3	-45 $f = f_c$
-7	-64
-10	-71
-20	-84
-40	-90

The complete Bode diagram for the circuit of Figure 16-5 is shown in Figure 16-7.

One of the more significant reasons for the popularity of the Bode diagram is that it permits one to examine quickly and easily the transfer function of a *combination system* composed of two or more networks in series, that is, networks whose transfer functions are known. An example will illustrate the procedure.

Example 28 A network consisting of a series inductance and a shunt resistance is to be connected after the network of Example 26 (Figure 16-5). The resulting circuit is shown in Figure 16-8. Construct the Bode diagram for the transfer function of the combination network. The components of the RL network have been chosen such that the f_c for the circuit is 10 kHz.

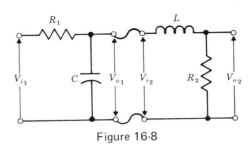

Figure 16-8

The first step is to construct, on the same set of axes, the transfer-function plots for the two separate transfer circuits. The plot for the RC circuit can be constructed using the procedure just described in the solution of Example 27. The Bode diagram for the RL circuit is similarly constructed, as follows.

1. The transfer function of the RL circuit is obtained from

$$V_{o2} = \frac{R_2}{R_2 + jX_L} V_{i2}$$

Thus,

$$G_2 = \frac{V_{o2}}{V_{i2}} = \frac{R_2}{R_2 + jX_L} = \frac{1}{1 + j\dfrac{X_L}{R_2}} = \frac{1}{1 + j\omega \dfrac{L}{R_2}}$$

$$= \frac{1}{1 + j\omega T_2}$$

$$= \frac{1}{\sqrt{1 + (\omega T_2)^2}}\ \underline{/-\arctan \omega T_2}$$

From this equation, we can write

$$|G_2| = \left| \frac{V_o}{V_i} \right| = \frac{1}{\sqrt{1 + (\omega T_2)^2}}$$

or

$$G_2 \text{ (in dB)} = 20 \log \sqrt{\frac{1}{1 + (\omega T_2)^2}} \quad (5)$$

$$\phi_2 = -\arctan \omega T_2 \quad (6)$$

Equation (5) shows that $G_2 = 0$ dB when $\omega T_2 \ll 1$ and that the curve of $G_{2\,dB}$ will have a -20-dB/decade slope when $\omega T_2 \gg 1$, that is, for $f > f_c$. Equation (6) shows that ϕ_2 approaches $0°$ when $\omega T_2 \ll 1$, $-90°$ when $\omega T_2 \gg 1$. Therefore, the Bode plot for the RL circuit has the same slope as that for the RC circuit. The -20-dB/decade line is drawn from the corner frequency, $f_{c2} = 10$ kHz.

2.   Since the input of the second network (the RL circuit) is the output of the first circuit (the RC circuit), it follows that

$$V_{o_2} = G_2 V_{i_2}$$

But $V_{i_2} = V_{o_1} = G_1 V_{i_1}$, where G_1 and G_2 are the gains of the first and second circuits, respectively. Hence, $V_{o_2} = G_1 G_2 V_{i_1}$, and the overall transfer function is

$$G_T = V_{o_2}/V_{i_1} = G_1 G_2$$

or

$$|G_T| = |G_1| \cdot |G_2| \qquad (7)$$

and

$$\phi_T = \phi_1 + \phi_2$$

If we take the logarithm of equation (7), we have

$$\log |G_T| = \log |G_1| + \log |G_2|$$

Multiplying both sides by 20, we get

$$20 \log |G_T| = 20 \log |G_1| + 20 \log |G_2|$$

Hence,

$$G_{T_{dB}} = G_{1_{dB}} + G_{2_{dB}}$$

The amplitude of the overall transfer function, in decibels, is equal to the sum of the individual amplitudes, in decibels.

The graph of $G_{T_{dB}}$ is obtained simply by adding the values of the individual amplitudes point by point. The result of this procedure is shown in Figure 16-9.

The curve of the total phase angle ϕ_T of the overall transfer function is similarly obtained by adding the values of ϕ_1 and ϕ_2, point by point, from the curves of these quantities (Figure 16-9).

As we can see on the diagram of Figure 16-9, the resultant gain curve has a slope of -40 dB/decade for $f > f_{c_2}$, and ϕ_T approaches $-180°$ when $f \gg f_{c_2}$.

It is important to realize that the technique of determining an overall transfer function for two transfer circuit blocks, as illustrated by Example 28, can be extended to any number of transfer blocks in series with the signal path.

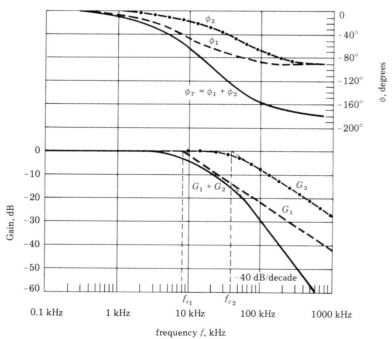

Figure 16-9

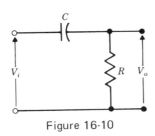

Figure 16-10

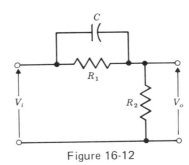

Figure 16-12

Exercises 16-10

1. Draw the graphs of the transfer characteristic (Bode diagram) of the transfer network of Figure 16-10. Construct the diagram using the general expression ωT as the variable for the horizontal axis. Plot the characteristics over the frequency range $0.01 < \omega T < 100$.

2. The circuit of Exercise 1 is to be used in series with an amplifier having a transfer function $G = 40$ dB $\underline{/0^\circ}$ over the entire frequency range of interest. Plot the Bode diagram for the amplifier on the same set of coordinates that you used for Exercise

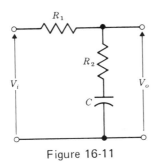

Figure 16-11

1. Show the overall transfer function for the combined system.

3. It can be shown that the transfer function of the network of Figure 16-11 is

$$\frac{V_o}{V_i} = \frac{1 + j\omega R_2 C}{1 + j\omega C(R_1 + R_2)}$$

Letting $R_2 C = T_2$ and $C(R_1 + R_2) = T_1$, show that

$$|G| \text{ dB} = 20 \log \sqrt{1 + (\omega T_2)^2} \\ - 20 \log \sqrt{1 + (\omega T_1)^2}$$

Construct the Bode diagram for this amplitude function. Assume that $T_1 = 10 T_2$. What is the value of G for values of $f \ll f_{c_1}$? of $f \gg f_{c_2}$?

4. Derive the expression for the phase-angle component of the transfer function for the network of Exercise 3. Plot the phase-angle curve on the same graph with the amplitude function of Exercise 3.

5. Develop an expression for the amplitude of the transfer function of Figure 16-12. Plot the Bode diagram.

Chapter 17
Exponential Equations

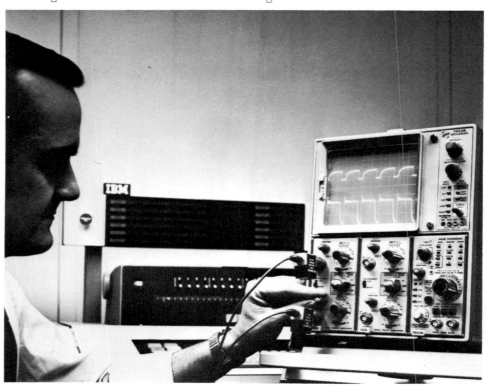

The transient characteristics of a circuit have an important influence on the wave forms of the signals it processes. In some instances these circuit properties are utilized to produce unusual wave shapes for the accomplishment of specific and desired purposes. In other cases the effect of these characteristics is the only objectionable distortion as depicted in the oscilloscope display in the photograph. (Courtesy of Tektronix, Inc.)

17-1 THE SIGNIFICANCE OF $y = y_0 e^{-x}$ AND $y = y_0(1 - e^{-x})$

The equations $y = y_0 e^{-x}$ and $y = y_0(1 - e^{-x})$ are called *exponential equations* because the independent variable x appears in them as an exponent. The relationships expressed by these two equations are also those of some important basic electric circuit responses and, hence, are important in the study of electronics. Let us consider their application to the circuit of Figure 17-1—a capacitor "charging" circuit. The capacitor C will store a charge Q if the switch (SW) is closed and allowed to remain closed long enough. Let us recall that charge is related to time and current,

$$Q = it$$

(The unit of Q is the coulomb—1 coulomb = 1 ampere $\times$ 1 second.)

When a capacitor (a circuit element which stores energy in the form of an accumulated excess or deficiency of electrons) contains a charge Q, a voltage v_C appears across its terminals which is related to its "size," C, and the charge,

$$v_C = \frac{Q}{C}$$

(For a given charge, v_C will be larger for a smaller capacitor, and so forth.) Referring again to Figure 17-1, if $Q = 0$ when we close the switch, then $v_C = 0$ at $t = 0$. And, since $v_C = Q/C$ and $Q = it$, v_C cannot change immediately—it takes time for Q to build up and cause v_C. Of course, Kirchhoff's voltage law applies to the situation,

$$v_C + v_R = V_s$$

The current i in the circuit can be determined using Ohm's law,

$$i = \frac{v_R}{R}$$

Therefore, at $t = 0+$ (an infinitesimal time after we close SW)

$$v_C = 0 \quad \text{and} \quad v_R = V_s$$

$$i = \frac{v_R}{R} = \frac{V_s - v_C}{R} = \frac{V_s}{R}$$

As time elapses the capacitor charges,

$$Q = it$$

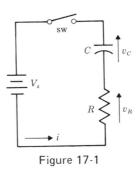

Figure 17-1

and v_C increases,

$$v_C = \frac{Q}{C} = \frac{it}{C}$$

The current is affected because

$$i = \frac{v_R}{R} = \frac{V_s - v_C}{R} = \frac{V_s - it/C}{R}$$

It is apparent that, with time, v_C grows larger and i grows smaller (we say i "decays"). As i grows smaller, v_C grows larger at a *slower rate*. The voltage across the capacitor approaches (but, theoretically, never equals) the battery voltage V_s. Similarly, the current approaches 0 A. These relationships are not simple linear relationships with respect to time. They are exponential relationships. They may be stated as follows:

$$v_C = V_s(1 - e^{-t/RC}) \qquad (1)$$

$$i = \frac{V_s}{R} e^{-t/RC} \qquad (2)$$

Let us note the similarity between these equations and the generalized exponential equations stated above. In equation 1, $y = v_C$ and $y_0 = V_s$; in equation (2), $y = i$ and $y_0 = V_s/R$. In both equations $x = -t/RC$. In this exponent—t/RC—t is the time variable and RC is a constant for a given circuit. However, the exponent must be a unitless number and therefore the product RC must have the unit of time, cancelling the time unit of the variable t. The RC product is called the *time constant* of the circuit and is sometimes represented by t_o, T, or τ (Greek tau).

$RC = \tau$ s when R is in ohms and C is in farads.

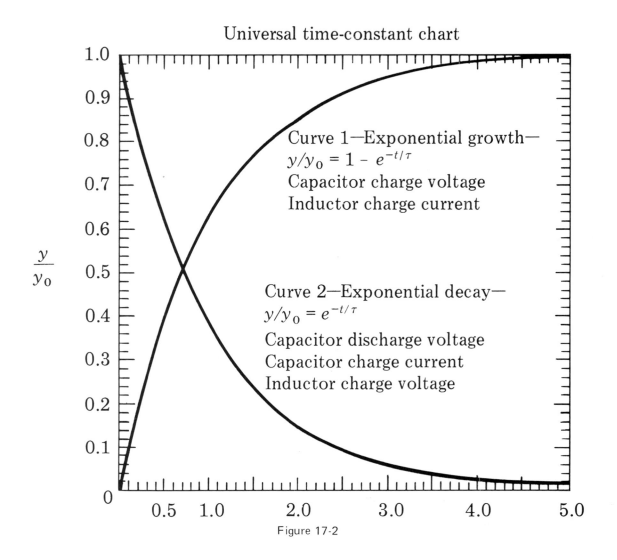

Figure 17-2

The chart labeled "Universal time-constant chart" shows:

Curve 1—Exponential growth—
$y/y_0 = 1 - e^{-t/\tau}$
Capacitor charge voltage
Inductor charge current

Curve 2—Exponential decay—
$y/y_0 = e^{-t/\tau}$
Capacitor discharge voltage
Capacitor charge current
Inductor charge voltage

Universal Time-Constant Chart Let us alter these basic exponential equations simply by dividing through by y_0. This produces the so-called *normalized forms* and ones which are useful for many applications because they are more general.

$$\frac{y}{y_0} = 1 - e^{-t/\tau} \qquad (3)$$

$$\frac{y}{y_0} = e^{-t/\tau} \qquad (4)$$

Graphs of equations (3) and (4) can be constructed by evaluating the equations for y/y_0 for a large number of values of t/τ, and plotting the values against t/τ. A chart showing the curves of both equations is shown in Figure 17-2. It is usually referred to as a *Universal Time-Constant Chart* or *UT-CC*. The graph of equation (3) is often referred to as the "exponential growth curve" and that of equation (4) as the "exponential decay curve."

The Universal Time-Constant Chart (UT-CC) can be used to determine the current and voltages in a capacitor-charging circuit (see Figure 17-1). Examples will illustrate the process.

Example 1   In a circuit like that of Figure 17-1, $R = 1$ MΩ, $C = 1$ μF, and $V_s = 100$ V. Find (a) i, (b) v_C, and (c) v_R, at 1-s intervals for $0 \leqslant t \leqslant 5$ s. (d) Sketch i and v_C versus t.

First, we find $\tau = RC$,

$$\tau = 1 \times 10^6 \times 1 \times 10^{-6} = 1 \text{ s}$$

Therefore, at $t = 0$, $t/\tau = 0$; at $t = 1$, $t/\tau = 1$; and so forth. (a) We refer to the chart of Figure 17-2 and recall that for the current in a charging circuit [equation (2)],

$$i = \frac{V_s}{R} e^{-t/\tau}$$

Dividing by V_s/R,

$$\frac{i}{V_s/R} = e^{-t/\tau}$$

But curve 2 on the chart is for $y/y_0 = e^{-t/\tau}$. Thus,

$$\frac{i}{V_s/R} = \frac{y}{y_0}$$

We can obtain values for y/y_0 for various values of t/τ from curve 2 of the UT-CC, Figure 17-2. Thus, since

$$i = \frac{V_s}{R} \cdot \frac{y}{y_0}$$

$$i = \left(\frac{100}{1} \times 10^6\right) \left(\frac{y}{y_0}\right) = 100 \left(\frac{y}{y_0}\right) \mu A$$

(curve 2)

Then, since $v_R = iR$,

$$v_R = (1 \times 10^6)i = 100 \left(\frac{y}{y_0}\right) \text{ V} \qquad \text{(curve 2)}$$

We find v_C using the relations of equation (3),

$$\frac{y}{y_0} = 1 - e^{-t/\tau}$$

and

$$v_C = V_s(1 - e^{-t/\tau})$$

That is,

$$v_C = V_s \cdot \frac{y}{y_0} \qquad \text{(curve 1)}$$

We construct a table of values for i, v_R, and v_C by reading y/y_0 values corresponding to the desired values of t/τ from curve 1 or 2 of the UT-CC. This is the table at the bottom of this page.

Example 2   If the equation for the voltage v_C across a capacitor during discharge in a circuit like that of Figure 17-3 (Switch in position 2) is

$$v_C = V_0 e^{-t/\tau}$$

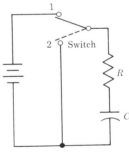

Figure 17-3

t (=t/τ)	0	1	2	3	4	5
$y/y_0 = e^{-t/\tau}$	1	0.37	0.14	0.05	0.018	0.007
(a) i, μA	100	37	14	5	1.8	0.7
(b) v_R, V	100	37	14	5	1.8	0.7
$y/y_0 = 1 - e^{-t/\tau}$	0	0.63	0.86	0.95	0.98	0.99
(c) v_C, V	0	63	86	95	98	99

(d)

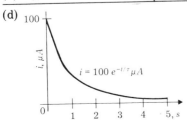

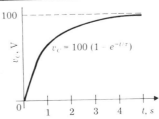

find v_C for $t = 0$, 0.25τ, 0.5τ, 1.0τ, and 3τ using the UT-CC. Sketch the graph of v_C versus t/τ. The initial voltage across C, due to a residual charge, is $V_0 = 80$ V.

Since

$$\frac{y}{y_0} = e^{-t/\tau}$$

and

$$\frac{v_C}{V_0} = e^{-t/\tau}$$

then

$$v_C = \left(\frac{y}{y_0}\right) V_0 = 80 \left(\frac{y}{y_0}\right)$$

t/τ	0	0.25	0.5	1.0	3.0
$y/y_0 = e^{-t/\tau}$	1	0.78	0.61	0.37	0.05
v_C, V	80	62	49	30	4

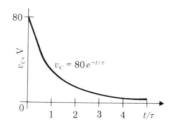

The response of a dc circuit containing a capacitor during the time when the capacitor is either charging or discharging is called its *transient response*. *Transient analysis* is the analysis of such responses. These analyses always make use of the expressions $e^{-t/\tau}$ or $1 - e^{-t/\tau}$. A summary of these so-called *RC* transients is included in Table 17-1.

Transients in Circuits Containing *R* and *L*
An inductance L is also an energy-storing circuit component and has a response in a dc circuit very similar to that of capacitance. For example, the equation for the transient current in an RL "charging" circuit (see Figure 17-4) is

$$i = \frac{V_s}{R}(1 - e^{-t/\tau})$$

In this case the time constant τ of the circuit is

$$\tau = \frac{L}{R}$$

(τ is in seconds when L is in henries and R is in ohms.) A summary of RL transients is included in Table 17-1. Graphs showing how the various voltages and currents in such circuits vary with t/τ are also shown. Examples illustrate how to use the Universal Time-Constant Chart (see Figure 17-2) to analyze RL circuits.

Example 3 An RL circuit is to be energized by a source with $V_s = 24$ V (see Figure 17-4). The circuit contains $R = 800\ \Omega$ and $L = 1.6$ mH. Determine (a) τ, and (b) the current at $t = 0$ s, $0.5\ \mu$s, $1\ \mu$s, $2\ \mu$s, $5\ \mu$s, and $10\ \mu$s. (c) Sketch i versus time.

(a) $\tau = \dfrac{L}{R} = \dfrac{1.6 \times 10^{-3}}{800} = \underline{2\ \mu s}$

(b) From Table 17-1, the current in an RL charging circuit is

$$i = \frac{V_s}{R}(1 - e^{-t/\tau})$$

For this circuit

$$\frac{V_s}{R} = \frac{24}{800} = 30 \text{ mA}$$

Thus,

$$i = 30(1 - e^{-t/\tau}) = 30 \left(\frac{y}{y_0}\right) \text{ mA}$$

We read values for y/y_0 from the UT-CC and tabulate:

t, μs	0	0.5	1	2	5	10
t/τ	0	0.25	0.5	1	2.5	5
$y/y_0 = 1 - e^{-t/\tau}$	0	0.22	0.39	0.63	0.92	0.99
$i = 30\dfrac{y}{y_0}$, mA	0	6.6	12	19	28	30

(c) The graph for i is like curve 1 of Figure 17-2.

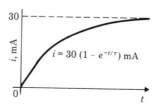

Table 17-1
Summary of RL and RC Circuit Transients

Circuit	Equation	Graph of function
RC charging circuit	$\tau = RC$	
	$i = \dfrac{V_s}{R} e^{-t/\tau}$	
	$v_C = V_s(1 - e^{-t/\tau})$	
	$v_R = V_s e^{-t/\tau}$	
RC discharging circuit	$i = -\dfrac{V_0}{R} e^{-t/\tau}$	
	$v_C = V_0 e^{-t/\tau}$	
	$v_R = -V_0 e^{-t/\tau}$	

Table 17-1 (cont.)

Circuit	Equation	Graph of Function
RL charging circuit	$\tau = L/R$ $i = \dfrac{V_s}{R}(1 - e^{-t/(\tau)})$ $v_L = V_s e^{-t/(\tau)}$ $v_R = V_s(1 - e^{-t/(\tau)})$	
RL discharging circuit	$i = I_0 e^{-t/(\tau)}$ where I_0 = current flowing when discharge starts $v_L = -I_0 R e^{-t/(\tau)}$ $v_R = I_0 R e^{-t/(\tau)}$	

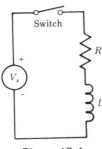

Figure 17-4

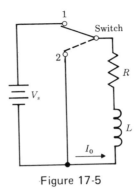

Figure 17-5

Example 4   In an RL charging circuit, $L = 250$ mH, $R = 1.2$ kΩ, and $V_s = 12$ V. How long will it take for the current to reach 3.5 mA?

$$I = i_{max} = \frac{V_s}{R} = \frac{12}{1200} = 10 \text{ mA},$$

$$\frac{i}{I} = \frac{y}{y_0} = \frac{3.5}{10} = 0.35$$

On the $1 - e^{-t/\tau}$ curve, $y/y_0 = 0.35$ when $t/\tau = 0.45$, approximately. Thus, since

$$\tau = \frac{L}{R} = \frac{0.25}{1200} = 208 \text{ } \mu s,$$

$$t = 0.45\tau = (0.45)(208) = \underline{94 \text{ } \mu s}$$

That is, i will reach 3.5 mA, which is 35% of its final current, approximately 94 μs after the switch is closed.

Example 5   In an RL discharging circuit (see Figure 17-5 and Table 17-1) containing $L = 2.5$ H and $R = 5$ Ω, there is a current $I_0 = 5$ A when the circuit starts to discharge. What will be the current i in the circuit 750 ms after discharge begins? What will v_L be at that time?

$$\tau = \frac{L}{R} = \frac{2.5}{5} = 0.5 \text{ s} = 500 \text{ ms}$$

At $t = 750$ ms, $t/\tau = 750/500 = 1.5$. From the UT-CC, we have at $t/\tau = 1.5$,

$$\frac{y}{y_0} = e^{-t/\tau} = 0.225$$

That is,

$$i/I_0 = 0.225,$$

and

$$i = 0.225 \text{ } I_0 = (0.225)(5 \text{ A}) = \underline{1.125 \text{ A}}$$

(I_0 is the current in the circuit at $t = 0$.)
   Then

$$v_L = -iR = (-1.125)(5) = -\underline{5.625 \text{ V}}$$

Example 6   An RL discharging circuit (see Figure 17-5) is conducting 10 mA, and 25 ms later it conducts 2.7 mA. The resistance of the circuit is known to be 150 Ω. What is L?

Here we use the 10 mA as I_0. Thus,

$$\frac{i}{I_0} = \frac{y}{y_0} = \frac{2.7}{10} = 0.27$$

From the UT-CC curve for $e^{-t/\tau}$ (the equation for i is $i = I_0 e^{-t/\tau}$) when $y/y_0 = 0.27$, we find that $t/\tau = 1.3$, approximately. Hence,

$$t = 25 \text{ ms} = 1.3\tau, \qquad \tau = \frac{25}{1.3} = 19.2 \text{ ms}$$

and since $\tau = L/R$,

$$L = R\tau = (150)(19.2)(10^{-3}) = \underline{2.88 \text{ H}}$$

Exercises 17-1   In Exercises 1–6 the given values are for R and C, respectively. Determine τ for the RC combinations.

1. 100 kΩ, 0.001 μF   2. 1.2 MΩ, 2.5 μF
3. 470 kΩ, 200 μF   4. 0.01 Ω, 220 pF
5. 0.56 MΩ, 5 pF   6. 100 MΩ, 2 pF

In Exercises 7–12 are listed L- and R-values. Find τ for these combinations.

7. 1 H, 2 Ω 8. 500 mH, 50 Ω
9. 10 H, 20 kΩ 10. 250 mH, 1.2 kΩ

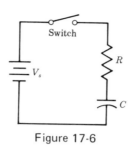

Figure 17-6

11. 375 μH, 100 kΩ 12. 875 mH, 3.3 kΩ

13. Calculate the value of the unknown component, R or C.
    (a) $\tau = 10$ ms, $C = 68$ pF
    (b) $R = 27$ kΩ, $\tau = 50$ μs

14. Repeat Exercise 13 for the following:
    (a) $R = 270$ Ω, $\tau = 73$ ms
    (b) $C = 5$ pF, $\tau = 55$ μs

15. Calculate the value of the unknown component, R or L.
    (a) $L = 250$ mH, $\tau = 10$ ms
    (b) $R = 1.2$ kΩ, $\tau = 500$ μs

16. Repeat Exercise 15 for the following:
    (a) $L = 2$ H, $\tau = 5$ s
    (b) $R = 1.2$ MΩ, $\tau = 2.5$ ns

In Exercises 17–20 the given values refer to the charging circuit of Figure 17-6. Determine the following (use the Universal Time-Constant Chart, Figure 17-2, in your solutions):
(a) τ
(b) i and v_C for $t/\tau = 0.1, 0.5, 1.0, 2.5$, and 5
(c) Sketch the graphs of i and v_C versus t/τ.

17. 100 kΩ, 10 μF, 100 V

18. 5 kΩ, 0.1 μF, 30 V

19. 0.9 MΩ, 100 pF, 1.5 V

20. 1 kΩ, 2000 μF, 10 V

21. For the circuit of Exercise 17, how much time will be required to charge the capacitor from $v_C = 0$ to the point that $v_C = 30$ V?

22. Refer to Exercise 19. Determine the charging time t required to obtain $v_C = 0.15$ V if $v_C = 0$ at $t = 0$.

23. In the circuit of Exercise 18 the current at $t = 0+$ (immediately after switch is closed) will be $I_0 = V_s/R = 6$ mA.
    (a) What is t/τ when $i = 2$ mA?
    (b) What is t for this value of i?

24. Refer to Exercise 20. (a) What is the value of t/τ when $i = 8$ mA? (b) What is t for this value of i?

In Exercises 25–30 values are given for R, C, and V_0, the voltage across C at the instant when SW is switched to position 2 (v_C at $t = 0$) for the capacitor discharging circuit of Figure 17-7. Use the Universal Time-Constant Chart (Figure 17-2) and the summary of Table 17-1 to aid you in your solutions.

25. (20 kΩ, 5 μF, 75 V) Find v_C when $t = 0$, 0.25τ, 0.5τ, 1.0τ, and 3τ. Sketch the graph of v_C versus t.

26. (1.5 kΩ, 0.05 μF, 200 V) Find v_C when $t = 0, 0.1\tau, 1\tau$, and 5τ. Sketch v_C versus t.

27. (1.5 MΩ, 50 pF, 10 V) Find i when $t = 0$, 1τ, and 5τ. Sketch i versus t.

28. (500 kΩ, 0.01 μF, 200 mV) Find i when $t = 0, 0.1\tau, 2\tau$, and 4τ. Sketch i versus t.

29. (100 kΩ, 1.0 μF, 50 V) The capacitor has been discharging for 50 ms. What are v_C and i?

30. (10 kΩ, 100 pF, 10 V) The capacitor has been discharging for 100 ns. What are v_C and i?

In Exercises 31–34 the given values are for the circuit of Figure 17-8 with SW in position 1, the "charging" position.

31. (10 kΩ, 1 H, 100 V) Determine (a) τ, and (b) the current i at $t = 0$ s, 10 μs, 100 μs, and 500 μs after SW is closed to position 1. (c) Sketch i versus t.

32. Determine v_L for the circuit of Exercise 31, for the values of t specified. Sketch v_L versus t.

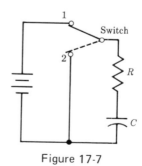

Figure 17-7

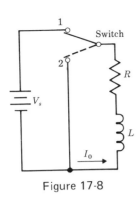

Figure 17-8

33. (100 Ω, 250 μH, 6 V) Determine (a) τ, and (b) the current i at $t = 0$, 1 μs, 2.5 μs, and 12.5 μs after SW is closed to position 1. Sketch i versus t.

34. Determine v_L for the circuit of Exercise 33, for the values of t specified. Sketch v_L versus t.

In Exercises 35–38 values are given for the circuit of Figure 17-8 with SW in position 2, the "discharging" position.

35. (100 kΩ, 5 mH, $I_0 = 10$ mA) Determine (a) τ, and (b) v_L for $t = 0$, 10 ns, 50 ns, and 200 ns. Sketch v_L versus t.

36. Determine i for the circuit of Exercise 35, for the values of t specified. Sketch i versus t.

37. (100 Ω, 2.5 H, $I_0 = 10$ A) Determine (a) τ, and (b) v_L for $t = 0$, 2.5 ms, 25 ms, and 100 ms. Sketch v_L versus t.

38. Determine i for the circuit of Exercise 37 for the values of t specified. Sketch i versus t.

39. An RL discharging circuit at one instant is conducting 100 mA and 50 μs later $i = 50$ mA. If $R = 10$ kΩ, what is L?

40. For an RL discharging circuit $I_0 = 10$ mA at $t = t_o$. At $t = t_o + 1$ ms, $i = 3$ mA. If $L = 500$ mH, what is R?

17-2 TRANSIENT ANALYSIS WITH THE CALCULATOR

A knowledge of natural logarithms (see Section 16-5) and a calculator with the $\ln x$ and e^x functions enable one to make rapid evaluations of the equations of the form $y = y_0 e^{-t/\tau}$

and $y = y_0 (1 - e^{-t/\tau})$. Transient analyses (see Section 17-1) of considerable complexity can be performed with relative ease.

Example 7 A capacitor is being charged (see Table 17-1) in a circuit in which $C = 0.1$ μF, $R = 50$ kΩ, and $V_s = 100$ V. Determine the time required to obtain $v_C = 28$ V if $v_C = 0$ at $t = 0$.

We start with the formula for v_C from Table 17-1, $v_C = V_s(1 - e^{-t/\tau})$. Then, dividing by V_s,

$$1 - e^{-t/\tau} = \frac{v_C}{V_s}$$

Solving for $e^{-t/\tau}$,

$$e^{-t/\tau} = 1 - \frac{v_C}{V_s}$$

Since $\log_b b^x = x$ (see Chapter 16), we simplify the exponential expression by taking the natural logarithms of both members of the equation,

$$\ln e^{-t/\tau} = \ln (1 - v_C/V_s),$$
$$-t/\tau = \ln (1 - 28/100) = -0.3285$$

[$\ln (1 - 28/100) = \ln 0.72 = -0.3285$ using $\ln x$ key on the calculator.]

Therefore,

$$t = 0.3285\tau = 0.3285\,RC$$
$$= (0.3285) (50 \times 10^3) (0.1 \times 10^{-6})$$
$$t = 1.643 \times 10^{-3} \text{ s} = \underline{1.643 \text{ ms}}$$

Example 8 The charging voltage in an RC charging circuit is $V_s = 300$ V. If $R = 75$ kΩ and $C = 0.5$ μF, what are v_C and v_R after 1 ms?

We calculate τ,

$$\tau = RC = (75 \times 10^3) (0.5 \times 10^{-6}) = 37.5 \text{ ms}$$

Thus, for $t = 1$ ms,

$$\frac{t}{\tau} = \frac{1}{37.5} = 0.02667$$

From Table 17-1 we obtain the formula for v_C,

$$v_C = V_s(1 - e^{-t/\tau}) = 300(1 - e^{-0.02667})$$

We evaluate $e^{-0.02667}$ on the calculator using the e^x key; thus,

$$v_C = 300(1 - 0.9737) = \underline{7.89 \text{ V}}$$

and

$$v_R = V_s - v_C = 300 - 7.89 = \underline{292.1 \text{ V}}$$

or, using the exponential equation,

$$v_R = V_s e^{-t/\tau} = 300 e^{-0.02667} = 292.1 \text{ V}$$

Example 9   In the RC charging circuit of Example 8 the capacitor has been previously charged so that it already has a voltage $V_0 = 75$ V when a new charging cycle begins. Determine t for $v_C = 125$ V.

The formula $v_C = V_s(1 - e^{-t/\tau})$ must now be modified since it represents only the change in the capacitor voltage. A more general formula is

$$v_C = (V_s - V_0)(1 - e^{-t/\tau}) + V_0$$

in which the expression $V_s - V_0$ represents the net potential in the circuit available to charge the capacitor. Thus,

$$v_C = 125 = (300 - 75)(1 - e^{-t/\tau}) + 75$$

$$1 - e^{-t/\tau} = \frac{125 - 75}{300 - 75} = 0.22222$$

$$e^{-t/\tau} = 1 - 0.22222 = 0.77778$$

Taking natural logarithms

$$-t/\tau = \ln 0.77778 = -0.25131$$

$$t = 0.25131\tau = (0.25131)(37.5 \text{ ms})$$

$$= \underline{9.424 \text{ ms}}$$

Example 10   Refer to Figure 17-9. A half-wave rectifier is connected to the 120-V, 60-Hz line. It supplies a transistor radio which draws 20 mA. Determine the value of the filter capacitor which would be required to limit the peak-to-peak ripple voltage [Δv in Figure 17-9(b)] to not more than 8 V. Assume that the discharge time of C is equal to the period of the ac cycle (1/60 s).

The peak-to-peak ripple voltage, Δv, is equal to the amount of voltage the capacitor "loses" between rectifier pulses when it discharges through R_L; thus,

$$\Delta v = V_0 - v_C$$

where V_0 = voltage on capacitor at peak of rectified waveform = $1.414\, V_{\text{line}}$

v_C = voltage across capacitor at end of discharge time between pulses of rectified line voltage

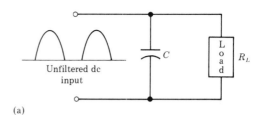

(a)

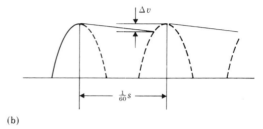

(b)

Figure 17-9

Thus,

$$v_C = V_0 - \Delta v = (1.414)(120) - 8$$

$$= 161.7 \text{ V}$$

but

$$v_C = 161.7 = V_0 e^{-t/\tau}$$

$$= (1.414)(120) e^{-(1/60)/\tau}$$

$$e^{-0.01667/\tau} = \frac{161.7}{(1.414)(120)} = 0.9530$$

Taking the natural logarithms,

$$-\frac{0.01667}{\tau} = \ln 0.9530 = -0.04814$$

$$\tau = \frac{0.01667}{0.04814} = 0.3463$$

The discharge circuit includes C and R_L. But

$$R_L = \frac{V_{DC}}{I_L} = \frac{(1.414)(120)}{0.020} = 8484\ \Omega$$

and, thus

$$C = \frac{\tau}{R_L} = \frac{0.3463}{8484} = \underline{40.8\ \mu F}$$

Example 11   In digital computers and other electronics equipment utilizing rectangular pulses [see Figure 17-10(a)] an in-

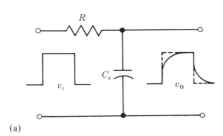

(a)

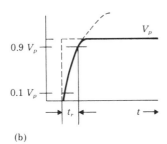

(b)

Figure 17-10

herent problem is the distortion of the pulse caused by the charging and discharging of the interlead and interelectrode capacitances of the circuit and devices. Pulses cannot change circuit voltage levels instantaneously—the charging and discharging require a rise time t_r and a fall time t_f. Rise time is defined as the time required for v_{pulse} to change from 0.10 to 0.90 of its maximum value V_p [see Figure 17-10(b)]. If the equivalent circuit to which a pulse is applied [see Figure 17-10(a)] contains $R = 2$ kΩ and $C_s = 50$ pF, what will be the rise time of applied pulses?

We find the solution by using the standard formula for the charging voltage of a capacitor,

$$v_C = V_s(1 - e^{-t/\tau})$$

and the relation

$$t_r = t_{0.9} - t_{0.1}$$

That is, the rise time is equal to the difference in times required for v_C to equal 0.9 V_p and 0.1 V_p.

$$\tau = RC_s = 2 \times 10^3 \times 50 \times 10^{-12} = 100 \times 10^{-9}$$
$$= 10^{-7} \text{ s}$$

For $t_{0.9}$

$$\frac{v_C}{V_p} = 0.9 = 1 - e^{-t_{0.9}/10^{-7}}$$
$$e^{-t_{0.9}/10^{-7}} = 1 - 0.9 = 0.1$$

Taking natural logarithms,

$$-t_{0.9}/10^{-7} = \ln 0.1 = -2.303$$
$$t_{0.9} = (2.303)(10^{-7}) = 230.3 \times 10^{-9} \text{ s}$$
$$= 230.3 \text{ ns}$$

For $t_{0.1}$,

$$\frac{v_C}{V_p} = 0.1 = 1 - e^{-t_{0.1}/10^{-7}}$$
$$e^{-t_{0.1}/10^{-7}} = 1 - 0.1 = 0.9$$
$$-t_{0.1}/10^{-7} = \ln 0.9 = -0.1054$$
$$t_{0.1} = (0.1054)(10^{-7}) = 10.54 \times 10^{-9} \text{ s}$$
$$= 10.54 \text{ ns}$$

Then

$$t_r = t_{0.9} - t_{0.1} = 230.3 - 10.54 = \underline{219.8 \text{ ns}}$$

It will be useful for you to remember that for any circuit

$$t_r/\tau = 2.198$$

Exercises 17-2 In Exercises 1–39 use the e^x and $\ln x$ functions of an electronic calculator to evaluate appropriate expressions leading to solutions. Refer to Table 17-1 to obtain equations relating to the given circuits.

1. A capacitor is being charged in a circuit in which $C = 0.1$ μF, $R = 50$ kΩ, and $V_s = 100$ V. Determine the time required to obtain $v_C = 35$ V. The capacitor is initially uncharged.

2. Determine the time required to charge the capacitor of Exercise 1 from $v_C = 0$ to $v_C = 65$ V.

3. In an RC charging circuit $R = 1$ kΩ, $C = 50$ μF, and $V_s = 30$ V. Determine t corresponding to $v_C = 100$ mV (at $t = 0$, $v_C = 0$).

4. For the circuit of Exercise 3 determine $t_{25 \text{ v}}$ if $v_0 = 0$.

5. In an RL charging circuit $R = 200$ Ω, $L = 0.5$ H, and $V_s = 100$ V. Determine the time required for the current to increase from 0, at $t = 0$, to 10 mA.

6. Repeat Exercise 5 for $i = 400$ mA.

7. The charging voltage in an RC circuit is 300 V, $R = 75$ kΩ, and $C = 0.5$ μF. What are v_C and i after 15 ms?

8. Find v_C and i for the circuit of Exercise 7 for $t = 100$ ms.

9. An RC charging circuit consists of $R =$ 1.5 MΩ and $C = 200$ pF. If $V_s = 5$ V, find v_C for $t = 3$ ms.

10. In the circuit of Exercise 9 the resistance is replaced with $R = 1.5$ kΩ. Find v_C for $t = 3$ μs.

11. In an RL charging circuit, $R = 400$ Ω, $L = 200$ μH, and $V_s = 5$ V. Find i for $t = 10$ ns.

12. Find i for $t = 2$ μs for the circuit of Exercise 11.

13. Refer to Exercise 1. If $v_C = V_0 = 17$ V at $t = 0$, what is t for $v_C = 80$ V? (See Example 9.)

14. Refer to Exercise 3. If $v_C = V_0 = 23$ V at $t = 0$, what is t for $v_C = 29$ V?

15. In the circuit of Exercise 7, $v_C = 125$ V when charging begins. Determine v_C for $t = 5$ ms.

16. Find v_C for $t = 50$ ms for the circuit of Exercise 9 if $V_0 = 2.75$ V.

17. Refer to the power-supply filter problem of Example 10, above. Find the value of C required to limit Δv to 5 V if $I_L = 10$ mA and all other circuit conditions are unchanged.

18. If the load current in the circuit of Example 10 is 15 mA and $C = 100$ μF, what is the peak-to-peak ripple voltage across the load?

19. Refer to the discussion of pulse rise time in Example 11. A computer timing pulse (rectangular) "sees" an equivalent circuit containing $R = 1.5$ kΩ and $C = 70$ pF. Determine t_r.

20. Refer to Exercise 19. Find t_r if $R = 10$ kΩ and $C = 50$ pF.

21. Refer to Exercise 19. Find t_r if $R = 5.6$ kΩ and $C = 30$ pF.

22. Refer to Exercise 19. Find t_r if $R = 800$ Ω and $C = 25$ pF.

23. Many persons working in electronics memorize the value of v_C/V_s at $t = \tau$ ("one time constant"). This knowledge provides a convenient rule-of-thumb guide to circuit action involving capacitor-charging transients. Calculate this ratio and express as a percent. Memorize it for future reference. To what other transient responses does this percent figure apply?

24. Repeat Exercise 23 for v_C/V_0 at $t = 1$ time constant during the discharge transient.

25. In the circuit of Figure 17-11 the voltage at point A is used as a triggering voltage to initiate a function in a circuit connected to A. A voltage of -5 V is required for the triggering function. If the voltage across Q_1 is effectively 0 V when Q_1 is conducting, how long will it take to achieve the triggering voltage at A after Q_1 is switched off? Assume that $I_C = 0$ when Q_1 is off.

26. Repeat Exercise 25 but assume that Q_1 discharges the capacitor to -1.0 V only ($V_0 = -1.0$ V).

27. In the circuit of Figure 17-12 the voltage v_0 drops from 12 V to 8.4 V in 150 μs after the switch is opened. What is the value of R_2?

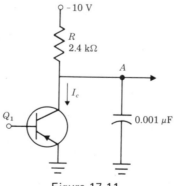

Figure 17-11

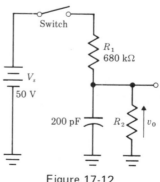

Figure 17-12

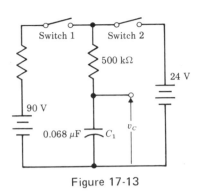

Figure 17-13

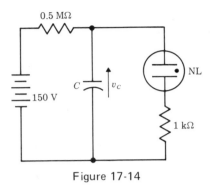

Figure 17-14

28. What is τ for the charging circuit of Figure 17-12? Assume that $R_2 = 2.7$ MΩ. (*Hint:* Derive the Thévenin equivalent for the circuit seen by C.) What is the maximum voltage to which C can be charged?

29. In the circuit of Figure 17-13 C_1 is charged to 90 V when switch 1 is closed. If switch 1 is opened and switch 2 closed, what will v_C be after 17 ms? (*Hint:* Refer to Example 9 and assume that the circuit is charging with $V_s = 24$ V and $V_0 = 90$ V.)

30. In the circuit of Figure 17-13, if the capacitor is to be discharged to approximately 24 V (say 24.447 V) from an initial v_C of 90 V by closing switch 2 for 17 ms, opening it for 1 ms, closing it for 17 ms, opening it for 1 ms, and so on, what is the total time required to achieve the desired result?

31. A certain purely dissipative circuit normally draws 10 A when energized by a 100-V source. We want to avoid any sudden current in-rush when we energize the circuit. If we want to delay the current to less than 80% of its final value for not less than 0.5 s, what value of L will we need?

32. Inductance is often used in computer memory circuits to limit the magnitude of currents generated by noise voltages that are induced in the circuits when magnetic cores are "flipped." Estimate the maximum noise current which will be generated in a circuit consisting of $L = 2.5$ mH in series with a 0.5-kΩ R. The $V_{noise} = 125$ V persists for 2 μs. What would be the value of I_{noise} if L were not present?

33. A voltage pulse $V_p = 10$ V is applied to a series circuit consisting of $C = 1$ μF and $R = 1.5$ MΩ. The pulse persists for $\frac{1}{2}$ s. Estimate the voltage across R at the instant the voltage starts to return to zero.

34. Refer to Exercise 33. If R is reduced to 680 kΩ, what must C be to produce v_R greater than or equal to 80% of $v_{R\,max}$ when the pulse is terminated? Assume that the desired C is to be obtained by adding to the 1-μF C already in the circuit. How much must be added and how should it be connected?

35. The circuit of Figure 17-14 is a simplified version of a neon "flasher" device. When v_C reaches 94.82 V, the neon lamp (NL) will conduct and discharge the capacitor. (a) If we want the time between flashes to be 10 s, what value of C should we use? Assume that C is discharged virtually to zero each time NL flashes ($t = 5\tau$). (b) On the basis of this assumption, how long does NL remain on each time it flashes? Assume that the effective resistance of the lamp when conducting is zero.

36. Repeat Exercise 35, assuming that NL extinguishes when v_C reaches 10 V.

37. In the popular jargon of electronics the circuit of Figure 17-15(a) is called a "differentiating circuit." It derives this title from the operation "to differentiate" in calculus. A rigorous analysis of the circuit requires the use of differention from the calculus, a branch of higher mathematics. If a square-wave pulse [see Figure 17-15(b)] is applied as v_{in}, we can consider the circuit simply as an RC

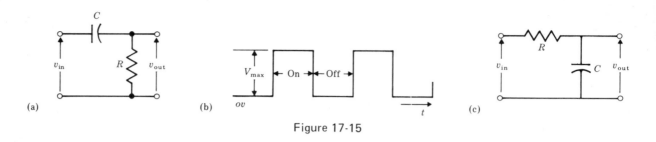

Figure 17-15

charging circuit when the pulse is "on" and a discharging circuit when the pulse is "off." That is, $v_{out} = v_R$. What is the equation for v_{out} during (a) on time of pulse? (b) off time? Sketch v_{out} for (c) $t_{on} = 10\tau$, (d) $t_{on} = 0.1\tau$. Align the sketches of v_{out} vertically with a sketch of v_{in}.

38. Refer to Exercise 37 and Figure 17-15. The circuit of Figure 17-15(c) is sometimes called an "integrating circuit" because in a complete and rigorous analysis of it the operation of integration from calculus is used. Again, if a square-wave pulse like that of Figure 17-15(b) is ap-

plied, we can consider the circuit as an RC charging and discharging circuit in which $v_{out} = v_C$. State the equation for v_{out} during the time when (a) $v_{in} = V_{max} = 10$ V; (b) $v_{in} = 0$ V. Sketch v_{out} for (c) $t_{on} = 5\tau$; (d) $t_{on} = 0.2\tau$.

39. It is often useful to have a Universal Time-Constant Chart which shows the relationships of the expressions $y/y_0 = e^{-t/\tau}$ and $y/y_0 = 1 - e^{-t/\tau}$ in greater detail for values of t between 0 and 0.2τ. Use Table II in the Appendix or the $\ln x$ function of your calculator to construct such a chart. (See UT-CC, Figure 17-2.)

Chapter 18
Concept of Slope

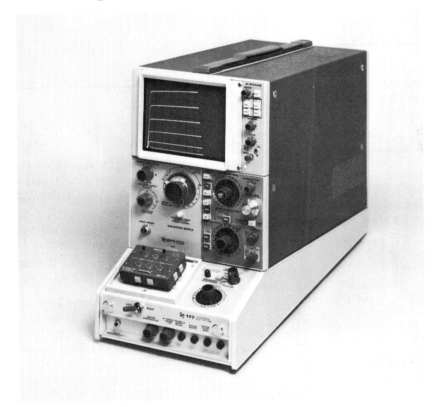

Finding the slope of the characteristic curves of an active device such as a transistor yields important information about its circuit properties. (Courtesy of Tektronix, Inc.)

18-1 SLOPE OF A STRAIGHT LINE

If we study the graph of an equation such as $y = 1.5x$ [see Figure 18-1(a)] we observe the similarity of the "rise" of the line to the rise of an inclined surface or hill, as seen in a side view. The concept of the "slope" of a hill or an inclined surface (a "sloping" surface) is a familiar one. We also say that the graph of a straight line possesses a mathematical characteristic called *slope*. The mathematical meaning of slope is basically the same as that of our everyday experience, namely, "inclination with respect to the horizontal." Defined formally:

> *Slope is the vertical distance between two points on the same straight line divided by the horizontal distance between those two points.*

In mathematics texts, slope is generally represented by the letter m. The slope of the line in Figure 18-1 can be given symbolically as

$$m = \frac{y_A - y_B}{x_A - x_B} \qquad (1)$$

where y_A and y_B = ordinates of points A and B, respectively, and

$$x_A, x_B = \text{Abscissas of } A \text{ and } B$$

Example 1   Determine the slope of the line through the points $(-3, -5)$ and $(2, -2)$. See Figure 18-2.

$$m = \frac{-5 - (-2)}{-3 - 2} = \frac{-3}{-5} = \underline{0.6}$$

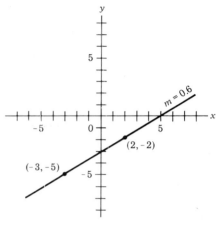

Figure 18-2

The slope of a line provides information about its direction:

1.   The larger the absolute value of the slope of a line, the more nearly vertical is the line.

2.   A line with a positive slope rises to the right. A line has a positive slope when the dependent variable is directly proportional to the independent variable.

3.   A line which falls to the right has a negative slope.

When a straight line passes through the origin (see Figure 18-3) we may choose the origin as one of the points for determining the slope. Formula (1) then becomes

$$m = \frac{y_A - y_B}{x_A - x_B} = \frac{y_A - 0}{x_A - 0} = \frac{y_A}{x_A}$$

When we apply this formula to a linear I-versus-V characteristic which passes through the origin, the characteristic of some electric

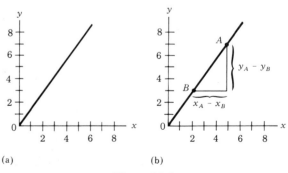

(a)   (b)

Figure 18-1

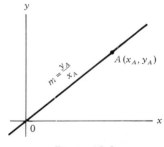

Figure 18-3

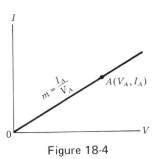

Figure 18-4

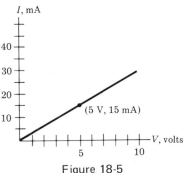

Figure 18-5

circuit device, we find that the slope has a special meaning related to a so-called physical *parameter* of the device. Let us consider the graph of Figure 18-4. It is apparent that the slope of the characteristic is given by

$$m = \frac{I_A}{V_A}$$

But, note that the ratio of the current in a device to the voltage across it is the definition of the physical concept of *conductance*, G (unit: siemens). Further, since $G = 1/R$, the reciprocal of the slope is the equivalent of the *resistance* of the device:

$$R = \frac{1}{\text{slope of } I\text{-versus-}V \text{ characteristic}}$$

Example 2   Determine G and R for the device whose I-V characteristic is shown in Figure 18-5.

G = Slope of I-V curve

$$= \frac{15 \times 10^{-3}}{5} = 3 \times 10^{-3} \text{ S} = 3 \text{ mS}$$

$$R = \frac{1}{G} = \frac{1}{3 \times 10^{-3}} = 0.3333 \times 10^3 \text{ } \Omega$$

$$= \underline{0.3 \text{ k}\Omega}$$

Note that even though the calculator gives us many digits in calculating the value for R, it is not realistic to retain more than one or two significant digits since the accuracy of the values taken from a graph is very limited.

In the light of the above information it is not difficult for us to accept the idea that the I-V characteristic (graph of I-versus-V) of a linear, passive resistance is a straight line through the origin and with a positive slope. Conversely, if a device has such a characteristic it will behave like a linear, passive resistance.

Exercises 18-1   In Exercises 1–10 find the slope of the straight line through the given pairs of points.

1. (1, 3) and (3, 5)
2. (−3, −2) and (1, 4)
3. (1, 3) and (3, −5)
4. (2, −5) and (−3, 4)
5. (−1, −2) and (−5, 3)
6. (−6, 3) and (0, 0)
7. (0, 0) and (4, −5)
8. (−1, −1) and (4, 1)
9. (2, −3) and (5, −7)
10. (2, 8) and (6, 2)

In Exercises 11–16 determine G and R for the devices whose I-versus-V characteristics are shown. Be sure to observe the units used on the graph and give the correct units with your answers.

11.

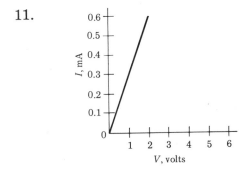

12.

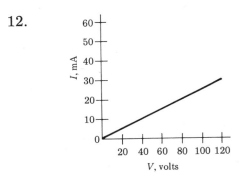

13.

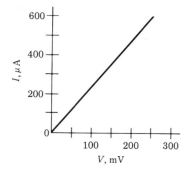

14.

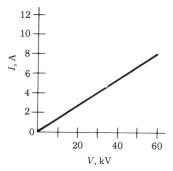

15.

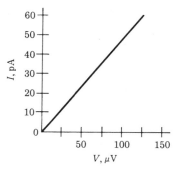

16

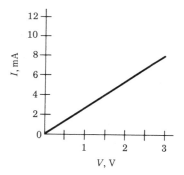

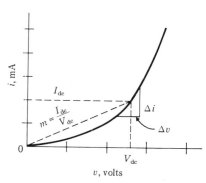

Figure 18-6

18-2 THE SLOPE OF NONLINEAR *I-V* CHARACTERISTIC CURVES

In the preceding section we say that finding the slope of a linear *I-V* characteristic provides the numerical value of the conductance of the device represented by the characteristic. The reciprocal of the slope is equivalent to the value of the resistance of the device. Let us see how the concept of slope is applied to obtain related information about devices with nonlinear *I-V* characteristics.

Consider the *I-V* characteristic of a semiconductor diode shown in Figure 18-6. It is apparent that since the graph is not a straight line it does not simply have one value of slope or resistance. Indeed, the slope is different for different parts of the curve. It is common to consider two types of resistances for such devices: *static forward resistance, R_F,* and *dynamic resistance, r_p.*

Static Forward Resistance, R_F   If a static (unchanging) dc voltage V_{dc} is connected across the device, a corresponding static current I_{dc} will result, as indicated by the *I-V* curve. The static forward resistance is the reciprocal of the slope of a straight line drawn from the origin through the point (V_{dc}, I_{dc}) on the characteristic curve of the device. That is,

$$R_F = \frac{V_{dc}}{I_{dc}}$$

Dynamic Resistance r_p   The dynamic resistance of a nonlinear device is equal to the reciprocal of the slope of a very short length of the curve centered at a point specified by V_{dc}. The point is called the *operating point.* Dynamic resistance is defined

$$r_p = \frac{\Delta v}{\Delta i}$$

(Δv is read "delta v" and means "a small change in v" or "an increment of v." The symbol Δ is the Greek letter "delta.")

Example 3   The curve shown in Figure 18-7 is the *I-V* characteristic of a semiconductor diode. Determine R_F and r_p at the following

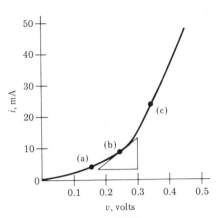

Figure 18-7

operating points: (a) $V_{dc} = 0.16$ V, (b) $V_{dc} = 0.24$ V, and (c) $V_{dc} = 0.34$ V.

(a) At the point labeled (a) on the curve of Figure 18-7, $V_{dc} = 0.16$ V and $I_{dc} = 2.4$ mA (approx.). Therefore,

$$R_F = \frac{V_{dc}}{I_{dc}} = \frac{0.16}{2.4 \times 10^{-3}} = \underline{67\ \Omega}$$

We find r_p by assuming that the curve is linear if we consider only a short length of it. For example, we could look at the curve between $V_{dc} = 0.12$ V, $I_{dc} \simeq 1.5$ mA and $V_{dc} = 0.20$ V, $I_{dc} \simeq 4$ mA. Thus,

$$r_p = \frac{\Delta V_{dc}}{\Delta I_{dc}} = \frac{0.20\ V - 0.12\ V}{4\ mA - 1.5\ mA} = \underline{32\ \Omega}$$

(b) At the point (b) on the curve, $V_{dc} = 0.24$ V and $I_{dc} \simeq 6.4$ mA. Therefore,

$$R_F = \frac{V_{dc}}{I_{dc}} = \frac{0.24}{6.4 \times 10^{-3}} = \underline{38\ \Omega}$$

To determine the value of r_p at $V_{dc} = 0.24$ V let us draw a tangent to the curve at $V_{dc} = 0.24$ V and consider the slope of this tangent (see Figure 18-7). (A tangent is a line which touches a curve at one point only. It therefore has the same slope as that of the curve at that point.)

$$r_p = \frac{1}{\text{slope of tangent}} = \frac{0.3\ V - 0.18\ V}{10\ mA - 2\ mA}$$

$$= \frac{0.12}{8 \times 10^{-3}} = \underline{15\ \Omega}$$

(c) For the curve at (c),

$$R_F = \frac{0.34}{22 \times 10^{-3}} = \underline{15\ \Omega}$$

$$r_p = \frac{0.36\ V - 0.32\ V}{26.4\ mA - 18\ mA} = \frac{0.04}{8.4 \times 10^{-3}} = \underline{5\ \Omega}$$

The results of Example 3 demonstrate an important fact to be remembered about the relationship of the slope of the *I-V* characteristic of a device and its resistance:

The greater the slope of an I-V characteristic, the smaller the resistance of the device.

Exercises 18-2   In Exercises 1–6 find R_F and r_p at the specified operating points on the given *I-V* characteristics.

1. For the characteristic of Figure 18-7 at $V_{dc} = 0.2$ V, 0.26 V, and 0.3 V.

2. For the characteristic of Figure 18-7 at $V_{dc} = 0.1$ V, 0.26 V, and 0.4 V.

3. For the characteristic of Figure 18-8 at $V_{dc} = 0.2$ V, 0.4 V, and 0.6 V.

4. For the characteristic of Figure 18-8 at $V_{dc} = 0.3$ V, 0.6 V, and 0.7 V.

5. For the characteristic of Figure 18-9 at $V_{dc} = 6$ V, 10 V, and 16 V.

6. For the characteristic of Figure 18-9 at $V_{dc} = 8$ V, 14 V, and 24 V.

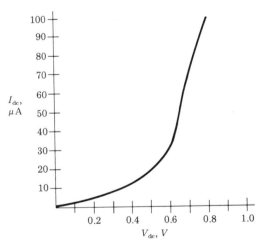

Figure 18-8

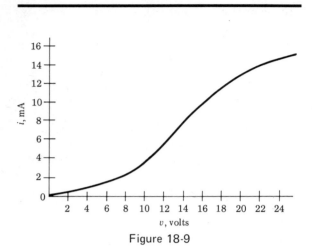

Figure 18-9

7. A device is tested and found to have the following $I\text{-}V$ characteristic:

v, V	0	5	10	20	40	60	80	100	120
i, mA	0	100	160	215	300	370	415	470	510

Plot this characteristic and determine R_F and r_p at the operating points $V = 20$ V, 60 V, and 80 V.

8. Find R_F and r_p for the device of Exercise 7 at the points $V = 10$ V, 50 V, and 120 V.

Chapter 19
Computer Number Systems; Binary Arithmetic

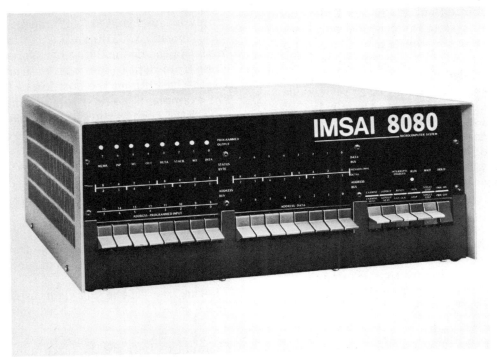

*Try to imagine a computer without numbers. (Courtesy of IMSAI
Manufacturing Corporation)*

19-1 FUNDAMENTALS OF NUMBER SYSTEMS

All of the numerical calculations presented in this book up to this point have utilized the *decimal* or *base-10* number system. However, working with computers or other digital equipment generally requires a knowledge of one or more other number systems—*binary (base 2)* and *octal (base 8)* or *hexadecimal (base 16)*, for example. In this chapter you will learn (a) the basic ideas which make these *positional number systems* work, (b) how to convert quantities from one number system to another, and (c) how to perform basic arithmetic operations in the binary number system.

As was pointed out in Chapter 2 the decimal number system (a positional number system) works because of its use of two important basic ideas: (1) a set of symbols or characters called *digits*, and (2) the *place value system*. You are urged to review Section 2-1 at this time to refresh your knowledge of these concepts because the other number systems which you will learn and make use of in digital electronics or computer technology also make use of these same ideas.

Each positional number system is constructed around a specific number of characters or digits. This number (of symbols) is called the *base* or *radix* of the system. Table 19-1 lists the symbols used in several common positional number systems.

You are familiar with the meaning or "count" of the digits of all the systems, except perhaps the higher-valued digits of the hexadecimal system. For example, if you were counting something, triangles perhaps, in the octal or decimal systems you would know that

1 triangle = △	4 triangles = △△△△
2 triangles = △△	5 triangles = △△△△△
3 triangles = △△△	and so on.

Note that in each of the systems represented above there are only enough symbols to represent a count of radix – 1. To count higher (or represent greater quantities) requires the use of the place value concept in which a digit acquires a *weight*—a value other than its basic value—as determined by its *place* or *position* in the number. For example, in Chapter 2 we reviewed the idea that in the decimal number system, "5" simply means five, but "50" means 5×10, "500" means 5×100, or 5×10^2, and so forth. In the decimal number system, then, going from right to left each position is worth 10 (the base) times the position to its right. For example, if we refer to a decimal counter or register, such as the six-place readout on a calculator, the position weights have the values shown in Figure 19-1.

The value of any number displayed can be analyzed as follows:

$$N = (D_5 \times 10^5) + (D_4 \times 10^4) + (D_3 \times 10^3)$$
$$+ (D_2 \times 10^2) + (D_1 \times 10^1) + (D_0 \times 10^0)$$

The symbols D_5, D_4, D_3, and so forth, represent any of the digits of this particular system: 0–9.

Let us transfer the idea demonstrated here (place value) to other numbers systems.

Table 19-1

Name of system	Symbols or digits used	Base or radix (No. of symbols)
Decimal	0, 1, 2, 3, 4, 5, 6, 7, 8, 9	10
Binary	0, 1	2
Ternary	0, 1, 2	3
Quinary	0, 1, 2, 3, 4	5
Octal	0, 1, 2, 3, 4, 5, 6, 7	8
Hexadecimal	0, 1, 2, 3, 4, 5, 6, 7, 8, 9, A, B, C, D, E, F	16

Example 1 Indicate the place values (decimal weights) of the registers shown.

(a) Base 8:

?	?	?	?	?	?
D_5	D_4	D_3	D_2	D_1	D_0.

8^5 8^4 8^3 8^2 8^1 8^0

(a)

D_5	D_4	D_3	D_2	D_1	D_0.

32,768's   4096's   512's   64's   8's   units

(b) Base 5:

?	?	?	?	?	?
D_5	D_4	D_3	D_2	D_1	D_0.

5^5 5^4 5^3 5^2 5^1 5^0

(b)

D_5	D_4	D_3	D_2	D_1	D_0.

3125's   625's   125's   25's   5's   units

Place values $\longrightarrow$ 10^5 10^4 10^3 10^2 10^1 10^0
as powers
of base (10)

D_5	D_4	D_3	D_2	D_1	D_0.

100,000's 10,000's 1000's 100's 10's units —Decimal point

Figure 19-1

(c) Base 2:

? ? ? ? ? ?

D_5	D_4	D_3	D_2	D_1	D_0.

2^5 2^4 2^3 2^2 2^1 2^0

(c)

D_5	D_4	D_3	D_2	D_1	D_0.

32's 16's 8's 4's 2's units

We summarize the basic facts that apply to any positional number system.

For any positional number system:
1. The number of symbols (digits) used in the system is equal to the base (radix), B.
2. The digit which represents the largest count is 1 less than the base ($=B - 1$).
3. Each position multiplies the digit in that position by a power of the base (B^0, B^1, B^2, . . .).
4. The weight (base multiplier) of positions increases by the first power of the base between positions, going from right to left.

Example 2 Analyze the number 5362_{10} showing the real value of each digit.

$5362 = (5 \times 10^3) + (3 \times 10^2) + (6 \times 10^1)$
$+ (2 \times 10^0)$
$= $ five 1000's + three 100's + six 10's
+ two 1's

Example 3 Express the following numbers in the format of Example 2 and find their decimal equivalents. (*Note:* The subscripts indicate the base of the number. The subscript is not written if the base is obvious from the context.)

(a) 1101_{10}
(a) $1101_{10} = (1 \times 10^3) + (1 \times 10^2)$
$+ (0 \times 10^1) + (1 \times 10^0)$
$= $ one 1000 + one 100 + no 10's
+ one 1 = $\underline{1101_{10}}$

(b) 1101_2
(b) $1101_2 = (1 \times 2^3) + (1 \times 2^2) + (0 \times 2^1)$
$+ (1 \times 2^0)$
$= (1 \times 8) + (1 \times 4) + (0 \times 2)$
$+ (1 \times 1)$
$= $ one 8 + one 4 + no 2's + one 1
$= 13_{10}$

(c) 1101_8
(c) $1101_8 = (1 \times 8^3) + (1 \times 8^2) + (0 \times 8^1)$
$+ (1 \times 8^0)$
$= (1 \times 512) + (1 \times 64) + (0 \times 8)$
$+ (1 \times 1)$
$= $ one 512 + one 64 + no 8's
+ one 1
$= \underline{577_{10}}$

Exercises 19-1 In Exercises 1–10 the given sets of symbols represent all of the digits of a positional number system. (a) Give the base or radix of the system. (b) State the decimal weight or position value, expressed as a power of the base, of each of eight positions, going from right to left and starting with the lowest value position of a whole number (see Exercise 1 below).

1. 0, 1, 2, 3, 4
Answer: (a) Radix (base) = 5
(b) $5^0, 5^1, 5^2, 5^3, 5^4, 5^5, 5^6, 5^7$

2. 0, 1, 2
3. 0, 1, 2, 3, 4, 5, 6, 7
4. 0, 1, 2, 3
5. 0, 1
6. 0, 1, 2, 3, 4, 5, 6, 7, 8, 9
7. 0, 1, 2, 3, 4, 5, 6, 7, 8, 9, A, B
8. 0, 1, 2, 3, 4, 5, 6, 7, 8, 9, A, B, C, D

9. $0, 1, 2, 3, 4, 5, 6, 7, 8, 9, A, B, C, D, E, F$

10. $0, 1, 2, A, B, C, \triangle$

In Exercises 11–20 write the numbers, given in number systems of various radices, in the format of Example 2 and find their decimal equivalents.

11. 1011_{10} 12. 2356_{10}

13. 1011_2 14. 1110_2

15. 1011_5 16. 2310_5

17. 1011_8 18. 1267_8

19. 1011_{16} 20. $10AE_{16}$

19-2 OCTAL NUMBER SYSTEM

Let us apply the principles of the previous section to the *octal* (base 8) number system, an important system in computer technology.

There are eight digits in the octal system and the largest is 7. When counting from zero the sequence is 0 through 7. On the next count we return to 0 in the units (8^0) position and generate a *carry* of 1 creating the number 10_8. We must remember to read this "one zero" and not "ten" since "ten" is a number in the decimal system only. The subscript is used to identify the number system being used. As the count continues the units position goes from 0 to 7, returns to zero while generating a carry, and so forth.

When quantities to be counted are large, additional positions are required. As each position reaches and exceeds the count of 7, a carry is generated (a 1 is added to the next position to the left) and the count in the position returns to 0. The decimal values of several octal positions are shown in Figure 19-2. (The letters SD mean "significant digit," LSD = least significant digit, MSD = most significant digit, and so on. See Section 2-5.)

Let us consider in detail a specific octal number, say 275_8. Because of our habitual and extensive use of decimal numbers we will

8^4	8^3	8^2	8^1	8^0
MSD	4SD	3SD	2SD	LSD.
4096's	512's	64's	8's	units

Figure 19-2

be inclined to read 275_8 as "two hundred seventy-five." This is incorrect since we have used words that apply only to decimal numbers. We should read 275_8 as "two-seven-five to the base eight" or simply "two-seven-five." Since we are not accustomed to "thinking in octal" we may convert it into its decimal equivalent in order to get a sense of the quantity involved:

$$275_8 = (2 \times 8^2) + (7 \times 8^1) + (5 \times 8^0)$$
$$= (2 \times 64) + (7 \times 8) + (5 \times 1)$$
$$= 128 + 56 + 5 = 189_{10}$$

A calculator, especially one with an accumulator memory function ($M+$), is very helpful in performing such conversions. Powers of a number can be found by repeated multiplication: $8^3 = 8 \times 8 \times 8$.

Example 4 Convert the given octal numbers to their decimal equivalents.

(a) 36_8
(a) $36_8 = (3 \times 8) + (6 \times 1) = 24 + 6 = \underline{30_{10}}$
(b) 122_8
(b) $122_8 = (1 \times 8^2) + (2 \times 8) + (2 \times 1)$
$= 64 + 16 + 2 = \underline{82_{10}}$
(c) 2076_8
(c) $2076_8 = (2 \times 8^3) + (0 \times 8^2) + (7 \times 8)$
$+ (6 \times 1) = 1024 + 56 + 6$
$= \underline{1086_{10}}$
(d) $56,312_8$
(d) $56,312_8 = (5 \times 8^4) + (6 \times 8^3) + (3 \times 8^2)$
$+ (1 \times 8) + (2 \times 1) = \underline{23,754_{10}}$

Decimal-to-Octal Conversion It is sometimes necessary to convert a decimal number into its octal equivalent. The procedure we will use is an algorithm referred to as "repeated division." (An algorithm is a specific, special method for solving a certain kind of problem. Often, the mathematical basis for the algorithm is not apparent in the application of the method.)

Example 5 Convert the number 189_{10} to its octal equivalent.

Step 1: Divide the number by the new base and write the remainder as the LSD of the equivalent.

Quotient   Remainder

$$\frac{23}{8\overline{)189}_{10}} + 5\text{———}$$

Step 2: Divide first quotient by the new base and write the remainder as 2SD.

Quotient   Remainder

$$\frac{2}{8\overline{)23}} + 7\text{———} \quad \text{(first quotient)}$$

Step 3: Divide again and write the remainder as 3SD.

Quotient   Remainder

$$\frac{0}{8\overline{)2}} + 2\text{———} \quad \text{(second quotient)}$$

$$2 \quad 7 \quad 5_8$$

The process is completed when the quotient is zero. The remainder for the final step is the MSD.

Streamlined, the process appears as follows:

Remainder

$$8\lfloor 189_{10} \qquad \text{LSD}$$
$$8\lfloor 23 \qquad 5 \uparrow$$
$$8\lfloor 2 \qquad 7$$
$$0 \qquad 2$$

MSD $189_{10} = 275_8$

Example 6   Convert the given decimal numbers to their octal equivalents.

(a) 30

(a)   Remainder

$$8\lfloor 30$$
$$8\lfloor 3 \qquad \uparrow 6$$
$$0 \qquad 3$$

$$30_{10} = 36_8$$

(b) 82

(b)   Remainder

$$8\lfloor 82$$
$$8\lfloor 10 \qquad \uparrow 2$$
$$8\lfloor 1 \qquad 2$$
$$0 \qquad 1$$

$$82_{10} = 122_8$$

(c) 1086

(c)   Remainder

$$8\lfloor 1086$$
$$8\lfloor 135 \qquad 6 \uparrow$$
$$8\lfloor 16 \qquad 7$$
$$8\lfloor 2 \qquad 0$$
$$0 \qquad 2$$

$$1086_{10} = 2076_8$$

(d) 23,754

(d)   Remainder

$$8\lfloor 23,754$$
$$8\lfloor 2,969 \qquad 2 \uparrow$$
$$8\lfloor 371 \qquad 1$$
$$8\lfloor 46 \qquad 3$$
$$8\lfloor 5 \qquad 6$$
$$0 \qquad 5$$

$$23,754_{10} = 56,312_8$$

Exercises 19-2   In Exercises 1–12 convert the given octal numbers to their decimal equivalents. (See Example 4.)

1. 36_8 2. 77_8
3. 156_8 4. 272_8
5. 307_8 6. 573_8
7. 1007_8 8. 1725_8
9. 2537_8 10. 4137_8
11. $11,321_8$ 12. $67,543_8$

In Exercises 13–24 convert the decimal numbers to octal equivalents. (See Examples 5 and 6.)

13. 63   14. 30
15. 186   16. 110
17. 379   18. 199
19. 981   20. 519
21. 2143   22. 1375
23. 28,515   24. 4817

19-3 HEXADECIMAL NUMBERS

The *hexadecimal* (base 16) number system is an important one in computer technology. Hexadecimal, as well as octal, numbers are frequently used as a shorthand notation for binary numbers, as we shall see.

The hexadecimal system requires 16 digits. The relationship between the hexadecimal digits and their equivalent decimal values is shown in Table 19-2.

Table 19-2
Decimal Equivalents of Hexadecimal Digits

Hexadecimal digit	0	1	2	3	4	5	6	7	8	9	A	B	C	D	E	F
Decimal value	0	1	2	3	4	5	6	7	8	9	10	11	12	13	14	15

16^4	16^3	16^2	16^1	16^0
MSD	4SD	3SD	2SD	LSD.
65,536's	4096's	256's	16's	units

Figure 19-3

The decimal equivalents or weights of several hexadecimal positions are shown in Figure 19-3.

It is frequently necessary to convert hexadecimal numbers to decimal numbers, and/or decimals to hexadecimals. The basic patterns of the procedures are the same as those demonstrated in the preceding section for octal-decimal, decimal-octal conversions. The calculator can be used to a very great advantage in performing the arithmetic involved in these conversions.

Example 7 Convert the given hexadecimal numbers to their equivalents in decimal notation.

(a) 39

(a) $39 = (3 \times 16) + (9 \times 1)$
$= 57_{10}$

(b) $4A$

(b) $4A = (4 \times 16) + (10 \times 1)$
$= 74_{10}$

(c) $1F9$

(c) $1F9 = (1 \times 16^2)$
$+ (15 \times 16) + (9 \times 1)$
$= (1 \times 256) + (15 \times 16)$
$+ (9 \times 1) = 505_{10}$

(d) $B89D$

(d) $B89D = (11 \times 16^3)$
$+ (8 \times 16^2) + (9 \times 16)$
$+ (13 \times 1)$
$= (11 \times 4096) + (8 \times 256)$
$+ (9 \times 16) + (13 \times 1)$
$= 47,261_{10}$

(*Note:* Any power of 16 can be found on a calculator by using the y^x key, if available, or simply by repeated multiplication: $16^3 = 16 \times 16 \times 16 = 4096$.)

Example 8 Convert the given decimal numbers to hexadecimal equivalents

(a) 57

(a) 16⌐57
16⌐3 9
0 3

$57_{10} = 39$

(*Note:* On the calculator $57 \div 16 = 3.5625$. You write the "3" as the quotient and find the remainder thus, $0.5625 \times 16 = 9$)

(b) 74

(b) 16⌐74
16⌐4 A
0 4

$74_{10} = 4A$

(Calculator steps:
$74 \div 16 = 4.625$
$0.625 \times 16 = 10_{10} = A$)

(c) 505

(c) 16⌐505
16⌐31 9
16⌐1 F
0 1

$505_{10} = 1F9$

(Calculator steps:
$505 \div 16 = 31.5625$
$0.5625 \times 16 = 9$
$31 \div 16 = 1.9375$
$0.9375 \times 16 = 15_{10} = F$

(d) 47,261

(d) 16⌐47,261
16⌐2,953 D
16⌐184 9
16⌐11 8
0 B

$47,261_{10} = B89D$

(Calculator steps:
1. $47,260 \div 16 = 2953.8125$
2. $0.8125 \times 16 = 13_{10} = D$
3. $2953 \div 16 = 184.5625$
4. $0.5625 \times 16 = 9$
5. $184 \div 16 = 11.5, 0.5 \times 16 = 8$)

Exercises 19-3 In Exercises 1–14 convert the given hexadecimal numbers to their decimal equivalents. (See Example 7.)

1. 42

2. $D5$

3. $15A$

4. $4C9$

5. $D8B$

6. ABC

7. 1132

8. $1A35$

9. *A27C*

10. *6BCA*

11. *12,5A7*

12. *10,9CD*

13. *A0,B8C*

14. *D1,2DF*

In Exercises 15–26 convert the given decimal numbers to hexadecimal equivalents. (See Example 8.)

15. 66

16. 213

17. 1360

18. 346

19. 2748

20. 3467

21. 6709

22. 4402

23. 27,594

24. 41,596

25. 68,045

26. 75,175

19-4 THE BINARY NUMBER SYSTEM

The *binary* or *radix 2* number system is used widely in the study, design, and operation of digital computer and other digital electronics equipment. This is true because, generally, the circuits used in such equipment are designed to operate only in one or the other of two states. The binary number system uses two digits —0 and 1—and these can represent the two states of a circuit.

In the binary system digits are called *bits* (from *bi*nary dig*its*). Examples of binary whole numbers are:

4 bit: 1101   8 bit: 10110011

16 bit: 1001110011110001

Although binary numbers may contain any number of bits, the binary numbers associated with computers generally contain 8, 16, 32, or 64 bits. The decimal equivalents or weights of several binary positions are shown in Figure 19-4.

As with other number systems, conversions to and from decimal numbers is a common requirement. The procedures are similar to those demonstrated in the preceding sections.

Example 9 Convert the given binary numbers to decimal numbers.

(a) 101

(b) 1101

(c) 10111

(d) 10111011

(a) $101 = (1 \times 2^2) + (0 \times 2) + (1 \times 1)$
$= 4 + 1 = 5$

2^5	2^4	2^3	2^2	2^1	2^0
MSD	5SD	4SD	3SD	2SD	LSD
32's	16's	8's	4's	2's	units

Figure 19-4

A convenient method of performing binary to decimal conversions is as follows:

1. Write the binary number with extra space between bits.
2. Write the weights, 1, 2, 4, 8, 16, and so on, beneath the bits.
3. Cross out any weights beneath a 0 bit.
4. Add up the remaining bits on your calculator.

(a) 101 Step 1: Write number. 1 0 1
Step 2: Write weights. 4 2 1
Step 3: Cross out weights. 4 2̸ 1
Step 4: Add weights. 4 + 1 = 5
101 = 5

(b) 1101   1 1 0 1
8 4 2̸ 1   8 + 4 + 1 = 13
1101 = 13

(c) 1 0 1 1 1
16 8̸ 4 2 1   16 + 4 + 2 + 1 = 23
10111 = 23

(d) 1 0 1 1 1 0 1 1
128 6̸4̸ 32 16 8 4̸ 2 1
128 + 32 + 16 + 8 + 2 + 1 = 187
10111011 = 187

Example 10 Convert the given decimal numbers to their binary equivalents.

(a) 5

(a)   Remainders
2|5   LSD
2|2 1
2|1 0
0 1 MSD
5 = 101

(b) 13

(b)   Remainders
2|13   LSD
2|6 1
2|3 0
2|1 1
0 1 MSD
13 = 1101

(c) 23

(c)   Remainders

$2\underline{|23}$ LSD
$2\underline{|11}$ 1 ↑
$2\underline{|5}$ 1
$2\underline{|2}$ 1
$2\underline{|1}$ 0
  0   1 | MSD

$$23 = 10111$$

(d) 187

(d) $2\underline{|187}$ LSD
$2\underline{|93}$ 1 ↑
$2\underline{|46}$ 1
$2\underline{|23}$ 0
$2\underline{|11}$ 1
$2\underline{|5}$ 1
$2\underline{|2}$ 1
$2\underline{|1}$ 0
  0   1 | MSD

$$187 = 10111011$$

Small decimal numbers can also be converted quickly by trial and error: (1) List possible weights. (2) Try crossing out weights until correct total is determined.

(a) $5 : 4\,\cancel{2}\,1$ $5 = 4 \text{ bit} + 1 \text{ bit} = 101$

(b) $13: 8\,4\,\cancel{2}\,1$ $13 = 8 + 4 + 1 = 1101$

(c) $23: 16\,\cancel{8}\,4\,2\,1$ $23 = 16 + 4 + 2 + 1$
$$= 10111$$

Binary-Octal Conversions   Because each position in a binary number represents only a power of two, a large number of bits is required to represent even moderately sized numbers. Although conversion from binary to decimal is relatively simple, it is tedious and time consuming for large numbers. Therefore, binary-octal conversions, which can be performed virtually by inspection, are popular.

Note that each group of three bits in a binary number can represent all the values between 0 and 7:

$$000 = 0 \quad 100 = 4$$
$$001 = 1 \quad 101 = 5$$
$$010 = 2 \quad 110 = 6$$
$$011 = 3 \quad 111 = 7$$

Therefore, since the octal system utilizes digits 0 through 7, we can convert directly from binary to octal simply by converting each group of three bits, starting from the LSD in whole numbers, to the equivalent octal digit. For example,

$$101111_2 = \underbrace{101}_{5} \ \underbrace{111}_{7_8} \qquad 101111_2 = 57_8$$

Similarly, octal-to-binary conversions are made by converting each octal digit into an appropriate set of three bits,

$$\underbrace{001}_{1} \ \underbrace{010}_{2} \ \underbrace{110}_{6_8} \qquad 126_8 = 1010110_2$$

Example 11   Convert the following binary numbers to their octal equivalents.

(a) 111
(a) $111 = 7_8$

(b) 1110
(b) $1110 = \underbrace{001}_{1} \ \underbrace{110}_{6}{}_2 = \underline{16_8}$

(c) 101101
(c) $101101 = \underbrace{101}_{5} \ \underbrace{101}_{5} = \underline{55_8}$

(d) 110110011
(d) $110110011 = \underbrace{110}_{6} \ \underbrace{110}_{6} \ \underbrace{011}_{3} = \underline{663_8}$

(e) 10111000101
(e) $10111000101 = \underbrace{010}_{2} \ \underbrace{111}_{7} \ \underbrace{000}_{0} \ \underbrace{101}_{5}$
$$= \underline{2705_8}$$

Example 12   Convert the following octal values to the binary equivalents.

(a) 17_8
(a) 17_8 $1_8 = 001$ $7_8 = 111$
$$17_8 = 001 \quad 111 = \underline{1111}$$

(b) 75_8
(b) $75_8 = \underbrace{111}_{7} \ \underbrace{101}_{5}$ or $\underline{111101}$

(c) 236_8
(c) $236_8 = \underbrace{010}_{2} \ \underbrace{011}_{3} \ \underbrace{110}_{6}$ or $\underline{10011110}$

(d) 1274_8
(d) $1274_8 = \underbrace{001}_{1} \ \underbrace{010}_{2} \ \underbrace{111}_{7} \ \underbrace{100}_{4}$ or
$$\underline{1010111100}$$

Binary-hexadecimal Conversions There is also a special relationship between binary and hexadecimal numbers which makes conversions between these two number systems simple and popular. A group of four bits can represent all values between 0_{10} and 15_{10}. Therefore, four bits can represent or be represented by the hexadecimal digits:

$$0000 = 0_{10} = 0_{16} \qquad 0001 = 1_{10} = 1_{16}$$

$$0010 = 2_{10} = 2_{16} \qquad 0011 = 3_{10} = 3_{16}$$

$$0100 = 4_{10} = 4_{16} \qquad 0101 = 5_{10} = 5_{16}$$

$$0110 = 6_{10} = 6_{16} \qquad 0111 = 7_{10} = 7_{16}$$

$$1000 = 8_{10} = 8_{16} \qquad 1001 = 9_{10} = 9_{16}$$

$$1010 = 10_{10} = A_{16} \qquad 1011 = 11_{10} = B_{16}$$

$$1100 = 12_{10} = C_{16} \qquad 1101 = 13_{10} = D_{16}$$

$$1110 = 14_{10} = E_{16} \qquad 1111 = 15_{10} = F_{16}$$

Example 13 Convert the following binary numbers to the corresponding hexadecimal values.

(a) 10110001

$$\text{(a) } 10110001 = \underset{B}{1011} \;\; \underset{1}{0001} = B1_{16}$$

(b) 1111010

$$\text{(b) } 1111010 = \underset{7}{0111} \;\; \underset{A}{1010} = 7A_{16}$$

(We add a leading 0-bit to complete the 4-bit group for the MSD.)

(c) 11100100111110

$$\text{(c) } 11100100111110 = \underset{3}{0011} \;\; \underset{9}{1001} \;\; \underset{3}{0011}$$

$$\underset{E}{1110} = 393E_{16}$$

(d) 1000111000110101

$$\text{(d) } 1000111000110101 = \underset{8}{1000} \;\; \underset{E}{1110}$$

$$\underset{3}{0011} \;\; \underset{5}{0101} = 8E35_{16}$$

Exercises 19-4 In Exercises 1-14 convert the given binary values to their decimal equivalents.

1. 101
2. 110
3. 1011
4. 1101
5. 1001
6. 1111
7. 101001
8. 110101
9. 1010101
10. 10011010
11. 1101001100
12. 1011110001
13. 100011101010
14. 1011100110101111

In Exercises 15-30 the values given are decimal numbers. Convert to binary equivalents.

15. 5
16. 7
17. 9
18. 12
19. 21
20. 30
21. 41
22. 61
23. 108
24. 121
25. 217
26. 1075
27. 4072
28. 5917
29. 20,073
30. 100,569

31-44. Convert the binary numbers given in Exercises 1-14 to their octal equivalents.

In Exercises 45-50 convert the sets of three octal numbers to their binary equivalents.

45. 3, 11, 115
46. 5, 21, 234
47. 37, 452, 1076
48. 74, 652, 4370
49. 150, 2507, 35,460
50. 761, 6130, 74,050

51-64. Convert the binary numbers given in Exercises 1-14 to their hexadecimal equivalents.

In Exercises 65-70 convert the sets of hexadecimal numbers to their binary equivalents.

65. $9, D, 10$
66. $7, E, C9$
67. $A, 27, 1A9$
68. $B, 9F, 79E$
69. $2A3F, 365A7B9C$
70. $90D5, 80B06C39$

19-5 FRACTIONAL NUMBERS

As in the decimal number system, binary numbers may also contain fractional parts. A point is used to separate the whole part from the fractional part, for example, 101101.0111. The bits to the right of the point have decimal weights which are equivalent to negative powers of two: $2^{-1}, 2^{-2}, 2^{-3}$, and so forth. Several fractional positions and their decimal weights are shown in Figure 19-5.

2^{-1}	2^{-2}	2^{-3}	2^{-4}	2^{-5}	2^{-6}
$.B_1$	B_2	B_3	B_4	B_5	B_6
0.5	0.25	0.125	0.0625	0.03125	0.015625

Figure 19-5

A fractional binary number (or the fractional part of a binary number) is converted to its decimal equivalent by summing the decimal weights of the bits present in the number.

Example 14   Convert the following binary numbers to the decimal equivalents.

(a) 11.011
(a) 11.011 = 2 + 1 + 0.25 + 0.125 = 3.375
(b) 101.101001
(b) 101.101001 = 4 + 1 + 0.5 + 0.125 + 0.015625 = 5.640625

Fractional binary numbers may be converted to octal and hexadecimal notation by separating the fractional bits into groups of three or four bits, respectively, starting at the point.

Example 15   Convert the following binary numbers into octal notation.

(a) 1101.01101
(b) 100111001.11001101
(a) 1101.01101: 001   101 . 011   010 = $\underline{15.32_8}$
          1   5 . 3   2

(Notice that we add a 0-bit in the sixth position to the right of the point to complete the group for the octal LSD.)

(b) 100111001.11001101: 100   111   001 .
                           4   7   1 .
110   011   010 = $\underline{471.632_8}$
 6   3   2

Example 16   Convert the binary numbers of Example 15 into hexadecimal notation.

(a) 1101.01101: 1101 . 0110   1000 = $\underline{D.68_{16}}$
          D   . 6   8

(We add three 0-bits to complete the 4-bit group for the hexadecimal LSD.)

(b) 100111001.11001101: 0001   0011
                                1   3
1001 . 1100   1101 = $\underline{139.CD_{16}}$
 9   . C   D

Exercises 19-5   In Exercises 1–10 convert the given binary numbers to decimal equivalents.

1. 0.11   2. 0.01
3. 0.101   4. 0.1011
5. 11.0011   6. 110.0001
7. 1001.101   8. 1100.0001
9. 1101.1101   10. 1111.0111

11–20. Convert the binary numbers of Exercises 1–10 to octal notation.

In Exercises 21–26 convert the given binary numbers to hexadecimal notation.

21. 0.001101   22. 0.100111
23. 1011.1011111   24. 10111.00110101
25. 1011011.101100011
26. 1100101.1010101

In exercises 27–30 we have the binary readouts of two indicating registers on the console of a large computer—Register A is for the whole part of a number, Register B for the fractional part. Translate the readouts into hexadecimal notation.

27. A:0000000101101011
     B:1101010000000000
28. A:0000000010111101
     B:0101110100000000
29. A:0000110010010011
     B:1110001110000000
30. A:1100000001111111
     B:1111000010010001

19-6 BINARY ARITHMETIC

The extensive calculations performed by modern computers are accomplished primarily through the processes of addition or subtraction of binary numbers. Since there are only two bits in the binary number system the "arithmetic tables" are simple and easily memorized (as well as implemented electronically).

Addition of Binary Numbers When we learned to add decimal numbers the process included learning (memorizing), by one method or another, the "addition table" of decimal digits—a table which indicated the sums of all possible combinations of two digits: $0 + 0, 0 + 1, \ldots, 1 + 1, 1 + 2, \ldots, 2 + 2, 2 + 3, \ldots$ and so on. Similarly, learning binary addition requires that we learn the binary "addition table," an extremely simple table consisting of only five basic combinations, detailed below. Memorize this table; it is used for all binary addition. The sums in lines 4 and 5 are read "one-zero" and "one-one." When more than two bits are involved we add them two at a time, using the above rules.

BINARY ADDITION TABLE

1. $0 + 0 = 0$
2. $0 + 1 = 1$
3. $1 + 0 = 1$
4. $1 + 1 = 10$, or 0 plus carry of 1
5. $1 + 1 + 1 = 11$, or 1 plus carry of 1

Example 17 Perform the indicated binary addition.

(a) $0 + 1 + 0 = ?$
(a) $0 + 1 = 1$, $1 + 0 = 1$, $0 + 1 + 0 = 1$
(b) $0 + 1 + 1 = ?$
(b) $0 + 1 = 1$, $1 + 1 = 10$, $0 + 1 + 1 = 10$
(c) $0 + 1 + 1 + 1 = ?$
(c) $0 + 1 = 1$, $1 + 1 + 1 = 11$, $0 + 1 + 1 + 1 = 11$

When adding large (multibit) binary numbers we must use *carry bits*. We are familiar with the process from decimal addition.

Example 18 Perform the indicated addition.

(a) $359_{10} + 865_{10}$

```
          111 ←——Carries
    (a)   359
          865
         ————
         1224
```

(b) $111 + 11$

(b)
```
    1   1   1   ←——Carries
    |   1   1   1
    ↓       1   1
    1   0   1   0
```

(c) $100011 + 011101$

```
111111 ←——Carries
```

(c)
```
    100011
    011101
   ———————
   1000000
```

Subtraction of Binary Numbers There are four lines in the "subtraction table" for binary numbers.

BINARY SUBTRACTION TABLE

1. $0 - 0 = 0$
2. $1 - 0 = 1$
3. $1 - 1 = 0$
4. $0 - 1 = 1$ and a "borrow" of 1

The *borrow* in binary subtraction is similar to that in the subtraction of decimal numbers.

Example 19 Perform the indicated subtraction. Check the binary operations by converting to decimal numbers and performing the operations.

(a) $(-)\begin{matrix}511_{10}\\379_{10}\end{matrix}$ (b) $(-)\begin{matrix}1000_2\\101_2\end{matrix}$ (c) $(-)\begin{matrix}110011_2\\101101_2\end{matrix}$

(a)
```
        4    10   11  ←—— Place values after
                         borrows accounted
                         for (in and out)
       -1   -1        ←—— Borrows out
   (-)  5   (¹)1  (¹)1 ←—— Minuend
        3    7    9   ←—— Subtrahend
       ———————————
        1    3    2   ←—— Difference
```

1. In the first column, $1 - 9 = 2$ ($11 - 9 = 2$) with a borrow from second column.

2. In the second column, $1 - 1$ (borrow from first column) $= 0$, then $0 - 7 = 3$ ($10 - 7 = 3$) with a borrow from third column. The 2SD of difference is 3.

3. In the third column, 5 − 1 (borrow from second column) = 4, then 4 − 3 (MSD of subtrahend) = 1, the MSD of difference.

(b)

```
                                  Place values
         0    1    1   10  ← after borrows
        -1   -1   -1        ← Borrows out
      1 (¹)0 (¹)0 (¹)0      ← Minuend
(−)     1    0    1         ← Subtrahend
      ──────────────
      0    0    1    1      ← Difference
```

1. In the first column, 0 − 1 = 1 with a borrow (from second column).

2. In the second column, 0 − 1 (borrow from first column) = 1 (with a borrow from third column); then 1 − 0 (2SD of subtrahend) = 1 (2SD of difference).

3. In the third column, 0 − 1 (borrow from second column) = 1 (with a borrow from fourth column); then 1 − 1 (3SD of subtrahend) = 0 (3SD of difference).

4. In the fourth column, 1 − 1 (borrow from third column) = 0; then 0 − 0 (4SD of subtrahend) = 0 (MSD of difference).

(c)

```
         -1   -1
       1  1  0  0  1  1      Minuend
(−)    1  0  1  1  0  1      Subtrahend
       ─────────────────
       0  0  0  1  1  0      Difference
```

Checks:

```
(b) (−) 1000        (−)  8
         101              5
        ─────            ───
          11               3
```

```
(c) (−)  1  1  0  0  1  1    (−) 51
         1  0  1  1  0  1        45
         ─────────────────      ───
                  1  1  0          6
```

Exercises 19-6 In Exercises 1–10 add the given binary numbers.

1. 101
   11

2. 1100
   111

3. 1011
   101

4. 1011
   1110

5. 10011
   11011

6. 11111
   10101

7. 100011
   110011

8. 110110
   101010

9. 1110001
   1010011

10. 11010101
    11010111

11–20. Check your solutions for Exercises 1–10 by performing the addition of the equivalent decimal numbers.

In Exercises 21–30 perform the indicated subtraction of binary numbers.

21. (−) 101
        10

22. (−) 100
        11

23. (−) 1001
        110

24. (−) 1011
        111

25. (−) 10111
        1110

26. (−) 10000
        1011

27. (−) 100100
        100011

28. (−) 10011000
        10001111

29. (−) 10000110
        10000011

30. (−) 11100001
        10111110

31–40. Check your solutions for Exercises 21–30 by performing the operations on the equivalent decimal numbers.

19-7 SUBTRACTION USING COMPLEMENTS

The subtraction of binary numbers in modern computers is usually performed by the addition of *complements* because it permits a much more economical hardware design.

One's Complement The one's complement of a binary number is obtained by subtracting the number from a number of the same bit count with all bits one's. We find the one's complement of 10110110:

```
       1 1 1 1 1 1 1 1
(−)    1 0 1 1 0 1 1 0
       ────────────────
       0 1 0 0 1 0 0 1 = one's complement of
                          10110110
```

You will be pleased to know that the one's complement can also be obtained simply by changing all 1's to 0's and all 0's to 1's. The computer also finds this very easy to accomplish.

Example 20 Find the one's complements of the given binary numbers.

(a) 10110010
(a) 10110010 number
    01001101 one's complement

(b) 11010101
(b) 11010101 number
    00101010 one's complement

We state the procedure for performing subtraction using the one's complement.

To subtract S (the subtrahend) from M (the minuend) using the one's complement:
1. Find the one's complement of S.
2. Add the one's complement of S to M.
3. Add the carry from MSB to the LSB (end-around carry). B = bit.

Example 21 Perform the indicated subtractions using the one's complement method.

(a) $(-)$ $\begin{array}{r} 11001 \\ 10101 \end{array}$ (b) $(-)$ $\begin{array}{r} 11011101 \\ 1101101 \end{array}$

(a)
$$(+) \begin{array}{r} 11001 \\ 01010 \end{array} \leftarrow \text{(one's complement}$$
$$\text{①} \overline{00011} \quad \text{of 10101)}$$
$$\longrightarrow 1 \quad \text{(add end-around}$$
$$\text{difference} \quad \overline{100} \quad \text{carry)}$$

(b)
$$(+) \begin{array}{r} 11011101 \\ 10010010 \end{array} \leftarrow \text{(one's comple-}$$
$$\text{①} \overline{01101111} \quad \text{ment)}$$
$$\longrightarrow 1 \quad \text{(add end-around}$$
$$\text{difference} \quad \overline{01110000} \quad \text{carry)}$$

Two's Complement The two's complement is equal to the one's complement plus one (one's complement + 1). We describe the procedure for subtraction with the two's complement.

To subtract S from M using the two's complement:
1. Find the one's complement of S, and add 1.
2. Add the two's complement of S to M.
3. Ignore the carry out of the MSB.

Example 22 Perform the indicated subtractions using the two's complement method.

(a) $(-)$ $\begin{array}{r} 11001 \\ 10101 \end{array}$

(a) Change subtrahend to one's complement and add one:

$$(\text{Subtrahend}) \ 10101$$
$$(\text{one's complement}) \ 01010 + 1 = 01011$$
$$\text{Add to minuend:} \ (+) \begin{array}{r} 11001 \\ 01011 \end{array}$$
$$(1) \ \overline{00100} \quad \text{(Answer)}$$
$$\qquad\qquad \nwarrow \text{(Ignore carry out}$$
$$\qquad\qquad\qquad \text{of MSB)}$$

(b) $(-)$ $\begin{array}{r} 11011101 \\ 1101101 \end{array}$

(Change subtrahend to one's complement.)

(b) 01101101
$$10010010 + 1 = 10010011 \ (\text{one's complement of S, plus 1})$$

Then
$$(+) \begin{array}{r} 11011101 \\ 10010011 \end{array}$$
$$(1) \ \overline{01110000} \quad \text{(Answer)}$$
$$(\text{ignore}) \searrow$$

Exercises 19-7 As Exercises 1–10, perform the subtractions of Exercises 21–30 in Exercises 19-6, using the one's complement method.

As Exercises 11–20, perform the subtractions of Exercises 21–30 in Exercises 19-6, using the two's complement method.

Chapter 20

Boolean Algebra

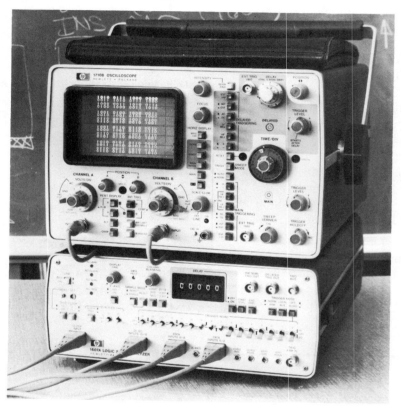

A logic analyzer is a mechanized truth table constructor. (Courtesy of Hewlett-Packard Company)

Although the title of this chapter sounds strange, the subject matter is relatively simple and easy to learn. *Boolean algebra* was named after George Boole, an Englishman, who developed its concepts and presented them in a paper entitled *The Mathematical Analysis of Logic* in 1847. In 1938 an American, Claude E. Shannon, a graduate student at the time, demonstrated how Boolean algebra could be used to analyze two-state or switching circuits. Boolean algebra is sometimes referred to as the "algebra of logic," the "algebra of switching circuits," "digital algebra," or "binary algebra."

20-1 LOGIC: VARIABLES AND FUNCTIONS

Logic is the science of correct reasoning, the reasoning which is required to interpret the outcome of a situation determined by two or more conditions. Each condition may be *present* or *not present*, or in the terminology of logic, *true* or *false*. Consider the following situations.

1. A tape recorder will be placed in the RECORD mode if the RECORD and FORWARD keys are depressed.
2. The tape advances if either the FORWARD or FAST FORWARD keys are depressed.

The situations described are relatively uncomplex and the outcomes can be interpreted intuitively (without significant conscious effort). Boolean algebra provides a direct, mathematical process for simplifying and interpreting the outcomes of these as well as extremely complex situations where simple reasoning might prove to be inadequate in the face of numerous, and possibly interrelated and redundant, conditions.

Logic Variables In Boolean algebra, variables such as $A, B, C, \ldots, X, Y, Z$ are used to represent logic statements or conditions. In regular algebra variables can usually have any value. In logic, however, statements are either true or false, or conditions are present or not present. A "1" is used to represent the "value" of a true statement or condition present, and a "0" is used to represent the "value" of a false statement or a condition not present. Thus, if we use A to represent

the statement "the recorder is in RECORD mode," then A can be true and we say that the "value" of A is 1, written $A = 1$. The statement can also be false; then the "value" of A is 0, or $A = 0$. The only two possible values for a variable, then, are 1 and 0. There is no quantitative significance to these values. Their significance is logical only even though we make statements like "A equals 1" or "A equals 0."

Logic Operations In attempting to interpret the outcome of situations describable with logic statements there are three key words that make all the difference in the analysis. These three words become the *operators* or *functions* of Boolean algebra:

1. *AND* This operation is referred to as *conjunction* or the *logical product*. The symbol of the operation is $(\cdot)$.
2. *OR* This is the operation of *disjunction* or *logical sum*. The symbol is $(+)$.
3. *NOT* *Negation* is symbolized by $(^-)$ or sometimes $(')$.

The AND Function The *AND* function can be illustrated by Statement 1.

A tape recorder will be placed in RECORD mode if the RECORD *AND* FORWARD keys are depressed.

Let us use X to represent the statement "the recorder is in RECORD mode." A to represent "the RECORD key is depressed," and B to represent "the FORWARD key is depressed." The situation is then symbolized as

$$X = A \cdot B$$

Note that this Boolean equation means that X is true only if both A and B are true. Just as in ordinary algebra, the $(\cdot)$ symbol may be omitted. The equation may be written

$$X = AB$$

The *AND* function may also be illustrated by the electric circuit of Figure 20-1. The light-emitting diode (LED) X is not energized unless both switches A *AND* B are closed: $X = AB$.

A modern computer as well as many other equipments employing so-called digital devices are made up of a multitude of electronic circuits which perform various logic operations. The circuits are called *gates* and are

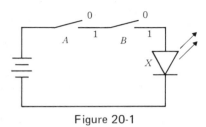

Figure 20-1

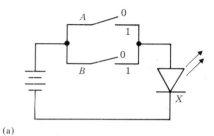

(a)

(b)

C	D	Y
0	0	0
0	1	1
1	0	1
1	1	1

(c)

Figure 20-3

represented by a set of standardized *logic symbols*. The gates are said to *implement* the logic (that is, cause it to happen with hardware). A truth table is a chart which shows all the logic conditions which may exist at the input and output of a logic gate. The logic symbol and truth table for a two-input *AND* gate are shown in Figures 20-2(a) and (b), respectively.

MEMORY JOGGER

A 1 means "true" (or condition present) and a 0 means "false" (condition not present).

The OR Function   The logic *OR* function can be illustrated with Statement 2.

The tape advances if either the FORWARD *OR* FAST-FORWARD key is depressed.

Let us use the variable Y to represent "the tape is advancing," C to represent "the FORWARD key is depressed," and D to represent "the FAST-FORWARD key is depressed."

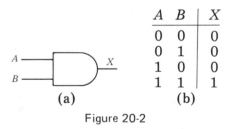

A	B	X
0	0	0
0	1	0
1	0	0
1	1	1

(a)   (b)

Figure 20-2

The logic equation is

$$Y = C + D$$

Note that Y is true if either C or D is true. A simple electric switching circuit to implement the function (allow it to happen or be *realized*) is shown in Figure 20-3(a). The gate symbol and truth table are shown in Figures 20-3(b) and (c), respectively.

The NOT Function   The *NOT* function is one which inverts a condition: a TRUE is changed to FALSE; a FALSE is changed to TRUE. Consider the two statements:

3. The recorder is in RECORD mode.
4. The recorder is *NOT* in RECORD mode.

Obviously, both Statements 3 and 4 cannot be true at the same time. If we let A represent Statement 3 and *NOT A* or $\overline{A}$ represent Statement 4, then

A	NOT A
True	False
False	True

or

A	$\overline{A}$
1	0
0	1

Two opposite statements or conditions are called *complements*.

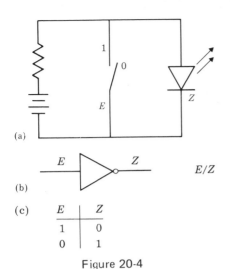

(a)

(b) E ⊳ Z   E/Z

(c)
E	Z
1	0
0	1

Figure 20-4

A circuit to implement the *NOT* function is shown in Figure 20-4(a). Notice that if the switch is in its normal position, the LED is on; if the switch is operated, the LED goes off. Thus,

$$Z = \overline{E}$$

which we read "Z equals *NOT E.* The gate circuit for this function is called an *inverter.* Its symbol and truth table are shown in Figures 20-4(b) and (c), respectively.

We summarize the three Boolean functions: their equations, symbols, and truth tables.

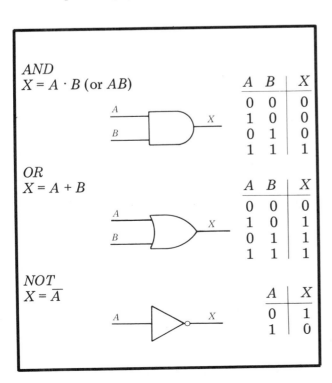

Example 1 Write a logic equation and construct the logic symbol and truth table to represent the following statement: "The car may be operated if the hand brake is released and the seat belt fastened."

Let M represent "the car is operable," let B represent "the hand brake is released," and let S represent "the seat belt is fastened."

$$M = B \cdot S$$

B	S	M
0	0	0
0	1	0
1	0	0
1	1	1

Example 2 Write a logic equation and construct the logic symbol and truth table to represent the statement: "The right front window may be moved up or down by the passenger's switch or the driver's switch."

Let W represent "the window is moved," let D represent "the driver's switch is activated," and let P represent "the passenger's switch is activated."

$$W = D + P$$

D	P	W
0	0	0
1	0	1
0	1	1
1	1	1

Example 3 Illustrate the following situation with a logic equation, a logic symbol, and a truth table: "A player wins if his move does not place him on square X."

Let W represent "the player wins," and let X represent "the player lands on X."

$$W = \overline{X}$$

X	W
0	1
1	0

Exercises 20-1

1. Write a Boolean or logic equation to illustrate an *AND* function with two input variables and one output variable.
2. Repeat Exercise 1 for the *OR* function.
3. The variables P and Q are related by the *NOT* function. Write an equation showing the relationship.
4. Draw and label the logic symbols that correspond to these equations: (a) $X = A + B$; (b) $X = \overline{A}$; and (c) $X = AB$.

5. Construct the truth tables for the equations of Exercise 4.

6. Write equations that state the relationships shown by the following logic symbols:

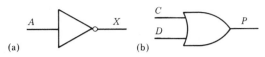

(a)

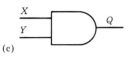

(c)

7. Write the equations that state the relationships illustrated in the following truth tables:

(a)

A	B	C
0	0	0
1	0	1
0	1	1
1	1	1

(b)

P	Y
0	1
1	0

(c)

R	S	Z
0	0	0
0	1	0
1	0	0
1	1	1

8. Construct the truth tables for the relationships illustrated by the symbols of Exercise 6.

9. Draw the logic symbols that correspond to the truth tables of Exercise 7.

In Exercises 10–12 for the given situations, (a) choose variables and write an equation, (b) draw the logic diagram, and (c) construct the truth table.

10. An automobile engine may be started if the ignition system is turned on and the starter actuated.

11. In a transceiver, the receiver circuit is normally connected to the antenna. Actuating the push-to-talk switch disconnects the receiver.

12. A stop-light lamp is energized by the brake switch or the turn signal.

20-2 BOOLEAN EXPRESSIONS

The basic operations (or functions)—*AND*, *OR*, and *NOT*—discussed in the preceding section may be combined in a limitless number of ways. The *AND* and *OR* functions may be used for more than two conditions (inputs); the outputs of functions may become the inputs of other functions, and so on. The expressions containing the variables and operation symbols used to describe these compound relationships are called *Boolean expressions*. Boolean algebra provides a set of rules for interpreting these expressions and manipulating them so as to arrive at the simplest form and thereby save hardware. Let us consider several examples of more complex Boolean expressions before proceeding to simplification procedures.

Example 4   Write Boolean expressions to correspond to the given logic diagrams.

(a)

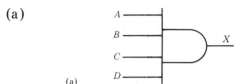

(a)

(a) $X = A \cdot B \cdot C \cdot D$,   or $X = ABCD$

(b)

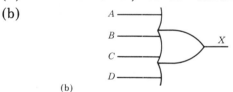

(b)

(b) $X = A + B + C + D$

(c)

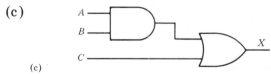

(c)

(c) $X = AB + C$ (Read "*A AND B, OR C.*")

(d)

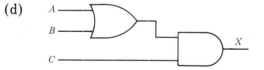

(d) $X = C(A + B)$ [Read "*C AND (A OR B).*"]

(Note that we use parentheses to indicate "addition" is to occur before "multiplicacion," as in ordinary algebra.)

(e)

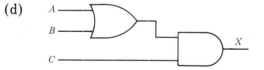

(e) $X = A\overline{B}C$ (Read "*A AND NOT B AND C.*")

(f)

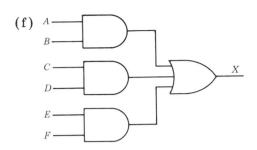

(f) $X = AB + CD + EF$

(g)

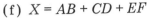

(g) $X = \overline{ABC}$

(h)

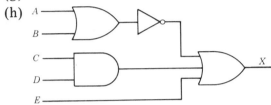

(h) $X = \overline{A + B} + CD + E$

It is important that you be able to verbalize the logic equations of various logic combinations. Study carefully the combinations illustrated in Example 4. (The samples do not include all possible combinations by any means, but are sufficient to demonstrate basic ideas of combinational logic.) Next, cover up the equations and state or write the relationship by inspecting the diagrams. Check your results. Finally, cover up the diagrams, and construct diagrams which will provide the logic described by the equations. Again, check your work.

Example 5 The Exclusive-*OR* Function In the expression $X = A + B$, X will be true if either A or B or both A and B are true. It is desired to exclude the condition AB. (a) Write Boolean equation(s) for the exclusive-*OR* function. (b) Construct the logic diagram(s) for realizing the function. (c) Construct the exclusive-*OR* truth table.

(a) The condition may be achieved in two ways:

$$X = (A + B)\overline{AB} \quad \text{or} \quad X = A\overline{B} + \overline{A}B$$

"X is true when either A or B is true *AND NOT* when both A *AND* B are true." Or, "X is true when A *AND NOT* B are true, *OR* when *NOT* A, *AND* B are true."

(b)

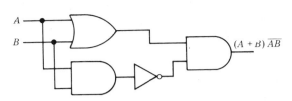

(a)

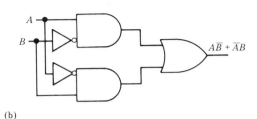

(b)

Figure 20-5

(c)

A	B	X
0	0	0
1	0	1
0	1	1
1	1	0 (AB excluded)

The exclusive-*OR* (*XOR*) function illustrated in Example 5 is so common it has a special logic operator symbol ($\oplus$) and logic symbol (diagram) of its own (see Figure 20-6.)

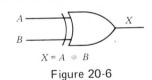

$$X = A \oplus B$$

Figure 20-6

Exercises 20-2 For Exercises 1–10 write Boolean equations for the given logic diagrams (refer to Example 4).

1.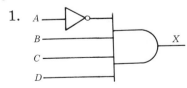

2.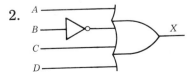

3.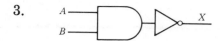

4.

5.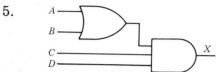

6.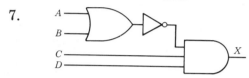

7.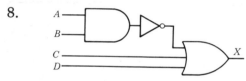

8.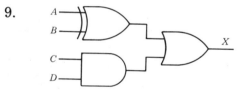

9.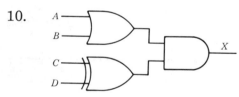

10.

In Exercises 11–20 sketch the logic diagrams for the given Boolean equations.

11. $X = A + B + \overline{C}$ 12. $X = A\overline{B}CD$

13. $X = \overline{\overline{A}B}$ 14. $X = \overline{A + \overline{B}}$

15. $X = AB + \overline{C} + D$ 16. $X = A\overline{B}(C + D)$

17. $X = AB(\overline{C} + D)$ 18. $X = A + B + \overline{CD}$

19. $X = (A \oplus B)(C + D)$

20. $X = A \oplus B + \overline{CD}$

21. Show with a truth table that the expression $X = (A + B)\overline{AB}$ excludes AB; that is, $X = 0$ when $AB = 1$.

22. Repeat Exercise 21 for $X = A\overline{B} + \overline{A}B$.

23. The conditions for tape advance (see Statement 2 in Section 20-1) are more completely described as follows: The tape advances if either the FORWARD or FAST-FORWARD key is depressed *AND* a cassette is in place *AND* power is on. Write the Boolean equation and draw the logic diagram.

20-3 DEFINITIONS, POSTULATES, AND THEOREMS

Variables   As previously indicated, Boolean *variables* are letters such as $A, B, \ldots, X, Y$, and so on.

Constants   There are only two Boolean *constants*—0 and 1. These have logical significance but no numerical significance.

Equivalence   Two expressions are equivalent if one expression is true (= 1) only when the other expression is true, and one is false (= 0) only when the other is false.

Complements   Two expressions are complements of each other if one is true (= 1) when the other is false (= 0), and vice versa. The *NOT* function generates the complement of (negates) an expression.

Postulates   A statement that describes a self-evident truth is called a *postulate*. The Boolean postulates are summarized in the box below. Postulates form the basis of theorems which are useful in manipulating and simplifying logical expressions. It is essential that you learn these thoroughly.

POSTULATES		
1a. $A = 1$ or else		1b. $A = 0$
2a. $1 \cdot 1 = 1$		2b. $0 + 0 = 0$
3a. $1 \cdot 0 = 0 \cdot 1 = 0$		3b. $0 + 1 = 1 + 0 = 1$
4a. $0 \cdot 0 = 0$		4b. $1 + 1 = 1$
5a. $\overline{1} = 0$		5b. $\overline{0} = 1$

In Figure 20-7 the postulates are demonstrated with logic diagrams.

POSTULATE SYMBOLIZATION

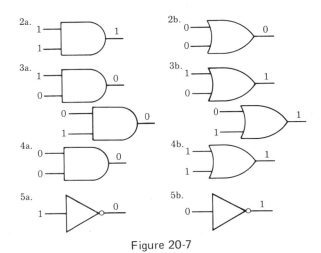

Figure 20-7

Duals Note the special relationship between the a and b parts of each postulate: the a parts utilize the *AND* operator (·), and the b parts utilize the *OR* operator (+). Furthermore, the constants in one part are the *complements* of those in the other. This kind of special relationship is called the *dual relation-*

ship. The *dual* of an expression is created by

changing all ·'s to +'s

changing all +'s to ·'s

changing all 1's to 0's

changing all 0's to 1's

but not complementing any variable.

Boolean theorems will also be studied in dual pairs.

Theorems Some frequently occurring relationships, although true, are not obviously true by simple inspection (as are postulates). These relationships are called *theorems* and are provable by the use of the postulates. Once proven they are accepted as laws or principles and enable us to simplify or transform expressions into more useful expressions. The first three theorems concern the properties of *commutation, association,* and *distribution* of Boolean expressions. These properties are similar to those of ordinary algebra (review Sections 1-5, 2-3, and 3-4). We illustrate the theorems relating to these properties in the box below. You are encouraged to become thoroughly familiar with these properties because they are essential to the process of manipulating logic expressions.

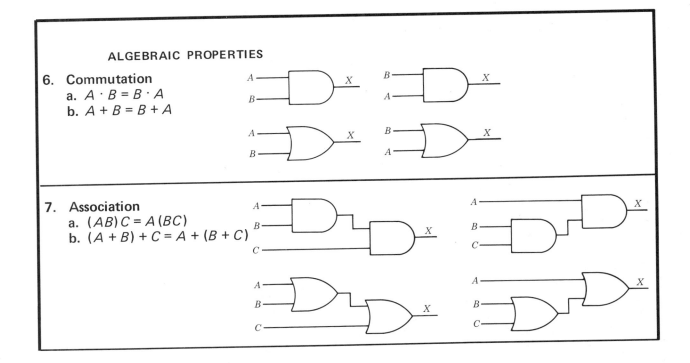

ALGEBRAIC PROPERTIES

6. **Commutation**
    a. $A \cdot B = B \cdot A$
    b. $A + B = B + A$

7. **Association**
    a. $(AB)C = A(BC)$
    b. $(A + B) + C = A + (B + C)$

8. **Distribution**
   a. $A(B + C) = AB + AC$
   b. $A + BC = (A + B)(A + C)$

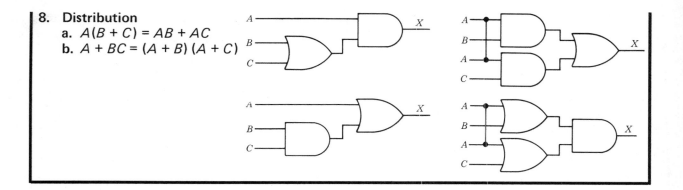

Example 6 Using the postulates and truth tables demonstrate that the distribution theorem (8a and b) is true.

8a. $A(B + C) = AB + AC$

A	B	C	B + C	A(B + C)	AB	AC	AB + AC
0	0	0	0	0	0	0	0
0	0	1	1	0	0	0	0
0	1	0	1	0	0	0	0
0	1	1	1	0	0	0	0
1	0	0	0	0	0	0	0
1	0	1	1	1	0	1	1
1	1	0	1	1	1	0	1
1	1	1	1	1	1	1	1

LM ———— = ———— RM (over columns $A(B+C)$ and $AB+AC$)

8b. $A + BC = (A + B)(A + C)$

A	B	C	BC	A + BC	A + B	A + C	(A + B)(A + C)
0	0	0	0	0	0	0	0
0	0	1	0	0	0	1	0
0	1	0	0	0	1	0	0
0	1	1	1	1	1	1	1
1	0	0	0	1	1	1	1
1	0	1	0	1	1	1	1
1	1	0	0	1	1	1	1
1	1	1	1	1	1	1	1

LM ———— = ———— RM (over columns $A + BC$ and $(A+B)(A+C)$)

Note that constructing a truth table (see Example 6) is a process of systematically substituting constants (0 and 1) for all variables (see columns headed A, B, and C in Example 6). Next we evaluate various parts of the expressions (see columns headed AB, AC, $B + C$, and so on). We evaluate by applying the postulates. If either B or C or both equals 1 and $A = 1$, then $B + C = 1$ because of Postulate 3b: $0 + 1 = 1$, and $A(B + C) = 1$ because of Postulate 2a: $1 \cdot 1 = 1$. Or if $A = 1$ or $B = 1$, then $A + B = 1$; and if $A = 1$ or $C = 1$, then $A + C = 1$, and therefore,

$(A + B)(A + C) = 1$, and so on. An equation is true (the two members are *equivalent*) when the columns representing its two members are identical (pattern of 0's and 1's is the same). Compare the columns labeled LM and RM in each of the truth tables of Example 6.

Study the truth tables of Example 6 in detail until you thoroughly understand the procedure for constructing and interpreting a truth table, as well as understand the meaning of the distributive property.

The next five theorems are concerned with the relationship of a variable with itself or a constant. They are presented, along with their logic diagrams, in the box below.

THEOREMS

9. a. $0 \cdot A = 0$

   b. $1 + A = 1$

$0 \cdot$ anything $= 0$ $1 +$ anything $= 1$

10. a. $1 \cdot A = A$

   b. $0 + A = A$

ANDing an expression with 1, or ORing an expression with 0 does not change the expression.

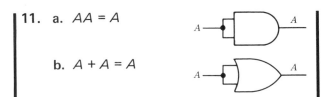

11. a. $AA = A$

    b. $A + A = A$

12. a. $A\overline{A} = 0$

    b. $A + \overline{A} = 1$

ANDing (or the product of) an expression with its complement always equals 0. ORing (or the sum of) an expression with its complement always equals 1.

13. $\overline{\overline{A}} = A$

Double negation

We illustrate these theorems with worked-out examples.

Example 7 Simplify the following expressions using Theorem 9.

(a) $0 \cdot (A + BC)$
(a) $0 \cdot (A + BC) = 0$ because $0 \cdot X = 0$, where X is any variable or expression.
(b) $1 + AB + C(D + E)$
(b) $1 + AB + C(D + E) = 1$ because $1 + X = 1$

Example 8 Simplify the following expressions. State which theorem is used to justify the simplification.

(a) $(A + BCD) \cdot 1$
(a) $(A + BCD) \cdot 1 = A + BCD$, Theorem 10a
(b) $0 + AB + C(D + E)$
(b) $0 + AB + C(D + E) = AB + C(D + E)$, Theorem 10b

Example 9 Simplify the following expressions. Which theorem(s) justifies(y) the simplification?

(a) ABB
(a) $ABB = AB$ (11a)
(b) $A + C + C + A$
(b) $A + C + C + A = A + C$ (11b)
(c) $CDD + CD + E$
(c) $CDD + CD + E = CD + E$ (11a and 11b)

Example 10 Simplify the following expressions.

(a) $A\overline{A}B$
(a) $A\overline{A}B = 0 \cdot B = 0$ (12a, 9a)
(b) $A + \overline{A} + B$
(b) $A + \overline{A} + B = 1 + B = 1$ (12b, 9b)
(c) $A\overline{A} + C$
(c) $A\overline{A} + C = 0 + C = C$ (12a, 10b)

Example 11 Simplify the following expressions.

(a) $\overline{\overline{AB}}$
(a) $\overline{\overline{AB}} = AB$ (13)
(b) $\overline{\overline{A + B}}$
(b) $\overline{\overline{A + B}} = A + B$ (13)
(c) $\overline{\overline{AB}}(\overline{C} + D)$
(c) $\overline{\overline{AB}}(\overline{C} + D) = AB(\overline{C} + D)$ (13)

Exercises 20-3A In Exercises 1–20 simplify the given expressions.

1. $ABBAC$ 2. $(AB + AB)C$
3. $ABA + C + BA$ 4. $AB + B(A + C)$
5. $\overline{A}ABC$ 6. $AB(\overline{A} + B)$
7. $(A + \overline{A})C + A\overline{A}C$ 8. $B + \overline{B} + (A + \overline{A})B$
9. $A\overline{A}(A + B + \overline{C})$ 10. $(\overline{C} + D)C\overline{D}$
11. $E + \overline{E}F$ (*Hint:* Apply Theorem 8b: $(E + \overline{E}F) = (E + \overline{E})(E + F)$)
12. $E + \overline{E}F + \overline{F}$ 13. $(A + \overline{A}B)\overline{A}B$
14. $C\overline{D}(C + \overline{C}D)$ 15. $D + E + \overline{F} + \overline{E}F$
16. $A(B + \overline{B}C + \overline{C})$ 17. $\overline{A} + B\overline{C}(\overline{B} + C)$
18. $A(A + B) + \overline{E}F + \overline{F} + E$
19. $\overline{\overline{A}\overline{A}B}$ 20. $AC(\overline{\overline{C} + D})$

DeMorgan's Theorem One of the more important theorems for simplification is known as *DeMorgan's theorem*, named after a friend of Boole's. It enables us to find the complement of (or invert) an expression. The result is to change an expression, which is basically an *AND* function, for example, into a complementary one, which is an *OR* function, and vice versa. It is applied to a complete expression or part of an expression by performing two steps:

1. Change all *OR*s (+) to *AND*s (·) and all *AND*s (·) to *OR*s (+).
2. Complement all the variables.

DEMORGAN'S THEOREM

14a. $\overline{A \cdot B \cdot \ldots \cdot C} = \overline{A} + \overline{B} + \ldots + \overline{C}$

14b. $\overline{A + B + \cdots + C} = \overline{A} \cdot \overline{B} \cdot \ldots \cdot \overline{C}$

(Variables may be single variables or expressions.)

Example 12　Change the following expressions into their complements and draw the logic diagrams for the two forms of each.

(a) AB　　(b) $\overline{A} + B$　　(c) $A + B\overline{C}$

(a) $\overline{AB} = ?$

Step 1: $AB \longrightarrow A + B$

Step 2: $\overline{A + B} \longrightarrow \overline{A} + \overline{B}$

Therefore, $\overline{AB} = \overline{A} + \overline{B}$

(*Note:* The small circle at the output or inputs of the symbols above is an alternate symbol for an inverter.)

(b) $\overline{\overline{A} + B} = ?$

Step 1: $\overline{A} + B \longrightarrow \overline{A}B$

Step 2: $\overline{\overline{A}B} \longrightarrow A\overline{B}$

Therefore, $\overline{\overline{A} + B} = A\overline{B}$

(c) $\overline{A + B\overline{C}} = ?$

Step 1: $A + B\overline{C} \longrightarrow A(B + \overline{C})$

Step 2: $\overline{A}(\overline{B} + \overline{\overline{C}}) \longrightarrow \overline{A}(\overline{B} + C)$

Therefore, $\overline{A + B\overline{C}} = \overline{A}(\overline{B} + C)$

Example 13　By the use of truth tables prove that the two forms of the expressions in Example 12 are equivalent.

(a) $\overline{AB} = \overline{A} + \overline{B}$

A	B	AB	$\overline{AB}$	$\overline{A}$	$\overline{B}$	$\overline{A} + \overline{B}$
0	0	0	1	1	1	1
1	0	0	1	0	1	1
0	1	0	1	1	0	1
1	1	1	0	0	0	0

(b) $\overline{\overline{A} + B} = A\overline{B}$

A	B	$\overline{A}$	$\overline{B}$	$\overline{A} + B$	$\overline{\overline{A} + B}$	$A\overline{B}$
0	0	1	1	1	0	0
1	0	0	1	0	1	1
0	1	1	0	1	0	0
1	1	0	0	1	0	0

(c) $\overline{A + B\overline{C}} = \overline{A}(\overline{B} + C)$

A	B	C	$\overline{C}$	$B\overline{C}$	$A + B\overline{C}$	$\overline{A + B\overline{C}}$	$\overline{A}$	$\overline{B}$	$\overline{B} + C$	$\overline{A}(\overline{B} + C)$
0	0	0	1	0	0	1	1	1	1	1
0	0	1	0	0	0	1	1	1	1	1
0	1	0	1	1	1	0	1	0	0	0
0	1	1	0	0	0	1	1	0	1	1
1	0	0	1	0	1	0	0	1	1	0
1	0	1	0	0	1	0	0	1	1	0
1	1	0	1	1	1	0	0	0	0	0
1	1	1	0	0	1	0	0	0	1	0

NAND and NOR Logic　Because of the nature of certain circuits which have been found both highly effective as switching circuits as

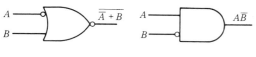
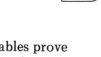
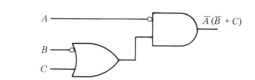

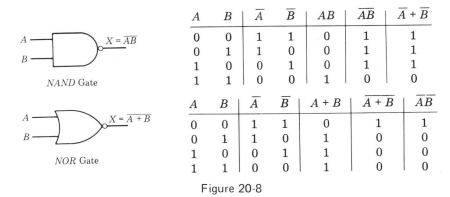

A	B	$\overline{A}$	$\overline{B}$	AB	$\overline{AB}$	$\overline{A}+\overline{B}$
0	0	1	1	0	1	1
0	1	1	0	0	1	1
1	0	0	1	0	1	1
1	1	0	0	1	0	0

A	B	$\overline{A}$	$\overline{B}$	$A+B$	$\overline{A+B}$	$\overline{A}\,\overline{B}$
0	0	1	1	0	1	1
0	1	1	0	1	0	0
1	0	0	1	1	0	0
1	1	0	0	1	0	0

Figure 20-8

well as relatively easy to manufacture, the majority of available integrated circuit (IC) gates perform the functions of *NOT AND* or *NOT OR* and are called *NAND* and *NOR* gates, respectively. The logic symbols and truth tables of these functions are shown in Figure 20-8.

Let us note a most significant feature of *NAND* and *NOR* logic: By DeMorgan's theorem, $\overline{AB} = \overline{A} + \overline{B}$ and $\overline{A+B} = \overline{A}\,\overline{B}$. Therefore, a *NAND* gate is the equivalent of an *OR* gate with inverted inputs; a *NOR* gate is identical to an *AND* gate with inverted inputs. The truth tables verify this conclusion (see Figure 20-8).

Example 14 A two-input *AND* gate is required for a certain application to perform the logic $X = AB$. Only *NOR* gates are available. Design a combination that will perform the desired function.

Since we know the final gate includes an inherent inversion (*NOT OR*) a good procedure is (1) write the final function, (2) invert, and (3) invert again to obtain the expression which can be implemented with *NOR*.

$$(1)\ AB \quad (2)\ \overline{AB} = \overline{A} + \overline{B}$$
$$(3)\ \overline{\overline{AB}} = AB = \overline{\overline{A} + \overline{B}}$$

It is evident that a *NOR* gate $(\overline{A + B})$ with inverted inputs $(\overline{\overline{A} + \overline{B}})$ would perform the required function. And since $(\overline{A + A}) = \overline{A}$ by Theorem 11b, *NOR* gates with their inputs tied together can serve as inverters. The logic

diagram and truth table are shown in Figure 20-9.

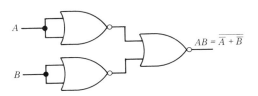

A	B	$\overline{A}$	$\overline{B}$	$\overline{A}+\overline{B}$	$\overline{\overline{A}+\overline{B}}$	AB
0	0	1	1	1	0	0
0	1	1	0	1	0	0
1	0	0	1	1	0	0
1	1	0	0	0	1	1

Figure 20-9

Exercises 20-3B In Exercises 21–30 change the given expressions into their complements and draw the logic diagrams for the two forms of each.

21. $\overline{A}B$

22. $A + \overline{B}$

23. $\overline{A}BC$

24. $A + B + \overline{C}$

25. $AB + C$

26. $A + \overline{B}C$

27. $AB(C + D)$

28. $A + \overline{B}(C + \overline{D})$

29. $AB + A(C + \overline{DE})$

30. $\overline{A}\overline{B}C(D + E\overline{F}\overline{G})$

In Exercises 31–34 the given logic diagrams contain *NAND* and/or *NOR* gates only.
(a) Write the Boolean equation for the dia-

grams. (b) Write the equivalent inverted equations. (Example: $\overline{AB} = A + \overline{B}$.) (c) Construct the truth tables for the given diagrams.

31.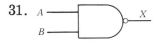

32.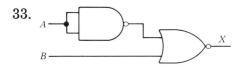

33.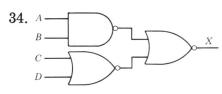

34.

In Exercises 35–38 draw logic diagrams utilizing only *NAND* and/or *NOR* gates to implement the given Boolean equations (see Example 14).

35. $\overline{\overline{AB}}$ 36. $A + B$

37. $A\overline{B}$

38. $AB(C + D)$ (*Hint:* $\overline{AB(C + D)} = \overline{\overline{AB} + \overline{C + D}}$)

Redundancy, and Absorption Theorems In many logical expressions the appearance of some literals is *redundant* (unnecessary to the interpretation of the situation represented). The absorption theorems (see box below) show how these expressions may be simplified by *absorbing* the redundant literals.

ABSORPTION THEOREMS

15a. $A + AB = A$ 15b. $A(A + B) = A$
16a. $A + \overline{A}B = A + B$ 16b. $A(\overline{A} + B) = AB$

Proving the absorption theorems makes them more understandable and meaningful.

Example 15 Prove the following identities.

(a) $A + AB = A$
(a) $X = A + AB$
    $= A(1 + B)$ (8a and 10a)
    $= A \cdot 1$ (9b)
Therefore, $A + AB = A$ (10a)

(b) $A(A + B) = A$
(b) $X = A(A + B)$
    $= (A + 0)(A + B)$ (10b)
    $= A + 0 \cdot B$ (8b)
    $= A + 0$ (9a)
Therefore, $A(A + B) = A$ (10b)

(c) $A + \overline{A}B = A + B$
(c) $X = A + \overline{A}B$
    $= (A + \overline{A})(A + B)$ (8b)
    $= 1 \cdot (A + B)$ (12b)
Therefore, $A + \overline{A}B = A + B$ (10a)

(d) $A(\overline{A} + B) = AB$
(d) $X = A(\overline{A} + B)$
    $= A\overline{A} + AB$ (8a)
    $= 0 + AB$ (12a)
Therefore, $A(\overline{A} + B) = AB$ (10a)

Examples of simplification demonstrate the application of the absorption theorems.

Example 16 Simplify the following expressions.

(a) $AB\overline{C} + ABD + AB + FAB$
(a) $AB\overline{C} + ABD + AB + FAB = AB + AB(\overline{C} + D + F) = AB$ (15a)

(b) $CD(ABC + CD + EFG)$
(b) $CD(ABC + CD + EFG) = CD$ (15b)

(c) $ABC + \overline{AB} + ABD$
(c) $ABC + \overline{AB} + ABD = \overline{AB} + AB(C + D) = \overline{AB} + C + D$ (*Remember:* $AB = \overline{\overline{AB}}$) (16a)

(d) $\overline{AB}(CD + AB + EF)$
(d) $\overline{AB}(CD + AB + EF) = \overline{AB}(CD + EF)$ (16b)

SUMMARY OF POSTULATES AND THEOREMS

1a. $A = 1$ or else   1b. $A = 0$

2a. $1 \cdot 1 = 1$ 2b. $0 + 0 = 0$

3a. $1 \cdot 0 = 0 \cdot 1 = 0$ 3b. $0 + 1 = 1 + 0 = 1$

4a. $0 \cdot 0 = 0$ 4b. $1 + 1 = 1$

5a. $\overline{1} = 0$ 5b. $\overline{0} = 1$

6a. $A \cdot B = B \cdot A$

6b. $A + B = B + A$

7a. $(AB)C = A(BC)$

7b. $(A + B) + C = A + (B + C)$

8a. $A(B + C) = AB + AC$

8b. $A + BC = (A + B)(A + C)$

9a. $0 \cdot A = 0$ 9b. $1 + A = 1$

10a. $1 \cdot A = A$ 10b. $0 + A = A$

11a. $A \cdot A = A$ 11b. $A + A = A$

12a. $A \cdot \overline{A} = 0$ 12b. $A + \overline{A} = 1$

13. $\overline{\overline{A}} = A$

14a. $\overline{A \cdot B \cdot \ldots \cdot C} = \overline{A} + \overline{B} + \cdots + \overline{C}$
(DeMorgan's theorem)

14b. $\overline{A + B + \cdots + C} = \overline{A} \cdot \overline{B} \cdot \ldots \cdot \overline{C}$

15a. $A + AB = A$

15b. $A(A + B) = A$

16a. $A + \overline{A}B = A + B$

16b. $A(\overline{A} + B) = AB$

Exercises 20-3C In Exercises 39–48 simplify the given expressions (see Example 16).

39. $B + ABC + BCD$

40. $AC + AC(B + D)$

41. $X(X + Y)$

42. $DE(AB + CD + DE)$

43. $P + N\overline{P}$

44. $AB + \overline{A}BC + DEF$

45. $F(D + \overline{F})$

46. $\overline{XY}(AB + CD + XY)$

47. $AB + A\overline{B}$

48. $(A + B)(A + \overline{B})$

In Exercises 49–54 use Boolean algebra or truth tables to show that the given simplifications are valid.

49. $ABC + A\overline{B}C + AB\overline{C} + A\overline{B}\,\overline{C} = A$

50. $(A + B + C)(A + \overline{B} + C)(A + B + \overline{C})$
    $(A + \overline{B} + \overline{C}) = A$

51. $AB + A\overline{B}C = AB + AC$

52. $(A + B)(A + \overline{B} + C) = (A + B)(A + C)$

53. $AB + \overline{A}C + BC = AB + \overline{A}C$

54. $(A + B)(\overline{A} + C)(B + C) = (A + B)(\overline{A} + C)$

20-4 APPLICATIONS

Setting up combinations of logic gates to perform specific functions can be systematized to a few straightforward steps. The procedure assumes that a design using a single type of commercially available gate (*NAND* or *NOR*) is preferred. A procedure for utilizing *NAND* gates is as follows:

1. Construct a truth table showing the relationship between inputs and the desired output(s).

2. From the truth table write a Boolean expression of the form

$$X = P_1 + P_2 + \cdots + P_n$$

where P_1, P_2, ... are products of the input variables in the rows in which $X = 1$.

3. Negate (place a bar over) each variable in each instance where it appears as a 0 in the truth table.

4. Use DeMorgan's theorem to convert the expression to the product form:

$$X = \overline{\overline{P_1}\,\overline{P_2} \ldots \overline{P_3}}$$

5. Connect *NAND* gates as required by the expression of Step 4.

We illustrate the procedure with several examples.

Example 17 Design a gate connection using only *NAND* gates which will perform the function of a half-adder (HA) for two binary bits. Assume that complements of variables, as well as the variables, are available.

Step 1:

A	B	S	C
0	0	0	0
0	1	1	0
1	0	1	0
1	1	0	1

Step 2: Sum: $\overline{A}B + A\overline{B}$
          Carry: AB

Step 3: $S = \overline{A}B + A\overline{B}$
          $C = AB$

Step 4: $S = \overline{\overline{\overline{A}B}} + \overline{\overline{A\overline{B}}} = \overline{\overline{\overline{A}B} \cdot \overline{A\overline{B}}}$
          $C = \overline{\overline{AB}} = \overline{\overline{AB} \cdot \overline{AB}}$

Step 5: $S = \overline{\overline{\overline{AB}} \cdot \overline{\overline{AB}}}$

$C = \overline{\overline{\overline{AB}}}$

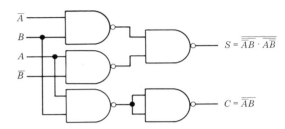

$$S = \overline{\overline{\overline{AB}} \cdot \overline{\overline{AB}}}$$

$$C = \overline{\overline{AB}}$$

Example 18   A data selector is required which will provide an output when only two of the three bits A, B, or C are present, and only when the C-bit is one of the two. Design the logic circuit using only $NAND$ gates.

Step 1:

A	B	C	X
0	1	1	1
1	0	1	1

Step 2: $\overline{A}BC + A\overline{B}C$

Step 3: $\overline{A}BC + A\overline{B}C$

Step 4: $X = \overline{\overline{\overline{A}BC + A\overline{B}C}} = \overline{\overline{\overline{A}BC} \cdot \overline{A\overline{B}C}}$

Step 5:

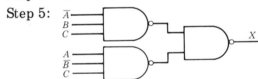

Example 19   Design the logic circuit which will produce the SUM (S) output of a binary full adder (FA). Use $NAND$ gates only. Assume that both variables and complements are available.

Step 1:

A	B	C	S
0	0	1	1
0	1	0	1
1	0	0	1
1	1	1	1

Step 2: $\overline{A}\,\overline{B}C + \overline{A}B\overline{C} + A\overline{B}\,\overline{C} + ABC$

Step 3: $\overline{\overline{A}\,\overline{B}C} + \overline{\overline{A}B\overline{C}} + \overline{A\overline{B}\,\overline{C}} + \overline{ABC}$

Step 4: $\overline{\overline{\overline{A}\,\overline{B}C}} \cdot \overline{\overline{\overline{A}B\overline{C}}} \cdot \overline{\overline{A\overline{B}\,\overline{C}}} \cdot \overline{\overline{ABC}}$

Step 5:

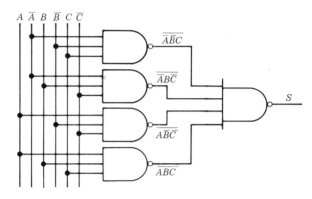

Exercises 20-4   In Exercises 1–6 design logic circuits to perform the given logic functions using $NAND$ gates only.

1. AND 2. NOR

3. $Exclusive$-OR

4. Equality comparator: $X = AB + \overline{A}\,\overline{B}$

5. Two, or more, out of three bits present

6. Full adder: $S = \overline{A}\,\overline{B}C + \overline{A}B\overline{C} + A\overline{B}\,\overline{C} + ABC$

In Exercises 7–12 design logic circuits to perform the given logic functions using NOR gates only.

7. $NAND$ 8. OR

9. Inequality comparator: $A\overline{B} + \overline{A}B$

10. $X = AB + \overline{A}\,\overline{B}$

11. At least two out of three bits present

12. Full subtractor: $D = M\overline{S}\,\overline{B} + \overline{M}S\overline{B} + \overline{M}\,\overline{S}B + MSB$

Chapter 21

Mathematics of Polyphase Systems

Three-phase transmission systems are the "freeways" of electrical energy. (Julie Lundquist, Van Cleve Photography)

21-1 INTRODUCTION

A polyphase electrical system is one in which an ac generator (alternator) is constructed so as to provide two or more equal voltages at fixed phase relationships with each other. The most common type of polyphase system utilizes three equal voltages $120°$ apart in phase. Such systems are called *three-phase systems.*

The windings of three-phase generators may be connected in *wye* [Figure 21-1(a)] or in *delta* [Figure 21-1(b)]. The voltages of the three windings—a-a', b-b', and c-c' (see Figure 21-1)—reach their maximums 120 electrical degrees apart. (The electrical degrees are based on the common frequency—$\theta = 2\pi ft$.) *Phase sequence* or *rotation* is a term used to describe the order in which a generator's terminal voltages reach their maximum. For example, if phase rotation is counterclockwise (CCW) for the generators in Figure 21-1, the terminal voltages reach peak values in the order ABC. Figure 21-2 shows the phasor diagram and the sine-wave pattern of the voltages from the three generator coils for the phase sequence ABC.

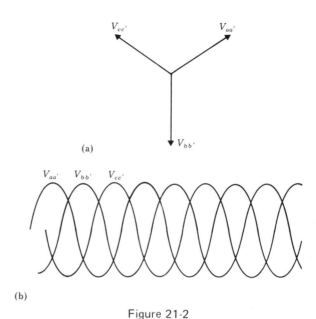

(a)

(b)

Figure 21-2

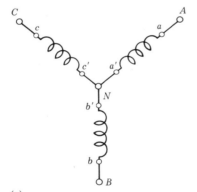

(a)

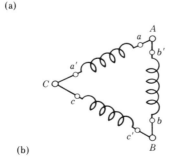

(b)

Figure 21-1

21-2 THREE-PHASE VOLTAGES

We present examples to illustrate important basic concepts concerning three-phase voltages and methods for working with three-phase circuits.

Example 1   The three windings of a 3ϕ generator produce the voltages $V_{aa'} = 120\underline{/30°}$ V, $V_{bb'} = 120\underline{/-90°}$ V, $V_{cc'} = 120\underline{/150°}$ V [see Figure 21-3(b)]. The winding terminals are connected a' to c, and b to c' [see Figure 21-3(b)]. Is it safe to "close the delta?" That is, is there a voltage between terminals a-b' which will cause a current to circulate in the windings if these terminals are connected? (See Section 8-2 for a review of the meaning of double-subscript notation.)

The voltage $V_{ab'}$ is found by tracing the circuit [see Figure 21-3(c)] from a to b' (through windings A, C, and B) adding the voltages vectorally.

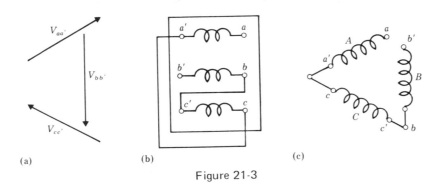

Figure 21-3

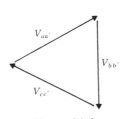

Figure 21-4

$V_{ab'} = V_{aa'} + V_{cc'} + V_{bb'}$
$= 120\underline{/30°} + 120\underline{/150°} + 120\underline{/-90°}$
$= 103.92 + j60 - 103.92 + j60 + 0 - j120$
$= \underline{0}$

A phasor diagram demonstrates the same result (see Figure 21-4). It is safe to close a properly connected delta circuit. (Answer)

Example 2 (a) If the windings of the 3ϕ generator of Example 1 are connected in a wye circuit (see Figure 21-5), what are the line-to-line voltages? (Line-to-line voltages are called the *system voltage* or simply the *line voltage.*) (b) What is the phase relationship of line voltages and the related winding or phase voltages?

(a) To find V_{AC} we start at line terminal A and trace the circuit to C:

$$V_{AC} = V_{AN} + V_{NC} = V_{AN} - V_{CN}$$
$$= 120\underline{/30°} - 120\underline{/150°}$$
$$= 120 (\cos 30° + j \sin 30°)$$
$$- 120 (\cos 150° + j \sin 150°)$$

but

$$\cos 150° = -\cos 30°, \quad \text{and} \quad \sin 150° = \sin 30°$$

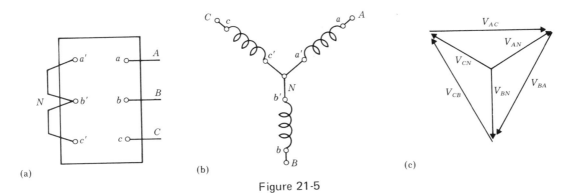

Figure 21-5

Therefore,

$$V_{AC} = 120 \ (2 \cos 30° - j0) = 120 \left(2 \ \frac{\sqrt{3}}{2} \right) \underline{/0°}$$

$$\left(\cos 30° = \frac{\sqrt{3}}{2} \right)$$

$$= \underline{120\sqrt{3}\underline{/0°}}$$

The phasor diagram of Figure 21-5(c) demonstrates the same result. Similarly,

$$V_{BA} = V_{BN} + V_{NA} = V_{BN} - V_{AN}$$
$$= 120\underline{/-90°} - 120\underline{/30°}$$
$$= -j120 - 120\frac{\sqrt{3}}{2} - j\frac{120}{2}$$
$$= -120\frac{\sqrt{3}}{2} - j180$$
$$= \underline{120\sqrt{3}\underline{/-240°} \text{ V}}$$

and

$$V_{CB} = V_{CN} + V_{NB} = V_{CN} - V_{BN}$$
$$= 120\underline{/150°} - 120\underline{/-90°}$$
$$= -120\frac{\sqrt{3}}{2} + j\frac{120}{2} + j120$$
$$= -120\frac{\sqrt{3}}{2} + j180$$
$$= \underline{120\sqrt{3}\underline{/120°} \text{ V}}$$

(b) The phasor diagram shows that V_{AC} lags V_{AN} by 30°, V_{BA} lags V_{BN} by 30°, and V_{CB} lags V_{CN} by 30°.

Examples 1 and 2 demonstrate several important basic facts about the voltages of a symmetrical (all phase voltages are equal, all line voltages are equal) three-phase system. We summarize.

---

1. The phasor sum of the coil or line voltages of a properly connected delta or wye system is equal to zero.
2. The line or system voltage of a delta-connected system is equal to the coil (or phase) voltage:

$$V_{line} = V_{phase}$$

3. The line or system voltage of a wye-connected system is equal to $\sqrt{3}$ (or 1.732) times the phase or coil voltage:

$$V_{line} = \sqrt{3} \ V_{phase}$$

$$V_{phase} = \frac{V_{line}}{\sqrt{3}}$$

Example 3   The line-to-neutral (or phase) voltage of a symmetrical wye-connected 3ϕ system is 120 V. What is the line voltage?

For a wye-connected system,

$$V_{line} = \sqrt{3} \ V_{phase} = \sqrt{3} \cdot 120 = \underline{207.8 \text{ V}}$$

Example 4   What would be the system voltage for the system of Example 3 if the generator windings were connected in delta?

For a delta-connected system,

$$V_{line} = V_{phase}$$

Therefore,

$$V_{line} = \underline{120 \text{ V}}$$

Exercises 21-2   In Exercises 1-4 the given sets of values represent line-to-neutral voltages of wye-connected, symmetrical, three-phase systems. Find the system (line) voltages.

1. 120 V, 240 V, 480 V
2. 2400 V, 12 kV, 30 kV
3. 65 kV, 80 kV, 130 kV
4. 170 kV, 230 kV, 370 kV

In Exercises 5 and 6 the given values represent the system voltages of symmetrical, delta-connected three-phase systems. Determine the phase voltages.

5. 120 V, 240 V, 480 V
6. 2400 V, 12 kV, 30 kV

In Exercises 7-10 the voltages are given for delta-connected systems. If the generator coils were reconnected in wye, what would be the system voltages?

7. 2400 V, 65 kV, 480 V
8. 170 kV, 240 V, 120 V
9. 12 kV, 80 kV, 30 kV
10. 120 V, 230 kV, 370 kV

In Exercises 11-14 line voltages for wye-connected systems are given. If the generator coils were reconnected in delta, what would be the system voltages?

11. 208 V, 416 V, 831 V
12. 4.16 kV, 20.8 kV, 52 kV
13. 112.6 kV, 138.6 kV, 225 kV
14. 294 kV, 398 kV, 641 kV

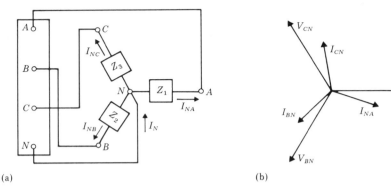

(a)

(b)

Figure 21-6

21-3 BALANCED, THREE-PHASE LOADS

A three-phase load is said to be *balanced* if in the three branches of the load the resistive components are equal, and the reactive components are equal. The three branches may be treated as three single-phase circuits. Ohm's law is used to find the currents in the branches. Kirchhoff's current law is used to determine currents at junction points. Examples demonstrate the procedure.

Example 5 A three-phase wye-connected load is connected to a three-phase, four-wire system (see Figure 21-6). $Z_1 = Z_2 = Z_3 = 6\underline{/15°}$ Ω, $V_{AN} = 120\underline{/0°}$ V, $V_{BN} = 120\underline{/-120°}$ V, $V_{CN} = 120\underline{/-240°}$ V. Determine (a) the phase currents, (b) the line currents, and (c) the current in the neutral, I_N. (d) Sketch the phasor diagram of phase voltages and line currents.

(a) The phase currents are determined using Ohm's law:

$$I_{NA} = \frac{V_{AN}}{Z_1} = \frac{120\underline{/0°}}{6\underline{/15°}} = \underline{20\underline{/-15°}}$$

$$I_{NB} = \frac{V_{BN}}{Z_2} = \frac{120\underline{/-120°}}{6\underline{/15°}} = \underline{20\underline{/-135°}}$$

$$I_{NC} = \frac{V_{CN}}{Z_3} = \frac{120\underline{/-240°}}{6\underline{/15°}} = \underline{20\underline{/-255°}}$$

The current directions marked on the diagram indicate the directions of actual currents during the time when the related line-to-neutral voltages are positive.

(b) Line currents are equal to phase currents because there is only one path for cur-

rent at the point where a line connects to a phase impedance.

(c) At N,

$$\begin{aligned} I_N &= I_{NA} + I_{NB} + I_{NC} \\ &= 20\underline{/-15°} + 20\underline{/-135°} + 20\underline{/-255°} \\ &= 19.318 - j5.176 - 14.142 - j14.142 \\ &\quad -5.176 + j19.318 \\ &= 19.318 - 19.318 + j19.318 - j19.318 \end{aligned}$$

$$I_N = \underline{0}$$

(d) The phasor diagram is shown in Figure 21-6(b). The line currents lag their corresponding line-to-neutral voltages by an angle which is equal to the angle of the impedance.

For a balanced, wye-connected load:
1. Line currents are equal to phase currents, $I_\phi = I_{\text{line}}$.
2. The neutral current is zero, $I_N = 0$.
3. Line voltage is equal to $\sqrt{3}$ times line-to-neutral voltage, $V_L = \sqrt{3}\, V_\phi$.

Example 6 A balanced delta-connected load [see Figure 21-7(a)] is connected to a system in which $V_{AB} = 120\underline{/0°}$ V, $V_{BC} = 120\underline{/-120°}$ V, and $V_{CA} = 120\underline{/-240°}$ V. For the load, $Z_1 = Z_2 = Z_3 = 6\underline{/15°}$ Ω. Determine (a) the phase currents in the load, and (b) the line

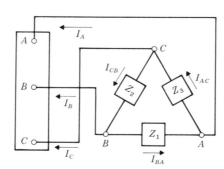

(a)

(b)

Figure 21-7

At node *B*:

$$I_B = I_{CB} - I_{BA} = -14.142 - j14.142$$
$$- 19.318 + j5.176$$
$$= -33.460 - j8.966 = 34.64\underline{/-165°} \text{ A}$$

At node *C*:

$$I_C = I_{AC} - I_{CB} = -5.176 + j19.318 + 14.142$$
$$+ j14.142$$
$$= 8.966 + j33.460 = 34.64\underline{/-285°} \text{ A}$$

(c) The phasor diagram is shown in Figure 21-7(b). Let us note that

$$\frac{|I_{\text{line}}|}{|I_\phi|} = \frac{34.64}{20} = 1.732 = \sqrt{3}$$

That is,

$$|I_{\text{line}}| = \sqrt{3}\,|I_{\text{phase}}|$$

Furthermore,

$$\theta_{I_L} = \theta_{I_\phi} - 30°$$

For a balanced, delta-connected load:
1. Line currents $=\sqrt{3} \times$ phase currents.
2. Line voltage = phase voltage.

---

currents. (c) Sketch the phasor diagram for line voltage, and phase and line currents.

(a) $I_{BA} = \dfrac{V_{AB}}{Z_1} = \dfrac{120\underline{/0°}}{6\underline{/15°}} = 20\underline{/-15°}$ A

$\qquad = 19.318 - j5.176$ A

$I_{CB} = \dfrac{V_{BC}}{Z_2} = \dfrac{120\underline{/-120°}}{6\underline{/15°}} = 20\underline{/-135°}$ A

$\qquad = -14.142 - j14.142$ A

$I_{AC} = \dfrac{V_{CA}}{Z_3} = \dfrac{120\underline{/-240°}}{6\underline{/15°}} = 20\underline{/-255°}$ A

$\qquad = -5.176 + j19.318$ A

(b) Line currents are found by applying Kirchhoff's current law at the three nodes: *A*, *B*, and *C*.

At node *A*:

$$I_A = I_{BA} - I_{AC} = 19.318 - j5.76 + 5.176$$
$$- j19.318 \text{ A}$$
$$= 24.494 - j24.494 = 34.64\underline{/-45°} \text{ A}$$

Exercises 21-3 In Exercises 1–6 the given values are for line-to-neutral voltages and phase impedance, respectively, for balanced, wye-connected three-phase loads. Determine (a) I_{line}, and (b) $|V_{\text{line}}|$. (c) Sketch the phasor diagrams for I_{line} and V_ϕ.

1. $V_{AN} = 120\underline{/0°}$ V, $V_{BN} = 120\underline{/-120°}$ V, $V_{CN} = 120\underline{/-240°}$ V, $Z_\phi = 10\underline{/0°}$ Ω $(Z_\phi = Z_1 = Z_2 = Z_3)$

2. $240\underline{/60°}$ V, $240\underline{/-60°}$ V, $240\underline{/-180°}$ V, $12\underline{/30°}$ Ω

3. $V_{AN} = 480\underline{/90°}$ V, phase sequence = *ABC*; $24\underline{/-30°}$ Ω

4. $2400\underline{/120°}$ V, phase sequence = *ABC*; $1.2\underline{/25°}$ kΩ

5. $7.2\underline{/0°}$ kV, phase sequence = *ACB*; $3\underline{/-20°}$ kΩ

6. $20\underline{/0°}$ kV, phase sequence = *ACB*; $8.5\underline{/18°}$ kΩ

In Exercises 7–12 the given values are for system voltage and phase impedance, respectively, for balanced, delta-connected three-phase loads. Determine (a) I_ϕ, and (b) I_{line}. (c) Sketch the phasor diagram.

7. 120 V, $8\underline{/15°}$ Ω. Assume ABC phase sequence.
8. 208 V, $18\underline{/-24°}$ Ω
9. 4160 V, $1.8\underline{/12°}$ kΩ
10. 12 kV, $2.4\underline{/30°}$ kΩ
11. 34 kV, $8.2\underline{/-18°}$ kΩ
12. 120 kV, $24\underline{/-12°}$ kΩ

21-4 UNBALANCED, THREE-PHASE LOADS

Unbalanced, Delta-Connected Loads Since the voltages across the branches of a delta-connected load are equal to the line voltage,

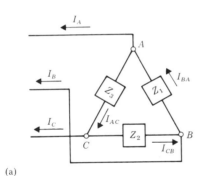

(a)

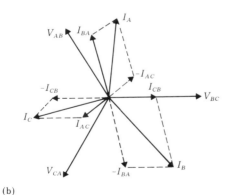

(b)

Figure 21-8

phase currents are found by applying Ohm's law. Line currents are found by applying Kirchhoff's current law at the nodes of the system.

Example 7 The branches of a delta-connected load consist of $Z_1 = 5\underline{/15°}$ Ω, $Z_2 = 8\underline{/0°}$ Ω, and $Z_3 = 12\underline{/25°}$ Ω [see Figure 21-8(a)]. The voltage $V_{AB} = 208\underline{/120°}$ V, and the phase sequence is ABC. Find (a) the phase currents, and (b) the line currents. (c) Construct a phasor diagram.

(a) Applying Ohm's law to find phase currents,

$$I_{BA} = \frac{V_{AB}}{Z_1} = \frac{208\underline{/120°}}{5\underline{/15°}} = \underline{41.6\underline{/105°} \text{ A}}$$

$$I_{CB} = \frac{V_{BC}}{Z_2} = \frac{208\underline{/0°}}{8\underline{/0°}} = 26\underline{/0°} \text{ A}$$

$$I_{AC} = \frac{V_{CA}}{Z_3} = \frac{208\underline{/-120°}}{12\underline{/25°}} = \underline{17.33\underline{/-145°} \text{ A}}$$

(b) We use KCL to determine line currents.

$$\begin{aligned}
I_A &= I_{BA} - I_{AC} = 41.6\underline{/105°} - 17.33\underline{/-145°} \\
&= -10.767 + j40.182 + 14.196 + j9.94 \\
&= 3.429 + j50.122 \text{ A} = 50.24\underline{/86.09°} \text{ A} \\
I_B &= I_{CB} - I_{BA} = 26\underline{/0°} - 41.6\underline{/105°} \\
&= 26 + j0 + 10.767 - j40.182 \\
&= 36.767 - j40.182 \\
&= 54.46\underline{/-47.54°} \text{ A} \\
I_C &= I_{AC} - I_{CB} = 17.33\underline{/-145°} - 26\underline{/0°} \\
&= -14.196 - j9.94 - 26 + j0 \\
&= -40.196 - j9.94 \\
&= 41.41\underline{/-166.11°} \text{ A}
\end{aligned}$$

(c) The phasor diagram is shown in Figure 21-8(b).

Unbalanced, Wye-connected Load with Neutral Connected When the neutral of a wye-connected load is connected to the neutral of the generator the system is referred to as a *three-phase, four-wire system*. Line currents are identical with their related phase currents and are found using Ohm's law.

Example 8 The impedances of Example 7 are connected in wye in a three-phase, four-

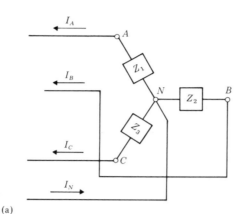

(a)

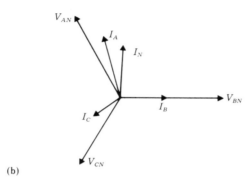

(b)

Figure 21-9

Unbalanced, Three-Wire, Wye-Connected Load   When the common point of an unbalanced, wye-connected, three-phase load is not connected to the system the voltages from the lines to this common point will not be equal. The common point is labeled 0 instead of N. It is necessary to utilize Kirchhoff's voltage law in, for example, the loop-current method to analyze loads of this type (review Section 10-6). A technique employing a concept referred to as the "displaced neutral" is also a popular method. Examples will illustrate each of these techniques.

Example 9   Analyze the circuit of Example 8 for the condition when the neutral is not connected. That is, determine (a) the line currents, and (b) the voltage across each impedance. (c) Construct a phasor diagram. Use the loop (or mesh)-current method. Assume that the line voltages are balanced and $\theta_{V_{AB}} = 150°$.

(a) Using the labeled diagram of Figure 21-10(a) we write KVL equations for the

wire system (see Figure 21-9). System voltage is 208 V. Phase sequence is ABC. Assume that $\theta_{V_{AN}} = 120°$. Calculate (a) the line currents, and (b) the current in the neutral, I_N. (c) Construct a phasor diagram.

(a) Applying Ohm's law.

$$I_A = I_{NA} = \frac{V_{AN}}{Z_1} = \frac{V_L/\sqrt{3}}{Z_1} = \frac{120/120°}{5/15°}$$
$$= \underline{24/105°} \text{ A}$$

$$I_B = I_{NB} = \frac{V_{BN}}{Z_2} = \frac{120/0°}{8/0°} = \underline{15/0°} \text{ A},$$

$$I_C = I_{NC} = \frac{120/-120°}{12/25°} = \underline{10/-145°} \text{ A}$$

(b) We find I_N by applying KCL at N,

$$I_{NA} + I_{NB} + I_{NC} = I_N$$
$$I_N = 24/105° + 15/0° + 10/-145° = -6.212$$
$$+ j23.182 + 15 - 8.192 - j5.736$$
$$= 0.596 + j17.446 \text{ A} = \underline{17.46/88.04°} \text{ A}$$

(c) The phasor diagram is shown in Figure 21-8(b).

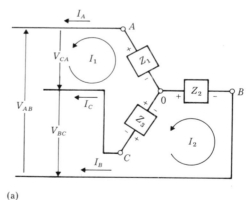

(a)

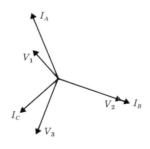

(b)

Figure 21-10

loops of currents I_1 and I_2, tracing CW:

$$I_1(Z_1 + Z_3) - I_2 Z_3 = -V_{CA}$$
$$-I_1 Z_3 + I_2(Z_2 + Z_3) = -V_{BC}$$

Substituting known values,

$$I_1(5\underline{/15^\circ} + 12\underline{/25^\circ}) - 12\underline{/25^\circ}I_2 = -208\underline{/-90^\circ}$$
$$= 208\underline{/90^\circ}$$
$$-12\underline{/25^\circ}I_1 + I_2(8\underline{/0^\circ} + 12\underline{/25^\circ}) = -208\underline{/30^\circ}$$
$$I_1(4.830 + j1.294 + 10.876 + j5.071)$$
$$- 12\underline{/25^\circ}I_2 = 208\underline{/90^\circ}$$
$$-12\underline{/25^\circ}I_1 + I_2(8 + 10.876 + j5.071)$$
$$= -208\underline{/30^\circ}$$

Simplifying,

$$16.95\underline{/22.06^\circ}I_1 - 12\underline{/25^\circ}I_2 = 208\underline{/90^\circ}$$
$$-12\underline{/25^\circ}I_1 + 19.55\underline{/15.04^\circ}I_2 = -208\underline{/30^\circ}$$

Solving for I_1 and I_2 by determinants,

$$I_1 = \frac{\begin{vmatrix} 208\underline{/90^\circ} & -12\underline{/25^\circ} \\ -208\underline{/30^\circ} & 19.55\underline{/15.04^\circ} \end{vmatrix}}{\begin{vmatrix} 16.95\underline{/22.06^\circ} & -12\underline{/25^\circ} \\ -12\underline{/25^\circ} & 19.55\underline{/15.04^\circ} \end{vmatrix}}$$

$$= \frac{4066.4\underline{/105.04^\circ} - 2496\underline{/55^\circ}}{331.37\underline{/37.1^\circ} - 144\underline{/50^\circ}}$$

$$= \frac{-1079.86 + j3920.4 - 1431.65 - j2044.6}{264.3 + j199.89 - 92.56 - j110.31}$$

$$= \frac{3134.69\underline{/143.24^\circ}}{193.7\underline{/27.55^\circ}}$$

$$= 16.18\underline{/115.69^\circ}\ A = -7.014 + j14.581\ A$$

$$I_2 = \frac{\begin{vmatrix} 16.95\underline{/22.06^\circ} & 208\underline{/90^\circ} \\ -12\underline{/25^\circ} & -208\underline{/30^\circ} \end{vmatrix}}{193.7\underline{/27.55^\circ}}$$

$$= \frac{3264.94\underline{/189.13^\circ}}{193.7\underline{/27.55^\circ}}$$

$$= 16.86\underline{/161.58^\circ}\ A = -16.0 + j5.33\ A$$

A study of the circuit diagram of Figure 21-10 shows that

$$I_A = I_1 = 16.18\underline{/115.69^\circ}\ A$$
$$I_B = -I_2 = -16.86\underline{/161.58^\circ} = 16.86\underline{/-18.42^\circ}\ A$$
$$I_C = I_2 - I_1 = 16.86\underline{/-18.42^\circ} - 16.18\underline{/115.69^\circ}$$
$$= -16.0 + j5.33 + 7.014 - j14.581$$
$$= -8.986 - j9.251 = 12.9\underline{/-134.17^\circ}\ A$$

(b) The voltages across the impedances are found by applying Ohm's law:

$$V_1 = V_{AO} = I_A Z_1 = 16.18\underline{/115.69^\circ} \cdot 5\underline{/15^\circ}$$
$$= 80.9\underline{/130.69^\circ}\ V$$

$$V_2 = V_{BO} = I_B Z_2 = 16.86\underline{/-18.42^\circ} \cdot 8\underline{/0^\circ}$$
$$= 134.9\underline{/-18.42^\circ}\ V$$
$$V_3 = \overline{V_{CO}} = I_C Z_3 = 12.9\underline{/-134.17^\circ} \cdot 12\underline{/25^\circ}$$
$$= 154.8\underline{/-109.17^\circ}\ V$$

(c) The phasor diagram is shown in Figure 21-10(b).

Example 10 Analyze the circuit of Example 8 for the condition when neutral connection is not used. Specifically, determine (a) the voltage across each phase impedance, and (b) the phase and line currents. (c) Construct a phasor diagram. Use the displaced neutral method for the analysis.

(a) A diagram such as that of Figure 21-11 is of great assistance in demonstrating the logic of the displaced neutral method. Writing KCL loop equations reveals the following:

$$V_{AN} + V_{NO} + V_{OA} = 0$$
$$V_{BN} + V_{NO} + V_{OB} = 0$$
$$V_{CN} + V_{NO} + V_{OC} = 0$$

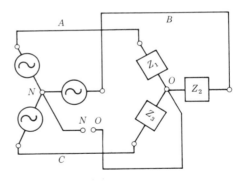

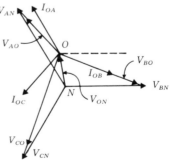

Figure 21-11

From these three equations are derived the equations for determining the voltages across the load impedances,

$$V_{A0} = V_{AN} + V_{N0}, \qquad V_{B0} = V_{BN} + V_{N0},$$
$$V_{C0} = V_{CN} + V_{N0}$$

The voltages V_{AN}, V_{BN}, and V_{CN} are all equal in magnitude (if system line voltages are maintained balanced) and equal to $V_{\text{line}}/\sqrt{3}$. Their respective phase angles are determined by the phase sequence of the system.

The voltage V_{N0}, as the voltage between two nodes between which are connected two or more branches, may be found with the aid of a form of Millman's theorem,

$$V_{N0} = \frac{V_{NA} Y_1 + V_{NB} Y_2 + V_{NC} Y_3}{Y_1 + Y_2 + Y_3} \qquad (1)$$

After finding V_{N0} and then V_{A0}, V_{B0}, and V_{C0} we will be able to find load currents using Ohm's law formulas, such as

$$I_1 = \frac{V_{A0}}{Z_1}$$

and so forth. From Example 8,

$$V_{AN} = \frac{208}{\sqrt{3}} \underline{/120°} = 120\underline{/120°} \text{ V}$$

$$V_{NA} = 120\underline{/-60°} \text{ V}$$
$$V_{BN} = 120\underline{/0°} \text{ V}, \qquad V_{NB} = 120\underline{/180°} \text{ V}$$
$$V_{CN} = 120\underline{/-120°} \text{ V}, \qquad V_{NC} = 120\underline{/60°} \text{ V}$$

The admittances are (see Example 7)

$$Y_1 = \frac{1}{Z_1} = \frac{1}{5\underline{/15°}} = 0.2\underline{/-15°}$$
$$= 0.1932 - j0.0518 \text{ S}$$

$$Y_2 = \frac{1}{Z_2} = \frac{1}{8\underline{/0°}} = 0.125\underline{/0°}$$
$$= 0.125 + j0 \text{ S}$$

$$Y_3 = \frac{1}{Z_3} = \frac{1}{12\underline{/25°}} = 0.0833\underline{/-25°}$$
$$= 0.0755 - j0.0352 \text{ S}$$
$$Y_1 + Y_2 + Y_3 = 0.3937 - j0.087 \text{ S}$$

The displaced neutral voltage V_{N0} is found by substituting into equation (1),

$$V_{N0} = \frac{\begin{array}{c}(120\underline{/-60°})(0.2\underline{/-15°}) + (120\underline{/180°}) \cdot \\ (0.125) + (120\underline{/60°})(0.0833\underline{/-25°})\end{array}}{0.3937 - j0.087}$$

$$= \frac{-0.596 - j17.446}{0.4032\underline{/-12.46°}} = \frac{17.456\underline{/-91.96°}}{0.4032\underline{/-12.46°}}$$

$$= 44.29\underline{/-79.5°} = 8.071 - j43.548 \text{ V}$$

Now, we find the load voltages,

$$V_{A0} = V_{AN} + V_{N0} = 120\underline{/120°} + 44.29\underline{/-79.5°}$$
$$= -60 + j103.923 + 8.071 - j43.548$$
$$= -51.929 + j60.375 = \underline{79.64\underline{/130.7°} \text{ V}}$$
$$V_{B0} = V_{BN} + V_{N0} = 120 + j0 + 8.071$$
$$- j43.548$$
$$= 128.071 - j43.548 = \underline{135.3\underline{/-18.78°} \text{ V}}$$
$$V_{C0} = V_{CN} + V_{N0} = -60 - j103.923 + 8.071$$
$$- j43.548$$
$$= -51.929 - j147.471 = \underline{156.3\underline{/-109.4°} \text{ V}}$$

(b) The phase and line currents can be found using Ohm's law.

$$I_A = I_1 = \frac{V_{A0}}{Z_1} = \frac{79.64\underline{/130.7°}}{5\underline{/15°}}$$
$$= 15.93\underline{/115.7°} \text{ A}$$
$$I_B = I_2 = \frac{V_{B0}}{Z_2} = \frac{135.3\underline{/-18.78°}}{8\underline{/0°}}$$
$$= 16.91\underline{/-18.78°} \text{ A}$$
$$I_C = I_3 = \frac{V_{C0}}{Z_3} = \frac{156.3\underline{/-109.4°}}{12\underline{/25°}}$$
$$= 13.03\underline{/-134.4°} \text{ A}$$

The values check with those found in Example 9 with a satisfactory degree of accuracy.

(c) The phasor diagram is shown in Figure 21-11(b).

Exercises 21-4 In Exercises 1–4 the given values are the phase load impedances and line voltage, respectively, for a delta-connected three-phase load. (See Figure 21-8.) Assume that $\theta_{V_{AB}} = 0°$ and an *ABC* phase sequence in each case. For each exercise (a) find the phase currents, (b) find the line currents, and (c) construct a phasor diagram.

1. $Z_1 = 5\underline{/0°}$ Ω, $Z_2 = 8 + j0$ Ω, $Z_3 = 10 + j0$ Ω, 240 V
2. $Z_1 = 5 + j2$ Ω, $Z_2 = 8 + j2$ Ω, $Z_3 = 10 + j2$ Ω, 240 V
3. $Z_1 = 5 - j2$ Ω, $Z_2 = 8 - j2$ Ω, $Z_3 = 10 - j2$ Ω, 240 V
4. $Z_1 = 40 + j30$ Ω, $Z_2 = 60 - j20$ Ω, $Z_3 = 80 + j10$ Ω, 720 V

In Exercises 5–8 the given values are the phase load impedances and system (line-to-line) voltages for wye-connected, four-wire, three-phase systems (see Figure 21-9). Phase sequence is ABC and $\theta_{V_{AN}} = 120°$. (a) Calculate the line/phase currents, (b) find the neutral current, I_N, and (c) construct a phasor diagram.

5. $Z_1 = 12 + j0 \ \Omega$, $Z_2 = 9 + j0 \ \Omega$,
   $Z_3 = 15 + j0 \ \Omega$, 208 V
6. $Z_1 = 12 + j4 \ \Omega$, $Z_2 = 9 + j4 \ \Omega$,
   $Z_3 = 15 + j4 \ \Omega$, 208 V
7. $Z_1 = 12 - j4 \ \Omega$, $Z_2 = 9 - j4 \ \Omega$,
   $Z_3 = 15 - j4 \ \Omega$, 208 V
8. $Z_1 = 15 + j3 \ \Omega$, $Z_2 = 8 - j6 \ \Omega$,
   $Z_3 = 6 + j8 \ \Omega$, 208 V

In Exercises 9–12 the given values are the phase load impedances and system (line-to-line) voltages for wye-connected, three-wire, three-phase systems. (See Figure 21-10.) Phase sequence is ABC and $\theta_{V_{AB}} = 150°$. For each exercise, using the loop-current method, (a) calculate the line/phase currents, (b) find the voltage across each impedance, and (c) construct a phasor diagram.

9. $Z_1 = 12 + j0 \ \Omega$, $Z_2 = 9 + j0 \ \Omega$,
   $Z_3 = 15 + j0 \ \Omega$, 208 V
10. $Z_1 = 12 - j4 \ \Omega$, $Z_2 = 9 + j4 \ \Omega$,
    $Z_3 = 15 + j4 \ \Omega$, 208 V
11. $Z_1 = 12 - j4 \ \Omega$, $Z_2 = 9 - j4 \ \Omega$,
    $Z_3 = 15 - j4 \ \Omega$, 208 V
12. $Z_1 = 15 + j3 \ \Omega$, $Z_2 = 8 - j6 \ \Omega$,
    $Z_3 = 6 + j8 \ \Omega$, 208 V

13–16. Repeat the analyses of Exercises 9–12 using the displaced-neutral method (see Example 10).

21-5 POWER IN THREE-PHASE LOADS

Balanced Three-Phase Loads The total power supplied to a balanced three-phase load, whether wye- or delta-connected, is three times the power for one phase ($P_T = 3P_\phi$). For wye connections,

$$P_\phi = V_\phi I \cos \theta_\phi, \quad \text{where } \theta_\phi = \arctan \frac{X_\phi}{R_\phi}$$

$$P_T = 3P_\phi = 3V_\phi I \cos \theta_\phi, \quad \text{but } V_\phi = \frac{V_{\text{line}}}{\sqrt{3}}$$

Thus,

$$P_T = \frac{3}{\sqrt{3}} V_{\text{line}} I \cos \theta_\phi = \sqrt{3} \, V_{\text{line}} I \cos \theta_\phi$$

Similarly, for balanced delta connections,

$$P_\phi = VI_\phi \cos \theta_\phi, \quad \text{and } P_T = 3VI_\phi \cos \theta_\phi$$

but, since

$$I_{\text{line}} = \sqrt{3} \, I_\phi \quad (I_\phi = I_{\text{line}}/\sqrt{3})$$

$$P_T = 3V\frac{I_{\text{line}}}{\sqrt{3}} \cos \theta = \sqrt{3}VI_{\text{line}} \cos \theta_\phi$$

Note: In formulas for power, V and I represent magnitudes only.

Example 11 Calculate the total power supplied to the wye-connected, three-phase load of Example 5. ($V_\phi = 120$ V, $I = 20$ A, $\theta_\phi = 15°$)

$$P_T = 3V_\phi I \cos \theta_\phi = 3 \times 120 \times 20 \times \cos 15°$$
$$= \underline{6.955 \text{ kW}}$$

Example 12 What power will be supplied to the delta-connected load of Example 6? ($V = 120$ V, $I_L = 34.64$ A, $\theta_\phi = 15°$)

$$P_T = \sqrt{3}VI_L \cos \theta = \sqrt{3} \times 120 \times 34.64$$
$$\times \cos 15° = \underline{6.954 \text{ kW}}$$

Example 13 A 3ϕ motor is rated at 15 kVA and is found to operate with $\theta_\phi = 12°$ when under normal load. Calculate the amount of power drawn from the system by the motor.

$$P_T = \sqrt{3}VI \cos \theta = 15 \times 10^3 \times \cos 12°$$
$$= \underline{14.67 \text{ kW}}$$

Example 14 A broadcast television transmitter draws approximately 25.5 A at each of the three terminals where it connects to a 2400-V, 3ϕ system. The phase angle is typically 18°. Determine P_T.

$$P_T = \sqrt{3}VI_L \cos \theta_\phi = \sqrt{3} \times 2400 \times 25.5$$
$$\times \cos 18°$$
$$P_T = \underline{100.813 \text{ kW}}$$

Unbalanced Three-Phase Loads The total power in three-phase systems with unbalanced loads can be determined using the formula

$$P_T = P_A + P_B + P_C$$

where P_A, P_B, and P_C represent the power dissipated in individual phase loads.

Example 15   Find P_T for the three-phase load of Example 7.

$$P_1 = VI_{BA} \cos \theta_1 = 208 \times 41.6 \times \cos 15°$$
$$= 8.36 \text{ kW}$$
$$P_2 = VI_{CB} \cos \theta_2 = 208 \times 26 \times \cos 0°$$
$$= 5.41 \text{ kW}$$
$$P_3 = VI_{AC} \cos \theta_3 = 208 \times 17.33 \cos 25°$$
$$= 3.27 \text{ kW}$$
$$P_T = P_1 + P_2 + P_3 = 8.36 + 5.41 + 3.27$$
$$= \underline{17.04 \text{ kW}}$$

Exercises 21-5   In Exercises 1–10 calculate P_T for the balanced, three-phase loads specified in the referred-to exercises of Section 21-3.

1. Exercise 1
2. Exercise 2
3. Exercise 3
4. Exercise 5
5. Exercise 6
6. Exercise 7
7. Exercise 8
8. Exercise 9
9. Exercise 11
10. Exercise 12

In Exercises 11–16 calculate P_T for the unbalanced, three-phase loads specified in the referred-to exercises of Section 21-4.

11. Exercise 1
12. Exercise 3
13. Exercise 5
14. Exercise 7
15. Exercise 9
16. Exercise 11
17. A short-wave radio transmitter is powered from a 2400-V, three-phase system. Its nameplate gives the rating: 105 kVA. The transmitter typically operates at the phase angle $\theta = 24°$. Calculate P_T.
18. A broadcast radio transmitter draws approximately 95-A line current (balanced) from a 480-V, 3ϕ system. The typical operating power factor ($\cos \theta$) is 0.92. Determine P_T.
19. A large computer installation operates with a balanced line current of 36 A on a 240-V, 3ϕ power line. The typical power factor is 0.85. Find (a) the three-phase kVA requirement of the system, and (b) P_T.
20. The sound system of a large athletic stadium typically draws a balanced line current of 15 A, at a power factor of 0.95, from a 208 V three-phase power line. Calculate P_T.

Chapter 22

A Mathematical Technique for Nonsinusoidal Waveforms

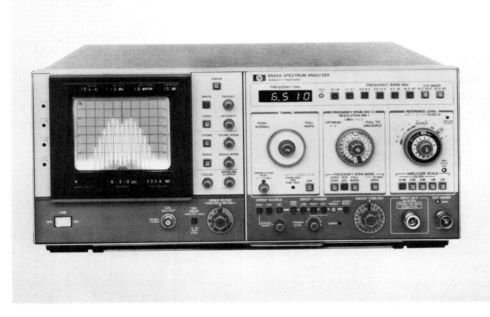

The spectrum analyzer is an effective means of determining the frequency components of a nonsinusoidal wave form. (Courtesy of Hewlett-Packard Company)

22-1 FUNDAMENTALS AND HARMONICS—FOURIER ANALYSIS

The mathematical concepts of the j-operator and phasor algebra (see Chapters 13 and 14) provide powerful techniques for the analysis of circuits excited by sinusoidal ac voltages (or currents). Unfortunately, these techniques are not directly applicable when the excitation is nonsinusoidal. And the signals processed in electronics equipment are almost all nonsinusoidal. The situation is not completely hopeless, however.

The French mathematician Jean Baptiste Joseph Fourier (1768–1830) (pronounced foo´ree ay) discovered the now well-known concept that any phenomenon which repeats itself in a pattern (we call this kind of phenomenon a *periodic function*) can be accurately represented by means of the sum of an infinite number of sine and cosine terms. The result, a polynomial, is one of a group of mathematical expressions called *infinite series*. This particular trigonometric series is named the *Fourier series* in honor of the discover.

A Fourier series, then, is a formula which tells us how we can synthesize (produce or generate) a periodic, nonsinusoidal signal such as a square wave, a sawtooth wave, or a rectified full or half-wave, and so on, by adding together some sinusoidal signals of the appropriate amplitudes and phase angles (strictly speaking, an infinite number of such signals). Or, the series tells us that if such a nonsinusoidal signal exists it can be considered to contain an infinite number of sinusoidal signals. The implication of this last statement is that it is possible to use phasor algebra to analyze, indirectly, the response of an electric circuit to a nonsinusoidal but periodic signal. We analyze the response to the individual sinusoidal signals and combine the results using the principle of superposition (see Chapter 8).

Periodic Functions   A periodic function is one whose pattern of variation (as might be observable on an oscilloscope) repeats itself after an amount of time (called its *period*) has elapsed. Period is typically designated T. We say the function satisfies the condition $f(t) = f(t + T)$. (Review Section 15-4 on the meaning of *function* and the use of function notation.)

The Fourier formula tells us that such a signal contains one sinusoidal component whose period is the same as that of the nonsinusoidal waveform. This component is called the *fundamental*. The frequency f_1 of the fundamental is

$$f_1 = \frac{1}{T}$$

In addition to the fundamental the signal contains an infinite number of sinusoidal signals whose frequencies are whole multiples of the fundamental. These components are called *harmonics*. Harmonics are related to the fundamental as follows:

$$f_2 = 2f_1, \quad f_3 = 3f_1, \quad f_4 = 4f_1$$
$$f_5 = 5f_1, \quad f_n = nf_1, \quad \text{and so on}$$

Example 1   Determine the frequencies of the fundamental and harmonic components through f_5 of the waveform shown in Figure 22-1.

$$f_1 = \frac{1}{T} = \frac{1}{0.5 \times 10^{-3}} = 2 \text{ kHz} \quad \text{(Fundamental)}$$
$$f_2 = 2f_1 = 4 \text{ kHz} \quad \text{(Second harmonic)}$$
$$f_3 = 3f_1 = 6 \text{ kHz} \quad \text{(Third harmonic)}$$
$$f_4 = 4f_1 = 8 \text{ kHz} \quad \text{(Fourth harmonic)}$$
$$f_5 = 5f_1 = 10 \text{ kHz} \quad \text{(Fifth harmonic)}$$

When stated in a very general form the Fourier series appears as follows:

$$f(x) = a_0 + a_1 \cos x + a_2 \cos 2x + a_3 \cos 3x$$
$$+ \cdots + a_n \cos nx + \cdots$$
$$+ b_1 \sin x + b_2 \sin 2x + b_3 \sin 3x + \cdots$$
$$+ b_n \sin nx + \cdots \qquad (1)$$

When this concept is applied to nonsinusoidal, but periodic, currents or voltages the series represents the instantaneous values of such quantities and appears as follows:

$$i = I_{dc} + I_{m_1} \cos \omega t + I_{m_2} \cos 2\omega t$$
$$+ \cdots + I_{m_n} \cos n\omega t$$
$$+ I'_{m_1} \sin \omega t + I'_{m_2} \sin 2\omega t$$
$$+ \cdots + I'_{m_n} \sin n\omega t + \cdots \qquad (2)$$
$$v = V_{dc} + V_{m_1} \cos \omega t + V_{m_2} \cos 2\omega t$$
$$+ V_{m_3} \cos 3\omega t + \cdots + V_{m_n} \cos n\omega t$$
$$+ V'_{m_1} \sin \omega t + V'_{m_2} \sin 2\omega t$$
$$+ V'_{m_3} \sin 3\omega t + \cdots + V'_{m_n} \sin n\omega t$$
$$\qquad (3)$$

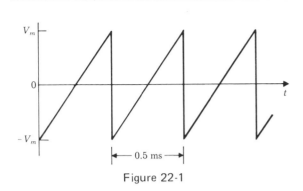

V_m

0

$-V_m$

t

|← 0.5 ms →|

Figure 22-1

Example 2   The Fourier analysis of a square wave is

$$v = \frac{4}{\pi} V_m \left(\sin \omega_1 t + \tfrac{1}{3} \sin 3\omega_1 t + \tfrac{1}{5} \sin 5\omega_1 t + \cdots \right)$$

Let $V_m = \pi/4$. (a) Make three sketches showing the synthesis of a square wave from sine-wave components: (i) $\sin \omega_1 t + \tfrac{1}{3} \sin 3\omega_1 t$, (ii) $\sin \omega_1 t + \tfrac{1}{3} \sin 3\omega_1 t + \tfrac{1}{5} \sin 5\omega_1 t$, and (iii) $\sin \omega_1 t + \tfrac{1}{3} \sin 3\omega_1 t + \tfrac{1}{5} \sin 5\omega_1 t + \tfrac{1}{7} \sin 7\omega_1 t$. (b) Construct a spectrum diagram

for the square wave showing the frequency components present and their relative amplitudes.

Let us take note of several important facts concerning the Fourier series:

1.   The first term, a constant, represents the *average value* of the function (phenomenon). In the case of electric current or voltage this term represents any dc component present in the signal.

2.   The angular velocity ω_1 is related to f_1, the fundamental frequency, and T, the period of the nonsinusoidal function,

$$\omega_1 = 2\pi f_1, \qquad f_1 = \frac{1}{T}$$

3.   The coefficients of the terms, $a_1, a_2, \ldots,$ $b_1, b_2, \ldots, I_{m_1}, I_{m_2}, \ldots, I'_{m_1}, I'_{m_2}, \ldots,$ and so forth, represent the *amplitudes* of the individual sinusoidal components. For some waveforms, some of these coefficients are zero. This means that for certain types of signals not all of the harmonic components will be present.

4.   Complete representation of a phenomenon requires the consideration of all terms for which the coefficients are not zero—an infinite number. For many

Solution for Example 2

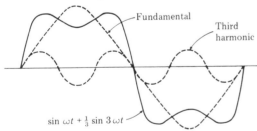

Fundamental

Third harmonic

$\sin \omega t + \tfrac{1}{3} \sin 3 \omega t$

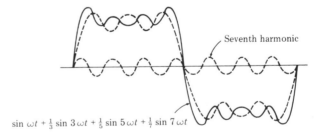

Seventh harmonic

$\sin \omega t + \tfrac{1}{3} \sin 3 \omega t + \tfrac{1}{5} \sin 5 \omega t + \tfrac{1}{7} \sin 7 \omega t$

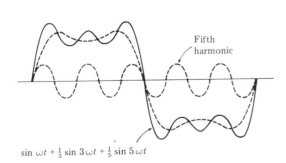

Fifth harmonic

$\sin \omega t + \tfrac{1}{3} \sin 3 \omega t + \tfrac{1}{5} \sin 5 \omega t$

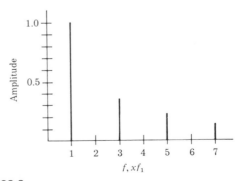

Amplitude

1.0

0.5

1   2   3   4   5   6   7

$f, x f_1$

Figure 22-2

phenomena, however, the amplitudes of the harmonics diminish rapidly with their order. For these, only a few, low-order harmonic terms are required to represent a function with a satisfactory degree of accuracy.

The coefficients of the Fourier series for any function which can be stated can be found using the process of *integration* (calculus). The coefficients of many common waveforms have been determined and are available in numerous texts and handbooks. (For example, see *Reference Data for Radio Engineers*, Howard W. Sams & Co., Inc., New York, 1968.)

Symmetry and the Fourier Coefficients The symmetry of a waveform refers to the similarity of the wave on either side of a dividing line (horizontal or vertical axis). There are three types of symmetry: *the axis symmetry of even functions, the point symmetry of odd functions, and half-wave or mirror symmetry.* Each type of symmetry is associated with a unique fact concerning the Fourier components.

A waveform exhibits axis symmetry if its shape is similar on either side of the vertical axis [see Figure 22-3(a)]. (If we fold the page along the vertical axis the two parts of the wave will coincide.) A function with axis symmetry is called an even function. An even function satisfies the condition $f(t) = f(-t)$.

> The coefficients of all sine terms are zero in the Fourier series of even functions.

An odd function is one which satisfies the condition $f(t) = -f(-t)$. Odd functions are symmetrical about the origin of the x, y-coordinate system [see Figure 22-3(b)].

> The coefficients of all cosine terms are zero in the Fourier series of odd functions.

A waveform can be tested visually for half-wave or mirror symmetry by shifting the negative portion by one-half of a period ($\frac{1}{2}T$) and testing for symmetry about the horizontal axis (see Figure 22-4). These functions satisfy the condition $f(t) = -f(t \pm T/2)$. Note that, in Figure 22-4, $f(t_x) = -f(t_x + T/2)$.

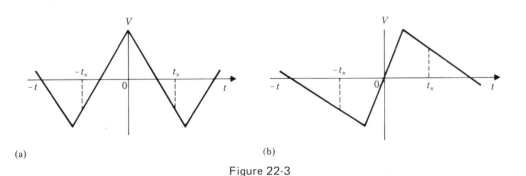

(a)   (b)

Figure 22-3

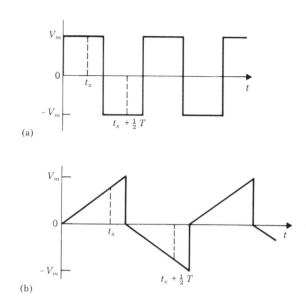

Figure 22-4

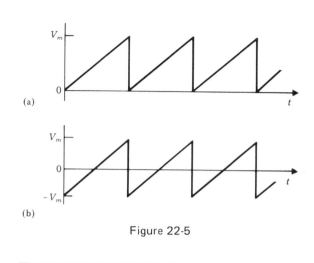

Figure 22-5

Functions with half-wave symmetry have only odd harmonics.

Average Value   Testing for the presence of a dc component (a net average value) in a nonsinusoidal voltage or current is performed by examining the areas under the curve (the area between the curve and the horizontal axis). If the areas under the curve are balanced (the area above the horizontal axis is equal to the area below the curve), there is no dc component. Refer to Figure 22-5; the curve in (a) has a dc component, the curve in (b) does not.

An oscilloscope with a switch for selecting either ac or dc coupling provides an effective and convenient method for measuring the dc component (average value) of a complex waveform. When a waveform is applied to an oscilloscope's vertical deflection circuit with ac coupling selected, a capacitor blocks the dc component and the waveform is displayed as if it had no dc component (areas above and below horizontal axis are equal). When coupling is direct the dc component is added

to the display. The amount of vertical displacement of the entire waveform when coupling is switched between "ac" and "dc" is equal to the dc component of the waveform.

Example 3   Refer to the waveform of Figure 22-6(a). (a) Describe the type of symmetry it exhibits (if any) and state the significance of the symmetry on the Fourier components. (b) Determine f_1, f_2, and f_3 of its sine-wave components. (c) Determine V_{dc}.

(a) Displaying the waveform without a dc component [see Figure 22-6(b)] demon-

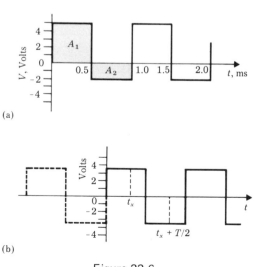

Figure 22-6

strates that it is symmetrical about the origin, and about the horizontal axis if displaced by $T/2$. It is, therefore, an odd function and has half-wave symmetry. The signal will contain no even harmonics or cosine terms [see equation 2].

(b) $f_1 = \dfrac{1}{T} = \dfrac{1}{1 \times 10^{-3}} = \underline{1 \text{ kHz}}, \quad f_2$ is

nonexistent

$f_3 = 3f_1 = \underline{3 \text{ kHz}}$

(c) We find V_{dc} by evaluating the formula,

$$V_{dc} = \frac{A_T}{T} = \frac{\text{total area under curve}}{\text{period}}$$

[see Figure 22-6(a)]

$$= \frac{A_1 + A_2}{T} = \frac{5 \times 0.5 + (-2 \times 0.5)}{1}$$

$$= \frac{2.5 - 1}{1} = \underline{1.5 \text{ V}}$$

Example 4 The output of a transistor amplifier appears as in Figure 22-7 even though the input is a pure sine wave. The waveform shifts +2.5 V when the oscilloscope input is switched from ac to dc coupling. (a) Does the distortion introduced by the amplifier include even harmonics—second, fourth, and so forth? Explain. (b) What is the dc component of the waveform?

(a) The waveform does not possess half-wave symmetry and therefore will include even harmonics.

(b) $V_{dc} = V_{\text{scope shift}} = \underline{+2.5 \text{ V}}$

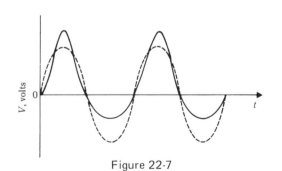

Figure 22-7

Exercises 22-1 In Exercises 1–8 write the solutions referring to the waveforms shown in Figure 22-8.

1. List all waveforms that represent even functions (that is, are symmetrical with respect to the vertical axis). (*Hint:* Imagine waveforms extended to left of zero.)

2. List all waveforms that represent odd functions (that is, are symmetrical about the origin).

3. List all waveforms that possess half-wave symmetry.

4. List all waveforms that have an average value.

5. List all waveforms (if any) that could be made even or odd by shifting the position of the vertical axis. State the type of function after the shift.

6. Sketch two waveforms (unlike any shown in Figure 22-8) that demonstrate even functions.

7. Sketch two waveforms (unlike any shown in Figure 22-8) that demonstrate odd functions.

8. Sketch two waveforms (unlike any in Figure 22-8) that demonstrate half-wave symmetry.

In Exercises 9–14, for the given waveforms, (a) state the significance of the symmetry of the waveform as it relates to the type of Fourier components which can be expected to be present (see Example 3); (b) determine f_1 and the first four harmonics that will be present; and (c) determine the average value of the waveform.

9. Waveform of Figure 22-8(a).

10. Waveform of Figure 22-8(b).

11. Waveform of Figure 22-8(c).

12. Waveform of Figure 22-8(d).

13. Waveform of Figure 22-8(e).

14. Waveform of Figure 22-8(f), $V_{\text{scope shift}} = -30$ mV when coupling changed from ac to dc.

15. Waveform of Figure 22-8(g), $V_{\text{scope shift}} = +18$ mV when coupling changed from dc to ac.

16. Consider that the waveforms of Figures 22-8(h) and 22-8(i) represent the distortion of pure sine waves by the addition of harmonics to the

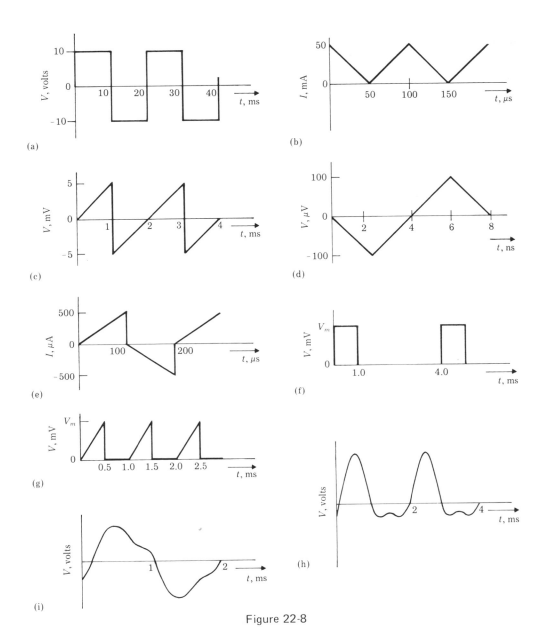

Figure 22-8

fundamental. On the basis of symmetry, which of the waves is distorted by a second harmonic content? Which by a third harmonic content?

17. Construct the spectrum diagram of a rectified full wave which has the following Fourier components (see Example 2):

$$v = \frac{2}{\pi} V_m \left(1 - \frac{2}{3} \cos 2\omega_1 t - \frac{2}{15} \cos 4\omega_1 t \right.$$
$$\left. - \frac{2}{35} \cos 6\omega_1 t \cdots \right)$$

18. Repeat Exercise 17 for a rectified half-wave. The Fourier analysis is

$$v = \frac{V_m}{\pi} \left(1 + \frac{\pi}{2} \sin \omega_1 t - \frac{2}{3} \cos 2\omega_1 t \right.$$
$$\left. - \frac{2}{15} \cos 4\omega_1 t - \frac{2}{35} \cos 6\omega_1 t - \cdots \right)$$

19. Repeat Exercise 17 for a sawtooth wave [see Figure 22-8(c)]:

$$v = \frac{2}{\pi} V_m \left(\sin \omega_1 t - \frac{1}{2} \sin 2\omega_1 t \right.$$
$$+ \frac{1}{3} \sin 3\omega_1 t - \frac{1}{4} \sin 4\omega_1 t$$
$$\left. + \frac{1}{5} \sin 5\omega_1 t + \cdots \right)$$

20. Repeat Exercise 17 for a triangle wave [see Figure 22-8(d)]:

$$v = \frac{8}{\pi^2} V_m \left(-\sin \omega_1 t + \frac{1}{9} \sin 3\omega_1 t \right.$$
$$- \frac{1}{25} \sin 5\omega_1 t$$
$$\left. + \frac{1}{49} \sin 7\omega_1 t + \cdots \right)$$

22-2 APPLICATIONS OF ANALYSIS OF NONSINUSOIDAL WAVEFORMS

The analysis techniques of Chapter 14 apply only for sinusoidal voltages or currents. However, those straightforward techniques can be used indirectly when the excitation is nonsinusoidal if use is also made of the theory embodied in the Fourier series and the superposition principle. In effect, a generator of a nonsinusoidal, periodic waveform can be replaced by a dc source plus an infinite number of sinusoidal ac generators (see Figure 22-9). The effect of the complex waveform is obtained by adding algebraically the effects of each of the individual sources. It may be necessary to include the effects of only two or three of the infinite number of sources to obtain the desired information in a practical situation.

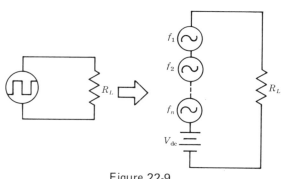

Figure 22-9

Effective Value of Nonsinusoidal Waveform
A frequently used application of these concepts is in the determination of the effective (or rms) values of the harmonics of a distorted sine wave. For example, one of the standard specifications of amplifiers for high-fidelity music systems is that of "percent total harmonic distortion" (THD). When a complex (or distorted) waveform appears across a resistor, the total power p_T in the resistor is found by considering the result of each frequency component acting independently. Thus,

$$p_T = p_1 + p_2 + p_3 + \cdots$$

But, since $p = V^2/R$, then

$$\frac{V_T^2}{R} = \frac{V_1^2}{R} + \frac{V_2^2}{R} + \frac{V_3^2}{R} + \cdots$$

Multiplying by R,

$$V_T^2 = V_1^2 + V_2^2 + V_3^2 + \cdots$$

where V_T, V_1, V_2, V_3, and so on, are effective values of voltages.

Taking the square root of the sum of the squares (called the quadratic sum) gives

$$V_T = \sqrt{V_1^2 + V_2^2 + V_3^2 + \cdots}$$

Or, if we are interested in the effective value of the harmonics only,

$$V_{\text{harmonics}} = \sqrt{V_2^2 + V_3^2 + V_4^2 + \cdots}$$

Total harmonic distortion is the ratio of the effective value of all harmonics to the effective value of the fundamental. That is,

$$\text{THD} = \frac{V_{\text{harmonics}}}{V_1} \times 100\%$$
$$= \frac{\sqrt{V_2^2 + V_3^2 + V_4^2 + \cdots}}{V_1} \times 100\%$$

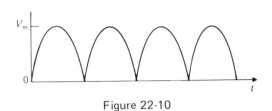

Figure 22-10

But

$$\text{Percent second} = \frac{V_2}{V_1} \times 100\%$$

$$\text{Percent third} = \frac{V_3}{V_1} \times 100\%, \text{ and so on}$$

Therefore,

THD

$$= \sqrt{(\text{percent second})^2 + (\text{percent third})^2 + \cdots}$$

Example 5   The fluctuation present in the output of an ac-to-dc power supply is called *ripple*. The ripple factor of a power supply is defined

$$\%V_{\text{ripple}} = \frac{\text{Effective value of ac components}}{\text{Average value}}$$
$$\times 100\%$$

The output of a full-wave rectified power supply (see Figure 22-10) can be represented as follows:

$$v = \frac{2V_m}{\pi} \left(1 - \tfrac{2}{3} \cos 2\omega_1 t - \tfrac{2}{15} \cos 4\omega_1 t\right.$$
$$\left. - \tfrac{2}{35} \cos 6\omega_1 t + \cdots \right)$$

where V_m = peak value of input ac sine wave
ω_1 = frequency of ac input wave

If V_{input} = 120 V (rms) and ω_1 = 377 rad/s, find (a) V_{dc}, (b) percent ripple in unfiltered output, and (c) frequency of the largest ripple component.

(a) The constant term in the Fourier series is the dc value of the wave. Thus, since

$$V_m = \sqrt{2}\, V$$
$$V_{\text{dc}} = \frac{2V_m}{\pi} = \frac{2 \times 1.414 \times 120}{\pi} = 108 \text{ V}$$

Now, since $V = V_m / \sqrt{2}$, the effective values of the ac components of the wave are

$$V_2 = \frac{2}{3} \times \frac{2}{\pi} V_{\text{input}}, \qquad V_4 = \frac{2}{15} \times \frac{2}{\pi} V_{\text{input}},$$

$$V_6 = \frac{2}{35} \times \frac{2}{\pi} V_{\text{input}}$$

Then

Effective value of all significant ac components

$$= \frac{2\, V_{\text{input}}}{\pi} \sqrt{(\tfrac{2}{3})^2 + (\tfrac{2}{15})^2 + (\tfrac{2}{35})^2 + \cdots}$$

$$= \frac{2 \times 120}{\pi} (0.6823) = 52.12 \text{ V}$$

(b) Hence,

$$\%V_{\text{ripple}} = \frac{52.12}{108} \times 100\% = \underline{48\%}$$

(c) The largest ripple component is contributed by the term $\tfrac{2}{3} \cos 2\omega_1 t$. Hence, the frequency of the largest ripple component is

$$f = \frac{2\omega_1}{2\pi} = \frac{377}{\pi} = \underline{120 \text{ Hz}}$$

Example 6   The output of a 30-W, hi-fi audio amplifier is analyzed with a spectrum analyzer and is found to have the following fundamental and harmonic components: V_1 = 1.256 V, V_2 = 11.31 mV, V_3 = 2.512 mV, and V_4 = 1.005 mV. Determine (a) percent second, third, and fourth harmonic distortion, and (b) THD.

(a) Percent second = $\dfrac{11.31 \times 10^{-3}}{1.256} \times 100\%$

$$= \underline{0.9\%}$$

Percent third = $\dfrac{2.512 \times 10^{-3}}{1.256} \times 100\%$

$$= \underline{0.2\%}$$

Percent fourth = $\dfrac{1.005 \times 10^{-3}}{1.256} \times 100\%$

$$= \underline{0.08\%}$$

(b) THD = $\sqrt{(0.9)^2 + (0.2)^2 + (0.08)^2}$
$$= \underline{0.925\%}$$

Exercises 22-2

1. The V_{input} for an unfiltered, full-wave rectified power supply is 12.6 (recall that ac voltages are effective values unless otherwise specified). The line frequency is 400 Hz. Determine (a) V_{dc}, (b) V_{ripple}, and (c) the frequency of the largest component of ripple. Refer to Example 5 for the Fourier representation for this waveform.

2. Repeat Exercise 1 for $V_{\text{input}} = 28$ V, $f_{\text{line}} = 50$ Hz.

3. Repeat Exercise 1 for $V_{\text{input}} = 36$ V, $f_{\text{line}} = 60$ Hz.

4. Repeat Exercise 1 for $V_{\text{input}} = 360$ V, $f_{\text{line}} = 400$ Hz.

5. The Fourier analysis of a half-wave rectified voltage (see Figure 22-11) is

$$v = \frac{V_m}{\pi}\left(1 + \frac{\pi}{2}\cos\omega_1 t - \frac{2}{3}\cos 2\omega_1 t \right.$$
$$\left. - \frac{2}{15}\cos 4\omega_1 t - \frac{2}{35}\cos 6\omega_1 t + \cdots\right)$$

where V_m = peak value of wave = $\sqrt{2}\, V_{\text{input}}$ (assuming no voltage drop across rectifier)

$$\omega_1 = 2\pi f_1$$
$$f_1 = \text{Line or input frequency}$$

Determine (a) V_{dc}, (b) V (rms) for each ac component, (c) $\%V_{\text{ripple}}$, and (d) f for largest component of ripple. Given: $V_{\text{input}} = 120$ V, $f_1 = 60$ Hz.

6. Repeat Exercise 5 for $V_{\text{input}} = 28$ V, $f_1 = 400$ Hz.

7. Repeat Exercise 5 for $V_{\text{input}} = 36$ V, $f_1 = 50$ Hz.

8. Repeat Exercise 5 for $V_{\text{input}} = 360$ V, $f_1 = 60$ Hz.

9. A testing laboratory report on an audio amplifier provides the following information on the output of the amplifier when it is processing a 1000-Hz test signal: $V_1 = 0.778$ V, $V_2 = 8.36$ mV, $V_3 = 2.75$ mV, and $V_4 = 0.937$ mV. Other harmonics were judged insignificant. Determine (a) the % distortion (see Example 6) and frequency of each of the reported harmonics and (b) THD.

10. Repeat Exercise 9 for the report: $f_1 = 10$ kHz, $V_1 = 40.0$ V, $V_2 = 0.365$ V, $V_4 = 93.6$ mV, $V_6 = 4.73$ μV.

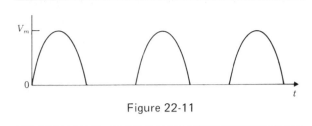

Figure 22-11

11. Repeat Exercise 9 for the report: $f_1 = 20$ Hz, $V_1 = 15.49$ V, $V_2 = 0.236$ V, $V_3 = 45.3$ mV, $V_4 = 6.36$ mV, $V_5 = 0.528$ mV.

12. Repeat Exercise 9 for the report: $f_1 = 5$ kHz, $V_1 = 2.65$ V, $V_2 = 0.575$ V, $V_3 = 2.56$ mV, $V_4 = 56.5$ μV.

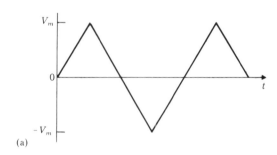

(a)

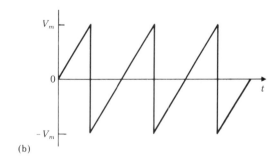

(b)

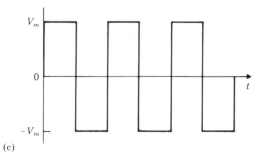

(c)

Figure 22-12

13. A proposed electronic unit is to extract and amplify the third harmonic of a triangular wave. The Fourier analysis of a triangular wave [see Figure 22-12(a)] is

$$v = \frac{8}{\pi^2} V_m \left(\sin \omega_1 t - \frac{1}{9} \sin 3\omega_1 t + \frac{1}{25} \sin 5\omega_1 t + \cdots \right)$$

If $V_m = 1$ mV, what gain must the unit have if the desired signal strength (rms) is $V_3 = 1.5$ V?

14. Repeat Exercise 13 for a sawtooth source wave [see Figure 22-12(b)]:

$$v = \frac{2}{\pi} V_m \left(\sin \omega_1 t - \frac{1}{2} \sin 2\omega_1 t + \frac{1}{3} \sin 3\omega_1 t - \frac{1}{4} \sin 4\omega_1 t + \cdots \right)$$

15. Repeat Exercise 13 for a square source wave [see Figure 22-12(c)]:

$$v = \frac{4}{\pi} V_m \left(\sin \omega_1 t + \frac{1}{3} \sin 3\omega_1 t + \frac{1}{5} \sin 5\omega_1 t + \cdots \right)$$

16. Repeat Exercise 13 for a rectified half-wave. Refer to Exercise 5 for the Fourier analysis.

APPENDIX A

Table I

FOUR-PLACE LOGARITHMS OF NUMBERS

N	0	1	2	3	4	5	6	7	8	9
10	0000	0043	0086	0128	0170	0212	0253	0294	0334	0374
11	0414	0453	0492	0531	0569	0607	0645	0682	0719	0755
12	0792	0828	0864	0899	0934	0969	1004	1038	1072	1106
13	1139	1173	1206	1239	1271	1303	1335	1367	1399	1430
14	1461	1492	1523	1553	1584	1614	1644	1673	1703	1732
15	1761	1790	1818	1847	1875	1903	1931	1959	1987	2014
16	2041	2068	2095	2122	2148	2175	2201	2227	2253	2279
17	2304	2330	2355	2380	2405	2430	2455	2480	2504	2529
18	2553	2577	2601	2625	2648	2672	2695	2718	2742	2765
19	2788	2810	2833	2856	2878	2900	2923	2945	2967	2989
20	3010	3032	3054	3075	3096	3118	3139	3160	3181	3201
21	3222	3243	3263	3284	3304	3324	3345	3365	3385	3404
22	3424	3444	3464	3483	3502	3522	3541	3560	3579	3598
23	3617	3636	3655	3674	3692	3711	3729	3747	3766	3784
24	3802	3820	3838	3856	3874	3892	3909	3927	3945	3962
25	3979	3997	4014	4031	4048	4065	4082	4099	4116	4133
26	4150	4166	4183	4200	4216	4232	4249	4265	4281	4298
27	4314	4330	4346	4362	4378	4393	4409	4425	4440	4456
28	4472	4487	4502	4518	4533	4548	4564	4579	4594	4609
29	4624	4639	4654	4669	4683	4698	4713	4728	4742	4757
30	4771	4786	4800	4814	4829	4843	4857	4871	4886	4900
31	4914	4928	4942	4955	4969	4983	4997	5011	5024	5038
32	5051	5065	5079	5092	5105	5119	5132	5145	5159	5172
33	5185	5198	5211	5224	5237	5250	5263	5276	5289	5302
34	5315	5328	5340	5353	5366	5378	5391	5403	5416	5428
35	5441	5453	5465	5478	5490	5502	5514	5527	5539	5551
36	5563	5575	5587	5599	5611	5623	5635	5647	5658	5670
37	5682	5694	5705	5717	5729	5740	5752	5763	5775	5786
38	5798	5809	5821	5832	5843	5855	5866	5877	5888	5899
39	5911	5922	5933	5944	5955	5966	5977	5988	5999	6010
40	6021	6031	6042	6053	6064	6075	6085	6096	6107	6117
41	6128	6138	6149	6160	6170	6180	6191	6201	6212	6222
42	6232	6243	6253	6263	6274	6284	6294	6304	6314	6325
43	6335	6345	6355	6365	6375	6385	6395	6405	6415	6425
44	6435	6444	6454	6464	6474	6484	6493	6503	6513	6522
45	6532	6542	6551	6561	6571	6580	6590	6599	6609	6618
46	6628	6637	6646	6656	6665	6675	6684	6693	6702	6712
47	6721	6730	6739	6749	6758	6767	6776	6785	6794	6803
48	6812	6821	6830	6839	6848	6857	6866	6875	6884	6893
49	6902	6911	6920	6928	6937	6946	6955	6964	6972	6981
50	6990	6998	7007	7016	7024	7033	7042	7050	7059	7067
51	7076	7084	7093	7101	7110	7118	7126	7135	7143	7152
52	7160	7168	7177	7185	7193	7202	7210	7218	7226	7235
53	7243	7251	7259	7267	7275	7284	7292	7300	7308	7316
54	7324	7332	7340	7348	7356	7364	7372	7380	7388	7396

Table I—continued

N	0	1	2	3	4	5	6	7	8	9
55	7404	7412	7419	7427	7435	7443	7451	7459	7466	7474
56	7482	7490	7497	7505	7513	7520	7528	7536	7543	7551
57	7559	7566	7574	7582	7589	7597	7604	7612	7619	7627
58	7634	7642	7649	7657	7664	7672	7679	7686	7694	7701
59	7709	7716	7723	7731	7738	7745	7752	7760	7767	7774
60	7782	7789	7796	7803	7810	7818	7825	7832	7839	7846
61	7853	7860	7868	7875	7882	7889	7896	7903	7910	7917
62	7924	7931	7938	7945	7952	7959	7966	7973	7980	7987
63	7993	8000	8007	8014	8021	8028	8035	8041	8048	8055
64	8062	8069	8075	8082	8089	8096	8102	8109	8116	8122
65	8129	8136	8142	8149	8156	8162	8169	8176	8182	8189
66	8195	8202	8209	8215	8222	8228	8235	8241	8248	8254
67	8261	8267	8274	8280	8287	8293	8299	8306	8312	8319
68	8325	8331	8338	8344	8351	8357	8363	8370	8376	8382
69	8388	8395	8401	8407	8414	8420	8426	8432	8439	8445
70	8451	8457	8463	8470	8476	8482	8488	8494	8500	8506
71	8513	8519	8525	8531	8537	8543	8549	8555	8561	8567
72	8573	8579	8585	8591	8597	8603	8609	8615	8621	8627
73	8633	8639	8645	8651	8657	8663	8669	8675	8681	8686
74	8692	8698	8704	8710	8716	8722	8727	8733	8739	8745
75	8751	8756	8762	8768	8774	8779	8785	8791	8797	8802
76	8808	8814	8820	8825	8831	8837	8842	8848	8854	8859
77	8865	8871	8876	8882	8887	8893	8899	8904	8910	8915
78	8921	8927	8932	8938	8943	8949	8954	8960	8965	8971
79	8976	8982	8987	8993	8998	9004	9009	9015	9020	9025
80	9031	9036	9042	9047	9053	9058	9063	9069	9074	9079
81	9085	9090	9096	9101	9106	9112	9117	9122	9128	9133
82	9138	9143	9149	9154	9159	9165	9170	9175	9180	9186
83	9191	9196	9201	9206	9212	9217	9222	9227	9232	9238
84	9243	9248	9253	9258	9263	9269	9274	9279	9284	9289
85	9294	9299	9304	9309	9315	9320	9325	9330	9335	9340
86	9345	9350	9355	9360	9365	9370	9375	9380	9385	9390
87	9395	9400	9405	9410	9415	9420	9425	9430	9435	9440
88	9445	9450	9455	9460	9465	9469	9474	9479	9484	9489
89	9494	9499	9504	9509	9513	9518	9523	9528	9533	9538
90	9542	9547	9552	9557	9562	9566	9571	9576	9581	9586
91	9590	9595	9600	9605	9609	9614	9619	9624	9628	9633
92	9638	9643	9647	9652	9657	9661	9666	9671	9675	9680
93	9685	9689	9694	9699	9703	9708	9713	9717	9722	9727
94	9731	9736	9741	9745	9750	9754	9759	9763	9768	9773
95	9777	9782	9786	9791	9795	9800	9805	9809	9814	9818
96	9823	9827	9832	9836	9841	9845	9850	9854	9859	9863
97	9868	9872	9877	9881	9886	9890	9894	9899	9903	9908
98	9912	9917	9921	9926	9930	9934	9939	9943	9948	9952
99	9956	9961	9965	9969	9974	9978	9983	9987	9991	9996

APPENDIX B

Table 11   Trigonometric functions of degrees

Angle°	sin	tan	cot	cos	Angle°	Angle°	sin	tan	cot	cos	Angle°
0.0	.00000	.00000	∞	1.00000	**90.0**	**4.5**	.07846	.07870	12.706	.99692	**85.5**
.1	.00175	.00175	572.96	1.00000	.9	.6	.08020	.08046	12.429	.99678	.4
.2	.00349	.00349	286.48	0.99999	.8	.7	.08194	.08221	12.163	.99664	.3
.3	.00524	.00524	190.98	.99999	.7	.8	.08368	.08397	11.909	.99649	.2
.4	.00698	.00698	143.24	.99998	.6	.9	.08542	.08573	11.664	.99635	.1
.5	.00873	.00873	114.59	.99996	.5	**5.0**	.08716	.08749	11.430	.99619	**85.0**
.6	.01047	.01047	95.489	.99995	.4	.1	.08889	.08925	11.205	.99604	.9
.7	.01222	.01222	81.847	.99993	.3	.2	.09063	.09101	10.988	.99588	.8
.8	.01396	.01396	71.615	.99990	.2	.3	.09237	.09277	10.780	.99572	.7
.9	.01571	.01571	63.657	.99988	.1	.4	.09411	.09453	10.579	.99556	.6
1.0	.01745	.01746	57.290	.99985	**89.0**	.5	.09585	.09629	10.385	.99540	.5
.1	.01920	.01920	52.081	.99982	.9	.6	.09758	.09805	10.199	.99523	.4
.2	.02094	.02095	47.740	.99978	.8	.7	.09932	.09981	10.019	.99506	.3
.3	.02269	.02269	44.066	.99974	.7	.8	.10106	.10158	9.8448	.99488	.2
.4	.02443	.02444	40.917	.99970	.6	.9	.10279	.10334	9.6768	.99470	.1
.5	.02618	.02619	38.188	.99966	.5	**6.0**	.10453	.10510	9.5144	.99452	**84.0**
.6	.02792	.02793	35.801	.99961	.4	.1	.10626	.10687	9.3572	.99434	.9
.7	.02967	.02968	33.694	.99956	.3	.2	.10800	.10863	9.2052	.99415	.8
.8	.03141	.03143	31.821	.99951	.2	.3	.10973	.11040	9.0579	.99396	.7
.9	.03316	.03317	30.145	.99945	.1	.4	.11147	.11217	8.9152	.99377	.6
2.0	.03490	.03492	28.636	.99939	**88.0**	.5	.11320	.11394	8.7769	.99357	.5
.1	.03664	.03667	27.271	.99933	.9	.6	.11494	.11570	8.6427	.99337	.4
.2	.03839	.03842	26.031	.99926	.8	.7	.11667	.11747	8.5126	.99317	.3
.3	.04013	.04016	24.898	.99919	.7	.8	.11840	.11924	8.3863	.99297	.2
.4	.04188	.04191	23.859	.99912	.6	.9	12014	.12101	8.2636	.99276	.1
.5	.04362	.04366	22.904	.99905	.5	**7.0**	.12187	.12278	8.1443	.99255	**83.0**
.6	.04536	.04541	22.022	.99897	.5	.1	.12360	.12456	8.0285	.99233	.9
.7	.04711	.04716	21.205	.99889	.3	.2	.12533	.12633	7.9158	.99211	.8
.8	.04885	.04891	20.446	.99881	.2	.3	.12706	.12810	7.8062	.99189	.7
.9	.05059	.05066	19.740	.99872	.1	.4	.12880	.12988	7.6996	.99167	.6
3.0	.05234	.05241	19.081	.99863	**87.0**	.5	.13053	.13165	7.5958	.99144	.5
.1	.05408	.05416	18.464	.99854	.9	.6	.13226	.13343	7.4947	.99122	.4
.2	.05582	.05591	17.886	.99844	.8	.7	.13399	.13521	7.3962	.99098	.3
.3	.05756	.05766	17.343	.99834	.7	.8	.13572	.13698	7.3002	.99075	.2
.4	.05931	.05941	16.832	.99824	.6	.9	.13744	.13876	7.2066	.99051	.1
.5	.06105	.06116	16.350	.99813	.5	**8.0**	.13917	.14054	7.1154	.99027	**82.0**
.6	.06279	.06291	15.895	.99803	.4	.1	.14090	.14232	7.0264	.99002	.9
.7	.06453	.06467	15.464	.99792	.3	.2	.14263	.14410	6.9395	.98978	.8
.8	.06627	.06642	15.056	.99780	.2	.3	.14436	.14588	6.8548	.98953	.7
.9	.06802	.06817	14.669	.99767	.1	.4	.14608	.14767	6.7720	.98927	.6
4.0	.06976	.06993	14.301	.99756	**86.0**	.5	.14781	.14945	6.6912	.98902	.5
.1	.07150	.07168	13.951	.99744	.9	.6	.14954	.15124	6.6122	.98876	.4
.2	.07324	.07344	13.617	.99731	.8	.7	.15126	.15302	6.5350	.98849	.3
.3	.07498	.07519	13.300	.99719	.7	.8	.15299	.15481	6.4596	.98823	.2
.4	.07672	.07695	12.996	.99705	.6	.9	.15471	.15660	6.3859	.98796	.1
4.5	.07846	.07870	12.706	.99692	**85.5**	**9.0**	.15643	.15838	6.3138	.98769	**81.0**
Angle°	cos	cot	tan	sin	Angle°	Angle°	cos	cot	tan	sin	Angle°

Table 11   Trigonometric functions of degrees (continued)

Angle°	sin	tan	cot	cos	Angle°	Angle°	sin	tan	cot	cos	Angle°
9.0	.15643	.15838	6.3138	.98769	**81.0**	**13.5**	.23345	.24008	4.1653	.97237	.5
.1	.15816	.16017	6.2432	.98741	.9	.6	.23514	.24193	4.1335	.97196	.4
.2	.15988	.16196	6.1742	.98714	.8	.7	.23684	.24377	4.1022	.97155	.3
.3	.16160	.16376	6.1066	.98686	.7	.8	.23853	.24562	4.0713	.97113	.2
.4	.16333	.16555	6.0405	.98657	.6	.9	.24023	.24747	4.0408	.97072	.1
.5	.16505	.16734	5.9758	.98629	.5	**14.0**	.24192	.24933	4.0108	.97030	**76.0**
.6	.16677	.16914	5.9124	.98600	.4	.1	.24362	.25118	3.9812	.96987	.9
.7	.16849	.17093	5.8502	.98570	.3	.2	.24531	.25304	3.9520	.96945	.8
.8	.17021	.17273	5.7894	.98541	.2	.3	.24700	.25490	3.9232	.96902	.7
.9	.17193	.17453	5.7297	.98511	.1	.4	.24869	.25676	3.8947	.96858	.6
10.0	.17365	.17633	5.6713	.98481	**80.0**	.5	.25038	.25862	3.8667	.96815	.5
.1	.17537	.17813	5.6140	.98450	.9	.6	.25207	.26048	3.8391	.96771	.4
.2	.17708	.17993	5.5578	.98420	.8	.7	.25376	.26235	3.8118	.96727	.3
.3	.17880	.18173	5.5026	.98389	.7	.8	.25545	.26421	3.7848	.96682	.2
.4	.18052	.18353	5.4486	.98357	.6	.9	.25713	.26608	3.7583	.96638	.1
.5	.18224	.18534	5.3955	.98325	.5	**15.0**	.25882	.26792	3.7321	.96593	**75.0**
.6	.18395	.18714	5.3435	.98294	.4	.1	.26050	.26982	3.7062	.96547	.9
.7	.18567	.18895	5.2924	.98261	.3	.2	.26219	.27169	3.6806	.96502	.8
.8	.18738	.19076	5.2422	.98229	.2	.3	.26387	.27357	3.6554	.96456	.7
.9	.18910	.19257	5.1929	.98196	.1	.4	.26556	.27545	3.6305	.96410	.6
11.0	.19081	.19438	5.1446	.98163	**79.0**	.5	.26724	.27732	3.6059	.96363	.5
.1	.19252	.19619	5.0970	.98129	.6	.6	.26892	.27921	3.5816	.96316	.4
.2	.19423	.19801	5.0504	.98096	.8	.7	.27060	.28109	3.5576	.96269	.3
.3	.19595	.19982	5.0045	.98061	.7	.8	.27228	.28297	3.5339	.96222	.2
.4	.19766	.20164	4.9594	.98027	.6	.9	.27396	.28486	3.5105	.96174	.1
.5	.19937	.20345	4.9152	.97992	.5	**16.0**	.27564	.28675	3.4874	.96126	**74.0**
.6	.20108	.20527	4.8716	.97958	.4	.1	.27731	.28864	3.4646	.96078	.9
.7	.20279	.20709	4.8288	.97922	.3	.2	.27899	.29053	3.4420	.96029	.8
.8	.20450	.20891	4.7867	.97887	.2	.3	.28067	.29242	3.4197	.95981	.7
.9	.20620	.21073	4.7453	.97851	.1	.4	.28234	.29432	3.3977	.95931	.6
12.0	.20791	.21256	4.7046	.97815	**78.0**	.5	.28402	.29621	3.3759	.95882	.5
.1	.20962	.21438	4.6646	.97778	.9	.6	.28569	.29811	3.3544	.95832	.4
.2	.21132	.21621	4.6252	.97742	.8	.7	.28736	.30001	3.3332	.95782	.8
.3	.21303	.21804	4.5864	.97705	.7	.8	.28903	.30192	3.3122	.95732	.2
.4	.21474	.21986	4.5483	.97667	.6	.9	.29070	.30382	3.2914	.95681	.1
.5	.21644	.22169	4.5107	.97630	.5	**17.0**	.29237	.30573	3.2709	.95630	**73.0**
.6	.21814	.22353	4.4747	.97592	.4	.1	.29404	.30764	3.2506	.95579	.9
.7	.21985	.22536	4.4373	.97553	.3	.2	.29571	.30955	3.2305	.95528	.8
.8	.22155	.22719	4.4015	.97515	.2	.3	.29737	.31147	3.2106	.95476	.7
.9	.22325	.22903	4.3662	.97476	.1	.4	.29904	.31338	3.1910	.95424	.6
13.0	.22495	.23087	4.3315	.97437	**77.0**	.5	.30071	.31530	3.1716	.95372	.5
.1	.22665	.23271	4.2972	.97398	.9	.6	.30237	.31722	3.1524	.95319	.4
.2	.22835	.23455	4.2635	.97358	.8	.7	.30403	.31914	3.1334	.95266	.3
.3	.23005	.23639	4.2303	.97318	.7	.8	.30570	.32106	3.1146	.95213	.2
.4	.23175	.23823	4.1976	.97278	.6	.9	.30736	.32299	3.0961	.95159	.1
13.5	.23345	.24008	4.1653	.97237	.5	**18.0**	.30902	.32492	3.0777	.95106	**72.0**
Angle°	cos	cot	tan	sin	Angle°	Angle°	cos	cot	tan	sin	Angle°

Table 11   Trigonometric functions of degrees (continued)

Angle°	sin	tan	cot	cos	Angle°	Angle°	sin	tan	cot	cos	Angle°
18.0	.30902	.32492	3.0777	.95106	**72.0**	**22.5**	.38268	.41421	2.4142	.92388	.5
.1	.31068	.32685	3.0595	.95052	.9	.6	.38430	.41626	2.4023	.92321	.4
.2	.31233	.32878	3.0415	.94997	.8	.7	.38591	.41831	2.3906	.92254	.3
.3	.31399	.33072	3.0237	.94943	.7	.8	.38752	.42036	2.3789	.92186	.2
.4	.31565	.33266	3.0061	.94888	.6	.9	.38912	.42242	2.3673	.92119	.1
.5	.31730	.33460	2.9887	.94832	.5	**23.0**	.39073	.42447	2.3559	.92050	**67.0**
.6	.31896	.33654	2.9714	.94777	.4	.1	.39234	.42654	2.3445	.91982	.9
.7	.32061	.33848	2.9544	.94721	.3	.2	.39394	.42860	2.3332	.91914	.8
.8	.32227	.34043	2.9375	.94665	.2	.3	.39555	.43067	2.3220	.91845	.7
.9	.32392	.34238	2.9208	.94609	.1	.4	.39715	.43274	2.3109	.91775	.6
19.0	.32557	.34433	2.9042	.94552	**71.0**	.5	.39875	.43481	2.2998	.91706	.5
.1	.32722	.34628	2.8878	.94495	.9	.6	.40035	.43689	2.2889	.91636	.4
.2	.32887	.34824	2.8716	.94438	.8	.7	.40195	.43897	2.2781	.91566	.3
.3	.33051	.35020	2.8556	.94380	.7	.8	.40355	.44105	2.2673	.91496	.2
.4	.33216	.35216	2.8397	.94322	.6	.9	.40514	.44314	2.2566	.91425	.1
.5	.33381	.35412	2.8239	.94264	.5	**24.0**	.40674	.44523	2.2460	.91355	**66.0**
.6	.33545	.35608	2.8083	.94206	.4	.1	.40833	.44732	2.2355	.91283	.9
.7	.33710	.35805	2.7929	.94147	.3	.2	.40992	.44942	2.2251	.91212	.8
.8	.33874	.36002	2.7776	.94088	.2	.3	.41151	.45152	2.2148	.91140	.7
.9	.34038	.36199	2.7625	.94029	.1	.4	.41310	.45362	2.2045	.91068	.6
20.0	.34202	.36397	2.7475	.93969	**70.0**	.5	.41469	.45573	2.1943	.90996	.5
.1	.34366	.36595	2.7326	.93909	.9	.6	.41628	.45784	2.1842	.90924	.4
.2	.34530	.36793	2.7179	.93849	.8	.7	.41787	.45995	2.1742	.90851	.3
.3	.34694	.36991	2.7034	.93789	.7	.8	.41945	.46206	2.1624	.90778	.2
.4	.34857	.37190	2.6889	.93728	.6	.9	.42104	.46418	2.1543	.90704	.1
.5	.35021	.37388	2.6746	.93667	.5	**25.0**	.42262	.46631	2.1445	.90631	**65.0**
.6	.35184	.37588	2.6605	.93606	.4	.1	.42420	.46843	2.1348	.90557	.9
.7	.35347	.37787	2.6464	.93544	.3	.2	.42578	.47056	2.1251	.90483	.8
.8	.35511	.37986	2.6325	.93483	.2	.3	.42736	.47270	2.1155	.90408	.7
.9	.35674	.38186	2.6187	.93420	.1	.4	.42894	.47483	2.1060	.90334	.6
21.0	.35837	.38386	2.6051	.93358	**69.0**	.5	.43051	.47698	2.0965	.90259	.5
.1	.36000	.38587	2.5916	.93295	.9	.6	.43209	.47912	2.0872	.90183	.4
.2	.36162	.38787	2.5782	.93232	.8	.7	.43366	.48127	2.0778	.90108	.3
.3	.36325	.38988	2.5649	.93169	.7	.8	.43523	.48342	2.0682	.90032	.2
.4	.36488	.39190	2.5517	.93106	.6	.9	.43680	.48557	2.0594	.89956	.1
.5	.36650	.39391	2.5386	.93042	.5	**26.0**	.43837	.48773	2.0503	.89879	**64.0**
.6	.36812	.39593	2.5257	.92978	.4	.1	.43994	.48989	2.0413	.89803	.9
.7	.36975	.39795	2.5129	.92913	.3	.2	.44151	.49206	2.0323	.89726	.8
.8	.37137	.39997	2.5002	.92849	.2	.3	.44307	.49423	2.0233	.89649	.7
.9	.37299	.40200	2.4876	.92784	.1	.4	.44464	.49640	2.0145	.89571	.6
22.0	.37461	.40403	2.4751	.92718	**68.0**	.5	.44620	.49858	2.0057	.89493	.5
.1	.37622	.40606	2.4627	.92653	.9	.6	.44776	.50076	1.9970	.89415	.4
.2	.37784	.40809	2.4504	.92587	.8	.7	.44932	.50295	1.9883	.89337	.3
.3	.37946	.41013	2.4383	.92521	.7	.8	.45088	.50514	1.9797	.89259	.2
.4	.38107	.41217	2.4262	.92455	.6	.9	.45243	.50733	1.9711	.89180	.1
22.5	.38268	.41421	2.4142	.92388	.5	**27.0**	.45399	.50953	1.9626	.89101	**63.0**
Angle°	cos	cot	tan	sin	Angle°	Angle°	cos	cot	tan	sin	Angle°

Table 11 Trigonometric functions of degrees (continued)

Angle°	sin	tan	cot	cos	Angle°	Angle°	sin	tan	cot	cos	Angle°
27.0	.45399	.50953	1.9626	.89101	**63.0**	**31.5**	.52250	.61280	1.6319	.85264	**59.5**
.1	.45554	.51173	1.9542	.89021	.9	.6	.52399	.61520	1.6255	.85173	.4
.2	.45710	.51393	1.9458	.88942	.8	.7	.52547	.61761	1.6191	.85081	.3
.3	.45865	.51614	1.9375	.88862	.7	.8	.52697	.62003	1.6128	.84989	.2
.4	.46020	.51835	1.9292	.88782	.6	.9	.52844	.62245	1.6066	.84897	.1
.5	.46175	.52057	1.9210	.88701	.5	**32.0**	.52992	.62487	1.6003	.84805	**58.0**
.6	.46330	.52279	1.9128	.88620	.4	.1	.53140	.62730	1.5941	.84712	.9
.7	.46484	.52501	1.9047	.88539	.3	.2	.53288	.62973	1.5880	.84619	.8
.8	.46639	.52724	1.8967	.88458	.2	.3	.53435	.63217	1.5818	.84526	.7
.9	.46793	.52947	1.8887	.88377	.1	.4	.53583	.63462	1.5757	.84433	.6
28.0	.46947	.53171	1.8807	.88295	**62.0**	.5	.53730	.63707	1.5697	.84339	.5
.1	.47101	.53395	1.8728	.88213	.9	.6	.53877	.63953	1.5637	.84245	.4
.2	.47255	.53620	1.8650	.88130	.8	.7	.54024	.64199	1.5577	.84151	.3
.3	.47409	.53844	1.8572	.88048	.7	.8	.54171	.64446	1.5517	.84057	.2
.4	.47562	.54070	1.8495	.87965	.6	.9	.54317	.64693	1.5458	.83962	.1
5	.47716	.54296	1.8418	.87882	.5	**33.0**	.54464	.64941	1.5399	.83867	**57.0**
.6	.47869	.54522	1.8341	.87798	.4	.1	.54610	.65189	1.5340	.83772	.9
.7	.48022	.54748	1.8265	.87715	.3	.2	.54756	.65438	1.5282	.83676	.8
.8	.48175	.54975	1.8190	.87631	.2	.3	.54902	.65688	1.5224	.83581	.7
.9	.48328	.55203	1.8115	.87546	.1	.4	.55048	.65938	1.5166	.83485	.6
29.0	.48481	.55431	1.8040	.87462	**61.0**	.5	.55194	.66189	1.5108	.83389	.5
.1	.48634	.55659	1.7966	.87377	.9	.6	.55339	.66440	1.5051	.83292	.4
.2	.48786	.55888	1.7893	.87292	.8	.7	.55484	.66692	1.4994	.83195	.3
.3	.48938	.56117	1.7820	.87207	.7	.8	.55630	.66944	1.4938	.83098	.2
.4	.49090	.56347	1.7747	.87121	.6	.9	.55775	.67197	1.4882	.83001	.1
.5	.49242	.56577	1.7675	.87036	.5	**34.0**	.55919	.67451	1.4826	.82904	**56.0**
.6	.49394	.56808	1.7603	.86949	.4	.1	.56064	.67705	1.4770	.82806	.9
.7	.49546	.57039	1.7532	.86863	.3	.2	.56208	.67960	1.4715	.82708	.8
.8	.49697	.57271	1.7461	.86777	.2	.3	.56353	.68215	1.4659	.82610	.7
.9	.49849	.57503	1.7391	.86690	.1	.4	.56497	.68471	1.4605	.82511	.6
30.0	.50000	.57735	1.7321	.86603	**60.0**	.5	.56641	.68728	1.4550	.82413	.5
.1	.50151	.57968	1.7251	.86515	.9	.6	.56784	.68985	1.4496	.82314	.4
.2	.50302	.58201	1.7182	.86427	.8	.7	.56928	.69243	1.4442	.82214	.3
.3	.50453	.58435	1.7113	.86340	.7	.8	.57071	.69502	1.4388	.82115	.2
.4	.50603	.58670	1.7045	.86251	.6	.9	.57215	.69761	1.4335	.82015	.1
.5	.50754	.58905	1.6977	.86163	.5	**35.0**	.57358	.70021	1.4281	.81915	**55.0**
.6	.50904	.59140	1.6909	.86074	.4	.1	.57501	.70281	1.4229	.81815	.9
.7	.51054	.59376	1.6842	.85985	.3	.2	.57643	.70542	1.4176	.81714	.8
.8	.51204	.59612	1.6775	.85895	.2	.3	.57786	.70804	1.4124	.81614	.7
.9	.51354	.59849	1.6709	.85806	.1	.4	.57928	.71066	1.4071	.71513	.6
31.0	.51504	.60086	1.6643	.85717	**59.0**	.5	.58070	.71329	.14019	.81412	.5
.1	.51653	.60324	1.6577	.85627	.9	.6	.58212	.71593	1.3968	.81310	.4
.2	.51803	.60562	1.6512	.85536	.8	.7	.58354	.71857	1.3916	.81208	.3
.3	.51952	.60801	1.6447	.85446	.7	.8	.58496	.72122	1.3865	.81106	.2
.4	.52101	.61040	1.6383	.85355	.6	.9	.58637	.72388	1.3814	.81004	.1
31.5	.52250	.61280	1.6319	.85264	**59.5**	**36.0**	.58779	.72654	1.3764	.80902	**54.0**
Angle°	cos	cot	tan	sin	Angle°	Angle°	cos	cot	tan	sin	Angle°

Table 11 Trigonometric functions of degrees (continued)

Angle°	sin	tan	cot	cos	Angle°	Angle°	sin	tan	cot	cos	Angle°
36.0	.58779	.72654	1.3764	.80902	**54.0**	**40.5**	.64945	.85408	1.1708	.76041	**49.5**
.1	.58920	.72921	1.3713	.80799	.9	.6	.65077	.85710	1.1667	.75927	.4
.2	.59061	.73189	1.3663	.80696	.8	.7	.65210	.86014	1.1626	.75813	.3
.3	.59201	.73457	1.3613	.80593	.7	.8	.65342	.86318	1.1585	.75700	.2
.4	.59342	.73726	1.3564	.80489	.6	.9	.65474	.86623	1.1544	.75585	.1
.5	.59482	.73996	1.3514	.80386	.5	**41.0**	.65606	.86929	1.1504	.75471	**49.0**
.6	.59622	.74267	1.3465	.80282	.4	.1	.65738	.87236	1.1463	.75356	.9
.7	.59763	.74538	1.3416	.80178	.3	.2	.65869	.87543	1.1423	.75241	.8
.8	.59902	.74810	1.3367	.80073	.2	.3	.66000	.87852	1.1383	.75126	.7
.9	.60042	.75082	1.3319	.79968	.1	.4	.66131	.88162	1.1343	.75011	.6
37.0	.60182	.75355	1.3270	.79864	**53.0**	.5	.66262	.88473	1.1303	.74896	.5
.1	.60321	.75629	1.3222	.79758	.9	.6	.66393	.88784	1.1263	.74780	.4
.2	.60460	.75904	1.3175	.79653	.8	.7	.66523	.89097	1.1224	.74664	.3
.3	.60599	.76180	1.3127	.79547	.7	.8	.66653	.89410	1.1184	.74548	.2
.4	.60738	.76456	1.3079	.79441	.6	.9	.66783	.89725	1.1145	.74431	.1
.5	.60876	.76733	1.3032	.79335	.5	**42.0**	.66913	.90040	1.1106	.74314	**48.0**
.6	.61015	.77010	1.2985	.79229	.4	.1	.67043	.90357	1.1067	.74198	.9
.7	.61153	.77289	1.2938	.79122	.3	.2	.67172	.90674	1.1028	.74080	.8
.8	.61291	.77568	1.2892	.79016	.2	.3	.67301	.90993	1.0990	.73963	.7
.9	.61429	.77848	1.2846	.78908	.1	.4	.67430	.91313	1.0951	.73846	.6
38.0	.61566	.78129	1.2799	.78801	**52.0**	.5	.67559	.91633	1.0913	.73728	.5
.1	.61704	.78410	1.2753	.78694	.9	.6	.67688	.91955	1.0875	.73610	.4
.2	.61841	.78692	1.2708	.78586	.8	.7	.67816	.92277	1.0837	.73491	.3
.3	.61978	.78975	1.2662	.78478	.7	.8	.67944	.92601	1.0799	.73373	.2
.4	.62115	.79259	1.2617	.78369	.6	.9	.68072	.92926	1.0761	.73254	.1
.5	.62251	.79544	1.2572	.78261	.5	**43.0**	.68200	.93252	1.0724	.73135	**47.0**
.6	.62388	.79829	1.2527	.78152	.4	.1	.68327	.93578	1.0686	.73016	.9
.7	.62524	.80115	1.2482	.78043	.3	.2	.68455	.93906	1.0649	.72897	.8
.8	.62660	.80402	1.2437	.77934	.2	.3	.68582	.94235	1.0612	.72777	.7
.9	.62796	.80690	1.2393	.77824	.1	.4	.68709	.94565	1.0575	.72657	.6
39.0	.62932	.80978	1.2349	.77715	**51.0**	.5	.68835	.94896	1.0538	.72537	.5
.1	.63068	.81268	1.2305	.77605	.9	.6	.68962	.95229	1.0501	.72417	.4
.2	.63203	.81558	1.2261	.77494	.8	.7	.69088	.95562	1.0464	.72294	.3
.3	.63338	.81849	1.2218	.77384	.7	.8	.69214	.95897	1.0428	.72176	.2
.4	.63473	.82141	1.2174	.77273	.6	.9	.69340	.96232	1.0392	.72055	.1
.5	.63608	.82434	1.2131	.77162	.5	**44.0**	.69466	.96569	1.0355	.71934	**46.0**
.6	.63742	.82727	1.2088	.77051	.4	.1	.69591	.96907	1.0319	.71813	.9
.7	.63877	.83022	1.2045	.76940	.3	.2	.69717	.97246	1.0283	.71691	.8
.8	.64011	.83317	1.2002	.76828	.2	.3	.69842	.97586	1.0247	.71569	.7
.9	.64145	.83613	1.1960	.76717	.1	.4	.69966	.97927	1.0212	.71447	.6
40.0	.64279	.83910	1.1918	.76604	**50.0**	.5	.70091	.98270	1.0176	.71325	.5
.1	.64412	.84208	1.1875	.76492	.9	.6	.70215	.98613	1.0141	.71203	.4
.2	.64546	.84507	1.1833	.76380	.8	.7	.70339	.98958	1.0105	.71080	.3
.3	.64679	.84806	1.1792	.76267	.7	.8	.70463	.99304	1.0070	.70957	.2
.4	.64812	.85107	1.1750	.76154	.6	.9	.70587	.99652	1.0035	.70834	.1
40.5	.64945	.85408	1.1708	.76041	**49.5**	**45.0**	.70711	1.00000	1.0000	.70711	**45.0**
Angle°	cos	cot	tan	sin	Angle°	Angle°	cos	cot	tan	sin	Angle°

List of Abbreviations

alternating current	ac	microsiemens	μS
ampere	A	microvolt	μV
angle whose tangent is	arctan, $\tan^{-1}$	microwatt	μW
average	ave	mile	mi
centimeter	cm	miles per hour	mi/hr, mph
cosine	cos	mil	mil
cotangent	cot	milliampere	mA
coulomb	C	millihenry	mH
cycles per second	cycles/s, cps	millijoule	mJ
decibel	dB	milliohm	mΩ
degree	deg	millisecond	ms
direct current	dc	millisiemens	mS
electromotive force	emf	millivolt	mV
farad	F	milliwatt	mW
foot	ft	minimum	min
gallon	gal	minute	min
gram	gm	nanofarad	nF
henry	H	nanosecond	ns
hertz	Hz	ohm	Ω
hour	hr	picofarad	pF
inch	in.	pound	lb
joule	J	power factor	PF
kilohm	kΩ	radian	rad
kilovolt	kV	reactive kVA	kVAr
kilovolt-ampere	kVA	reactive VA	VAr
kilowatt	kW	revolutions per minute	rev/min, rpm
kilowatt-hour	kWh	revolutions per second	rev/sec, rps
maximum	max	root-mean-square	rms
megawatt	MW	second	s
megohm	MΩ	siemens (formerly mho)	S
meter	m	sine	sin
microampere	μA	tangent	tan
microfarad	μF	volt	V
microhenry	μH	volt-ampere	VA
microsecond	μs	watt	W

ANSWERS TO SELECTED EXERCISES

Exercises 1-1

1. Two point seven five times ten to the third power = $2.75 \times 1000 = 2750$
3. four point nine six times ten to the eighteenth =
   4,960,000,000,000,000,000
5. three point one six times ten to the minus three = $3.16 \times 0.001 = 0.00316$
7. five point nine seven times ten to the minus sixteen =
   0.000000000000000597
9. ten to the third power (or ten cubed) = 1000
11. two cubed = $2 \cdot 2 \cdot 2 = 8$ 13. x squared = $x \cdot x$
15. x to the a power means "take x as a factor a times"

Exercises 1-2

1. $-59°, +75°$ 3. $+\$150, -\$75$ 5. $+15$ yd, -7 yd   7. $-3 < 2$
9. $-2 > -15$ 11. $0 > -15$ 13. $-100 < -40$ 15. $|-5| > 3$
17. $|-18| < |-30|$ 19. $|x| > 4$ 21. $-5C < 3C$ 23. $Q_1 > Q_2$

25. (a)

$-2 \leq x \leq 2$

(number line from -3 to 3)

(b) x may be $-2, -1, 0, 1$, or 2

27. (a)

x (at -1; number line from -2 to 0)

(b) x can only be -1

29. (a)

$-6 \leq x \leq -2$

(number line from -6 to 0)

(b) x may be $-6, -5, -4, -3$, or -2

Exercises 1-3

1. 11   3. 9.3   5. -11 7. -27 9. -6 11. 2   13. -9
15. -1 17. -4 19. 7   21. 22.2   23. 12.44   25. -9.21
27. 14.42   29. -70.33 31. 4 C   33. -18 C

Exercises 1-4

1. 3   3. -5 5. -12 7. -8 9. -8 11. 20   13. -7 15. 27
17. -9 19. 23   21. -5 23. 7   25. -8.09 27. -18.56
29. 70.04   31. 12.5 C   33. -50 C   35. 16 C

Exercises 1-5   1. -8 3. -15 5. -1 7. 7   9. 0   11. 0   13. -2 15. -12
17. -5.2 19. -6.7 21. -27 C   23. -9 C   25. -2.17 C

Exercises 1-6   1. 12   3. -12 5. -10 7. 21   9. 5   11. -3 13. -5 15. -4
17. -3 19. 2   21. 4   23. -5 25. 0   27. 24   29. -36
31. 24   33. -60 35. -36 37. 0   39. -240 41. -4 43. -4
45. -5.625 47. -2.667 49. 2   51. -30 C   53. -94.2 C
55. 15 C   57. -4.75 C

Exercises 2-1A   1. (a) $4.5 \times 10 \,(4.5 \times 10^1)$ (b) 4.5×10^2 (c) 4.5×10^3
(d) 4.5 (or 4.5×10^0) (e) 4.5×10^5 (f) 4.5×10^6
3. (a) 7.7×10^{-2} (b) 7.7×10^{-1} (c) 7.7×10^{-3} (d) 7.7×10^{-4}
(e) 7.7 (or 7.7×10^0) (f) 7.7×10^{-6}
5. (a) $8.13 \,(8.13 \times 10^0)$ (b) 8.13×10^{-3} (c) 8.13×10^6
(d) 8.13×10^8 (e) 8.13×10^{-8} (f) 8.13×10
7. (a) 36.84×10^2 (b) 368.4×10^3 (c) 0.3684×10^4 (d) 36.84×10^{-3}
(e) 0.3684×10^{-3} (f) 3684×10^{-5}
9. (a) 0.901×10^{-3} (b) 9010×10^{-5} (c) 90.1×10^{-2} (d) 90.1×10^2
(e) 901×10^3 (f) 0.901×10^5
11. (a) 312×10^2 (b) 463×10^{-4} (c) 572×10 (d) 695×10^{-6}
(e) 597×10^{-3} (f) 680×10^4
13. (a) 3560 (b) 356,000 (c) 0.0356 (d) 0.00000356
(e) 0.00000000356 (f) 3,560,000,000,000
15. (a) 0.00000912 (b) 9,120,000 (c) 0.0000912
(d) 9,120,000,000,000,000 (e) 912 (f) 0.00912
17. (a) 0.0000888 (b) 888,000 (c) 88,800 (d) 88,800,000
(e) 0.000888 (f) 0.000000888
19. (a) 62,800 (b) 0.0628 (c) 628,000 (d) 0.0628 (e) 0.00628
(f) 62.8
21. (a) 178 (b) 1.78 (c) 178,000 (d) 0.00178 (e) 0.00000178
(f) 178
23. (a) 0.0000603 (b) 7,500,000 (c) 673,000 (d) 205,000 (e) 0.0316
(f) 0.00761

Exercises 2-1B   25. (a) 4.57×10^3 (b) 0.0457×10^6 (c) 457×10^3 (d) 45.7×10^{-3}
(e) 4570×10^{-6} (f) 457×10^{-9}
27. (a) $810,000 \times 10^{-3}$ (b) 81×10^3 (c) 0.0081×10^6 (d) 8.0×10^{-3}
(e) 8100×10^{-6} (f) $81,000 \times 10^{-9}$
29. (a) 0.673×10^6 (b) 0.673×10^6 (c) 673×10^3 (d) 673×10^{-6}
(e) 67.3×10^{-9} (f) 0.673×10^{-3}
31. (a) 7.03×10^3 (b) 0.0703×10^6 (c) 70.3×10^{-3} (d) 7030×10^{-6}
(e) 70.3×10^{-9} (f) 0.0703×10^{-9}
33. (a) 7.55×10^{-3} (b) 7550×10^{-6} (c) 0.755×10^6 (d) 755×10^3
35. (a) 0.054×10^5 (b) 0.7404×10^6 (c) 630.5×10^{-6}
(d) 0.0472×10^{-3}

Exercises 2-2

1. (a) 65 mA (b) 256 ms (c) 375 mH (d) 37 mS (e) 1568 mA
(f) 0.739 ms (g) 1037 mV (h) 4.9 mV (i) 0.00425 mV
(j) 32,500 mA

3. (a) 5.6 kΩ (b) 0.75 kV (c) 12 kHz (d) 0.68 kΩ (e) 330 kV
(f) 1010 kHz (g) 1.6 km (h) 0.072 kΩ (i) 2.2 kΩ (j) 0.98 kHz

5. (a) 0.455 MHz (b) 2.7 MΩ (c) 0.77 MV (d) 0.68 MΩ (e) 10.7 MHz
(f) 0.068 MΩ (g) 48 MHz (h) 0.15 MΩ (i) 800 MHz (j) 0.41 MΩ

7. (a) 5.6 μF (b) 458 μS (c) 0.57 μs (d) 2755 μH (e) 15 μF
(f) 985 μH (g) 0.00015 μF (h) 87.7 μS (i) 78 μs (j) 32,780 μs

9. (a) 27 GHz (b) 150 pF (c) 450 am (d) 500 PHz (e) 2.5 nW
(f) 15 THz (g) 0.8 EHz (h) 55 pF (i) 0.18 am (j) 70 THz

11. (a) 1.5×10^{-2} A, 0.015 A (b) 3.7×10^5 Ω, 370,000 Ω
(c) 5.3×10^{-5} s, 0.000053 s (d) 7.1×10^7 Ω, 71,000,000 Ω
(e) 9.6×10^{-7} s, 0.00000096 s (f) 8.4×10^{11} Hz, 840,000,000,000 Hz
(g) 6.75×10^{-4} H, 0.000675 H (h) 4.8×10^5 Ω, 480,000 Ω
(i) 29 Ω (j) 1.6×10^{-16} W

13. (a) 10,700,000 Hz (b) 1.5×10^{-8} F (c) 41,000 Ω (d) 720,000 Ω
(e) 0.0037 S (f) 3.6×10^{-9} s (g) 5×10^{-15} W (h) 8.2×10^{-6} s
(i) 640,000 Hz (j) 1.5×10^{18} Hz

15. (a) 1.5 mA (b) 0.455 MHz (c) 15,000 pF (d) 0.0036 μs
(e) 480 kΩ (f) 50,000 MHz (g) 0.675 mH (h) 0.37 MΩ (i) 160 aW
(j) 0.053 mS

17. (a) 4500 GHz (b) 0.000027 μF (c) 3300 Ω (d) 0.68 MΩ
(e) 56,000 Ω (f) 0.25 mH (g) 2700 ps (h) 0.0027 μs
(i) 0.0043 fW (j) 5500 μA

19. (a) 0.2 A (b) 0.2 mV (c) 0.48 MΩ (d) 1.2×10^{-3} μF (e) 12 μs
(f) 25 GHz (g) 4.5 μA (h) 6500 Ω (i) 3.7 MΩ (j) 1.5 μA

Exercises 2-3

1. 10^7 3. 10^{-3} 5. 10^{-10} 7. 2^5 9. x^5 11. 10^3 13. 10^{x+y}
15. 10^3 17. 10^{-7} 19. $10 \, (10^1)$ 21. a^2 23. a^{-7} 25. I
27. 1 29. 1 31. -5 33. -6 35. $8 \times 10^5 = 800,000$
37. $1.2 \times 10^5 = 120,000$ 39. $6 \times 10^3 = 6000$ 41. $3 \times 10^4 = 30,000$
43. $3 \times 10^6 = 3,000,000$ 45. $2 \times 10^{10} = 20,000,000,000$
47. $1.15028 \times 10^{-1} = 0.115028$ 49. $2.9989 \times 10^5 = 299,890$
51. $7.41 \times 10^3 = 7410$

Exercises 2-4

1. (a) 5.05×10^3 (b) -3.22×10^{-3} (c) 5.55×10^4 (d) 2.45×10^{-2}
3. (a) 109 kΩ (b) 2.045 mA (c) 1.0723 MΩ (d) -29.5 mV
5. (a) 1.5×10^4 (b) -2.05×10^{-3} (c) 5.8×10^6 (d) 9.0×10^{-6}
7. (a) 6 kΩ (b) 4.0 mA (c) 1.8 mV (d) 750 mS

Exercises 2-5

1. 12 V 3. 117 V 5. 100.8 V 7. 19.6 V 9. 62.56 V
11. 5.5 V 13. 7.134 V 15. 12.48 V 17. 3.25 V 19. 1.65 V
21. 3 A 23. 2.9 A 25. 1.528 A 27. 17.92 mA 29. 2.788 mA
31. 6.944 μA 33. 395.6 nA 35. 91.67 nA 37. 848.2 pA

39. 3.573 pA   41. 4.667 Ω 43. 42.59 Ω 45. 5.714 Ω
47. 5.139 Ω 49. 275 Ω 51. 14.69 Ω 53. 1061 Ω 55. 4375 Ω
57. 48.95 Ω 59. 16.91 Ω 61. 24 W   63. 1260 W   65. 22.83 mW
67. 51.79 mW   69. 825 kW   71. 0.9576 mW   73. 144.1 nW
75. 19.44 kW   77. 33 μW   79. 50.78 W

Exercises 2-6

1. 10^{12} 3. 10^{-10} 5. 10^6 7. x^{ab} 9. x^{4ab} 11. $x^3 y^3$
13. $a^6 b^{18}$ 15. $a^2 \times 10^6$ 17. 1.225×10^{-7} 19. 2.256×10^7
21. 16 W   23. 10.63 kW   25. 1.287 MW   27. 2.419 W
29. 68.6 mW   31. 5.756 W   33. 1.375 mW   35. 620.7 μW
37. 3.285 μW   39. 2.873 μW   41. 36 W   43. 25.23 W
45. 420.1 W   47. 15.41 mW   49. 25.65 mW   51. 1.736 μW
53. 14.24 nW   55. 7.975 MW   57. 402.9 fW   59. 9.577 aW

Exercises 2-7

1. 10^3 3. 10^{-3} 5. 2×10^2 7. 2×10^3 9. 10^3 11. 6.758
13. 7.53×10^2 15. 0.1948   17. 888.6   19. 209   21. 2401
23. 169   25. 3.358   27. 3   29. 0.1429   31. 109.5 V
33. 201.2 V   35. 47.49 V   37. 63.19 V   39. 821.6 V
41. 70.71 mA   43. 791.6 μA   45. 205.3 μA   47. 1.18 μA
49. 74.26 μA   51. 1.049 V   53. 25.98 V   55. 25 V   57. 424.3 V
59. 547.7 V   61. 0.6455 A   63. 0.1118 A   65. 3.536 mA
67. 2.113 mA   69. 741.2 μA   71. 0.7416 V   73. 18.37 V
75. 17.68 V   77. 300 V   79. 387.3 V   81. 0.4564 A
83. 79.06 mA   85. 2.5 mA   87. 1.494 mA   89. 524.1 μA

Exercises 2-8

1. $3.6 \times 10^6, 3.458 \times 10^6$ 3. 1.8, 1.716   5. 5, 5.256
7. $4 \times 10^{-5}, 3.6904 \times 10^{-5}$ 9. $1.6 \times 10^7, 1.521 \times 10^7$
11. $6.4 \times 10^{-7}, 6.084 \times 10^{-7}$ 13. 20, 20.32   15. 0.9, 0.9033
17. 80 V   19. 7 μA   21. 3.5 Ω 23. 120 W

Exercises 3-1

1. 2   3. 4   5. -4   7. -2   9. -3   11. 4   13. 10   15. 10
17. -8   19. -28   21. 8   23. 35   25. -3   27. 12.6   29. b/a
31. V/I 33. V/I 35. Q/I 37. Q/I 39. P/V
41. $4I = 960, I = 240$ mA   43. $0.15x = 277.50, x = 1850$ mi
45. $I_T/4 = 5, I_T = 20$ mA   47. $\frac{1}{6}x = 105, x = \$630$

Exercises 3-2

1.

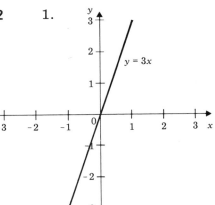

3.

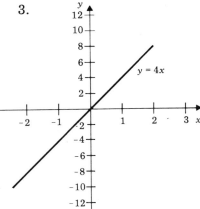

5.

7.

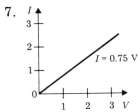

9.

11.

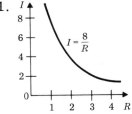

13. (a)

(b) The graph is linear.

15. (a)

(b) The graph is linear.

17. (a)

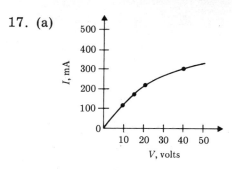

(b) The graph is nonlinear.

Exercises 3-3

1. $3a$ and $-2a$ are like terms.   3. $5i$ is like $-12i$, $-2v$ is like $7v$.
5. $a^2 bc^3$ and $-2a^2 bc^3$ are like terms.   7. $3i^3$ and $-5i^3$ are like terms.
9. $4V_1^2 G_1$ and $3V_1^2 G_1$ are like terms.   11. (a) $9ax$ (b) $-6R_1 R_2 R_3$
    (c) $-4V^2 I$ (d) $2I_1 R_1 - 4I_2 R_2$ (e) $3fL_1 - 7fL_2$
13. (a) $11d^2 ef$ (b) $-2IR$ (c) $-16R_1 R_2 R_3$ (d) $8I^2 R - 9I_1^2 R_1$
    (e) $-5V'I' + 14VI$ 15. $4x - 2y$ 17. $6xy - 16xz$
19. $27VI - 4I^2 R$

Exercises 3-4

1. $12x$ 3. $-20V$ 5. $6x$ 7. $-20V^2 IR$ 9. $16v^2$ 11. $9v^4$
13. $3x + 3y$ 15. $12I + 15i$ 17. $5VG_1 + 10VG_2 - 25VG_3$
19. $-2IR_1 + 4IR_2 + 14IR_3$ 21. $4I - 3$ 23. $-2i - 6$ 25. $-2V + 6v$
27. $V_1 + V_2 + V_3 - 5$ 29. $IR - 7iR - 3Ir + 8ir$ 31. $x^2 + 3x + 2$
33. $x^2 + 2x - 8$ 35. $IR + Ir + iR + ir$
37. $IR_1 + IR_2 + IR_3 + iR_1 + iR_2 + iR_3$ 39. $3x^2 - 3x - 18$

Exercises 3-5

1. 25 3. $5x + 10$ 5. $-2i - 6$ 7. $2V_1 + V_2 + V_3 - 2$
9. $-26R_1 + 7R_2$ 11. $5VI - 2V_1 I_1 - 2V_2 I_2$ 13. $2R_2 - 4$
15. $X_L - X_C + 18$ 17. (a) $40D_P + 40D_Q + 880$ (b) 1600 ns or 1.6 μs
19. $r + 4(R_1 + R_2 + R_3)$ 21. $6I_1 R_1 - 7[iR - 3(V_1 - 2v)]$
23. $4R - (6r - 7R')$ 25. $X_1 - (R_2 - X_2 + X_3) - Z_1$

Exercises 3-6

1. $3x$ 3. $-2y$ 5. $-4xy^2$ 7. $9b$ 9. $-4V$ 11. $2ab + ac$
13. $-3ab - bc$ 15. $-1 - 7b$ 17. $R_1 + R_2 + R_3 + R_4$
19. $2abcd + 6a^{-1} b^{-1} cd^2$, or $2abcd + 6cd^2/ab$
21. $C_3^{-1} + C_1^{-1} + C_2^{-1}$ or $1/C_3 + 1/C_1 + 1/C_2$ 23. $-4I + 3R - R^2$
25. $X_1^2 + X_2^2 + X_3^2$

Exercises 3-7

1. (a) $2 \cdot 2 \cdot 3$ (b) $3 \cdot 5$ (c) $2 \cdot 3 \cdot 3$ (d) $3 \cdot 3 \cdot 3$
3. (a) $2 \cdot 7$ (b) $3 \cdot 7$ (c) $2 \cdot 2 \cdot 7$ (d) $2 \cdot 3 \cdot 7$
5. $2 \cdot x \cdot y \cdot z$ 7. $2 \cdot 2 \cdot 2 \cdot I \cdot I \cdot R$ 9. $a(x + y)$ 11. $5I(R_1 + 3R_2)$
13. $B_0 (1 + R)$ 15. $I_L (r - R)$ 17. $R_1 (R_2 + R_3)$

19. $V(G_1 + G_2 + G_3 + G_4)$ 21. $I(V_1 + V_2 + V_3 + V_4)$
23. $10^3(3 + 5 + 2 + 10) = 20 \times 10^3$ 25. $m(a + b + c - d)$
27. $V - I(R_1 + R_2 + R_3)$ 29. $I_T - V(G_1 + G_2)$

Exercises 3-8

1. $c^2 + 2cd + d^2$ 3. $v^2 + 2iv + i^2$ 5. $x^2 - 4x + 4$ 7. $R^2 - 6R + 9$
9. $4x^2 + 4xy + y^2$ 11. $16v^2 - 8iv + i^2$ 13. $i^2 + 10ir + 25r^2$
15. $Z^2 - 16XZ + 64X^2$ 17. $4a^2 + 12ab + 9b^2$ 19. $25v^2 - 40iv + 16i^2$
21. $X^4 - 12X^2 + 36$ 23. $X^2 - 4f^2L^2X + 4f^4L^4$
25. $144 - 24v^2g^2 + v^4g^4$ 27. $\frac{1}{4}a^2 + \frac{2}{3}ab + \frac{4}{9}b^2$ 29. $\frac{9}{49} - \frac{6}{7}v + v^2$
31. $4v^2 - 2v + 0.25$ 33. $0.04i^2 - 0.12iv + 0.09v^2$ 35. 484
37. 1024   39. 1.44   41. 10,201   43. 4761   45. 6889

Exercises 3-9

1. $(x + 1)^2$ 3. $(v - 2)^2$ 5. $(r - 4)^2$ 7. $(2V + 1)^2$ 9. $(4gv + i)^2$
11. $(9 - 5g)^2$ 13. $[(R + X) - 2)]^2$ 15. $(2v - 6)^2$ 17. $(hi + 2)^2$
19. $(9s - 11)^2$ 21. $6x$ 23. i^2 25. $12vg$ 27. $6irv$ 29. $64Z^2$
31. $96v^2g$ 33. $\frac{8}{3}xy$ 35. $0.4ir$

Exercises 3-10

1. $x^2 - 1$ 3. $9 - i^2$ 5. $h^2i^2 - 100$ 7. $9c^2 - 9g^2$ 9. $9v^2 - 25i^2$
11. $4\beta^2i^2 - 49\alpha^2I^2$ 13. $\frac{4}{9}i^2 - \frac{9}{16}r^2$ 15. $0.25\omega^2L^2 - 0.09\omega^2C^2$
17. $4a^8x^{10} - 64b^6g^{14}$ 19. $36n^6p^4 - 1$ 21. 3584   23. 9984
25. 22,484   27. 159,964   29. 2496 V

Exercises 3-11

1. $(x - 3)(x + 3)$ 3. $(i - 4)(i + 4)$ 5. $(2R - 7)(2R + 7)$
7. $(11i^2r - 10)(11i^2r + 10)$ 9. $(1 - 7Z)(1 + 7Z)$ 11. $(\frac{3}{5}v - 1)(\frac{3}{5}v + 1)$
13. $(0.5x - 0.4)(0.5x + 0.4)$
15. $(I^2R - V^2I^2)(I^2R + V^2I^2)$ also $I^4(R - V^2)(R + V^2)$
17. $\left(\frac{1}{R} - \frac{1}{r}\right)\left(\frac{1}{R} + \frac{1}{r}\right)$ 19. $[(R + X)^2 - Z][(R + X)^2 + Z]$

Exercises 3-12

1. $x^2 + 7x + 10$ 3. $v^2 + 11v + 28$ 5. $r^2 - 5r + 6$ 7. $g^2 - 7g + 6$
9. $X^2 - 4X - 21$ 11. $h^2i^2 - 5hi - 14$ 13. $10v^2 - 26vV + 12V^2$
15. $12h^2r^2 + 32hrR + 16R^2$ 17. $v^2 + 2iv - 15i^2$
19. $v^2/6 - vV/6 + V^2/24$

Exercises 3-13

1. $(x + 2)(x + 1)$ 3. $(i + 3)(i + 2)$ 5. $(R - 2)(R - 3)$
7. $(Z - 4)(Z - 5)$ 9. $(I - 2)(I + 1)$ 11. $(i + 5)(i - 2)$
13. $(2s + 1)(s + 2)$ 15. $(2p + 3)(2p + 1)$ 17. $(3f - 5)(3f - 1)$
19. $(2r - 3)(2r - 1)$ 21. $(4i + 3)(2i - 3)$ 23. $(7hi - 2)(hi + 3)$
25. $(x + 7y)(x + 6y)$ 27. $(3i - 7r)(i - r)$ 29. $(5\beta + 2I)(\beta - 5I)$
31. $(3R^2 + 2X^2)(2R^2 - 5X^2)$

Exercises 3-14

1. $i(R - r)$ 3. $R(2I - 3i)(2I + 3i)$ 5. $4(v + 3)^2$ 7. $5(v - 4)^2$
9. $8(v + 2)(v + 8)$ 11. $10(r - 3)(r - 1)$ 13. $2(v - 3)(v + 2)$
15. $3(2x + 3)^2$ 17. $7(2X - 5)^2$ 19. $(3R - 2)(2R + 3)$
21. $(5v + 1)(4v + 3)$ 23. $(3v + 2i)(2v + 3i)$ 25. $(5v + 7g)(5v - 3g)$
27. $(v^2 + 3)(v^2 + 4)$ 29. $(7v^2 + 9i^2)(2v^2 + 5i^2)$ 31. $v(2v - 3i)(2v + 3i)$
33. $X(X - 2)(X - 4)$ 35. $R^3(4R - 7r)(3R + 5r)$ 37. $(x + y)(x + z)$
39. $(v + i)(V + I)$ 41. $(a + b)(c + d)$ 43. $(x + \frac{2}{3})^2$ 45. $(R + \frac{2}{3})(R - \frac{1}{2})$
47. $\left(3I - \dfrac{V^2}{R}\right)\left(3I + \dfrac{V^2}{R}\right)$ 49. $(-3v + 2i)(v - \frac{2}{3}i)$

Exercises 3-15

1. $a + 3$ 3. $R - 2$ 5. $2i + 3 + 2/(3i - 2)$ 7. $3x + 7$
9. $R^2 + 10R + 25$ 11. $4I^3 - 2I + 2$ 13. $V^2I^2 + 3VI + 9$ 15. $\omega L + 1$
17. $12I^4(R + r)^2 + 3I^2(R + r) + 2$ 19. $e - v$ 21. $R - 3X$ 23. $Av + V$
25. $R^2 + 2X^2$ 27. $A^2v^2 - 2Avir + 4i^2r^2$

Exercises 4-1

1. 2 3. 2.2 5. 6 7. 7.5 9. -10 11. -12.4 13. -14
15. 8.4 17. $\frac{1}{4}$ 19. $-\frac{13}{10}$ 21. (a) $V_x + 3 = 12$ (b) 9 V
23. (a) $h_R + 8 = 44$ (b) 36 hr 25. (a) $R_T - 15 = 39$ (b) 54 kΩ
27. (a) $V_{NL} - 2.35 = 10.85$ (b) 13.2 V 29. (a) $x + 18 = 23$ (b) 5 m

Exercises 4-2

1. 2 3. 2 5. 4 7. 2 9. 4 11. 3 13. 1 15. -0.5
17. 5 19. 2 21. 12 23. 9 25. $1/(c + d)$ 27. -2 29. 7

Exercises 4-3

1. 2 3. $R_T - R_1 - R_3$ 5. $V_C + IR_L$ 7. b/a 9. V/I 11. $X_L/2\pi f$
13. P/I^2 15. $V/(R + r)$ 17. ab 19. IR 21. βI_b 23. 7
25. V/I 27. I_c/β 29. 1 31. $(V_g - V_t)/r$ 33. $V_T/(R_1 + R_2 + R_3)$
35. $(V_{CC} - V_C)/R_L$ 37. $\sqrt{P/R}$

Exercises 4-4

1. $R_3 = R_T - (R_1 + R_2 + R_4)$ 3. $I_c = I_e - I_b$ 5. $R_T = R_1 R_2/(R_1 + R_2)$
7. $v = L \, \Delta i/\Delta t$ 9. $1/R_T = 1/R_1 + 1/R_2 + 1/R_3 + 1/R_4$ 11. 13 Ω, 39 Ω
13. 0.7 V, 2.94 V 15. 33 kΩ, 48 kΩ 17. 28 mA, 3 mA
19. 544.7 μA 21. 592.3 kHz 23. 4.286 Ω, 10.714 Ω
25. 3 A, 10.5 A 27. 20 resistors, 10 capacitors
29. two 22-kΩ resistors, three 15-kΩ resistors

Exercises 4-5

1. $P_x = I_x V_x$ 3. $P_T = P_1 + P_2 + P_3 + P_4 + P_5$ 5. $P_4 = P_T - (P_1 + P_2 + P_3)$
7. $V_x^2 = P_x R_x$, $V_x = \sqrt{P_x R_x}$ 9. $I = \sqrt{\dfrac{P_1 + P_2 + P_3}{R_T}}$

11. (a)

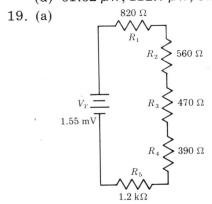

(b) 0.3333 A   (c) 3.333 V, 6.666 V   (d) 1.111 W, 2.222 W

13. (a)

(b) 138.6 μA   (c) 249.5 mV, 374.2 mV, 126.1 mV
(d) 34.58 μW, 51.87 μW, 17.48 μW

15. (a)

(b) 3.074 μA   (c) 209 μV, 172.1 μV, 368.9 μV
(d) 642.6 pW, 529.2 pW, 1134 pW

17. (a)

(b) 86.69 μA   (c) 0.7109 V, 1.30 V, 7.108 V, 3.381 V
(d) 61.62 μW, 112.7 μW, 616.2 μW, 293.1 μW

19. (a)

(b) 450.6 nA (c) 369.5 μV, 252.3 μV, 211.8 μV, 175.7 μV, 540.7 μV
(d) 166.5 pW, 113.7 pW, 95.43 pW, 79.18 pW, 243.6 pW
21. 6 W 23. 5.65 V 25. 1.5 mA 27. (a) 250 μA (b) 4.7 kΩ
29. 7.2 kΩ

Exercises 5-2

1. 9/15 3. 5/16 5. 12 V/3 V 7. R_1/R_T

9. $\dfrac{v^2 - 16}{v^2 + 4v + 4}$ or $(v^2 - 16)/(v^2 + 4v + 4)$ 11. $5 \div 11$ 13. $5vi \div 7gR$

15. $(v + V) \div (i + I)$ 17. 12/20 19. $(R^2 - 4)/(R^2 + R - 6)$ 21. 4/7

23. $(v + 2)/(i - 2)$ 25. 15/39 27. $4v^2/6i^2$ 29. $I^2 rR/(I^2 r + I^2 R)$

31. $(iR + 3R)/(iR + 5R)$ 33. $(gv^2 + 2gv)/(gv^2 - 2gv)$ 35. $\dfrac{\dfrac{2v}{R} + \dfrac{1}{R}}{\dfrac{V}{R} - \dfrac{v}{R}}$

Exercises 5-3

1. $\frac{2}{3}$ 3. $v/5r$ 5. $3x/4y$ 7. $3py^3$ 9. $2ab^{-2}$ or $2a/b^2$ 11. $-\frac{3}{7}$
13. 7 15. $-(i^2/r^2)$ 17. $3a^{-1}c^{-1}xy^{-1}$ or $3x/acy$
19. $\frac{11}{4}p^{-1}r^{-1}st$ or $11st/4pr$ 21. $x(x - 2)$ or $x^2 - 2x$ 23. $V/3$
25. $5(i + I)/g(i - I)$ 27. $\frac{5}{3}$ 29. $R/2r$ 31. $\frac{1}{7}$ 33. R_1

Exercises 5-4

1. $\frac{1}{3}$ 3. $\frac{16}{105}$ 5. 25 7. ax/by 9. $R_2 R_5/R_4 R_6$ 11. $3/4iV$

13. $Rr/3(r + R)$ 15. $a + 3$ 17. $iv - iV$ 19. $\dfrac{V + v}{R_1 - R_2}$ 21. $\frac{3}{4}$

23. $-\frac{1}{3}$ 25. 20 27. $3r/-5$ 29. V 31. $\dfrac{R_3 + R_4}{R_3 R_4 (R_1 + R_2)}$

33. $\dfrac{VR}{R + r}$ 35. $\dfrac{r}{R + r}$

Exercises 5-5

1. $\frac{19}{15}$ or $1\frac{4}{15}$ 3. $\frac{3}{4}$ 5. $\frac{34}{15}$ or $2\frac{4}{15}$ 7. $\frac{61}{24}$ or $2\frac{13}{24}$ 9. $\dfrac{4b - 3a}{ab}$

11. $\dfrac{ab - 2a^2}{4b^2}$ 13. $\dfrac{4a + 5b - 3c}{abc}$ 15. $\dfrac{R_1 + R_2}{R_1 R_2}$

17. $\dfrac{3xyz + 4x^2 + 5y^2 + 3z}{xyz}$ 19. $\dfrac{4 + 2(i + I)}{iR + IR}$ 21. $\dfrac{19 - v}{6(v - 3)}$ 23. $\dfrac{R^2 + X^2}{R^2 - X^2}$

25. $\dfrac{a - b}{a + b}$ 27. $\dfrac{x}{x - y}$ 29. $\dfrac{a + 3b}{a - 3b}$ 31. $\dfrac{5i + 1}{(i + 1)(i - 1)(i - 1)}$

Exercises 5-6

1. $\dfrac{5}{-7}$ or $-\dfrac{5}{7}$ 3. $\dfrac{y - x}{b - a}$ or $-\dfrac{y - x}{a - b}$ 5. $\dfrac{5V^2}{4R^2}$ or $\dfrac{-5V^2}{-4R^2}$

7. $\dfrac{V^2 + v^2}{r - R}$ or $\dfrac{-V^2 - v^2}{R - r}$ 9. -1 11. 1 13. $\dfrac{7V}{R - r}$

15. $\dfrac{7x + 2}{(x - 2)(x - 2)}$ 17. $-\dfrac{R - 4}{R + 3}$ 19. $-\dfrac{2V}{R_1 + R_2}$

Exercises 5-7 1. $\frac{31}{8}$ 3. $\frac{4+x}{2}$ 5. $\frac{R_1(I+1)+R_2(I-3)}{R_1+R_2}$ 7. $\frac{8a^2+ab+7b^2}{2a+2b}$ 9. 1

11. $\frac{r-v^2}{(v+1)^2}$ 13. V 15. $3\frac{2}{5}$ 17. $\frac{V}{R}-I$ 19. $h_{12}+R_L+\frac{30V_C}{I_b}$

21. $1+\frac{v^2}{V^2}$ 23. $v-6+\frac{7}{v}$ 25. False, $\frac{R+3}{3}$ 27. False, $\frac{I^2-4}{I^2-8}$

29. False, $\frac{V+v}{R}\cdot\frac{R}{V+v}=1$ 31. False, $1\div\left(\frac{1}{R_1}+\frac{1}{R_2}\right)=\frac{R_1R_2}{R_1+R_2}$

33. False, $\frac{IR+V}{R}=I+\frac{V}{R}$ 35. True

Exercises 5-8 1. $\frac{2}{3}$ 3. dxy/a 5. $-\frac{34}{31}$ 7. $\frac{6}{19}$ 9. x/b 11. $\frac{RrV}{R+r}$

13. $\frac{i(iR-V)}{R^2}$ 15. $v/2$ 17. $\frac{2i}{v+i}$ 19. $\frac{R_pR_m}{R_pR_m+R_sR_p+R_sR_m}$

21. $\frac{r_c+R_L}{r_cr_p+R_Lr_p+r_cR_L}$ 23. $\frac{rV+vR+vr}{rR}$ 25. $\frac{R-v}{R+v}$

Exercises 6-1 1. 1   3. 1   5. 12   7. $\frac{5}{2}$ 9. -3 11. 5   13. 2   15. 2

Exercises 6-2 1. 1   3. -3 5. -3 7. 1   9. $i=a$ is an extraneous root

Exercises 6-3 1. $IR,\ V/I$ 3. $V/R,\ IR$ 5. $1/f$ 7. $1/2\pi CX_C,\ 1/2\pi fX_C$

9. $\frac{I_2}{I_1}R_2,\ \frac{I_1}{I_2}R_1,\ \frac{R_2}{R_1}I_2$ 11. $\frac{R_TR_2}{R_2-R_T},\ \frac{R_TR_1}{R_1-R_T}$ 13. $\frac{R_1R_2}{R_1+R_2},\ \frac{R_TR_2}{R_2-R_T}$

15. $\frac{\beta}{1+\beta}$ 17. $V_L+I_Lr_i,\ (V_{NL}-V_L)/r_i$ 19. $R_L\left(\frac{\mu}{A}-1\right),\ \frac{A}{\mu-A}r_p$

21. 360 ft, $\theta/360t$ 23. $1/2\pi fQR$ 25. $\frac{1}{2}\frac{\omega_0}{\omega_0-\omega_1},\ 2Q\omega_1/(2Q-1)$

Exercises 6-4 1. $\frac{1}{R_T}=\frac{1}{R_1}+\frac{1}{R_2},\ \frac{1}{R_T}=\frac{R_2}{R_1R_2}+\frac{R_1}{R_1R_2},\ R_T=\frac{R_1R_2}{R_1+R_2}$

3. $R_T=\dfrac{1}{\dfrac{1}{R}+\dfrac{1}{R}+\dfrac{1}{R}+\cdots}=\dfrac{1}{\dfrac{n}{R}}=\dfrac{R}{n}$

5. $R_T=\dfrac{1}{\dfrac{1}{R_1}+\dfrac{1}{R_2}+\dfrac{1}{R_3}+\cdots}=\dfrac{R_1}{\dfrac{R_1}{R_1}+\dfrac{R_1}{R_2}+\dfrac{R_1}{R_3}+\cdots}=\dfrac{R_1}{1+\dfrac{R_1}{R_2}+\dfrac{R_1}{R_3}+\cdots}$

7. $V_T=I_1R_1,\ R_2=\dfrac{I_1R_1}{I_T-I_1},\ I_2=I_T-I_1$

9. $\dfrac{1}{R_T}=\dfrac{1}{R_1}+\dfrac{1}{R_2}+\dfrac{1}{R_3}+\cdots,\ G_T=G_1+G_2+G_3+\cdots$

11. $I_1 = \dfrac{I_T R_T}{R_1} = I_T \cdot \dfrac{R_1 R_2}{R_1 + R_2} \cdot \dfrac{1}{R_1} = \dfrac{R_2}{R_1 + R_2} I_T$

13. (a)

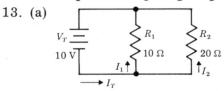

(b) 1 A, 0.5 A, 1.5 A   (c) 10 W, 5 W

15. (a)

(b) 416.7 μA, 277.8 μA, 824.2 μA, 1.519 mA
(c) 312.5 μW, 208.3 μW, 618.1 μW

17. (a)

(b) 11.03 μA, 13.39 μA, 6.25 μA, 30.67μA
(c) 8.273 nW, 10.04 nW, 4.688 nW

19. (a)

(b) 1.524 mA, 833.3 μA, 152.4 μA, 320.5 μA, 2.83 mA
(c) 19.05 mW, 10.42 mW, 1.905 mW, 4.006 mW

21. (a)

(b) 1.89 μA, 2.768 μA, 3.298 μA, 3.974 μA, 1.292 μA, 13.22 μA
(c) 2.93 nW, 4.29 nW, 5.112 nW, 6.16 nW, 2.002 nW

23. 14.43 Ω 25. 3.656 kΩ 27. 775.7 Ω 29. 138.2 kΩ

31. 8.3 mA   33. 8.998 V   35. (a) 1.463 V
(b) 81.28 μA, 21.51 μA, 140.3 μA,   (c) 10.43 kΩ

37. (a) Add another resistor in parallel.

$R_x = I_T R_T/(4.594 - 2.75)(10^{-3}) = 6.8$ kΩ (b) 2.73 kΩ

39. 120 mA

Exercises 6-5   1. $21.65\ \Omega$ 3. 2.027 W   5. $21.87\ \Omega$ 7. $154.3\ \mu$V   9. 2.343 mA
11. 7.343 V   13. $56.25\ \mu$W   15. 10.17 kΩ 17. 5.22 V
19. 0.4299 A

Exercises 6-6   1. $1/4\pi^2 f_r^2 L$ 3. $\dfrac{N_1^2}{N_2^2} Z_2$ 5. $CQ_s^2 R^2$

7. $n^2 r^4/(9r + 10l)$, $(n^2 r^4 - 9rL)/10L$ 9. $i^2(r + R)$, $p/i^2 - R$ 11. L/Z_0^2
13. $mV_z^2/2V_A$, $2eV_A/V_z^2$

Exercises 7-1   1. 25 3. 8 5. 5 7. 40 9. 15 11. -120 13. -150
15. 22.5 17. 3.846 19. 4 21. 15 mi/gal
23. 30 students/instructor   25. 0.06 A/V or 60 mS   27. 60
29. $133,333\ \Omega/$V   31. (a) 25 (b) 450 (c) 4×10^6 (d) 0.02
33. (a) 2500 (b) $25,000$ (c) 0.95 (d) 2500
35. (a) 300 (b) 100 (c) 400 (d) 80
37. (a) 8.333 mS (b) 5.556 mS (c) 3.75 mS (d) 21.43 mS
39. (a) 2 (b) 0.05 (c) 20 (d) 0.1714 41. voltage, resistance, V/Ω

43. current, voltage, A/V   45. $I_1 = \dfrac{R_2}{R_1 + R_2} I_T$

47. (a) 2.991 (b) 1.227 49. 250 cards/min

Exercises 7-2   1. 6 3. 800 5. 5.726 7. $\dfrac{R_1}{R_2} R_3$ 9. $\dfrac{I_2}{I_1} R_2$

11. $\dfrac{I_1' R_1}{I_1 R_2} = \dfrac{I_2' R_3}{I_2 R_4}$, I_1' and I_1 cancel because $I_1' = I_1$; I_2' and I_2 cancel because

$I_2' = I_2$. Thus, $\dfrac{R_1}{R_2} = \dfrac{R_3}{R_4}$
13. $27/3 = 9$ 15. $564.1\ \Omega$

Exercises 7-3   1. $I = kGV$ 3. $I_C = k\beta I_B$ 5. $\lambda = k(1/f)$ 7. $G = k(1/R)$
9. $C = k\,A/d$ 11. $L = k\,N^2 A/l$ 13. $30\ \mu$A   15. 6.25 mA
17. 468.75 m   19. 555.6 nS   21. 166.7 pF   23. 142.2 mH

Exercises 7-4   1. 5 V   3. 12 V   5. 0.4649 mV   7. 1.075 mV   9. 6.46 V
11. 158 mV   13. 1.431 V

Exercises 7-5   1. 3.569 mA   3. 131.5 kΩ 5. 0.3021 A, 0.1679 A

7. $1.324\ \mu$A, $0.5258\ \mu$A   9. $R_{sh} = \dfrac{I_m}{I_T - I_m} R_m$

Exercises 7-6
1. 25%, 33%, 50%   3. 15%, 30%, 90%   5. 200%, 500%, 1000%
7. 0.15%, 0.025%, 1.9%   9. 50%, 75%, 33.33%
11. 166.7%, 750%, 1375%   13. 0.25, 0.33, 0.5   15. 4.5, 10, 25
17. $7.57   19. $120.45   21. 0.0782 mA   23. 44%   25. 8.083%
27. 0.016 V, 1.026%   29. 3.333 V   31. 98.1 MHz   33. (a) 0.005%
 (b) 5070 Ω 35. (a) 0.07% (b) 17.22 kΩ

Exercises 7-7
1. $37.62   3. 517 k$\Omega$, 423 k$\Omega$ 5. No, 34.2 kΩ 7. 120 V
9. $296.68   11. 3.397%   13. 0.002193%   15. 6 V   17. 46.84%
19. 85.71%   21. 30.5 kW   23. (a) 0.0767 V (b) 0.4267V (c) 17.98%
25. (a) 2.522 V (b) 37.522 V (c) 6.722%   27. (a) 0.1741 mA
 (b) 1.828 mA (c) 9.523%   29. (a) 16.25 μA (b) 113.8 μA
 (c) 14.28%

Exercises 8-2

1. [circuit: I_1 arrow, resistor labeled 5 A]

3. [circuit: a, I_{ab}, b, resistor labeled 8 mA]

5. [circuit: d, I_{de}, e, resistor, I_1]

7. [circuit: arrow, 50 mA, resistor]

9. [circuit: $I_m = -40$ mA, resistor]

11. [circuit: a, $-$, I_1, $+$, b, resistor labeled V_1]

13.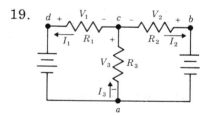

15. [circuit diagram with I_2, q, V_b, R_b, I_1, V_a, p, R_a, V_c, I_3, R_c, r, V_d, R_d, I_4, s]

17. [circuit diagram with V_1, R_1, V_2, R_2, V_3, R_3, R_4, V_4, R_5, V_5, V_T, I_1]

19. [circuit diagram with V_1, R_1, I_1, V_2, R_2, I_2, V_3, R_3, I_3]

21. [circuit: a, V_{ba}, b, resistor, I_{ab}]

23.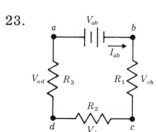

25.

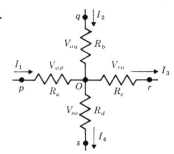

27.

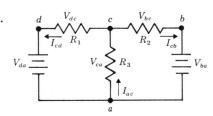

29.

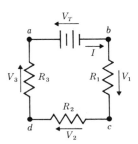

31.

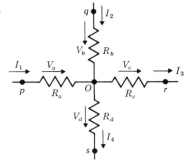

33.

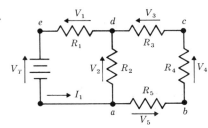

35.

37.

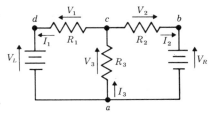

39.

41.

V_1

9 V

43.

V_a

15 V

45.

$V_1 = 15\ mV$

R_1

47.

V_{ab}

a   b

I 16 mV

49.

$V_{cd} = 24\ mV$

c   d

R_1

I

$V_1 = -24\ mV$

Exercises 8-3   1. (a)

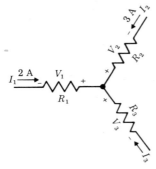

(b) $I_1 + I_2 + I_3 = 0$ (c) $I_3 = -5A$

3. (a)

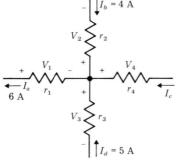

(b) $-I_a + I_b + I_c + I_d = 0$ (c) $I_c = -3A$

5. (a)

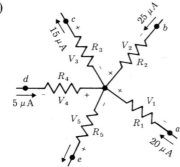

(b) $I_a + I_b - I_c + I_d - I_e = 0$ (c) $I_e = 35 \ \mu A$

7. (a)

(b) $I_1 + I_2 - I_3 = 0,$ $I_3 - I_4 - I_5 - I_6 = 0$
(c) $I_3 = 8 \ A,$ $I_6 = 3 \ A$

9. (a)

18 V

c d

2 A   6 A

V_3 R_3 $6\ \Omega$ R_1 $2\ \Omega$ V_1

$R_2 = 3\ \Omega$

b a

V_2

(b) $18 - V_1 - V_2 - V_3 = 0$

(c) $V_2 = -6$ V   [polarity is opposite of that indicated in diagram in (a)]

11. (a)

D $V_2 = 7$ V C

R_2

V_1 R_1 V_{s1} 20 V

10 V

A B

V_{s2}

(b) $-V_2 + V_{s1} - V_{s2} - V_1 = 0$ (c) $V_1 = 3$ V

13. (a)

1   2

6 V

V_{12} 12 V

9 V

18 V

3   4

(b) $6 + 12 + 9 + 18 - V_{12} = 0$ (c) $V_{12} = 45$ V

15. (a)

2 A V_{ps}

s p

$12\ \Omega$

V_{sr} 15 V V_{pq} 3 V

V_{qr}

r q

R_1

(b) $-V_{ps} + V_{pq} + V_{qr} - V_{sr} = 0$ (c) $V_{qr} = 36$ V

Exercises 8-4A   1. $V_{AC} = -9$ V   3. $V_{CA} = 15$ V   5. $V_{CA} = 25$ V   7. $V_{AB} = 14$ V

9. $V_{AD} = -28$ V, $V_{AE} = -31$ V, $V_{DA} = 28$ V

11. (a)

(b) $V_A = -18$ V, $V_B = -6$ V, $V_D = 8$ V, $V_E = 11$ V

Exercises 8-4B

13. (a) $V_1 = 5$ V, $V_2 = 7.5$ V, $V_3 = 2.5$ V, $V_4 = 2$ V, $V_5 = 3$ V, $V_6 = 5$ V
   (b) $V_{AC} = +20$ V, $V_{AE} = +15$ V, $V_{EB} = +10$ V, $V_{EA} = -15$ V,
   $V_{GC} = +15$ V

15. (a)

(b) $V_A = +5$ V, $V_B = -20$ V, $V_C = -15$ V, $V_D = -12$ V, $V_E = -10$ V,
   $V_F = -7.5$ V

17. (a) $V_1 = 20$ V, $V_2 = 40$ V, $V_3 = 20$ V, $V_4 = 50$ V, $V_5 = 25$ V, $V_6 = 5$ V,
   $V_7 = 8$ V, $V_8 = 22$ V, $V_9 = 50$ V
   (b) $V_{BD} = +30$ V, $V_{BE} = +55$ V, $V_{BF} = -12$ V, $V_{BH} = +10$ V,
   $V_{DF} = -42$ V, $V_{DH} = -20$ V, $V_{EF} = -67$ V, $V_{EH} = -45$ V
   (c) $V_B = 60$ V, $V_C = 20$ V, $V_D = 30$ V, $V_E = 5$ V, $V_F = 72$ V, $V_H = 50$ V

19. $V_{BC} = +0.6$ V, $V_{CB} = -0.6$ V, $V_{DC} = -3.9$ V, $V_{EC} = -7.9$ V, $V_{EB} = -8.5$ V

21. (a) $V_A = +6$ V, $V_B = +2.0$ V, $V_C = +1.4$ V, $V_D = 0$ V, $V_E = -4$ V
   (b) $V_{DC} = -1.4$ V, $V_{CE} = +5.4$ V, $V_{BE} = +6$ V   23. -3.64 V

25. $V_G = +1.98$ V, $V_{GS} = -2.52$ V

Exercises 8-4C

27. $V_A = +8$ V, $V_B = +15$ V, $V_C = +5$ V, $V_D = -9$ V, $V_{CA} = -3$ V,
   $V_{CD} = +14$ V, $V_{AB} = -7$ V, $I_8 = -3$ mA

Exercises 8-5

1. -0.5 V   3. $+30$ V, 0 V
5. $V_A = 1.333$ V, $I_A = 0.2222$ A (from ground toward A)   7. 2.444 A
9. 7 V, 1.4 A (from ground toward A)   11. $I_L = 21$ mA, $V_A = 0$ V

13. $V_A = +14.07$ V, $V_B = -3.73$ V, $I_{AB} = -1.78$ A

15. $V_A = -6.805$ V, $V_B = -5.879$ V, $I_{AB} = 385.8$ μA

Exercises 9-2A

1. (a) 13.33 V, 6.667 Ω;
   (b) 5.713 V, 1.143 A; 7.998 V, 0.7998 A; 10.52 V, 0.4209 A

3. (a) 10 V, 7 kΩ; (b) 0 V, 1.429 mA; 3 V, 1 mA; 5.882 V, 0.5882 mA

5. (a) −3 V, 3.75 kΩ;
   (b) 0 V, −0.8 mA; −2.182 V, −0.2182 mA; −2.892 V, −28.92 μA

7. (a) −25 V, 25 kΩ; (b) −16.67 V, 5.558 mW

9. (a) 20 V, 18 Ω; (b) 0.7407 A, 4.938 W, 0.3704 A, 4.938 W

11. (a) 100 mV, 9.545 Ω; (b) 5.238 mA, 3.492 mA

13. (a) −1.5 V, 3.75 kΩ; (b) −1.174 V, −86.96 μA

Exercises 9-2B

15. 4.667 V, 2 Ω; 3.111 V, 0.7778 A   17. 3.429 V, 7.714 Ω, −0.1852 A

19. +20 V, 40 kΩ   21. 9.091 V, 0.4545 mA

23. −0.3158 mA, 5.242 V, 3.663 V   25. 0.2985 mA, 6.716 V, 7.910 V

27. −2.594 mA, 55.43 V, 24.30 V   29. −2.252 mA

31. $V_A = +28.03$ V, $V_B = +29.11$ V, $V_C = +100$ V, $V_D = +160$ V,
   $V_E = -40$ V, $V_F = -60$ V, $V_{FC} = -160$ V, $V_{ED} = -200$ V, $V_{AB} = -1.08$ V,
   $V_{AC} = -71.97$ V, $V_{BD} = -130.9$ V, $V_{AE} = +68.03$ V, $V_{BF} = +89.11$ V,
   $V_{BE} = +69.11$ V, $V_{AF} = 88.03$ V, $V_{AD} = -131.97$ V, $V_{FE} = -20$ V,
   $I_{AB} = +180$ μA, $I_{AC} = +3.599$ mA, $I_{BD} = +6.545$ mA, $I_{AE} = -3.779$ mA,
   $I_{BF} = -6.365$ mA, $I_{FE} = +1.0$ mA

33. $V_A = -31.52$ V, $V_B = -2.647$ V, $V_C = +100$ V, $V_D = +160$ V,
   $V_E = -120$ V, $V_F = -160$ V, $V_{AB} = -28.87$ V, $V_{AC} = -131.5$ V,
   $V_{AE} = +88.48$ V, $V_{BD} = -162.6$ V, $V_{BF} = +157.4$ V, $V_{CE} = +220$ V,
   $V_{CF} = +260$ V, $V_{CD} = -60$ V, $V_{ED} = -280$ V, $V_{EF} = +40$ V,
   $I_{AB} = +1.444$ mA, $I_6 = -0.788$ mA, $I_{CA} = -2.192$ mA, $I_{EA} = +4.424$ mA,
   $I_{DB} = -4.066$ mA, $I_{FB} = +2.623$ mA

35. (a) $V_3 = \dfrac{R_3}{R_1 + R_2 + R_3} V_S$ or $V_3 = \dfrac{R_p}{R_p + R_s} V_s$

   (b) $V_r = \dfrac{R_p R_m}{R_s R_p + R_s R_m + R_m R_p} \cdot V_s$

Exercises 9-3

1. 16 V, 8 Ω; 1.333 A   3. 20 V, 1 kΩ; 3.333 mA

5. 40 V, 80 kΩ; 0.3333 mA, 83.33 μA   7. 0.6667 mA, 166.7 μA

9. 20 V, 10 kΩ; 0.6667 mA

Exercises 9-4

1. (a) 3 A, 3 Ω (b) 1.8 A, 3.6 V; 1 A, 6 V; 0.6 A, 7.2 V

3. (a) 10 mA, 3 kΩ (b) 2.308 mA

5. (a) 96 μA, 62.5 kΩ
   (b) 72.73 μA, 1.455 V; 53.33 μA, 2.667 V; 36.92 μA, 3.692 V

7. (a) 1.429 mA, 46.67 kΩ (b) $V_A = 41.28$ V, $V_B = 22.23$ V

9. (a) 16.67 μA, 15 kΩ   (b) 11.11 μA, 925.7 nW; 5.556 μA, 926.1 nW

11. (a) 50 μA, 400 kΩ   (b) 22.22 μA

13. (a) 1 mA, 80 kΩ   (b) 909.1 μA, 90.91 μA

15. (a) 375 μA, 3.2 kΩ   (b) 0.7429 V, 142.9 μA

17. (a) 1.167 A, 12 Ω   (b) 0.5833 A, 0.3889 A

Exercises 9-5   1. 4.091 V, 0.2727 A   3. 4.615 V   5. (a) 1.5 A, 2.727 Ω
(b) 0.5294 A   7. (a) 4.615 V, 9.231 Ω   (b) 93.74 mA
9. 10.24 V, 1.024 mA   11. (a) 11.14 V   (b) −4.3 A   (c) 5.57 A
13. (a) 13.622 V   (b) +1.1 A   (c) 0.6811 A   15. 7.458 V, 0.6215 mA
17. (a) 11 mA, 0.678 kΩ   (b) 0.8594 mA
19. −369 μA (toward A), 14.26 kΩ, −184.5 μA, −V_A = 2.631 V

Exercises 9-6   1. R_A = 0.5 Ω, R_B = 0.3333 Ω, R_C = 1.0 Ω
3. R_A = 19.23 kΩ, R_B = 11.54 kΩ, R_C = 11.54 kΩ
5. R_1 = 39 Ω, R_2 = 65 Ω, R_3 = 65 Ω
7. R_1 = 7.045 kΩ, R_2 = 15.05 kΩ, R_3 = 10.03 kΩ
9. 38.51 mA, −0.231 V   11. 7.679 V, 12.77 Ω, 0.3007 A, 3.84 V
13. 4.898 V, 7.804 Ω

Exercises 10-2   1. 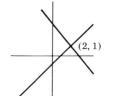   3. 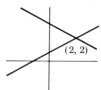   5.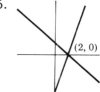

7.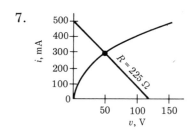

(a) 7V, 115 mA; 35 V, 280 mA; 50 V, 340 mA; 95 V, 455 mA; 110 V, 480 mA   (b) 0.06 kΩ; 0.1 kΩ; 0.15 kΩ; 0.2 kΩ; 0.22 kΩ

9. (a)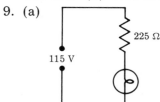

(b) See figure for Exercise 7.   (c) 45 V, 70 V   (d) 310 mA
(e) 14 W, 22 W

11. (a) 10 mA, 8 V   (b) 17 mA, 5.2 V   (c) 13 mA, 6.8 V

13. Load line is drawn between $V_C = 10$ V, $I_C = 0$ and $V_C = 0$, $I_C = 20$ mA
   (a) 9 mA, 5.2 V   (b) 15.5 mA, 2.3 V

Exercises 10-3

1. $u = 3$, $v = 2$ 3. $x = -25$, $y = 15$ 5. $r = 1$, $R = 2$ 7. $v = 2$, $V = -1$

9. $I_1 = 1$, $I_2 = 2$ 11. $I_1 = 1$, $I_2 = -1$ 13. $I_1 = 3$, $I_2 = -2$

15. $I_1 = 5$, $I_2 = -13$ 17. $I_1 = \frac{48}{11}$, $I_2 = \frac{10}{11}$ 19. $I_1 = 19$, $I_2 = \frac{7}{2}$

21. $u = 3$, $v = 2$ 23. $x = -25$, $y = 15$ 25. 1, 2   27. 2, -1   29. 1, 2

31. 1, -1   33. 3, -2   35. 5, -13   37. $\frac{48}{11}$, $\frac{10}{11}$ 39. 19, $\frac{31}{2}$

Exercises 10-4

1. 2, 4   3. 1, -1   5. 1, 2   7. 3, -4, 1   9. 4, 3, 2   11. 16, -16, 4

13. 5, 3, 12   15. 3, 1, -4   17. -28, -21, 3

Exercises 10-5

1. (a)

(b) $10I + 15I - 12.5 = 0$ (c) 0.5 A

3. (a)

(b) $2.5I_1 + (3 + 5)I_2 = 16$, $-(3 + 5)I_2 + (6 + 8)I_3 = 0$, $I_1 - I_2 - I_3 = 0$

(c) $I_1 = 2.108$ A, $I_2 = 1.341$ A, $I_3 = 0.7665$ A

5. (a)

(b) $I_1 - I_2 - I_3 = 0$, $4I_1 + 6I_2 = 4.5$, $-6I_2 + 5I_3 = -6$

(c) $I_1 = 0.1824$ A, $I_2 = 0.6284$ A, $I_3 = -0.4459$ A

7. (a)

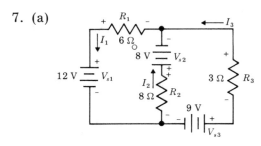

(b) $I_1 - I_2 - I_3 = 0$, $6I_1 + 8I_2 = 20$, $-8I_2 + 3I_3 = -17$

(c) $I_1 = 0.9333$ A, $I_2 = 1.80$ A, $I_3 = -0.8667$ A

9. (a)

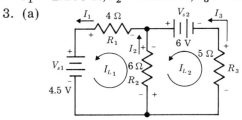

(b) $I_1 - I_2 - I_3 = 0$, $I_3 - I_4 - I_5 = 0$, $4I_1 + 3I_2 = 9$, $-3I_2 + 6I_3 + 7I_4 = 0$, $4I_1 + 6I_3 + 4I_5 = 0$. To reduce to three variables substitute $I_1 = I_2 + I_3$ into the third equation and $I_5 = I_3 - I_4$ into the fifth equation.

11. $I_1 = 1.208$ A, $I_2 = 1.39$ A, $I_3 = -0.1823$ A, $I_4 = 0.7519$ A, $I_5 = -0.9342$ A

Exercises 13–24. (a) Reverse the direction of the current arrows in the circuit diagrams of Exercises 1–12. Voltage polarity markings then remain the same as indicated for electron current. (b) Equations and solutions are the same as for Exercises 1–12.

Exercises 10-6
1. Loop currents: $I_{L_1} = 2.108$ A, $I_{L_2} = 0.7665$ A. Branch currents: $I_1 = 2.108$ A, $I_2 = 1.341$ A, $I_3 = 0.7665$ A

3. (a)

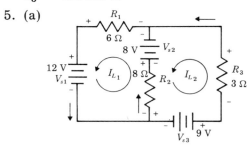

(b) $10I_{L_1} - 6I_{L_2} = 4.5$, $-6I_{L_1} + 11I_{L_2} = -6$

(c) $I_{L_1} = 0.1824$ A, $I_{L_2} = -0.4459$ A, $I_1 = 0.1824$ A, $I_2 = 0.6283$ A, $I_3 = -0.4459$ A

5. (a)

(b) $14I_{L_1} - 8I_{L_2} = 20, -8I_{L_1} + 11I_{L_2} = -17$

(c) $I_{L_1} = 0.9333$ A, $I_{L_2} = -0.8667$ A, $I_1 = 0.9333$ A, $I_2 = 1.80$ A, $I_3 = -0.8667$ A

7. (a)

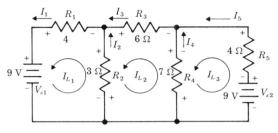

(b) $7I_{L_1} - 3I_{L_2} = 9, -3I_{L_1} + 16I_{L_2} - 7I_{L_3} = 0, -7I_{L_2} + 11I_{L_3} = -9$

9. $I_{L_1} = 1.208$ A, $I_{L_2} = -0.1823$ A, $I_{L_3} = -0.9342$ A, $I_1 = 1.208$ A, $I_2 = 1.3903$ A, $I_3 = -0.1823$ A, $I_4 = 0.7519$ A, $I_5 = -0.9342$ A

11.

$I_{L_1} = 0.6490$ A, $I_{L_2} = 0.2638$ A, $I_{L_3} = 0.2802$ A, $I_5 = I_{L_3} - I_{L_2} = 0.0164$ A

13. $I_5 = 0.0164$ A

Exercises 15–28. Reverse the direction of the current arrows on circuit diagrams. Equations and solutions will be identical to those for electron current.

Exercises 10-7
1. Use circuit as marked except label $V_a = -12$ V. $I_1 + I_2 - I_3 = 0$,
$$\frac{-12 - V_b}{6} + \frac{3 - V_b}{3} - \frac{V_b}{4} = 0 \quad I_1 = -I_{ab} = -1.778 \text{ A}$$
3. 2.068 A   5. 1.390 A   7. 0.1823 A   9. 11.20 W
11–20. See solutions to Exercises 1–10.

Exercises 11-1
1. 0.0698 rad   3. 0.4886 rad   5. 1.605 rad   7. 2.399 rad
9. 3.381 rad   11. 15.77°   13. 97.58°   15. 45.28°   17. 15.26°
19. 48.39°   21. 22.5°   23. 45°   25. 90°   27. 270°   29. 77.92°
31. 21.20°   33. 0.3209°   35. 149.54°   37. 327.16°   39. 690.99°
41. 1.571 in.   43. 7.5 cm   45. 0.3333 rad   47. 84 m
49. 120, 321 m   51. 0.00675 rad, 0.3867°   53. 4.735×10^{-5} rad

Exercises 11-2
1. 75°, 53°, 45°, 30°   3. 0.5708 rad, 0.4908 rad, 1.2108 rad, 1.1108 rad
5. 165°, 105°, 90°, 53°
7. 2.1416 rad, 1.3916 rad, 0.4116 rad, 0.09159 rad

9.

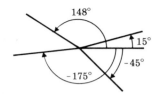

11.

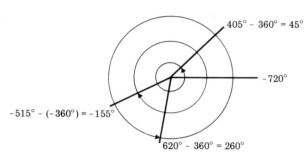

Exercises 11-3

1. (a) 12.53
   (b)

3. (a) 289.5
   (b)

5. (a) 43.37
   (b)
   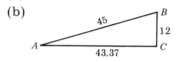

7. (a) 0.9177
   (b)

9. (a) 38,056
   (b)
   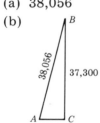

11. (a) 6.403 Ω   (b)

    (c) 51°

13. (a) 237.2 Ω   (b)

    (c) −72°

15. (a) 23.64 Ω   (b)

    (c) 58°
    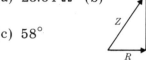

17. (a) 3.688 kΩ   (b)

    (c) 57°

19. (a) 90.14 kΩ   (b)

    (c) 87°
    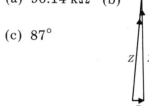

21. 28.62 ft   23. 9.37 mi   25. 98.54 m

Exercises 11-4
1. 0.8, 0.6, 1.333   3. 0.5039, 0.8638, 0.5833   5. 0.9449, 0.3273, 2.887
7. 0.7746, 0.6325, 1.225   9. 0.4472, 0.8944, 0.5   11. 0.6726, 1.10
13. 0.4146, 0.4556   15. 0.1483, 0.9889   17. 0.7071, 0.7071
19. 0, ∞

Exercises 11-5
1. 0.25882, 0.96593, 0.26795; 0.5, 0.86603, 0.57735; 0.70711, 0.70711, 1.0
3. 0.01745, 0.99985, 0.01746; 0.12187, 0.99255, 0.12278;
   0.20791, 0.97815, 0.21256
5. 0.06105, 0.99813, 0.06116; 0.40035, 0.91636, 0.43689;
   0.60321, 0.79758, 0.75629
7. 0.01222, 0.99993, 0.01222; 0.02094, 0.99978, 0.02095;
   0.04013, 0.99919, 0.04016
9. 0.21388, 0.97686, 0.21895; 0.73904, 0.67366, 1.0971;
   0.98058, 0.19612, 5.0
11. 30°   13. 15.29°   15. 38.8°   17. 70.3°   19. 84.9°

Exercises 11-6
1. $B = 60°, b = 8.66, c = 10$ 3. $A = 75°, a = 22.39, c = 23.18$
5. $B = 61.5°, a = 3.340, b = 6.152$ 7. $A = 48.3°, a = 26.13, b = 23.28$
9. $A = 36.87°, B = 53.13°, c = 5.0$ 11. $A = 41.81°, B = 48.19°, b = 11.18$
13. $A = 55.15°, B = 34.85°, a = 57.45$ 15. 39.78 m   17. 17.5 m

Exercises 11-7
1. 30°   3. 22.02°   5. 30°   7. 60°   9. 7.745°   11. 74.68°
13. 0.6614   15. 0.6895   17. 0.7071   19. 0.4841   21. 0.3822
23. 0.0822   25. $A = 36.87°, B = 53.13°, c = 5$
27. $A = 41.81°, B = 48.19°, b = 11.18$
29. $A = 55.15°, B = 34.85°, a = 57.45$

Exercises 11-8
1. $\theta = 41.81°, R = 13.42 \ \Omega$ 3. $\theta = 48.59°, R = 13.23 \ \Omega$
5. $\theta = 36.87°, X = 3 \ \Omega$ 7. $\theta = 55.62°, X = 7.015 \ k\Omega$
9. $\theta = 36.87°, Z = 5 \ \Omega$ 11. $\theta = 60.75°, Z = 11.46 \ k\Omega$
13. $Z = 13.86 \ \Omega, X = 6.928 \ \Omega$ 15. $Z = 96.59 \ \Omega, R = 93.30 \ \Omega$
17. $R = 49.24 \ \Omega, X = 8.682 \ \Omega$

Exercises 11-9
1. +, -, -   3. -, -, +   5. +, -, -   7. +, -, +   9. +, +, +   11. -, -, +
13. -, +, -   15. -, -, +

Exercises 11-10
1. 0.8192, -0.5736, -1.428   3. -0.7071, 0.7971, -1.0
5. -0.50, -0.8660, 0.5774   7. -0.1908, -0.9816, 0.1944
9. -0.50, 0.8660, -0.5774   11. -0.4226, -0.9063, 0.4663
13. -0.7660, -0.6428, 1.192   15. 0.2419, 0.9703, 0.2493
17. -0.2419, -0.9703, 0.2493   19. -0.3907, 0.9205, -0.4245
21. 0.1219, -0.9925, -0.1228   23. -0.4540, -0.8910, 0.5095

25. (a) 38.30 Ω, −32.14 Ω (b) capacitive
27. (a) 12.29 kΩ, 8.604 kΩ (b) inductive
29. (a) 17 Ω, −61.93° (b) capacitive
31. (a) 2.846 kΩ, 18.43° (b) inductive
33. (a) 3.906 Ω, 6.345 Ω (b) inductive
35. (a) 70.57 Ω, 72.15 Ω (b) inductive
37. 0.8415, 0.5403, 1.557   39. −0.3827, 0.9239, −0.4142
41. 0.3817, −0.9243, −0.4129   43. −0.7126, 0.7016, −1.016
45. 0, 1, 0   47. 0.9785, 0.2061, 4.747

Exercises 12-1

1.

x	$-\pi$	$-\dfrac{3\pi}{4}$	$-\dfrac{\pi}{2}$	$-\dfrac{\pi}{4}$	0	$\dfrac{\pi}{4}$	$\dfrac{\pi}{2}$	$\dfrac{3\pi}{4}$	π	$\dfrac{5\pi}{4}$	$\dfrac{3\pi}{2}$	$\dfrac{7\pi}{4}$	2π	$\dfrac{9\pi}{4}$	$\dfrac{5\pi}{2}$	$\dfrac{11\pi}{4}$	3π
y	0	−0.7	−1	−0.7	0	0.7	1	0.7	0	−0.7	−1	−0.7	0	0.7	1	0.7	0

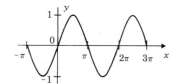

3.

x	$-\pi$	$-\dfrac{3\pi}{4}$	$-\dfrac{\pi}{2}$	$-\dfrac{\pi}{4}$	0	$\dfrac{\pi}{4}$	$\dfrac{\pi}{2}$	$\dfrac{3\pi}{4}$	π	$\dfrac{5\pi}{4}$	$\dfrac{3\pi}{2}$	$\dfrac{7\pi}{4}$	2π	$\dfrac{9\pi}{4}$	$\dfrac{5\pi}{2}$	$\dfrac{11\pi}{4}$	3π
y	−1	−0.7	0,	0.7	1	0.7	0	−0.7	−1	−0.7	0	0.7	1	0.7	0	−0.7	−1

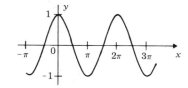

5.

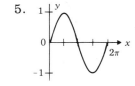

7.

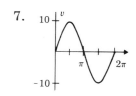

9.

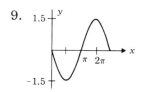

11.

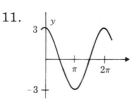

13.

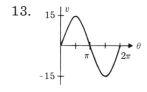

15. 1   17. 10 V   19. 1.5   21. 3   23. 15 μV
25. 8.716 V, 21.64 V, −100 V   27. −3.897 V, −3.182 V, 3.897 V
29. −125.6 mV, 125.6 mV, −102.5 mV

Exercises 12-2

1.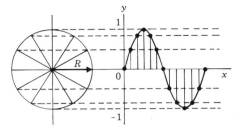

3.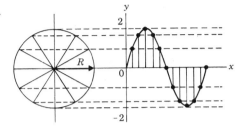

5.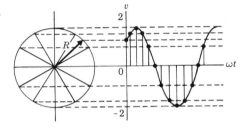

7.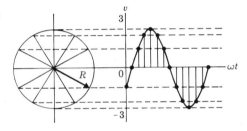

9.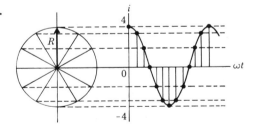

Exercises 12-3

1.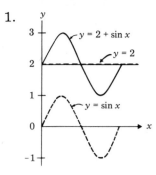

3.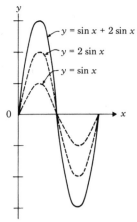

5.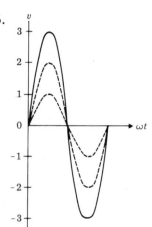

7.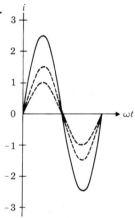

9.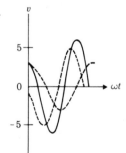

Exercises 12-4

1.

$S = 5, \theta = 53°$

3.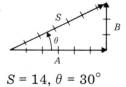

$S = 14, \theta = 30°$

5.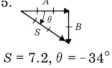

$S = 7.2, \theta = -34°$

7.

$S = 3, \theta = -30°$

9.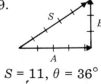

$S = 11, \theta = 36°$

11. 5, 53.13°   13. 13.89, 30.26°   15. 7.211, –33.69°
17. 2.915, –30.96°   19. 10.75, 35.54°   21. 9.979, 42.43°
23. 10.45, 57.14°   25. 21.22, 0°   27. 161.8, –24.64
29. 0.04319, 17.0°

Exercises 12-5   1. 3.606 V, 56.31°   3. 4.366 A, 66.37°   5. 3.905 A, 9.8°
7. 157.4 mV, 27.26°   9. 40.02 mA, 8.98°   11. 347.6 kV, 62.76°
13. 0.8585 V, –53.51°

Exercises 12-6   1. 0.01667 s, 8.333 ms, 0.8333 ms   3. 33.33 μs, 20 μs, 5.0 μs
5. 0.10 μs, 40 ns, 20 ns   7. 100 Hz, 60 Hz, 200 Hz
9. 4 Hz, 13.33 Hz, 62.5 Hz   11. 28.57 MHz, 55.56 MHz, 285.7 MHz
13. 250 ms, 4.0 Hz   15. 0.5 ms, 2000 Hz   17. 25 μs, 40 kHz
19. 5 s, 0.2 Hz   21. 1.25 ms, 800 Hz   23. 25 μs, 40 kHz
25. 306.1 m   27. 3.308 m   29. 7.576 MHz

Exercises 12-7   1. (a) 8 mV (b) 4 mV (c) 2.828 mV (d) $v = 4 \sin 6280t$ mV
3. (a) 0.4 V (b) 0.2 V (c) 0.1414 V (d) $v = 0.2 \sin 6.28 \times 10^5 t$ V
5. (a) 2.0 V (b) 1.0 V (c) 0.7071 V (d) $v = 1.0 \sin 6.28 \times 10^6 t$ V
7. (a) 8.0 V (b) 4.0 V (c) 2.828 V (d) $v = 4.0 \sin 628t$ V
9. (a) 40 V (b) 20 V (c) 14.14 V (d) $v = 20 \sin 125.6t$ V
11. 50.63 mV   13. 28.93 mV   15. 2.87°   17. 10.81°
19. (a) $V = 103$ V   (b) $V_{ave} = 74.5$ V
21. (a) 61.3 V   (b) $V_{ave} = 33$ V

Exercises 13-1   1. $j3$ 3. $j4$ 5. $j\sqrt{2}$ 7. $j\sqrt{7}$ 9. ja^2 11. (a) –1 (b) 1
13. (a) j (b) $-j$ 15. (a) 1 (b) 1   17. 1   19. 1

Exercises 13-2   1. 2, –3   3. –2, –4   5. 3, 1

Exercises 13-4   1. 5.831   3. 6.083   5. 114   7. 4.549 kΩ

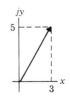   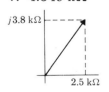

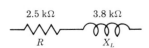

9. 1235 Ω

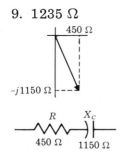

Exercises 13-5

1. $5 + j7$ 3. $15 - j12$ 5. $3 + j$ 7. $-3 + j11$ 9. $13 + j2$
11. $9 - j8$ 13. $30 - j18$ V   15. $11 - j21$ μV   17. $10 + j7$ A
19. $9 + j3$ Ω   21. $42 + j18$ kΩ   23. $8 - j4$ mS

Exercises 13-6

1. $5\underline{/53.13°}$ 3. $3.606\underline{/56.31°}$ 5. $9.22\underline{/-49.4°}$ 7. $10\underline{/-36.87°}$
9. $9.22\underline{/49.4°}$ Ω   11. $6.325\underline{/-55.3°}$ kΩ   13. $1.71\underline{/28.66°}$ MΩ
15. $6.325\underline{/-18.43°}$ V   17. $10\underline{/0°}$ mV   19. $3.157\underline{/-29.41°}$ μV
21. $2.236\underline{/63.43°}$ A   23. $44.6\underline{/70.35°}$ mA   25. $175.6\underline{/4.9°}$ μA
27. $3 + j4$ 29. $8.693 - j2.329$ 31. $13.52 + j6.505$ V
33. $33.47 + j10.23$ mV   35. $6.71 - j9.948$ A   37. $24.41 + j5.411$ mA
39. $14.49 + j3.882$ Ω   41. $3.05 - j4.697$ kΩ   43. $1.499 - j0.03927$ MΩ
45. $2.898 + j0.7765$ mS   47. $39.92 - j20.78$ μS

Exercises 13-7

1. $20\underline{/22°}$ 3. $11.5\underline{/-50°}$ 5. $13.7\underline{/77.31°}$ 7. $171.9\underline{/9.84°}$
9. $26.08\underline{/32.47°}$ 11. $12.5\underline{/15°}$ V   13. $5.46\underline{/49.05°}$ V
15. $5.534\underline{/17.57°}$ V   17. $1.0\underline{/3°}$ A   19. $87.5\underline{/-58°}$ mA
21. $(ac - bd) + j(ad + bc)$ 23. $R^2 + X^2$
25. $(R_1 R_2 - X_1 X_2) + j(R_1 X_2 + R_2 X_1)$ 27. $26 - j7$
29. $-33 + j56$ 31. $-6 + j15$

Exercises 13-8

1. $1.667\underline{/35°}$ 3. $1.846\underline{/20°}$ 5. $0.3816\underline{/-9.31°}$
7. $0.7555\underline{/-30.41°}$ 9. $3.2\underline{/10°}$ 11. $3.305\underline{/28.52°}$ kΩ
13. $2.667\underline{/25°}$ A   15. $266\underline{/85°}$ μA
17. $[ac + bd + j(bc - ad)] \div (c^2 + d^2)$

19. $(R^2 - X^2 + 2jRX) \div (R^2 + X^2)$ 21. $1 - j\dfrac{R}{X}$ 23. $-2 - j5$

25. $2.8 - j0.4$ 27. $\frac{9}{13} + j\frac{7}{13}$ 29. $-\frac{1}{2} + j\frac{3}{2}$

Exercises 14-1

1. (a) $6.766\underline{/32.19°}$ V, $5.726 + j3.605$ V
   (b)

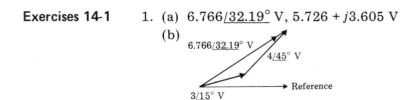

3. (a) $4.406\underline{/27.48°}$ V, $3.909 + j2.033$ V
   (b)
   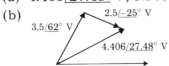

5. (a) $5.622\underline{/21.29°}$ V, $5.238 + j2.041$ V
   (b)
   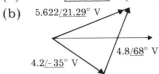

7. (a) $4.721\underline{/-9.58°}$ V, $4.655 - j0.786$ V
   (b)
   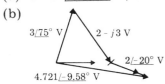

9. (a) $5.406\underline{/39.76°}$ V, $4.156 + j3.458$ V
   (b)
   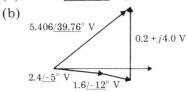

11. (a) $57.33\underline{/3.89°}$ mA, $57.2 + j3.89$ mA
    (b)

13. (a) $945.3\underline{/-18.59°}$ μA, $896 - j301.4$ μA
    (b)
    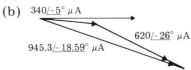

15. (a) $1.211\underline{/-1.81°}$ mA, $1.210 - j0.0383$ mA
    (b)
    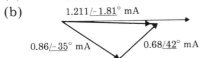

17. (a) $6.408\underline{/5.85°}$ mA, $6.375 + j0.653$ mA
    (b)
    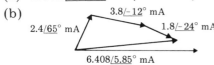

19. (a) $52.07\underline{/-1.36°}$ mA, $52.06 - j1.24$ mA
    (b)
    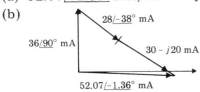

Exercises 14-2

1. (a) $5 + j3.016$ kΩ, $5.839\underline{/31.1°}$ kΩ

   (b)
   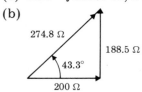

3. (a) $200 + j188.5$ Ω, $274.8\underline{/43.3°}$ Ω

   (b)
   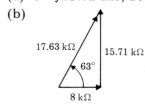

5. (a) $8 + j15.71$ kΩ, $17.63\underline{/63.01°}$ kΩ

   (b)
   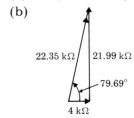

7. (a) $4 + j21.99$ kΩ, $22.35\underline{/79.69°}$ kΩ

   (b)
   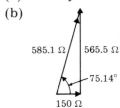

9. (a) $150 + j565.5$ Ω, $585.1\underline{/75.14°}$ Ω

   (b)
   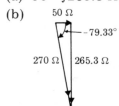

11. (a) $50 - j265.3$ Ω, $270\underline{/-79.33°}$ Ω

    (b)

13. (a) $2 - j1.592$ kΩ, $2.556\underline{/-38.52°}$ kΩ

    (b)
    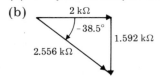

15. (a) $1.5 - j0.7958$ kΩ, $1.698\underline{/-27.95°}$ kΩ

(b)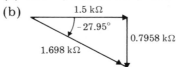

17. (a) $48 - j39.79$ kΩ, $62.35\underline{/-39.66°}$ kΩ

(b)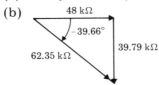

19. (a) $82 - j132.6$ Ω, $155.9\underline{/-58.27°}$ Ω

(b)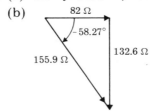

21. (a) $f = \dfrac{X_L}{2\pi L}$ (b) $f = \dfrac{1}{2\pi X_C C}$ 23. 4.33 kΩ, 2.5 kΩ, 397.9 mH

25. 1.449 kΩ, 0.3882 kΩ, 103 μH   27. 4.243 kΩ, 4.243 kΩ, 0.03751 μF

29. 1.128 kΩ, 0.4104 kΩ, 517.1 pF

Exercises 14-3

1. $96.05\underline{/38.66°}$ Ω, $1.197\underline{/-38.66°}$ A, $89.78\underline{/-38.66°}$ V, $71.82\underline{/51.34°}$ V, 107.5 W

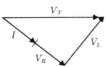

3. $27.46\underline{/10.49°}$ kΩ, $1.821\underline{/-10.49°}$ μA, $49.17\underline{/-10.49°}$ mV, $9.105\underline{/79.51°}$ mV, 89.53 nW

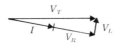

5. $477.6\underline{/-83.99°}$ Ω, $0.7852\underline{/83.99°}$ A, $39.26\underline{/83.99°}$ V, $373\underline{/-6.01°}$ V, 30.83 W

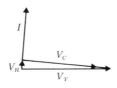

7. $174.9\underline{/-30.96°}$ kΩ, $1.572\underline{/30.96°}$ nA, $235.8\underline{/30.96°}$ μV, $141.5\underline{/-59.04°}$ μV, 0.3707 pW

9. $400\underline{/-53.13°}$ Ω, $0.3\underline{/53.13°}$ A

11. $1.467\underline{/62.39°}$ MΩ, $306.7\underline{/-62.39°}$ μA

13. $509.9\underline{/-78.69°}$ kΩ, $3.138\underline{/78.69°}$ mA

15. $15.53\underline{/33.18°}$ Ω, $7.727\underline{/-33.18°}$ A

17. $2.385\underline{/33.02°}$ MΩ, $1.593\underline{/-33.02°}$ pA   19. $187.2\underline{/15°}$ V, $243.3\underline{/-60°}$ V

Exercises 14-4

1. $89.8\underline{/-38.66°}$ V, $71.84\underline{/51.34°}$ V

3. $49.16\underline{/-10.49°}$ mV, $9.104\underline{/79.51°}$ mV

5. $39.26\underline{/83.99°}$ V, $372.9\underline{/-6.01°}$ V

7. $235.8\underline{/30.96°}$ μV, $141.5\underline{/-59.04°}$ μV

9. $72\underline{/53.13°}$ V, $39\underline{/143.1°}$ V, $135\underline{/-36.87°}$ V

11. $208.6\underline{/-62.39°}$ V, $766.9\underline{/27.61°}$ V, $368.1\underline{/-152.39°}$ V

13. $313.8\underline{/78.69°}$ V, $1569\underline{/168.69°}$ V, $3138\underline{/-11.31°}$ V

15. $7.508\underline{/-2.22°}$ V   17. $9.017\underline{/46.44°}$ μV

Exercises 14-5

1. $2.901\underline{/6.9°}$ Ω $= 2.88 + j0.3485$ Ω

3. $1.041\underline{/-28.57°}$ kΩ $= 0.9142 - j0.4978$ kΩ

5. $1.576\underline{/17.06°}$ kΩ $= 1.507 + j0.4624$ kΩ

7. $24\underline{/-25°}$ A, $20\underline{/15°}$ A, $41.37\underline{/-6.9°}$ A

9. $8.571\underline{/-27°}$ mA, $25\underline{/45°}$ mA, $28.82\underline{/28.57°}$ mA

11. $118.75\underline{/-48°}$ μA, $158.3\underline{/-5°}$ μA, $52.78\underline{/15°}$ μA, $301.4\underline{/-17.06°}$ μA

13. $5.947\underline{/-8.73°}$ Ω, $5.878 - j0.9026$ Ω

Exercises 14-6

1. $1.667\underline{/15°}$ A, $3.448\underline{/-6.89°}$ A   3. $1.5\underline{/-48°}$ mA, $3.262\underline{/-23.28°}$ mA

5. $1.933\underline{/-25.57°}$ mA, $5.637\underline{/46.43°}$ mA

7. $1.452\underline{/-30.28°}$ A, $1.045\underline{/44.46°}$ A

9. $2.138\underline{/-2.87°}$ mA, $7.145\underline{/37.94°}$ mA

Exercises 14-7

1. $0.1163\underline{/-54.46°}$ S   3. $202.1\underline{/14.04°}$ μS   5. $3.636\underline{/18°}$ mS

7. $172.5\underline{/75°}$ Ω   9. 4 Ω   11. $4.86 - j4.82$ mS, $6.845\underline{/-44.76}$ mS

13. $49.45\underline{/-18.5°}$ mA, $2.917\underline{/42°}$ mS, $3.847\underline{/-18.5°}$ mS

15. $22.42\underline{/-80.96°}$ mA, $35.72\underline{/84.97°}$ mA

17. $0.5635 - j1.208$ μS, $0.8781 - j0.2353$ μS, $4.256 + j3.571$ μS

Exercises 14-9

1. $53.19 + j53.79$ Ω   3. $178.1\underline{/0°}$ V   5. $191.3\underline{/-10.32°}$ V, $\theta = -6.91°$

7. $156.8\underline{/28.35°}$ mV

Exercises 14-10

1. 4.896 mV, 9.793 μA   3. $8.029\underline{/56.65°}$ kΩ, $326.7\underline{/-48.32°}$ mA

5. $7\sqrt{10/7}/20\pi RC$

Exercises 14-11

1. $1.554\underline{/169.15°}$ A, $0.5029\underline{/-104.14°}$ A, $1.661\underline{/6.75°}$ A
3. $1.555\underline{/169.14°}$ A, $0.5028\underline{/-104.13°}$ A, $1.661\underline{/6.75°}$ A
5. $7.196\underline{/6.75°}$ V, $1.333\underline{/0°}$ Ω, $1.661\underline{/6.75°}$ A
7. $4.982\underline{/6.75°}$ V, 0.9231 Ω, $0.8892\underline{/-38.81°}$ A
9. $1.833\underline{/-5.83°}$ mA, $0.5241\underline{/68.33°}$ mA, $2.039\underline{/8.49°}$ mA
11. $1.832\underline{/-5.83°}$ mA, $0.5242\underline{/68.33°}$ mA, $2.039\underline{/8.49°}$ mA
13. $15.31\underline{/3.59°}$ V, $2.829\underline{/28.73°}$ kΩ
15. $5.412\underline{/-25.14°}$ mA, $2.829\underline{/28.73°}$ kΩ, $2.039\underline{/8.48°}$ mA

Exercises 15-2

1. ±2 3. ±9 5. $\pm\frac{5}{4}$ 7. $\pm2\sqrt{c}$ 9. $\pm3\sqrt{b+3c}$ 11. $I = \pm\sqrt{P/R}$
13. $V = \pm\sqrt{P/G}$

Exercises 15-3

1. $-1, -5$ 3. $\frac{2}{3}, \frac{3}{5}$ 5. $6.405, 1.405$ 7. $\dfrac{3 \pm \sqrt{-23}}{4}$ 9. 4 cm, 10 cm
11. 2 cm, 5 cm   13. $1.055\ \mu F$ or $0.01067\ \mu F$ 15. $7.842\ \mu H$, $0.1162\ \mu H$

17. $X_C = \dfrac{R_L^2 + X_L^2 \pm \sqrt{(R_L^2 + X_L^2)^2 - 4R_C^2 X_L^2}}{2X_L}$

Exercises 15-4

1. 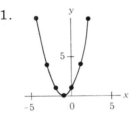   3. 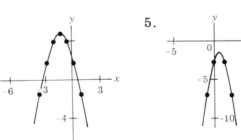   5.

Exercises 15-5

1. 2, max, 1   3. -4, min, -1 5. $\frac{2}{3}$, min, $\frac{2}{3}$ 7. 5 W, 1.0 A, 5 Ω
9. 1.667 mW, 1.667 mA, 600 Ω
11. Apply $x_{y_{ex}} = -b/2a$ to $P_L = -I_L^2 R_i + V_s I_L$

Exercises 16-1

1. $\log_2 4 = 2$, $\log_3 27 = 3$ 3. $\log_{10} 3.1623 = 0.5$, $\log_{16} 4 = 0.25$
5. $\log_{10} 3 = 0.4771$, $\log_{10} 300 = 2.4771$ 7. $\log_p V = x$, $\log_q I = t$
9. $\log_e 5 = 1.609$, $\log_e 50 = 3.912$ 11. $\log_2 \frac{1}{4} = -2$, $\log_4 \frac{1}{16} = -2$
13. $\log_{10} 0.01 = -2$, $\log_{10} 0.001 = -3$ 15. $4^{0.5} = 2$, $9^{0.5} = 3$
17. $25^{0.5} = 5$, $10^{3.6021} = 4000$ 19. $10^{0.3010} = 2$, $10^{0.4771} = 3$
21. $10^{0.7782} = 6$, $10^{3.7782} = 6000$ 23. $10^{-0.3010} = 0.5$, $10^{-0.6021} = 0.25$
25. $e^{-0.6931} = 0.5$, $e^{-1.3863} = 0.25$ 27. $p^x = V$, $q^t = I$

Exercises 16-2

1. $\log_b p + \log_b q$ 3. $\log_2 V + \log_2 I$ 5. $\log_{10} x - \log_{10} y$
7. $\log_2 V - \log_2 5$ 9. $2\log_5 V$ 11. $5\log_b x$ 13. $2\log_b x + \log_b y$
15. $2\log_2 I + \log_2 R$ 17. $\frac{1}{2}(\log_2 P + \log_2 R)$
19. $2(\log_2 V_1 - \log_2 V_2) + \log_2 R_2 - \log_2 R_1$ 21. $\log_b IR$
23. $\log_2 Rr$ 25. $\log_2 b/a$ 27. $\log_{2.7} I/i$ 29. $\log_{10} \sqrt{PR}$
31. $\log_{2.7} x^{4/3} y^{2/3}$ 33. $V = IR$ 35. $P = V^2/R$ 37. $v = Vb^{-tx}$

Exercises 16-3

1. 0.65896, 3.65896, 5.65896   3. $-0.32975, -2.32975, 1.67025$
5. 0.79099, 2.79099, -2.20901 7. $-2.08302, 3.91698, 4.91698$
9. 0.97909, 1.97909, -1.02091 11. $-4.71897, -7.96257, 3.03782$
13. $45,860 = 10^{4.6614}$ 15. $0.000468 = 10^{-3.3298}$
17. $2,358,000 = 10^{6.3725}$ 19. $6059 = 10^{3.7824}$ 21. $605,900 = 10^{5.7824}$
23. $975.8 = 10^{2.9894}$ 25. $\log 458,600 = 5.6614$
27. $\log 0.000468 = -3.3298$ 29. $\log 2.79 \times 10^{-3} = -2.5544$
31. $\log 72.35 \times 10^{-8} = -6.1406$ 33. $\log 9.758 \times 10^1 = 1.9894$

Exercises 16-4

1. $5679; 5.679 \times 10^{-5}$ 3. $10,595; 1.0595 \times 10^{-3}$
5. $555.5; 5.555 \times 10^{-3}$ 7. 2.3195; 0.4311   9. $1.139; 4.393 \times 10^{-2}$
11. $6059; 7.386 \times 10^{-3}$ 13. $45,860; 4.586 \times 10^{-3}$
15. $84,740; 8.474 \times 10^{-4}$ 17. 1.230; 0.1230   19. $8888; 8.888 \times 10^{-6}$

Exercises 16-5

1. $6.0234; -5.4895$ 3. $12.2968; -3.8213$ 5. $-6.8345; 4.6784$
7. 9.0728; 4.4676   9. $6.4489; -2.7615$ 11. 377.1; 37,710
13. 53,500; 5.350   15. 0.07017; 701.7   17. 0.159; 159
19. 5610; 0.5610   21. 1.5441   23. -0.25181 25. 3.3617
27. 4.0074   29. -3.8397 31. -7.1272 33. 3   35. 2

Exercises 16-6

1. $495.7\,\Omega$ 3. $434.4\,\Omega$ 5. $578.6\,\Omega$ 7. $144.2\,\Omega$ 9. $160\,\Omega$
11. $91.48\,\Omega$ 13. 0.996 cm   15. 0.2429 cm

Exercises 16-7

1. 13.01 dB, 20.79 dB, 26.53 dB, 31.76 dB
3. -13.01 dB, -3.01 dB, 7 dB, 17 dB
5. 26.02 dB, 41.58 dB, 53.06 dB, 63.52 dB
7. -26.02 dB, -6.02 dB, 14 dB, 34 dB   9. (a) 31.62, 5.62
   (b) 10,000; 100   11. (a) 0.0316, 0.1778 (b) 0.2818, 0.5309
13. 17 dB   15. -7.27 dB   17. -4.54 dB   19. 83.84 dB
21. -21.65 dB   23. 83.84 dB   25. -10.2 dB   27. 80.64 mV
29. 1.888 mV   31. 95.32 mV   33. 88.29 mA   35. (a) 30 W
   (b) 9.487 W (c) 3 W (d) 0.03 W   37. 13 dB   39. -36 dB
41. Output is reduced by 6 dB each time frequency changes by a factor of
   two. Power ratio = 0.25, 2 mW : 8 mW; voltage ratio =
   0.5, 30 mV : 60 mV.

Exercises 16-8   1. 10 mW, 31.62 kW, 10 μW   3. 0.7079 mW, 5.011 W, 17.78 mW
5. 7 dBm, 40 dBm, -1.249 dBm   7. -23.01 dBm, -38.24 dBm, 71.76 dBm
9. 3.01 dBm, 2 mW   11. 30.79 dBm, 1.2 W   13. 30.79 dBm, 1.2 W
15. $+22.02$ dB   17. -99.21 dBm   19. 1.585×10^{-9} W

Exercises 16-9   1. 30 dBm, 1 W   3. 53.01 dBm, 96 dB   5. 0.5 μW   7. 8 dB
9. 55 dB

Exercises 16-10   1.

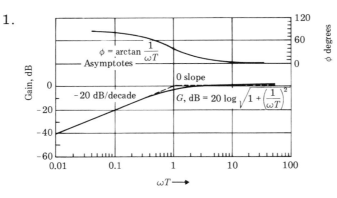

3. & 4.

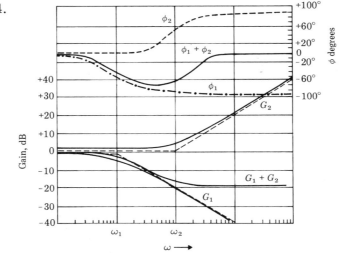

At $f \ll f_{c1}$, $G = 0$ dB; at $f \gg f_{c2}$, $G = -20 \log \left(\dfrac{R_1 + R_2}{R_2} \right)$

$\phi_1 = -\arctan \omega T_1$, $\phi_2 = \arctan \omega T_2$, $\phi_T = \phi_1 + \phi_2$

Exercises 17-1   1. 100 μs   3. 94 s   5. 2.8 μs   7. 0.5 s   9. 0.5 ms   11. 3.75 ns
13. 147.1 MΩ, 0.001852 μF   15. 25 Ω, 0.6 H
17. (a) 1 s   (b)

t/τ	0.1	0.5	1.0	2.5	5.0
i, mA	0.92	0.61	0.37	0.09	0.01
v_C, V	8	39	63	91	99

(c)

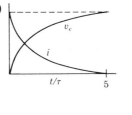

19. (a) 90 μs   (b)

t/τ	0.1	0.5	1.0	2.5	5.0
i, μA	1.5	1.02	0.62	0.15	0.02
v_C, V	0.12	0.6	0.9	1.4	1.49

(c) See Exercise 17(c)   21. 0.36 s   23. 1.1, 550 μs

25. 75 V, 60 V, 45.75 V, 27.75 V, 4.1 V; see Table 17-1 for a typical curve

27. 6.667 μA, 2.47 μA, 0.067 μA   29. 30.5 V, 0.305 mA

31. 100 μs, 0, 0.8 mA, 6.3 mA, 9.9 mA; see curve in Table 17-1

33. 2.5 μs, 0, 19.8 mA, 37.8 mA, 59.4 mA

35. 50 ns, 1000 V, 840 V, 370 V, 20 V

37. 25 ms, 1000 V, 920 V, 370 V, 20 V   39. 0.7143 H

Exercises 17-2

1. 2.154 ms   3. 0.1669 ms   5. 50.51 μs   7. 98.9 V, 2.681 mA

9. 4.9998 V   11. 247.5 μA   13. 7.116 ms   15. 146.8 V

17. 32.85 μF   19. 230.8 ns   21. 369.3 ns   23. 63%

25. 1.664 μs   27. 2.103 MΩ 29. 64.03 V   31. 3.107 H

33. 7.165 V   35. 20 μF, 100 ms   37. (a) $v_{out} = (V_{max} - V_0)e^{-t/\tau}$

     (b) $v_{out} = -V_C e^{-t/\tau}$

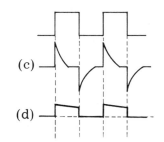

39.

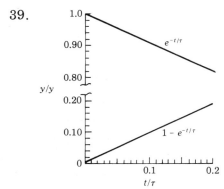

Exercises 18-1

1. 1   3. -4   5. -1.25   7. -1.25   9. -1.333   11. 0.3 mS, 3.3 kΩ

13. 2.5 mS, 0.4 kΩ 15. 0.5 μS, 2.0 MΩ

Exercises 18-2

1. 33 Ω, 20 Ω; 26 Ω, 13 Ω; 20 Ω, 6.7 Ω

3. 40 kΩ, 33 kΩ; 33 kΩ, 21 kΩ; 20 kΩ, 6 kΩ

5. 3.8 kΩ, 3.1 kΩ; 2.9 kΩ, 1.2 kΩ; 1.6 kΩ, 1.1 kΩ

7. 93 Ω, 220 Ω; 160 Ω, 330 Ω; 190 Ω, 340 Ω

Exercises 19-1

3. 8; $8^0, 8^1, 8^2, 8^3, 8^4, 8^5, 8^6, 8^7$ 5. 2; $2^0, 2^1, 2^2, 2^3, 2^4, 2^5, 2^6, 2^7$

7. 12; $12^0, 12^1, 12^2, 12^3, 12^4, 12^5, 12^6, 12^7$

9. 16; $16^0, 16^1, 16^2, 16^3, 16^4, 16^5, 16^6, 16^7$

11. $(1 \times 10^3) + (1 \times 10) + (1 \times 10^0) = 1011_{10}$

13. $(1 \times 2^3) + (1 \times 2) + (1 \times 2^0) = 11_{10}$
15. $(1 \times 5^3) + (1 \times 5) + (1 \times 5^0) = 131_{10}$
17. $(1 \times 8^3) + (1 \times 8) + (1 \times 8^0) = 521_{10}$
19. $(1 \times 16^3) + (1 \times 16) + (1 \times 16^0) = 4113_{10}$

Exercises 19-2   1. 30   3. 110   5. 199   7. 519   9. 1375   11. 4817   13. 77_8
15. 272_8 17. 573_8 19. 1725_8 21. 4137_8 23. $67,543_8$

Exercises 19-3   1. 66   3. 346   5. 3467   7. 4402   9. 41,596   11. 75,175
13. 658,316   15. 42_{16} 17. 550_{16} 19. ABC_{16} 21. $1A35_{16}$
23. $6BCA_{16}$ 25. $10,9CD_{16}$

Exercises 19-4   1. 5   3. 11   5. 9   7. 41   9. 85   11. 844   13. 2282
15. 101   17. 1001   19. 10101   21. 101001   23. 1101100
25. 11011001   27. 111111101000   29. 100111001101001   31. 5_8
33. 13_8 35. 11_8 37. 51_8 39. 125_8 41. 1514_8 43. 4352_8
45. 011, 1 001, 1 001 101   47. 11 111, 100 101 010, 1 000 111 110
49. 1 101 000, 10 101 000 111, 11 101 100 110 000   51. 5_{16} 53. B_{16}
55. 9_{16} 57. 29_{16} 59. 55_{16} 61. $34C_{16}$ 63. $8EA_{16}$
65. 1001, 1101, 1 0000   67. 1010, 0010 0111, 1 1010 1001
69. 10 1010 0011 1111, 11 0110 0101 1010 0111 1011 1001 1100

Exercises 19-5   1. 0.75   3. 0.625   5. 3.1875   7. 9.625   9. 13.8125   11. 0.6_8
13. 0.5_8 15. 3.14_8 17. 11.5_8 19. 15.64_8 21. 0.34_{16}
23. $B.BE_{16}$ 25. $5B.B18_{16}$ 27. $16B.D4_{16}$ 29. $C93.E38_{16}$

Exercises 19-6   1. 1000   3. 10000   5. 101110   7. 1010110   9. 11000100
11. 8   13. 16   15. 46   17. 86   19. 196   21. 011   23. 011
25. 1001   27. 1   29. 11   31. 3   33. 3   35. 9   37. 1   39. 3

Exercises 19-7   1. 011   3. 011   5. 1001   7. 1   9. 011   11. 011   13. 011
15. 1001   17. 1   19. 011

Exercises 20-1   1. $C = A \cdot B$ 3. $P = \overline{Q}$ 5.

(a)

A	B	X
0	0	0
0	1	1
1	0	1
1	1	1

(b)

A	X
0	1
1	0

(c)

A	B	X
0	0	0
0	1	0
1	0	0
1	1	1

7. (a) $C = A + B$ (b) $Y = \overline{P}$ (c) $Z = R \cdot S$

9. (a)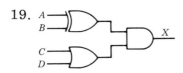

(c)

11. (a) $R = \overline{S}$ (b) (c)

S	R
0	1
1	0

Exercises 20-2

1. $X = \overline{A} \cdot B \cdot C \cdot D$ 3. $X = \overline{AB}$ 5. $X = (A + B)CD$

7. $X = (\overline{A + B})CD$ 9. $X = (A \oplus B) + CD$ 11.

13.

15.

17.

19.

21.

A B	AB	$\overline{AB}$	A + B	$(A + B)\overline{AB} = X$
0 0	0	1	0	0
0 1	0	1	1	1
1 0	0	1	1	1
1 1	1	0	1	0

23. $Y = (C + D)KP$

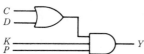

Exercises 20-3A

1. ABC 3. $AB + C$ 5. 0   7. C 9. 0   11. $E + F$ 13. $\overline{AB}$

15. 1   17. $\overline{A}$ 19. AB

Exercises 20-3B

21. $A + \overline{B}$

23. $A + \overline{B} + \overline{C}$

25. $(\overline{A} + \overline{B})\overline{C}$

27. $\overline{A} + \overline{B} + \overline{CD}$

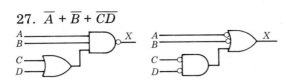

29. $(\overline{A} + \overline{B})[\overline{A} + \overline{C}(D + \overline{E})]$

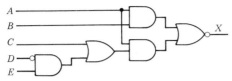

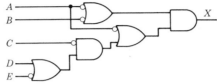

31. (a) $X = \overline{AB}$ (b) $X = \overline{A} + \overline{B}$

A	B	AB	$\overline{AB}$ = X
0	0	0	1
0	1	0	1
1	0	0	1
1	1	1	0

33. (a) $X = \overline{\overline{A} + B}$ (b) $X = A \cdot \overline{B}$

A	B	$\overline{A} + B$	X
0	0	1	0
0	1	1	0
1	0	0	1
1	1	1	0

35.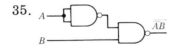

37.

Exercises 20-3C

39. B 41. X 43. $P + N$ 45. DF 47. A

49. Factor A, then show that $\overline{BC} = \overline{B}C + B\overline{C} + \overline{B}\overline{C}$.

51. $A(B + \overline{B}C) = A(B + C) = AB + AC$ [8(a), 16(a)]

53. Use truth table, or use $BC = 1 \cdot BC = (A + \overline{A})BC = ABC + \overline{A}BC$.
    Therefore, $AB + ABC + \overline{A}C + \overline{A}BC = AB + \overline{A}C$ [15(a)].

Exercises 20-4

1.

3.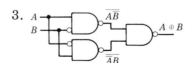

5. $\overline{\overline{\overline{ABC} \cdot \overline{A\overline{B}C} \cdot \overline{AB\overline{C}} \cdot \overline{\overline{A}BC}}}$

7.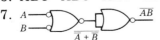

9.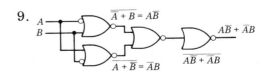

11. $A \ \overline{A} \ B \ \overline{B} \ C \ \overline{C}$

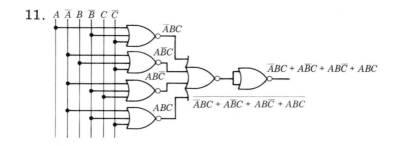

$\overline{A}BC$

$A\overline{B}C$

$AB\overline{C}$

ABC

$\overline{A}BC + A\overline{B}C + AB\overline{C} + ABC$

$\overline{\overline{A}BC + A\overline{B}C + AB\overline{C} + ABC}$

Exercises 21-2   1. 207.8 V, 415.7 V, 831.4 V   3. 112.6 kV, 138.6 kV, 225.2 kV
5. 120 V, 240 V, 480 V   7. 4157 V, 112.6 kV, 831.4 V
9. 20.78 kV, 138.6 kV, 51.96 kV   11. 120 V, 240.2 V, 479.8 V
13. 65 kV, 80 kV, 130 kV

Exercises 21-3   1. (a) $I_L = 12\underline{/\theta_{V_\phi} + 0°}$ A   (c)

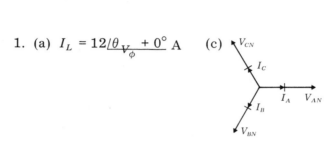

(b) $|V_L| = 207.8$ V
(For example, $V_{AB} = 207.8\underline{/\theta_{V_{AN}} + 30°}$, and so on.)

3. (a) $I_L = 20\underline{/\theta_{V_\phi} + 30°}$ A   (c)

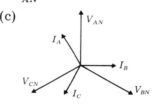

(b) $|V_L| = 831.4$ V

5. (a) $I_L = 2.4\underline{/\theta_{V_\phi} + 20°}$ A   (c)

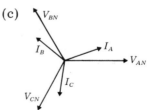

(b) $|V_L| = 12.5$ kV

7. (a) $15\underline{/\theta_V - 15°}$ A   (c)

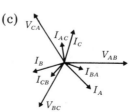

(b) $25.98\underline{/\theta_{I_\phi} - 30°}$ A

9. (a) $2.311\underline{/\theta_V - 12°}$ A

(c)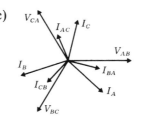

(b) $4.0\underline{/\theta_{I_\phi} - 30°}$ A

11. (a) $4.146\underline{/\theta_V + 18°}$ A

(c)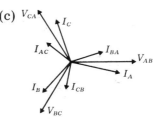

(b) $7.181\underline{/\theta_{I_\phi} - 30°}$ A

Exercises 21-4

1. (a) $48\underline{/0°}$ A, $30\underline{/-120°}$ A, $24\underline{/120°}$ A
   (b) $63.5\underline{/-19.1°}$ A, $68.15\underline{/-157.57°}$ A, $46.88\underline{/86.33°}$ A
   (c)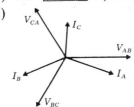

3. (a) $44.57\underline{/21.8°}$ A, $29.11\underline{/-105.96°}$ A, $23.53\underline{/131.31°}$ A
   (b) $56.92\underline{/-1.13°}$ A, $66.5\underline{/-137.95°}$ A, $46.28\underline{/99.36°}$ A
   (c)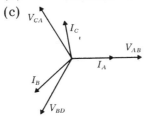

5. (a) $10\underline{/120°}$ A, $13.33\underline{/0°}$ A, $8\underline{/-120°}$ A
   (b) $4.664\underline{/21.8°}$ A

(c)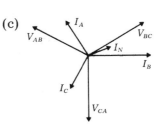

7. (a) $9.486\underline{/138.43°}$ A, $12.18\underline{/23.96°}$ A,
       $7.732\underline{/-105.07°}$ A
   (b) $4.282\underline{/61.81°}$ A

(c)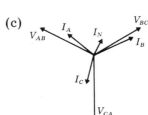

9. (a) $10.33\underline{/128.22°}$ A, $11.52\underline{/-3.67°}$ A, $8.973\underline{/-124.73°}$ A   (c)

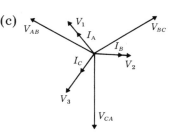

(b) $123.96\underline{/128.22°}$ V, $103.68\underline{/-3.67°}$ V, $134.6\underline{/-124.73°}$ V

11. (a) $9.302\underline{/146.79°}$ A, $10.93\underline{/17.7°}$ A, $8.818\underline{/-107.33°}$ A
    (b) $117.7\underline{/128.36°}$ V, $107.6\underline{/-6.26°}$ V, $136.9\underline{/-122.26°}$ V
    (c)

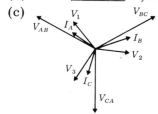

13. (a) $10.31\underline{/128.21°}$ A, $11.51\underline{/-3.67°}$ A, $8.96\underline{/-124.72°}$ A
    (b) $123.8\underline{/128.21°}$ V, $103.6\underline{/-3.67°}$ V, $134.4\underline{/-124.72°}$ V,
         $V_{NO} = 17.86\underline{/-158.2°}$ V
    (c) See Exercise 9.
15. (a) $9.296\underline{/146.81°}$ A, $10.93\underline{/17.71°}$ A, $8.808\underline{/-107.34°}$ A
    (b) $117.6\underline{/128.38°}$ V, $107.6\underline{/-6.25°}$ V, $136.7\underline{/-122.27°}$ V
         $V_{NO} = 17.51\underline{/-137.99°}$ V
    (c) See Exercise 11.

Exercises 21-5   1. 4.32 kW   3. 24.94 kW   5. 134.3 kW   7. 6.59 kW   9. 402.2 kW
11. 24.48 kW   13. 3.76 kW   15. 3.683 kW   17. 95.92 kW
19. 14.96 kVA, 12.72 kW

Exercises 22-1   1. b   3. a, b (if average value subtracted), d, e, i
5. a, even; d, even; f, even; h, even
9. (a) no even harmonics or cosine terms
   (b) 50 Hz, 150 Hz, 250 Hz, 350 Hz, 450 Hz   (c) 0
11. (a) no cosine terms   (b) 500 Hz, 1 kHz, 1.5 kHz, 2 kHz, 2.5 kHz   (c) 0
13. (a) no even harmonics   (b) 5 kHz, 15 kHz, 25 kHz, 35 kHz, 45 kHz
    (c) 0   15. (a) all components
    (b) 1 kHz, 2 kHz, 3 kHz, 4 kHz, 5 kHz   (c) −18 V
17. $\frac{2}{\pi}V_m$

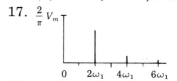

19. $\frac{2}{\pi}V_m$

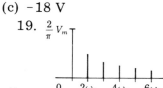

Exercises 22-2   1. (a) 11.34 V   (b) 5.47 V   (c) 800 Hz   3. (a) 32.41 V   (b) 15.64 V
(c) 120 Hz   5. (a) 54 V   (b) 60 V, 25.46 V, 5.09 V, 2.18 V
(c) 121%   (d) 60 Hz   7. (a) 16.2 V   (b) 18 V, 7.64 V, 1.53 V, 0.65 V
(c) 121%   (d) 50 Hz
9. (a) 1.07%, 2 kHz; 0.35%, 3 kHz; 0.12%, 4 kHz   (b) 1.13%
11. (a) 1.52%, 40 Hz; 0.29%, 60 Hz; 0.041%, 80 Hz; 0.0034%, 100 Hz
(b) 1.55%   13. 23,550 (87.4 dB)   15. 4998 (74 dB)

INDEX